MW00844713

Carpentry
Level One

Trainee Guide
Fifth Edition

PEARSON

Boston Columbus Indianapolis New York San Francisco Upper Saddle River
Amsterdam Cape Town Dubai London Madrid Milan Munich Paris Montreal Toronto
Delhi Mexico City São Paulo Sydney Hong Kong Seoul Singapore Taipei Tokyo

NCCER
President: Don Whyte
Director of Product Development: Daniele Stacey
Carpentry Project Manager: Rob Richardson
Senior Manager: Tim Davis
Quality Assurance Coordinator: Debie Ness
Desktop Publishing Coordinator: James McKay
Permissions Specialist: Amanda Werts
Production Specialist: Megan Casey
Editor: Chris Wilson

Writing and development services provided by S4Carlisle Publishing Services, Dubuque, IA
Project Manager: Barb Tucker
Writers: Michael B. Kopf
Art Development: S4Carlisle Publishing Services
Permissions Specialists: Kim Schmidt, Karyn Morrison
Media Specialist: Genevieve Brand
Copy Editor: Michael H. Toporek

Pearson Education, Inc.
Editorial Director: Vernon R. Anthony
Executive Editor: Alli Gentile
Editorial Assistant: Douglas Greive
Program Manager: Alexandrina B. Wolf
Operations Supervisor: Deidra M. Skahill
Art Director: Jayne Conte
Director of Marketing: David Gesell
Executive Marketing Manager: Derril Trakalo
Marketing Manager: Brian Hoehl
Marketing Coordinator: Crystal Gonzalez
Composition: NCCER
Printer/Binder: LSC Communications/Kendallville
Cover Printer: LSC Communications
Text Fonts: Palatino and Univers

Credits and acknowledgments for content borrowed from other sources and reproduced, with permission, in this textbook appear at the end of each module.

---

18 2020

PEARSON

Case bound ISBN-13: 978-0-13-340380-0
ISBN-10: 0-13-340380-7
Perfect bound ISBN-13: 978-0-13-340237-7
ISBN-10: 0-13-340237-1

Preface

To the Trainee

If you're ready to nail down a career in construction, consider carpentry. Carpenters make up the largest building trades occupation in the industry and those with all-around skills are in high demand. Carpenters are involved in many different kinds of construction activities, from building highways and bridges to installing kitchen cabinets.

Carpenters construct, erect, install, and repair structures and fixtures made from wood and other materials. Depending on the type of construction, size of company, and other factors, carpenters may specialize in one or two activities or may perform many different tasks. Each carpentry task is somewhat different, but most involve the same basic steps: working from blueprints, laying out the structure, assembling the structure, and checking the work afterward. Having good hand-eye coordination, an attention to detail, and the ability to perform math calculations will help you as you progress through your carpentry training.

We wish you success as you embark on your first year of training in the carpentry craft and hope that you will continue your training beyond this textbook. There are more than one million people employed in carpentry work in the United States, and as most of them can tell you, there are many opportunities awaiting those with the skills and desire to move forward in the construction industry.

New with *Carpentry Level One*

This fifth edition of *Carpentry Level One* presents a new instructional design and features a streamlined teaching approach to better prepare you for your career as a carpenter. In addition, *Carpentry Level One* contains updated artwork, detailed figures, and the latest tools and technology. Throughout this title, "Going Green" features emphasize ustainability and environmentally constructive procedures. *Building Materials, Fasteners, and Adhesives* contains more information on industrial materials and anchors, and *Orientation to the Trade* introduces students to SkillsUSA, a national nonprofit organization that serves teachers and high school and college students who are preparing for careers in trade, technical and skilled service occupations.

We invite you to visit the NCCER website at **www.nccer.org** for information on the latest product releases and training, as well as online versions of the *Cornerstone* newsletter and Pearson's NCCER product catalog.

Your feedback is welcome. You may email your comments to **curriculum@nccer.org** or send general comments and inquiries to **info@nccer.org**.

NCCER Standardized Curricula

NCCER is a not-for-profit 501(c)(3) education foundation established in 1996 by the world's largest and most progressive construction companies and national construction associations. It was founded to address the severe workforce shortage facing the industry and to develop a standardized training process and curricula. Today, NCCER is supported by hundreds of leading construction and maintenance companies, manufacturers, and national associations. The NCCER Standardized Curricula was developed by NCCER in partnership with Pearson, the world's largest educational publisher.

Some features of the NCCER Standardized Curricula are as follows:

- An industry-proven record of success
- Curricula developed by the industry for the industry
- National standardization providing portability of learned job skills and educational credits
- Compliance with the Office of Apprenticeship requirements for related classroom training *(CFR 29:29)*
- Well-illustrated, up-to-date, and practical information

NCCER also maintains a National Registry that provides transcripts, certificates, and wallet cards to individuals who have successfully completed modules of NCCER's Curricula. *Training programs must be delivered by an NCCER Accredited Training Sponsor in order to receive these credentials.*

Special Features

In an effort to provide a comprehensive, user-friendly training resource, we have incorporated many different features for your use. Whether you are a visual or hands-on learner, this book will provide you with the proper tools to get started in the Carpentry trade.

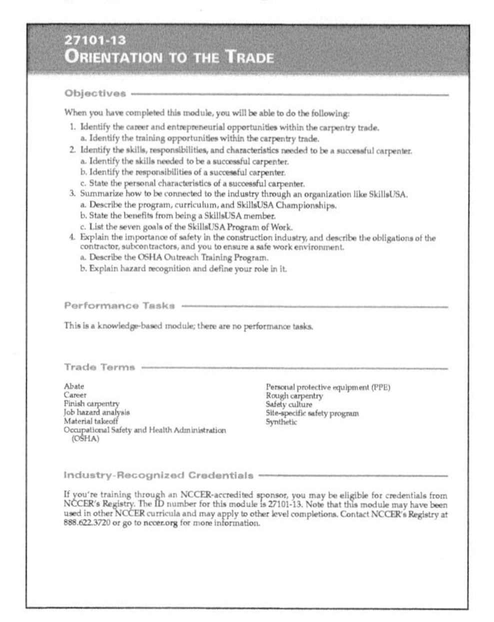

27101-13
ORIENTATION TO THE TRADE

Objectives

When you have completed this module, you will be able to do the following:

1. Identify the career and entrepreneurial opportunities within the carpentry trade.
   a. Identify the training opportunities within the carpentry trade.
2. Identify the skills, responsibilities, and characteristics needed to be a successful carpenter.
   a. Identify the skills needed to be a successful carpenter.
   b. Identify the responsibilities of a successful carpenter.
   c. State the personal characteristics of a successful carpenter.
3. Summarize how to be connected to the industry through an organization like SkillsUSA.
   a. Describe the program, curriculum, and SkillsUSA Championships.
   b. State the benefits from being a SkillsUSA member.
   c. List the seven goals of the SkillsUSA Program of Work.
4. Explain the importance of safety in the construction industry, and describe the obligations of the contractor, subcontractors, and you to ensure a safe work environment.
   a. Describe the OSHA Outreach Training Program.
   b. Explain hazard recognition and define your role in it.

Performance Tasks

This is a knowledge-based module; there are no performance tasks.

Trade Terms

Abate
Career
Finish carpentry
Job hazard analysis
Material takeoff
Occupational Safety and Health Administration (OSHA)
Personal protective equipment (PPE)
Rough carpentry
Safety culture
Site-specific safety program
Synthetic

Industry-Recognized Credentials

If you're training through an NCCER-accredited sponsor, you may be eligible for credentials from NCCER's Registry. The ID number for this module is 27101-13. Note that this module may have been used in other NCCER curricula and may apply to other level completions. Contact NCCER's Registry at 888.622.3720 or go to nccer.org for more information.

Introduction

This page is found at the beginning of each module and lists the Objectives, Performance Tasks, Trade Terms, and Required Trainee Materials for that module. The Objectives list the skills and knowledge you will need in order to complete the module successfully. The Performance Tasks give you an opportunity to apply your knowledge to the real-world duties that carpenters perform. The list of Trade Terms identifies important terms you will need to know by the end of the module. Required Trainee Materials list the materials and supplies needed for the module.

Special Features

Features provide a head start for those entering the Carpentry field by presenting technical tips and professional practices from craftworkers in various disciplines. These features often include real-life scenarios similar to those you might encounter on the job site.

Teamwork

Many of us like to follow all sorts of different teams: racing teams, baseball teams, football teams, and soccer teams. Just as in sports, a job site is made up of a team. As a part of that team, you have a responsibility to your teammates. What does teamwork really mean on the job? Craftworkers must sincerely do everything they can to build strong, professional working relationships with

Color Illustrations and Photographs

Full-color illustrations and photographs are used throughout each module to provide vivid detail. These figures highlight important concepts from the text and provide clarity for complex instructions. Each figure reference is denoted in the text in *italics* for easy reference.

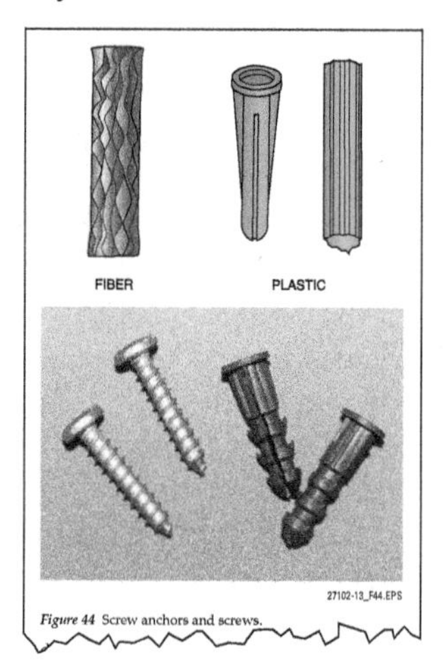

Figure 44 Screw anchors and screws.

Notes, Cautions, and Warnings

Safety features are set off from the main text in highlighted boxes and are organized into three categories based on the potential danger of the issue being addressed. Notes simply provide additional information on the topic area. Cautions alert you of a danger that does not present potential injury but may cause damage to equipment. Warnings stress a potentially dangerous situation that may cause injury to you or a co-worker.

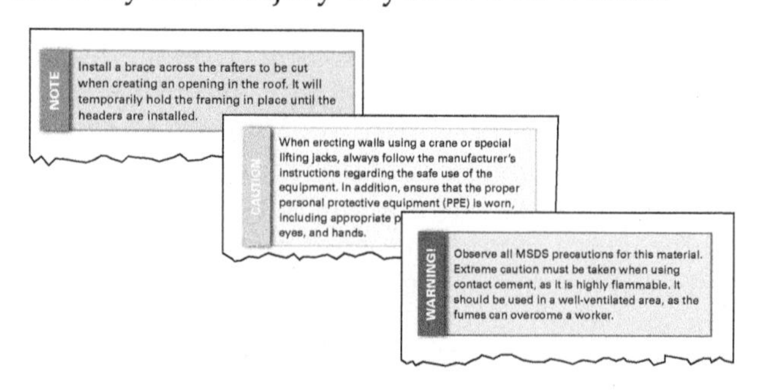

Going Green

Going Green looks at ways to preserve the environment, save energy, and make good choices regarding the health of the planet. Through the introduction of new construction practices and products, you will see how the "greening of America" has already taken root.

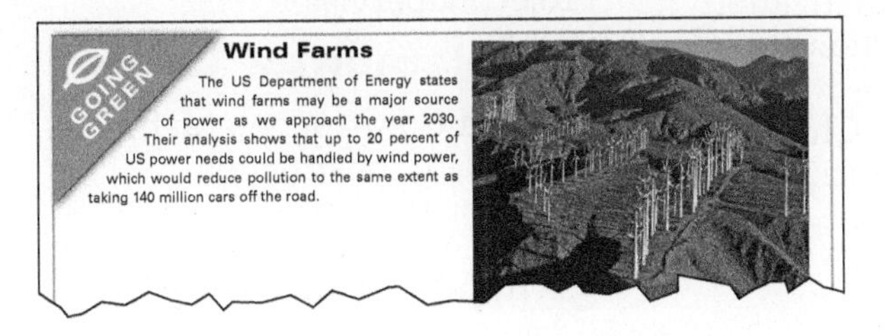

Did You Know?

The Did You Know? features offer hints, tips, and other helpful bits of information from the trade.

> **Did You Know?**
> **Balloon Framing**
> Balloon framing is frequently used in hurricane-prone areas for gable ends. In fact, this type of framing may be required by local building codes.

Step-by-Step Instructions

Step-by-step instructions are used throughout to guide you through technical procedures and tasks from start to finish. These steps show you not only how to perform a task but also how to do it safely and efficiently.

Step 1 Select the proper-size toggle bolt and drill bit for the job.

Step 2 Check the toggle bolt for damaged or dirty threads or a malfunctioning wing mechanism.

Step 3 Drill a hole completely through the surface to which the part is to be fastened.

Step 4 Insert the toggle bolt through the opening in the item to be fastened.

Step 5 Screw the wings onto the end of the toggle bolt, ensuring that the flat side of the wing is facing the bolt head.

Step 6 Fold the win... comp... bac...

Trade Terms

Each module presents a list of Trade Terms that are discussed within the text and defined in the Glossary at the end of the module. These terms are denoted in the text with **bold, blue type** upon their first occurrence. To make searches for key information easier, a comprehensive Glossary of Trade Terms from all modules is located at the back of this book.

> on at least 1½" of wood. In platform construction, the ends of all the joists are fastened to a **header joist**, also called a band joist.
> Joists must be doubled where extra loads require additional support. When a partition runs parallel to the joists, a double joist is placed underneath. Joists must also be doubled around openings in the floor frame for stairways, chimneys, etc., to reinforce the rough opening in the floor. These additional joists used at such openings are called **trimmer joists**. They support the headers that carry short joists called **tail joists**. Double joists should be spread where necessary to accommodate plumbing.
> In residential construction, floors traditionally

Review Questions

Review Questions reinforce the knowledge you have gained and are a useful tool for measuring what you have learned.

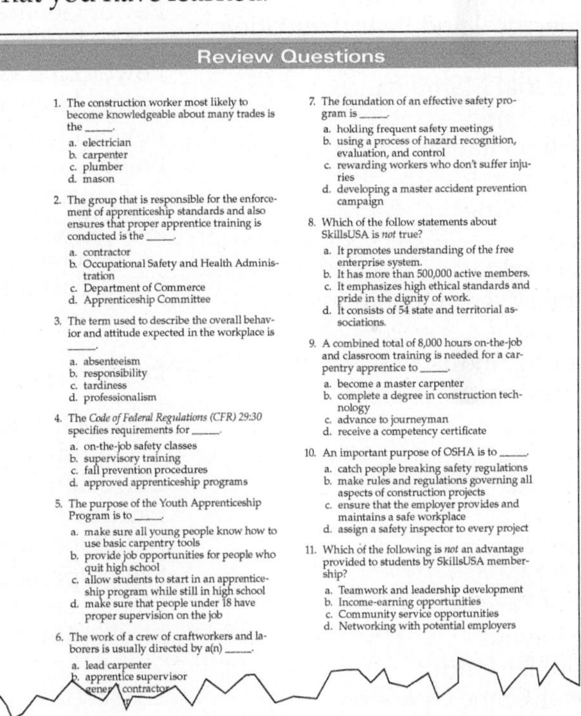

Review Questions

1. The construction worker most likely to become knowledgeable about many trades is the _____.
   a. electrician
   b. carpenter
   c. plumber
   d. mason
2. The group that is responsible for the enforcement of apprenticeship standards and also ensures that proper apprentice training is conducted is the _____.
   a. contractor
   b. Occupational Safety and Health Administration
   c. Department of Commerce
   d. Apprenticeship Committee
3. The term used to describe the overall behavior and attitude expected in the workplace is _____.
   a. absenteeism
   b. responsibility
   c. tardiness
   d. professionalism
4. The *Code of Federal Regulations (CFR) 29:30* specifies requirements for _____.
   a. on-the-job safety classes
   b. supervisory training
   c. fall prevention procedures
   d. approved apprenticeship programs
5. The purpose of the Youth Apprenticeship Program is to _____.
   a. make sure all young people know how to use basic carpentry tools
   b. provide job opportunities for people who quit high school
   c. allow students to start in an apprenticeship program while still in high school
   d. make sure that people under 18 have proper supervision on the job
6. The work of a crew of craftworkers and laborers is usually directed by a(n) _____.
   a. lead carpenter
   b. apprentice supervisor
   c. gener... contractor
7. The foundation of an effective safety program is _____.
   a. holding frequent safety meetings
   b. using a process of hazard recognition, evaluation, and control
   c. rewarding workers who don't suffer injuries
   d. developing a master accident prevention campaign
8. Which of the follow statements about SkillsUSA is *not* true?
   a. It promotes understanding of the free enterprise system.
   b. It has more than 500,000 active members.
   c. It emphasizes high ethical standards and pride in the dignity of work.
   d. It consists of 54 state and territorial associations.
9. A combined total of 8,000 hours on-the-job and classroom training is needed for a carpentry apprentice to _____.
   a. become a master carpenter
   b. complete a degree in construction technology
   c. advance to journeyman
   d. receive a competency certificate
10. An important purpose of OSHA is to _____.
    a. catch people breaking safety regulations
    b. make rules and regulations governing all aspects of construction projects
    c. ensure that the employer provides and maintains a safe workplace
    d. assign a safety inspector to every project
11. Which of the following is *not* an advantage provided to students by SkillsUSA membership?
    a. Teamwork and leadership development
    b. Income-earning opportunities
    c. Community service opportunities
    d. Networking with potential employers

NCCER Standardized Curricula

NCCER's training programs comprise more than 80 construction, maintenance, pipeline, and utility areas and include skills assessments, safety training, and management education.

Boilermaking
Cabinetmaking
Carpentry
Concrete Finishing
Construction Craft Laborer
Construction Technology
Core Curriculum:
Introductory Craft Skills
Drywall
Electrical
Electronic Systems Technician
Heating, Ventilating, and
Air Conditioning
Heavy Equipment Operations
Highway/Heavy Construction
Hydroblasting
Industrial Coating and Lining
Application Specialist
Industrial Maintenance
Electrical and Instrumentation
Technician
Industrial Maintenance
Mechanic
Instrumentation
Insulating
Ironworking
Masonry
Millwright
Mobile Crane Operations
Painting
Painting, Industrial
Pipefitting
Pipelayer
Plumbing
Reinforcing Ironwork
Rigging
Scaffolding
Sheet Metal
Signal Person
Site Layout
Sprinkler Fitting
Tower Crane Operator
Welding

Maritime

Maritime Industry Fundamentals

Green/Sustainable Construction

Building Auditor
Fundamentals of Weatherization
Introduction to Weatherization
Sustainable Construction
Supervisor
Weatherization Crew Chief
Weatherization Technician
Your Role in the Green
Environment

Energy

Alternative Energy
Introduction to the Power
Industry
Introduction to Solar
Photovoltaics
Introduction to Wind Energy
Power Industry Fundamentals
Power Generation Maintenance
Electrician
Power Generation I&C
Maintenance Technician
Power Generation Maintenance
Mechanic
Power Line Worker
Power Line Worker: Distribution
Power Line Worker: Substation
Power Line Worker:
Transmission
Solar Photovoltaic Systems
Installer
Wind Turbine Maintenance
Technician

Pipeline

Control Center Operations,
Liquid
Corrosion Control
Electrical and Instrumentation
Field Operations, Liquid
Field Operations, Gas
Maintenance
Mechanical

Safety

Field Safety
Safety Orientation
Safety Technology

Management

Fundamentals of Crew
Leadership
Project Management
Project Supervision

Supplemental Titles

Applied Construction Math
Careers in Construction
Tools for Success

Spanish Translations

Basic Rigging
(Principios Básicos de
Maniobras)
Carpentry Fundamentals
(Introducción a la
Carpintería, Nivel Uno)
Carpentry Forms
(Formas para Carpintería,
Nivel Trés)
Concrete Finishing, Level One
(Acabado de Concreto,
Nivel Uno)
Core Curriculum:
Introductory Craft Skills
(Currículo Básico:
Habilidades Introductorias del
Oficio)
Drywall, Level One
(Paneles de Yeso, Nivel Uno)
Electrical, Level One
(Electricidad, Nivel Uno)
Field Safety
(Seguridad de Campo)
Insulating, Level One
(Aislamiento, Nivel Uno)
Ironworking, Level One
(Herrería, Nivel Uno)
Masonry, Level One
(Albañilería, Nivel Uno)
Pipefitting, Level One
(Instalación de Tubería
Industrial, Nivel Uno)
Reinforcing Ironwork, Level One
(Herreria de Refuerzo,
Nivel Uno)
Safety Orientation
(Orientación de Seguridad)
Scaffolding
(Andamios)
Sprinkler Fitting, Level One
(Instalación de Rociadores,
Nivel Uno)

Acknowledgments

This curriculum was revised as a result of the farsightedness and leadership of the following sponsors:

Abbott Construction Inc.
ABC Eastern Pennsylvania Chapter
ABC Heart of America Chapter
ABC South Texas Chapter
Arkansas Department of Career Education
Bowden Contracting Company, Inc.
Brasfield & Gorrie
Construction Industry Training Council of Washington
Crossland Construction
The Haskell Company
HB Training & Consulting
JJA Construction & General Industry Training Consultants
Mid-Maine Technical Center
Mountain Home High School
PCL Construction
River Valley Technical Center
Robins & Morton
Santa Fe College
Suwannee-Hamilton Technical Center
Trenton High School
VisionQuest Academy

This curriculum would not exist were it not for the dedication and unselfish energy of those volunteers who served on the Authoring Team. A sincere thanks is extended to the following:

John Ambrosia
Keith Bennett
Owen Carpenter
Mark Champagne
Vincent Console
Howard Davis
Curtis Haskins
Hal Heintz
Jeff Henry
Erin Hunter
Kevin Kelley
Mark Knudson
Bob Makela
Tim Mosley
Mark Robinson
Tony Vazquez
John Yencho

NCCER Partners

American Fire Sprinkler Association
Associated Builders and Contractors, Inc.
Associated General Contractors of America
Association for Career and Technical Education
Association for Skilled and Technical Sciences
Carolinas AGC, Inc.
Carolinas Electrical Contractors Association
Center for the Improvement of Construction Management and Processes
Construction Industry Institute
Construction Users Roundtable
Construction Workforce Development Center
Design Build Institute of America
GSSC – Gulf States Shipbuilders Consortium
Manufacturing Institute
Mason Contractors Association of America
Merit Contractors Association of Canada
NACE International
National Association of Minority Contractors
National Association of Women in Construction
National Insulation Association
National Ready Mixed Concrete Association
National Technical Honor Society
National Utility Contractors Association
NAWIC Education Foundation
North American Technician Excellence
Painting & Decorating Contractors of America
Portland Cement Association
SkillsUSA®
Steel Erectors Association of America
U.S. Army Corps of Engineers
University of Florida, M. E. Rinker School of Building Construction
Women Construction Owners & Executives, USA

Contents

Module Nine

Basic Stair Layout

Introduces types of stairs and common building code requirements related to stairs. Focuses on techniques for measuring and calculating rise, run, and stairwell openings, laying out stringers, and fabricating basic stairways. (Module ID 27110-13; 12.5 Hours)

Glossary

Index

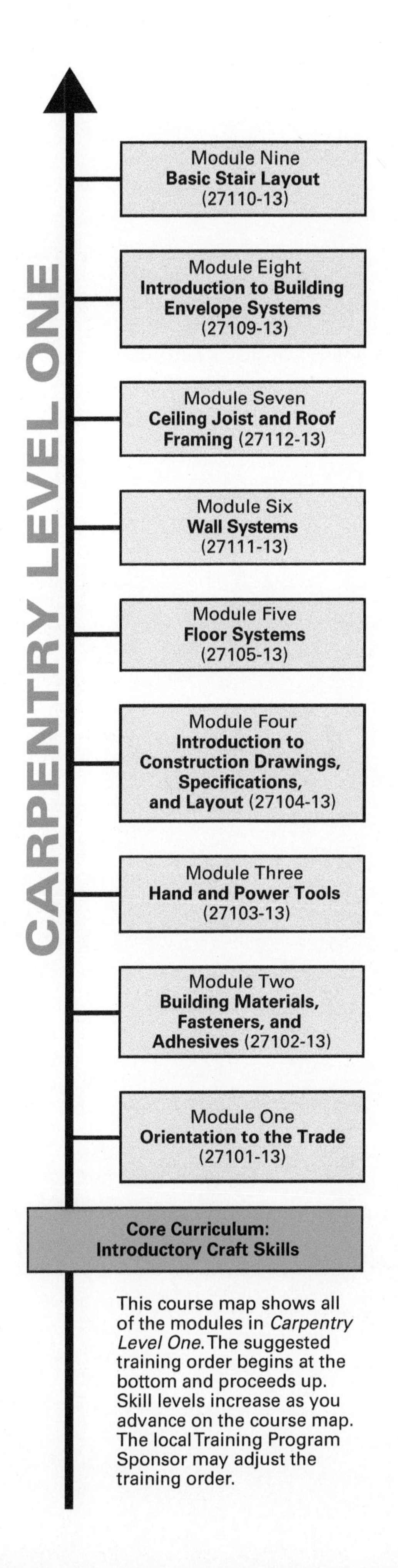

This course map shows all of the modules in *Carpentry Level One*. The suggested training order begins at the bottom and proceeds up. Skill levels increase as you advance on the course map. The local Training Program Sponsor may adjust the training order.

27101-13

Orientation to the Trade

Overview

The carpentry trade offers numerous career opportunities, from constructing concrete forms to creating fine cabinetry. Carpenters build beautiful structures that can last for centuries and have one of the highest job satisfaction rates of any career in the construction industry. They have opportunities to work in residential, commercial, and industrial construction. Carpenters are required to work safely to ensure their own personal safety and the safety of others on the job site.

Module One

Trainees with successful module completions may be eligible for credentialing through the NCCER Registry. To learn more, go to **www.nccer.org** or contact us at **1.888.622.3720**. Our website has information on the latest product releases and training, as well as online versions of our *Cornerstone* magazine and Pearson's product catalog.

Your feedback is welcome. You may email your comments to **curriculum@nccer.org,** send general comments and inquiries to **info@nccer.org**, or fill in the User Update form at the back of this module.

This information is general in nature and intended for training purposes only. Actual performance of activities described in this manual requires compliance with all applicable operating, service, maintenance, and safety procedures under the direction of qualified personnel. References in this manual to patented or proprietary devices do not constitute a recommendation of their use.

27101-13
ORIENTATION TO THE TRADE

Objectives

When you have completed this module, you will be able to do the following:

1. Identify the career and entrepreneurial opportunities within the carpentry trade.
   a. Identify the training opportunities within the carpentry trade.
2. Identify the skills, responsibilities, and characteristics needed to be a successful carpenter.
   a. Identify the skills needed to be a successful carpenter.
   b. Identify the responsibilities of a successful carpenter.
   c. State the personal characteristics of a successful carpenter.
3. Summarize how to be connected to the industry through an organization like SkillsUSA.
   a. Describe the program, curriculum, and SkillsUSA Championships.
   b. State the benefits from being a SkillsUSA member.
   c. List the seven goals of the SkillsUSA Program of Work.
4. Explain the importance of safety in the construction industry, and describe the obligations of the contractor, subcontractors, and you to ensure a safe work environment.
   a. Describe the OSHA Outreach Training Program.
   b. Explain hazard recognition and define your role in it.

Performance Tasks

This is a knowledge-based module; there are no performance tasks.

Trade Terms

Abate
Career
Finish carpentry
Job hazard analysis
Material takeoff
Occupational Safety and Health Administration (OSHA)
Personal protective equipment (PPE)
Rough carpentry
Safety culture
Site-specific safety program
Synthetic

Industry-Recognized Credentials

If you're training through an NCCER-accredited sponsor, you may be eligible for credentials from NCCER's Registry. The ID number for this module is 27101-13. Note that this module may have been used in other NCCER curricula and may apply to other level completions. Contact NCCER's Registry at 888.622.3720 or go to **nccer.org** for more information.

Contents

Topics to be presented in this module include:

Figures

SECTION ONE

1.0.0 CAREER AND ENTREPRENEURIAL OPPORTUNITIES

Objective

Identify the career and entrepreneurial opportunities within the carpentry trade.

a. Identify the training opportunities within the carpentry trade.

Trade Terms

Career: A profession for which an individual trains and considers to be a lifelong calling.

Opportunity is driven by knowledge and ability, which are in turn driven by education and training. This NCCER training program was designed and developed by the construction industry for the construction industry. It is the only nationally accredited, competency-based construction training program in the United States. A competency-based program requires that the trainee demonstrates the ability to safely perform specific job-related tasks in order to receive credit. This approach is unlike other apprentice programs that merely require a trainee to put in the required number of hours in the classroom and on the job.

The primary goal of NCCER is to standardize construction craft training throughout the country so that both contractors and craftworkers will benefit from the training, no matter where they are located. Trainees in an NCCER program will become part of a national registry and receive a certificate for each level of training completed. (See the examples in the *Appendix*.) When you apply for a job with any participating contractor in the country, a copy of your training transcript will be available to them. If your training is incomplete when making a job transfer, you can pick up where you left off because every participating contractor is using the same training program. In addition, many technical schools and colleges are using the program.

The construction industry employs more people and contributes more to the nation's economy than any other industry. Our society will always need new homes, roads, airports, hospitals, schools, factories, and office buildings. This means that there will always be a source of good-paying jobs and career opportunities for carpenters and other construction trade professionals.

As a construction worker, a carpenter can progress from apprentice through several levels:

- Journeyman carpenter
- Master carpenter
- Foreman/lead carpenter/crew leader
- Supervisor

Journeyman carpenter – After successfully completing an apprenticeship, a trainee becomes a journeyman. The term *journeyman* originally meant to journey away from the master and work alone. A person can remain a journeyman or advance in the trade. Journeymen may have additional duties such as supervisor or estimator. With larger companies and on larger jobs, journeymen often become specialists. For example, some carpenters may specialize in concrete formwork, while others may specialize in stair construction.

Master carpenter – A master craftworker is one who has achieved and continuously demonstrates

Careers in Carpentry

Apprentice training is the first step in a career that has endless possibilities. Carpenters gain knowledge of many different trades and skills while developing their craft. This broad set of skills is a valuable asset in the construction industry and will open the door to a wide variety of exciting career opportunities. In addition, the carpentry trade has long been rated number one for job satisfaction, perhaps because carpenters participate and take pride in creating beautiful and lasting structures, the results of which can often be seen at the end of each day.

27101-13_SA01.EPS

the highest skill levels in the trade. A master carpenter is a mentor (guide or coach) for journeymen and apprentice carpenters. Master carpenters often start their own businesses and become contractors/owners.

Foreman/lead carpenter/crew leader – This individual is a frontline leader who directs the work of a crew of craftworkers and laborers.

Supervisor – Large construction projects require supervisors who oversee the work of crews made up of foremen, journeymen, and apprentices. They are responsible for assigning, directing, and inspecting the work of crew members.

In addition, other careers may be suitable for an experienced carpenter, such as safety manager, project manager/administrator, estimator, architect, general contractor, construction manager, and contractor/owner.

More than any other construction worker, the carpenter is likely to become knowledgeable about many trades. This makes carpentry work interesting and challenging and creates a variety of career opportunities.

The important thing to understand is that a career is a lifelong learning process. To be an effective carpenter, you must keep current with new tools, materials, and methods. If you choose to work into a management role or to someday start a construction business, management and administrative skills will be needed in addition to keeping your carpentry skills honed. Every successful manager and business owner started the same way you are starting, and they all have one thing in common: a desire and willingness to continue learning. The learning process begins with apprentice training.

While developing your carpentry skills and gaining experience, you will have the opportunity to earn increased pay for your services. There is great financial incentive for learning and growing within the trade. You can't get to the top, however, without learning the basics.

1.1.0 Formal Construction Training

For many skilled carpenters, formal training started in high school while enrolled in building science or construction technology programs. These programs allow students to gain an understanding of basic carpentry methods and techniques while exploring the carpentry field to determine if it a good fit for them.

The federal government established registered apprenticeship training through the *Code of Federal Regulations (CFR) 29:29,* which dictates specific requirements for apprenticeship, and *CFR 29:30,* which dictates specific guidelines for recruitment, outreach, and registration into approved apprenticeship programs. The Department of Labor has established specific apprenticeship guidelines including a minimum numbers of hours required to complete an apprenticeship.

Education and training throughout the country is undergoing significant change. Educators and researchers have been learning and applying new techniques to adjust to how students learn and apply their education. New training delivery methods, such as mobile apps, are being developed so students can learn anywhere and at any time.

NCCER is an independent educational foundation founded and funded by the construction industry to provide quality instruction and instructional materials for a wide variety of crafts. The mission of NCCER is to offer training and credentialing of the construction workforce. NCCER has departed from traditional classroom learning and has adopted a competency-based training regimen. Competency-based training means that

Tools

Tools are an essential part of carpentry. Although modern tools have advanced beyond the primitive stone tools used by our ancestors, they serve the carpenter's needs in much the same way. Carpentry, like other trades, relies on tools to make difficult tasks easier. If you take the time to learn the proper way to handle and use your tools safely, you'll be able to work more effectively and produce a high-quality product.

27101-13_SA02.EPS

instead of requiring specific hours of classroom training and set hours of on-the-job learning (OJL), you are able to have to prove that you know what is required and can demonstrate that you perform the specific skill. NCCER also uses the latest technology, such as interactive computer-based training, to deliver the classroom portions of the training. All completion information for every trainee is sent to NCCER and kept within the National Registry. The National Registry can then confirm training and skills for workers as they move from company to company, state to state, or even within their own company.

The dramatic shortage of skills within the construction workforce, combined with the shortage of new workers coming into the industry, is providing an opportunity for the construction industry to design and implement new training initiatives. When enrolling in an NCCER program, it is critical that you work for a contractor who supports a national, standardized training program that includes credentials to confirm your skill development.

1.1.1 Apprenticeship Program

Apprentice training is a means for individuals entering a craft to learn from those who have mastered the craft. It focuses on learning by doing: real skills versus theory. Although some theory is presented in the classroom, it is always presented in a way that helps the trainee understand the purpose behind the required skill.

1.1.2 Youth Apprenticeship Program

A Youth Apprenticeship Program is also available that allows students to begin their apprentice training while still in high school. A student entering the program in eleventh grade may complete as much as two years of an NCCER program by high school graduation. In addition, the program, in cooperation with local craft employers, allows students to work in the trade and earn money while still in school. Upon graduation, the student can enter the industry at a higher level and with more pay than someone just starting the apprenticeship program.

This training program is similar to the one used by NCCER, learning centers, contractors, and colleges across the country. Students are recognized through official transcripts and can enter the next year of the program wherever it is offered. They may also have the option of applying the credits at a two-year or four-year college that offers degree or certification programs in the construction trades.

1.1.3 Apprenticeship Standards

All apprenticeship standards prescribe certain work-related or on-the-job learning. This OJL is broken down into specific tasks in which the apprentice receives hands-on training. In addition, a specified number of hours are required in each task. The total amount of OJL for a carpentry apprenticeship program is traditionally 8,000 hours, which amounts to four years of training. In a competency-based program, it may be possible to shorten this time by testing out of specific tasks through a series of performance exams.

In a traditional program, the required OJL may be acquired in increments of 2,000 hours per year. Layoffs or illness may affect the duration. The apprentice must log all work time and turn it in to the apprenticeship committee so that accurate time control can be maintained.

The classroom instruction and work-related training does not always run concurrently due to such reasons as layoffs, type of work needed to be done in the field, etc. Furthermore, apprentices with special job experience or coursework may obtain credit toward their classroom requirements. This reduces the total time required in the classroom while maintaining the total 8,000-hour OJL requirement. These special cases depend on the type of program and the regulations and standards under which it operates.

Informal OJL provided by employers is usually less thorough than that provided through a formal apprenticeship program. The degree of training and supervision in this type of program often depends on the size of the employing firm. A small contractor may provide training in only one area, while a large company may be able to provide training in several areas.

For those entering an apprenticeship program, a high school or technical school education is desirable. Courses in shop, mechanical drawing, and general mathematics are helpful. Manual dexterity, good physical condition, and quick reflexes are important. The ability to solve problems quickly and accurately and to work closely with others is essential. You must also have a high concern for safety.

The prospective apprentice must submit certain information to the apprenticeship committee. This may include the following:

- Aptitude test (General Aptitude Test Battery or GATB Form Test) results (usually administered by the local Employment Security Commission)
- Proof of educational background (candidate should have school transcripts sent to the committee)

- Letters of reference from past employers and friends
- Proof of age
- If the candidate is a veteran, a copy of Form DD214
- A record of technical training received that relates to the construction industry and/or a record of any pre-apprenticeship training
- High school diploma or General Equivalency Diploma (GED)

The apprentice must do the following:

- Wear proper safety equipment on the job
- Purchase and maintain tools of the trade as needed and required by the contractor
- Submit a monthly on-the-job learning report to the committee
- Report to the committee if a change in employment status occurs
- Attend classroom instruction and adhere to all classroom regulations such as attendance requirements

1.1.4 What to Expect from a Contractor

After an applicant has been selected for apprenticeship by the Committee, the contractor employing the apprentice agrees that the apprentice will be employed under conditions that will result in normal advancement. In return, the contractor requires the apprentice to make satisfactory progress in OJL and related classroom instruction. The contractor agrees that the apprentice will not be employed in a manner that may be considered in violation of the apprenticeship standards. The contractor also agrees to pay a prorated share of the cost of operating the apprenticeship program.

1.1.5 What to Expect from a Training Program

First and foremost, it is important that the contractor you select has a training program. The program should be comprehensive, standardized, and competency-based; not based on the amount of time spent in a classroom.

When contractors take the time and initiative to provide quality training, it is a sign that they are willing to invest in their workforce and improve the abilities of their workers. It is important that the training program is national in scope and that transcripts and completion credentials are issued to participants. Construction is unique in that the contractors share the workforce. A craftworker may work for several contractors throughout their time in the field. Therefore, it is critical that the training program help the worker move from company to company, city to city, or state to state without having to start at the beginning for each move. Ask how many contractors in the area use the same program before enrolling. Make sure that you will always have access to transcripts and certificates to verify your status and level of completion.

Training should be rewarded. The training program should have a well-defined compensation ladder attached to it. Successful completion and mastery of skill sets should be accompanied by increases in hourly wages.

Finally, the training curricula should be complete and up-to-date. Any training program has to be committed to maintaining its curricula, developing new delivery mechanisms (Internet, webinars, etc.), and being constantly vigilant for new techniques, materials, tools, and equipment in the workplace.

1.1.6 What to Expect from the Apprenticeship Committee

The Apprenticeship Committee is the local administrative body to which the apprentice is assigned and to which the responsibility is delegated for the appropriate training of the individual. Every apprenticeship program, whether state or federal, is covered by standards that have been approved by those agencies. The responsibility of enforcement is delegated to the Committee.

The Apprenticeship Committee is responsible not only for enforcement of standards, but must see to it that proper training is conducted so that a craftworker successfully completing from the program is fully qualified in those areas of training designated by the standards. Among the responsibilities of the Committee are the following:

- Screen and select individuals for apprenticeship and refer them to participating contractors for training.
- Place apprentices under written agreement for participation in the program.
- Establish minimum standards for OJL and related instruction, and monitor the apprentice to see that these criteria are adhered to during the training period.
- Hear all complaints of violations of apprenticeship agreements, whether by contractor or apprentice, and take action within the guidelines of the standards.
- Notify the registration agencies of all enrollments, completions, and terminations of apprentices.

History of Carpentry

27101-13_SA03.EPS

Primitive carpentry developed in forest regions during the latter years of the Stone Age, when early humans improved stone tools so they could shape wood for shelters, animal traps, and dugout boats. Between 4000 and 2000 BC, Egyptians developed copper tools, which they used to build vaults, bed frames, and furniture. Later in that period, they developed bronze tools and bow drills. An example of the Egyptians' skill in mitering, mortising, dovetailing, and paneling is the intricate furniture found in the tomb of Tutankhamen (King Tut). European carpenters did not produce such furniture until the Renaissance (1300 to 1500 AD), although they used timber to construct dwellings, bridges, and industrial equipment. In Denmark and ancient Germany, Neolithic people around 5000 BC built rectangular houses from timbers that were nearly 100-feet long. In England, the mortised and dovetailed joints of the stone structures at Stonehenge indicate that advanced carpentry techniques were known in ancient Britain. Before the Roman conquest of Britain (100 AD), its carpenters had already developed iron tools such as saws, hatchets, rasps, and knives. They even had turned-wood objects made on primitive pole lathes.

In the Middle Ages, carpenters began a movement toward specialization, such as shipwrights, wheelwrights, turners, and millwrights. However, general-purpose carpenters were still found in most villages and on large private estates. These carpenters could travel with their tools to outlying areas that had no carpenters or to a major building project that required temporary labor. During this period, European carpenters invented the carpenter's brace (a tool for holding and turning a drill bit). The plane, which the Romans had used centuries earlier, reappeared about 1200 AD. The progress of steelmaking also provided for advancements in the use of steel-edged tools and the advent of crude iron nails. (Wood pegs were used to hold wooden members together before the use of nails.) Screws were invented in the 1500s.

The first castles and churches in northern Europe were constructed of timber. When the great stone buildings replaced those made of timber, skilled carpenters built the floors, paneling, doors, and roofs. The erection of large stone buildings also led to the inventions of scaffolding for walls, framework for arch assemblies, and pilings to strengthen foundations. Houses and other smaller buildings were still made of timber and thinner wood. Clay was used to fill the gaps between the beams.

The art of carpentry contributed significantly to the grandeur of the great buildings of the Renaissance. Two noted masterpieces of timber construction are the outer dome of St. Paul's Cathedral in London and the Sheldonian Theater in Oxford, England. After the Renaissance, other examples of architecture requiring skilled carpentry appeared, including the mansard roof with its double slope (providing loftier attics), broad staircases, and sashed windows. These architectural features were incorporated into homes constructed in colonial America. In Chicago in 1840, George W. Snow introduced balloon-frame construction, which proved to be a more cost-effective and quicker method because it used machine-made studs and nails. In balloon framing, the studs run from the bottom floor to the uppermost rafters. This method gives the structure exceptional ability to handle strong winds, but requires very long studs that are difficult to manufacture, transport, and store. Because of these problems, balloon framing has almost disappeared, although it is used to some extent in Florida to frame the gable ends of buildings to provide protection from wind loads exerted during hurricanes. Today, platform (western) framing has almost completely replaced balloon-frame construction.

Another example of drawing upon other classical architecture is Thomas Jefferson's Monticello. Although Jefferson started work on Monticello in 1768, he left the estate in 1784 to serve as Minister of the United States to France. While spending several years in Europe, Jefferson had the opportunity to see many classical buildings. When he returned to Monticello in 1794, Jefferson began to remodel his house to include some of these classical architectural features.

Ancient Construction

Modern carpentry is a continuum from ancient times. Carpenters and other craftworkers were responsible for construction of the world's most celebrated examples of historical architecture. From the lavish interiors of ancient Egypt's great pyramids to the outer dome of London's St. Paul's Cathedral (shown here), the rich heritage of carpentry can be found across the world.

27101-13_SA04.EPS

1.0.0 Section Review

1. A person who has achieved and continuously demonstrates the highest skill levels in the carpentry trade is a(n) _____.
   a. estimator
   b. journeyman carpenter
   c. master carpenter
   d. apprentice

Section Two

2.0.0 Carpenter Skills, Responsibilities, and Characteristics

Objective

Identify the skills, responsibilities, and characteristics needed to be a successful carpenter.

a. Identify the skills needed to be a successful carpenter.
b. Identify the responsibilities of a successful carpenter.
c. State the personal characteristics of a successful carpenter.

Trade Terms

Finish carpentry: The portion of the carpentry trade associated with interior and exterior trim, cabinetry, siding, wall finishes, and decorative work.

Job hazard analysis: An approach to hazard recognition in which the task to be performed is broken down into its individual parts or steps, and then each step is analyzed for its potential hazard(s). When the hazard is identified, certain actions or procedures are recommended that will correct the hazard before an accident occurs.

Material takeoff: A list of building materials obtained by analyzing the project drawings (also known as a takeoff).

Rough carpentry: The portion of the carpentry trade associated with framing and other work that will be covered with finish materials.

Synthetic: Man-made (not naturally occurring); typically a result of a chemical reaction.

The scope of carpentry has expanded in modern times with the use of synthetic building materials and ever-improving tools. Today's carpenters must not only know about wood, but also about materials such as particleboard, gypsum board, plastics, and laminates. They must also know how to use many modern tools, fasteners, construction techniques, and safety equipment.

The duties of carpenters can vary significantly from one job to another. A carpenter who works for a commercial contractor may work primarily with concrete, steel, and preformed building materials (*Figure 1*). A carpenter who does residential work is more likely to work with wood-frame construction and wood-finish materials, but will also encounter an increasing variety of preformed and prefabricated building materials (*Figure 2*).

In the construction industry, carpentry is commonly divided into two categories: rough and finish. Examples of rough carpentry include:

- Erecting framing members
- Erecting concrete forms
- Building docks, bridges, and supports for tunnels and sewers

Examples of finish carpentry (*Figure 3*) include:

- Building stairs
- Installing doors, cabinets, wood paneling, and molding
- Installing acoustical tile

The duties of carpenters vary even within the broad categories of rough and finish carpentry. The type of construction, size of the company, skill of the carpenter, community size, and other

27101-13_F01.EPS

Figure 1 Typical commercial construction.

27101-13_F02.EPS

Figure 2 Typical residential construction.

27101-13_F03.EPS

Figure 3 Example of finish carpentry.

factors affect the carpenter's duties. Carpenters who are employed by a large contractor, for example, may specialize in one area, such as laying hardwood floors. Carpenters employed by a small firm may build wall frames, put in insulation, and install paneling, as well as perform concrete finishing, welding, and painting. The duties of carpenters also vary because each job is unique.

2.1.0 Carpenter Skills

In order to be successful, a professional carpenter must have the skills required to use building materials, tools, and equipment to produce a high-quality finished product in a minimum amount of time. A carpenter must be adept at adjusting methods to meet each situation. A successful carpenter must continuously remain current and knowledgeable about the technical advancements in building materials and equipment and gain the skills to use them. In addition, a professional carpenter never takes chances with regard to personal safety or the safety of others.

As in other building trades, a carpenter's work is active and sometimes strenuous. Prolonged standing, climbing, and squatting are often necessary. Many carpenters work outside under adverse weather conditions. Carpenters risk injury from slips or falls, from contact with sharp or rough materials, and from the use of sharp tools and power equipment. Being new to the trade increases the chance of being injured. Therefore, it is essential that you rely on the knowledge of more experienced workers, learn applicable safety procedures, and wear appropriate personal protective equipment.

In order to be successful in the carpentry trade, a person should possess the following:

- Physical strength and ability to lift and move materials
- Hand-eye coordination to use tools
- Ability to communicate clearly with co-workers
- Ability to perform math calculations to perform a **material takeoff** and lay out the structure
- Attention to detail to accurately measure and cut building materials

Carpenters often have great freedom in planning and performing their work. However, carpentry skills are standard, and most jobs involve the following skills to some extent:

- Use construction drawings to lay out the structure on the site.
- Use drawings and specifications to perform a material takeoff.
- Gather the materials, tools, and equipment needed for construction.
- Schedule the work.
- Prepare a **job hazard analysis**.
- Assemble the structure using hand and power tools.
- Check the work using levels, rules, and squares.

Carpenters also use powder-actuated and pneumatic tools and operate power equipment such as personnel lifts, equipment and material lifts, and small earth-moving equipment. More information about the various tools and construction methods used by carpenters is presented in other modules of this program.

2.2.0 Carpenter Responsibilities

A carpenter must be a responsible person with a high degree of concern for the safety of workers and the quality of the work.

2.2.1 Willingness to Take Responsibility

Every carpenter should take responsibility for working safely. In addition, most contractors expect their carpenters to see what needs to be done, then go ahead and do it. It is very tiresome to have to ask again and again that a certain job be done. Once responsibility has been delegated, you should continue to perform the duties without further direction.

2.2.2 Rules and Regulations

People can work together well only if there is some understanding about what work is to be done, when it will be done, and who will do it. Rules and regulations are a necessity in any work situation.

2.2.3 Tardiness and Absenteeism

Tardiness means being late for work and absenteeism means being off the job for one reason or another. Consistent tardiness and frequent absences are an indication of poor work habits, unprofessional conduct, and a lack of commitment to your contractor.

Work life is governed by the clock. All members of a carpentry crew are required to be at work at a specific time. Failure to get to work on time results in confusion, lost time, and resentment on the part of those who do come on time. In addition, frequent tardiness or absenteeism may lead to penalties, including dismissal. When accepting a job with a contractor, you agree to the terms of work. Perhaps it will allow you to see the picture more clearly if viewed from the supervisor's point of view. Supervisors cannot keep track of people if they come in any time they please. It is not fair to others to ignore tardiness. Failure to be on time may hold up the work of other carpenters and craftworkers. Better planning of your morning routine will often keep you from being delayed and so prevent a breathless, late arrival. In fact, arriving a little early indicates your interest and enthusiasm for your work, which is appreciated by contractors. The habit of being late is another one of those things that stand in the way of promotion.

It is sometimes necessary to take time off from work. No one should be expected to work when sick or when there is a serious issue at home. However, it is possible to get into the habit of letting unimportant and unnecessary matters keep you from the job. This results in lost production and hardship on those who try to carry on the work with less help. The contractor that hires you has a right to expect you to be on the job unless there is some very good reason for staying away. Certainly, do not let some trivial reason keep you home. Do not stay up late at night and be too tired to go to work the next day. If you are ill, spend the time at home to recover quickly. This, after all, is no more than what you would expect of a person you hired, and on whom you depended to do a certain job.

If it is necessary to stay home, then at least phone the office early in the morning so your supervisor can find another worker for the day, if needed. Some workers remain at home without contacting the contractor, which is the worst possible way to handle the matter. It leaves those at work uncertain about what to expect. They have no way of knowing whether you have merely been held up and will be in later, or whether immediate steps should be taken to assign your work to someone else. Courtesy alone demands that you let the supervisor know if you cannot come to work.

The most frequent causes of absenteeism are illness or death in the family, accidents, personal business, and dissatisfaction with the job. Some of the causes are legitimate and unavoidable, while others can be controlled. For most situations, you can carry on most personal business affairs after working hours. Frequent absences will reflect unfavorably on a worker when promotions are being considered.

Contractors sometimes resort to docking pay, demotion, and even dismissal in an effort to control tardiness and absenteeism. No contractor likes to impose penalties of this kind. However, in fairness to those workers who do come on time and who do not stay away from the job, a contractor is sometimes forced to discipline those who will not follow the rules.

The Customer

When you are on a job site, consider yourself to be working for both your contractor and the customer. If you are honest and maintain a professional attitude when interacting with customers, everyone will benefit. Your contractor will be pleased with your performance, and the customer will be happy with the work that is being done. Try seeing things from a customer's point of view. A good, professional attitude goes a long way toward ensuring repeat business.

Late for Work

Showing up on time is a basic requirement for just about every job. Your contractor is counting on you to be there at a set time, ready to work. While legitimate emergencies may arise that can cause you to be late for or even miss work, starting a bad habit of consistent tardiness is not something you want to do. What are the possible consequences that you could face as a result of tardiness and absenteeism?

2.3.0 Carpenter Characteristics

In addition to taking on the responsibilities assigned to a carpenter by a supervisor, the carpenter must also have strong personal characteristics to be successful on the job. These characteristics include:

- Professionalism
- Honesty
- Loyalty
- Willingness to learn
- Willingness to communicate
- Positive attitude

2.3.1 Professionalism

The word *professionalism* is a broad term that describes the desired overall behavior and attitude expected in the workplace. Most people would argue that professionalism must start at the top in order to be successful. It is true that management support of professionalism is important to its success in the workplace, but it is more important that individuals recognize the importance of professionalism.

Professionalism can be demonstrated in a variety of ways every minute you are on the job site. Most important is that you do not tolerate the unprofessional behavior of co-workers. This is not to say that you shun the unprofessional worker; instead, you work to demonstrate the benefits of professional behavior.

Professionalism is a benefit both to the contractor and the worker. It is a personal responsibility. Our industry is what each individual chooses to make of it; choose professionalism and the industry image will follow.

2.3.2 Honesty

Honesty and personal integrity are important characteristics of a successful professional. Professionals pride themselves on performing a job well and on being punctual and dependable. Each job is completed in a professional way, never by cutting corners or reducing materials. A valued professional maintains work attitudes and ethics that protect property such as tools and materials belonging to contractors, customers, and other trades from damage or theft at the shop or job site.

Honesty and success go hand-in-hand. It is not simply a choice between good and bad, but a choice between success and failure. Dishonesty will always catch up with you. Whether a person is stealing materials, tools, or equipment from the job site or simply lying about their work, it will

Smartphone Apps for Construction

Smartphones are becoming an increasingly popular form of communication, and also offer a great deal of versatility for craftworkers. Smartphone cameras can be used to document on-the-job activities or potential safety violations. Best practices can be communicated to crew members using video clips. Construction calculators can be downloaded, providing craftworkers with the same (or even greater) capabilities than a handheld calculator.

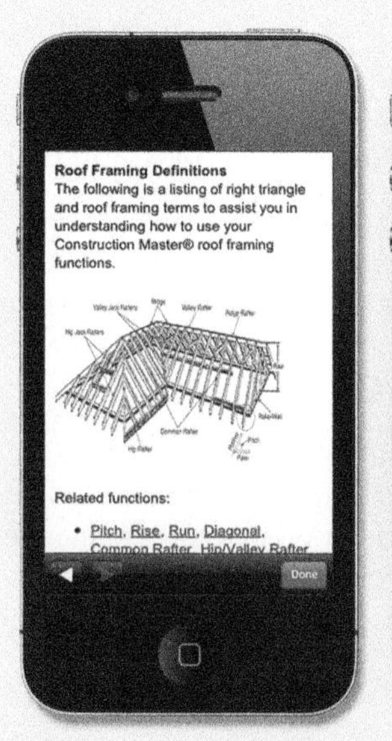

27101-13_SA05.EPS

not take long for the contractor to find out. Of course, a carpenter can always go and find another contractor, but this option will ultimately run out on them.

To be successful and enjoy continuous employment, consistency of earnings, and being sought after as opposed to seeking employment, then start out with the basic understanding of honesty in the workplace and you will reap the benefits. Honesty means more than giving a fair day's work for a fair day's pay; it means carrying out your side of a bargain; it means that your words convey true meanings and actual happenings. All thoughts as well as your actions should be honest. Contractors place a high value on an employee who is honest.

2.3.3 Loyalty

Craftworkers expect contractors to look out for their interests, to provide them with steady employment, and to promote them to better jobs as openings occur. Contractors feel that they, too, have a right to expect their workers to be loyal to them—to keep their interests in mind; to speak well of them to others; to keep any minor troubles strictly within the plant or office; and to keep absolutely confidential all matters that pertain to the business. Both contractors and workers should keep in mind that loyalty is not something to be demanded; rather, it is something to be earned.

2.3.4 Willingness to Learn

Every contractor has a unique way of doing things. Contractors expect their workers to be willing to learn these ways. Also, it is necessary to adapt to change and be willing to learn new methods and procedures as quickly as possible. Sometimes the purchase of new tools or equipment makes it necessary for even experienced carpenters to learn new methods and operations. It is often the case that craftworkers resent having to accept improvements because of the retraining that is involved. However, contractors will no doubt think they have a right to expect workers to put forth the necessary effort. Methods must be kept current to meet competition and show a profit. It is this profit that enables the contractor to continue in business and that provides jobs for the worker.

2.3.5 Willingness to Cooperate

To cooperate means to work together. In our modern business world, cooperation is the key to getting things done. Learn to work as a member of a team with your contractor, supervisor, and fellow workers in a common effort to get the work done efficiently, safely, and on time. Many people underestimate the importance of cooperating with others.

The term *human relations* is often associated with the willingness to cooperate. There is a tendency to pass off human relations as nothing more than common sense. What exactly is involved in

Ethical Principles for Members of the Construction Trades

Honesty: Be honest and truthful in all dealings. Conduct business according to the highest professional standards. Faithfully fulfill all contracts and commitments. Do not deliberately mislead or deceive others.

Integrity: Demonstrate personal integrity and the courage of your convictions by doing what is right even when there is great pressure to do otherwise. Do not sacrifice your principles for expediency, be hypocritical, or act in an unscrupulous manner.

Loyalty: Be worthy of trust. Demonstrate fidelity and loyalty to companies, contractors, fellow craftworkers, and trade institutions and organizations.

Fairness: Be fair and just in all dealings. Do not take undue advantage of another's mistakes or difficulties. Fair people display a commitment to justice, equal treatment of individuals, tolerance for and acceptance of diversity, and open-mindedness.

Respect for others: Be courteous and treat all people with equal respect and dignity regardless of sex, race, or national origin.

Law abiding: Abide by laws, rules, and regulations relating to all personal and business activities.

Commitment to excellence: Pursue excellence in performing your duties, be well informed and prepared, and constantly endeavor to increase your proficiency by gaining new skills and knowledge.

Leadership: By your own conduct, seek to be a positive role model for others.

human relations? One response would be to say that part of human relations is being friendly, pleasant, courteous, cooperative, adaptable, and sociable. Human relations is much more than just getting people to like you; it is also knowing how to handle difficult situations as they arise.

Human relations is knowing how to work with supervisors who are often demanding and may sometimes seem unfair. It is understanding the personality traits of others as well as yourself. Human relations is building sound working relationships in situations where others are forced on you. Human relations is knowing how to restore working relationships that have deteriorated for one reason or another. It is learning how to handle frustrations without hurting others. Human relations is building and maintaining relationships with all kinds of people, whether those people are easy to get along with or not.

Effective human relations is directly related to productivity. Productivity is the key to business success. Every craftworker is expected to produce at a certain level. Contractors quickly lose interest in a worker who has a great attitude but is able to produce very little. There are work schedules to be met and jobs that must be completed.

All craftworkers, both new and experienced, are measured by the amount of quality work they can safely turn out. The contractor expects all workers to do their fair share of the workload.

However, doing one's share in itself is not enough. To be productive, do your share (or more than your share) without antagonizing your fellow workers. Perform your duties in a manner that encourages others to follow your example. It makes little difference how ambitious you are or how capably you perform. You cannot become the kind of worker you want to be, or the type of worker a contractor wants you to be, without learning how to work with your peers.

2.3.6 Positive Attitude

A positive attitude is essential to a successful career. First, being positive means being energetic, highly motivated, attentive, and alert. A positive attitude is essential to safety on the job. Second, a worker with a positive attitude contributes to the productivity of others. Both negative and positive attitudes are transmitted to others on the job. A persistent negative attitude can spoil the positive attitudes of others. It is very difficult to maintain a high level of productivity while working next to a person with a negative attitude. Third, people favor a person who is positive. Being positive makes a person's job more interesting and exciting. Fourth, the kind of attitude transmitted to management has a great deal to do with a worker's future success in the company. Supervisors can determine a subordinate's attitude by their approach to the job, reactions to directives, and the way they handle problems.

A positive attitude is far more than a smile, which is only one example of an inner positive attitude. As a matter of fact, some people transmit a positive attitude even though they seldom smile. They do this by the way they treat others, the way they look at their responsibilities, and the approach they take when faced with problems.

Here are a few suggestions that will help you to maintain a positive attitude:

- Remember that your attitude follows you wherever you go. If you make a greater effort to be a more positive person in your social and personal lives, it will automatically help you on the job. The reverse is also true. One effort will complement the other.
- Negative comments are seldom welcomed by fellow workers on the job. Neither are they welcome on the social scene. The solution: talk about positive things and be complimentary. Constant complainers do not build healthy and fulfilling relationships.

Working with Other Trades

Cooperation among the various trades at the job site and respect for the work of other trades are essential to achieve a smooth-running project.

On many well-run jobs, a sense of togetherness (cooperation) soon develops and the trades work together in harmony. Many times, there is even a trade-off of activities that allows the project to progress at a uniform pace. The carpenter may, for example, build a storage locker for the electricians, who may in turn provide the carpenters with additional outlets. Such practices are fine as long as the material used does not short the project of materials needed for completion and can be done without additional cost.

The point to be made here, from the craftworker's standpoint, is that cooperation among the many trades on the job makes the work flow better and helps ensure timely completion of the project.

- Look for the good things in people on the job, especially your supervisor. Nobody is perfect, but almost everyone has a few worthwhile qualities. If you dwell on people's good features, it will be easier to work with them.
- Look for the good things where you work. What are the factors that make it a good place to work? Is it the hours, the physical environment, the people, the actual work being done, or is it the atmosphere? Keep in mind that you cannot be expected to like everything. No work assignment is perfect, but if you concentrate on the good things, the negative factors will seem less important and bothersome.
- Look for the good things in the contractor. Just as there are no perfect assignments, there are no perfect contractors. Nevertheless, almost all organizations have good features. Is the contractor progressive? What about promotional opportunities? Are there chances for self-improvement? What about the wage and benefit package? Is there a good training program? You cannot expect to have everything you would like, but there should be enough to keep you positive. In fact, if you decide to stick with a contractor for a long period of time, it is wise to look at the good features and think about them. If you think positively, you will act the same way.
- You may not be able to change the negative attitude of another worker, but you can protect your own attitude from becoming negative.

Trade Coordination

Carpenters are generally on the job before most other trades. Therefore, an important part of trade coordination is ensuring that the work done by the carpenters will accommodate the other trades. For example, plans for location of heating and air conditioning duct runs, plumbing fixtures and piping, and electrical components and wiring runs must be continually reviewed by the superintendent or supervisor to ensure that carpenters are allocating space, in the correct locations, for these facilities.

Teamwork

Many of us like to follow all sorts of different teams: racing teams, baseball teams, football teams, and soccer teams. Just as in sports, a job site is made up of a team. As a part of that team, you have a responsibility to your teammates. What does teamwork really mean on the job? Craftworkers must sincerely do everything they can to build strong, professional working relationships with fellow craftworkers, supervisors, and customers.

2.0.0 Section Review

1. Which of the following is *not* considered to be finish carpentry?
   a. Installing staircase railings
   b. Building a built-in bookcase
   c. Erecting concrete forms
   d. Installing baseboard trim

2. If you are absent from work due to illness, it is not necessary to contact your employer as they will assume you are not showing up.
   a. True
   b. False

3. The broad term that describes the desired overall behavior and attitude expected in the workplace is _____.
   a. integrity
   b. loyalty
   c. human relations
   d. professionalism

SECTION THREE

3.0.0 SKILLSUSA

Objective

Summarize how to be connected to the industry through an organization like SkillsUSA.

a. Describe the program, curriculum, and SkillsUSA Championships.
b. State the benefits from being a SkillsUSA member.
c. List the seven goals of the SkillsUSA Program of Work.

SkillsUSA is a partnership of students, teachers, and industry representatives working together to ensure America has a skilled workforce. SkillsUSA is a national organization serving teachers and high school and college students who are preparing for careers in technical, skilled, and service occupations, including carpentry and other building trades occupations. More than 320,000 students and advisers join SkillsUSA annually, who are organized into more than 17,000 sections and 54 state and territorial associations. Combining alumni and lifetime membership, the total number impacted is more than 320,000 students and advisers. SkillsUSA has served more than 10.5 million members in its history.

The mission of SkillsUSA is to assist its members in becoming world-class workers, leaders, and responsible citizens. SkillsUSA is an applied method of instruction for preparing America's high-performance workers in public career and technical programs. It provides quality education experiences for students in leadership, teamwork, citizenship, and character development. SkillsUSA builds and reinforces self-confidence, positive work attitudes, and communications skills. It emphasizes total quality at work: high ethical standards, superior work skills, lifelong education, and pride in the dignity of work. SkillsUSA also promotes understanding of the free-enterprise system and involvement in community service. SkillsUSA helps each student to excel.

3.1.0 Program, Curriculum, and SkillsUSA Championships

SkillsUSA programs help to establish industry standards for job skill training in the lab and classroom, and promote community service. SkillsUSA is recognized by the US Department of Education and is cited as a "successful model of employer-driven youth development training program" by the US Department of Labor.

SkillsUSA Championships include local, state, and national competitions in which students demonstrate occupational and leadership skills. The SkillsUSA Championships is the showcase for the best career and technical students in the nation (*Figure 4*). This is a multimillion-dollar event that occupies a space equivalent to 16 football fields. In 2011, there were more than 5,700 contestants in 94 separate events. Nearly 1,500 judges and contest organizers from labor and management make the national event possible. The philosophy of the Championships is to reward students for excellence, to involve industry in directly evaluating student performance and to keep training relevant to employers' needs. The national carpentry competition at the SkillsUSA Championships is sponsored by NCCER.

3.2.0 Membership

In 2011, more than 16,600 teachers and school administrators served as professional SkillsUSA members and advisers. More than 1,100 business, industry, and labor sponsors actively support SkillsUSA at the national level through financial aid, in-kind contributions, and involvement of their people in SkillsUSA activities. Many more work directly with state associations and local chapters. NCCER and SkillsUSA have a long-standing partnership, as both organizations share the goal of a skilled workforce.

3.2.1 The Value for Students

For many students, SkillsUSA is the first professional organization they will join. The experiences and knowledge gained provide an excellent platform for career development and success. SkillsUSA also sets the stage for involvement in other professional and service organizations. Advantages include:

- Teamwork and leadership development
- Reinforcement of employability skills
- Competition in a nationally recognized contest program

Learn more about the SkillsUSA Championships:

http://www.skillsusa.org/compete/skills.shtml

27101-13_F04.EPS

Figure 4 SkillsUSA Championships.

- Community service opportunities
- Access to scholarships
- Networking with potential employers

3.2.2 The Value for the Classroom and School

Great instructors are always looking for ways to engage students and build relationships. SkillsUSA provides the tools to do both. As a student-run organization, members feel a sense of empowerment and belonging. SkillsUSA is a motivator for students to put forth their best effort in the classroom, making daily lessons even more relevant to career success. As a SkillsUSA adviser, the activities, projects, and contests provide opportunities for instructors to build stronger relationships with students. Chapter activities and accomplishments can build a positive image for your program and your school. Benefits include:

- Recognition for the school within the community
- Opportunities to meet educational standards
- Development of career technical education (CTE) pathways
- Improved recruitment and enrollment
- More graduates equipped with essential skills

3.3.0 National Program of Work Standards

The heart of SkillsUSA is the Program of Work (POW), or what your chapter is going to do during the school year. It is the activities and projects—the plan of action—that your chapter will carry out.

The national Program of Work sets the pace for SkillsUSA nationwide. The expectation is that each chapter will carry out this Program of Work. All of the SkillsUSA programs are in some way related to the following seven major goals: professional development, community service, employment, ways and means, SkillsUSA Championships, public relations, and social activities.

Professional development – The goal of professional development is to prepare each SkillsUSA member for entry into the workforce and provide a foundation for success in a career. Becoming a professional does not stop with acquiring a skill, but involves an increased awareness of the meaning of good citizenship and the importance of labor and management in the world of work.

Community service – The goal of the community service standard is to promote and improve goodwill and understanding among all segments of the community through services donated by SkillsUSA chapters. In addition, SkillsUSA hopes to instill in its members a lifetime commitment to community service.

Employment – The goal of this standard is to increase student awareness of quality job practices and attitudes, and to increase the opportunities for employer contact and eventual employment.

Ways and means – The ways and means goal is to plan and participate in fundraising activities to allow all members to carry out the chapter's projects.

SkillsUSA Championships – The goal of the SkillsUSA Championships is to offer students the opportunity to demonstrate their skills and be recognized for them through competitive activities in occupational areas and leadership.

Public relations – The goal of the public relations standard is to make the general public aware of the good work that students in career and technical education are doing to better themselves and their community, state, nation, and world.

Social activities – The goal of this standard is to increase cooperation in the school and community through activities that allow SkillsUSA members to get to know each other in something other than a business or classroom setting.

Learn more about the Program of Work standards:

http://www.skillsusa.org/educators/chapmanage5.shtml
http://skillsusa.org/courses/07_Program/player.html

3.3.1 Chapter Activity Planner

Chapter members should discuss and develop their own program of work. Your instructor will assist in selecting activities that relate to your vocational training and will guide you in developing your personal skills in communications, organization, planning, and follow-through.

Chapter activities will provide some of the best opportunities for you to learn by doing. A successful program of work creates a positive learning atmosphere in the classroom, and allows you to learn how to accept responsibility, work as a team, manage a budget, and handle success and failure.

3.3.2 Chapter Elections and Training

Effective chapter officers ensure the chapter functions effectively and efficiently. Officers frequently are responsible for routine management tasks, such as organizing meetings, conducting meetings, scheduling work, and leading chapter activities. This helps you learn simple supervisory skills and creates responsible team spirit.

The election of chapter SkillsUSA officers is often one of the highlights of the SkillsUSA year. The outcome of the election affects the entire group's chances for having a successful program. Officers not only spark enthusiasm in the organization, but also carry on the routine business affairs that keep the program moving.

Being elected as a SkillsUSA officer gives you an opportunity to hone your leadership abilities. The officer selection process is an excellent way for you to learn valuable, practical lessons in leadership and teamwork.

3.3.3 Chapter Meetings

Valuable skills are learned and practiced through organized activities. Well-run meetings are a good example. Learning how to plan and work cooperatively with others is an important skill set. As individuals, we all have good ideas, but when people combine their ideas and efforts, great things can occur.

3.0.0 Section Review

1. The national carpentry competition at the SkillsUSA Championships is sponsored by _____.
   a. the International Brotherhood of Carpenters
   b. local contractors
   c. NCCER
   d. chapter officers

2. The number of business, industry, and labor sponsors that actively support SkillsUSA at the national level is _____.
   a. 60
   b. 600
   c. 900
   d. more than 1,100

3. Which of the following is *not* a goal of the SkillsUSA Program of Work?
   a. Community service
   b. Professional development
   c. Social activities
   d. Improved recruitment and enrollment

Section Four

4.0.0 Safety in the Construction Industry

Objective

Explain the importance of safety in the construction industry, and describe the obligations of the contractor, subcontractors, and you to ensure a safe work environment.

a. Describe the OSHA Outreach Training Program.
b. Explain hazard recognition and define your role in it.

Trade Terms

Abate: To reduce or minimize.

Occupational Safety and Health Administration (OSHA): An agency of the US Department of Labor whose mission is to set occupational safety and health standards for all places of employment, enforce these standards, ensure that employers provide and maintain a safe workplace for all employees, and provide research and educational programs to support safe working practices.

Personal protective equipment (PPE): Equipment or clothing designed to prevent or reduce injuries.

Safety culture: The culture created when the whole company sees the value of a safe work environment.

Site-specific safety program: A safety program developed for a job site that identifies and takes into account any specific potential hazards that may be encountered.

An obligation is like a promise or a contract. In exchange for the benefits of your employment and your own well-being, you agree to work safely. In other words, you are obligated to work safely. You are also obligated to make sure anyone you happen to supervise or work with is working safely. Your contractor is also obligated to maintain a safe workplace for all employees. Safety is everyone's responsibility (*Figure 5*).

Creating and maintaining a safety culture is an ongoing process that includes a sound safety structure and attitude, and relates to both organizations and individuals. While you are ultimately responsible for yourself, safety is everyone's responsibility.

27101-13_F05.EPS

Figure 5 Safety is everyone's responsibility.

Some contractors have safety committees. If you work for such a contractor, you are then obligated to that committee to maintain a safe work environment. This means three things:

- Follow the safety committee's rules for proper work procedures and practices.
- Report any unsafe equipment and conditions directly to the committee or your supervisor.
- Be willing to take an active role in implementing changes to the company safety policy, if necessary, and actively support changes.

Here is a basic rule to follow every working day: if you see something that is not safe, report it! Do not ignore it. It will not correct itself. While you may not have the authority to correct unsafe conditions encountered on the job site, you are obligated to report the issue.

Suppose you see a faulty electrical connection. You know enough to stay away from it, and you do. But then you forget about it. Why should you worry? It is not going to hurt you. Let somebody else deal with it. The next thing that happens is that a co-worker accidentally touches the live wire.

In the long run, even if you do not think an unsafe condition affects you—it does. Do not mess around; report what is not safe. Do not think your contractor will be angry because your productivity suffers while the condition is corrected. On the contrary, the contractor will be more likely to criticize you for not reporting a safety issue.

Contractors know that the short time lost in making conditions safe again is nothing compared with shutting down the whole job because of a major disaster. If that happens, you are out of work anyway. Do not ignore an unsafe condition. In fact, Occupational Safety and Health Administration (OSHA) regulations require you to report hazardous conditions. This applies to every

part of the construction industry. Whether you work for a large contractor or a small subcontractor, you are obligated to report unsafe conditions. The easiest way to do this is to tell your supervisor. If that person ignores the unsafe condition, report it to the next highest supervisor. If it is the owner who is being unsafe, let that person know your concerns. If nothing is done about it, report it to OSHA. If you are worried about your job being on the line, think about it in terms of your life, or someone else's life, being on the line.

The US Congress passed the Occupational Safety and Health Act in 1970. This act also created the Occupational Safety and Health Administration (OSHA), which is part of the US Department of Labor. The mission of OSHA is to set occupational safety and health standards for all places of employment, enforce these standards, ensure that employers provide and maintain a safe workplace for all employees, and provide research and educational programs to support safe working practices.

OSHA requires each employer to provide a safe and hazard-free work environment. OSHA also requires that employees comply with OSHA rules and regulations that relate to their conduct on the job. To gain compliance, OSHA can perform spot inspections of job sites, impose fines for violations, and even stop work from proceeding until the job site is safe.

According to OSHA standards, you are entitled to on-the-job safety training. As a new worker, you must be:

- Oriented to the company policies and procedures
- Trained how to do your job safely
- Oriented to the specific job that you are working on
- Provided with the required personal protective equipment (PPE)
- Warned about specific hazards pertaining to the project
- Supervised for safety while performing the work

OSHA was adopted with the stated purpose "to assure so far as possible every working man and woman in the Nation safe and healthful working conditions and to preserve our human resources."

Enforcement of the Occupational Safety and Health Act is provided by federal and state safety inspectors who have the legal authority to make employers pay fines for safety violations. The law allows states to have their own safety regulations and agencies to enforce them, but they must first be approved by the US Secretary of Labor. For states that do not develop such regulations and agencies, federal OSHA standards must be obeyed. These standards are listed in *OSHA Safety and Health Standards for the Construction Industry (29 CFR, Part 1926)*, sometimes called *OSHA Standards 1926*. Other safety standards that apply to the carpentry trade are published in *OSHA Safety and Health Standards for General Industry (29 CFR, Parts 1900 to 1910)*.

The most important general requirements that OSHA places on employers in the construction industry are:

- The employer must perform frequent and regular job site inspections of equipment and conditions of the job site.
- The employer must train all employees to recognize and avoid unsafe conditions, and to know the regulations that pertain to the job, so they may control or eliminate any hazards.
- No one may use any tools, equipment, machines, or materials that do not comply with *OSHA Standards 1926*.
- The employer must ensure that only qualified individuals operate tools, equipment, and machines. For some tools and equipment, such as powder-actuated tools (PATs) and construction forklifts, specialized training is required and only those workers that have successfully completed the training can use such tools or equipment.

4.1.0 OSHA Outreach Training Program

The OSHA Outreach Training Program for the Construction Industry is a volunteer, but recommended, program that teaches workers about their rights, contractor responsibilities, and complaint procedures, as well as how to identify, abate, avoid, and prevent job-related hazards. Even though OSHA considers this to be a voluntary program, some jurisdictions, contractors, and unions require this outreach training program

Drugs and Alcohol

When people use drugs and alcohol, they are putting both themselves and the people around them at serious risk. A construction site can be a dangerous environment, and it is important to be alert at all times. Using drugs and alcohol on the job is an accident waiting to happen. A positive drug screening may result in dismissal, denial of worker's compensation (if there is an injury), and the possibility of criminal charges (criminally negligent). You have an obligation to yourself, your employer, and your fellow employees to work safely. What should you do if you discover someone abusing drugs and/or alcohol at work?

before allowing anyone to work on construction sites, and to fulfill their safety training goals. The 10-hour construction industry program, also known as OSHA-10, is intended for entry-level workers. The OSHA-10 program covers a variety of construction safety and health hazards encountered on the job site, with emphasis on hazard identification, avoidance, control, and prevention, and not on the OSHA standards themselves. Mandatory topics included in the OSHA-10 program include:

- Focus Four Hazards (falls, electrocution, struck-by, and caught-in or between)
- Personal protective equipment (PPE) and lifesaving equipment
- Health hazards in construction

In addition, some of the following topics may be addressed, depending on the contractor and/or type of construction work the contractor is performing:

- Cranes, derricks, hoists, elevators, and conveyors
- Excavations
- Materials handling, storage, use, and disposal
- Scaffolds
- Stairways and ladders
- Hand and power tools

4.2.0 Site-Specific Safety Program

The process of hazard recognition, evaluation, and control is the foundation of an effective safety program. When hazards are identified and assessed, they can be addressed quickly, reducing the hazard potential. Simply put, the more aware you are of your surroundings and the dangers in them, the less likely you are to be involved in an accident. A site-specific safety program should be developed. The best approach in determining if a situation or equipment is potentially hazardous is to ask yourself these questions:

- How can this situation or equipment cause harm?
- What types of energy sources are present that can cause an accident?
- What is the magnitude of the energy?
- What could go wrong to release the energy?
- How can the energy be eliminated or controlled?
- Will I be exposed to any hazardous materials?
- What precaution or action plan can I put in place to eliminate the chance of injury or accident?

There are a number of ways to recognize hazards and potential hazards on a job site. Some techniques are more complicated than others. In order to be effective, they all must answer this question: what could go wrong with this situation or operation? No matter what hazard recognition technique you use, answering that question in advance will save lives and prevent equipment damage.

4.2.1 Job Hazard Analysis (JHA)

Performing a job hazard analysis (JHA), or job safety analysis (JSA), is one approach to hazard recognition. In a JHA, the task to be performed is broken down into its individual parts or steps and then each step is analyzed for its potential hazards. Once a hazard is identified, certain actions or procedures are recommended that will correct that hazard before an accident occurs. For example, during a JHA, it is determined that using workers will need to install flashing along the roof edge during the course of work. To alert workers of the roof edge, warning lines (yellow ropes and ribbons) will need to be installed at the roof edges. If workers are required to work outside the warning line, they must be tied off to an approved anchor and use a harness, rope, and rope grab. The rope grab should be positioned in such a way to prevent a fall from the leading edge whenever feasible. *Figure 6* shows an example of a form used to conduct a job safety analysis.

Learn more about the Occupational Safety and Health Administration:

http://www.osha.gov/law-regs.html

JOB HAZARD ANALYSIS FORM

Job Title: Date of Analysis:
Job Location: Conducted by:
PPE: Staffing:
Tools, Materials and Equipment: Duration:

Step	Hazards	Quality Concern	Environmental Concern	New Procedure or Protection

27101-13_F06.EPS

Figure 6 Job hazard analysis form.

4.0.0 Section Review

1. The Occupational Safety and Health Administration is part of the US Department of _____.
   a. Education
   b. Labor
   c. Health and Human Services
   d. Commerce

2. Which of the following is *not* a mandatory topic in the OSHA-10 program?
   a. Hand and power tools
   b. Personal protective equipment
   c. Focus Four Hazards
   d. Construction health hazards

3. When the task to be performed is broken down into its individual parts or steps and then each step is analyzed for its potential hazards, it is referred to as a(n) _____.
   a. material safety data sheet
   b. OSHA Form 300
   c. job hazard analysis
   d. safety plan

SUMMARY

There are many career opportunities for skilled carpenters. An apprenticeship program that combines competency-based, hands-on training with classroom instruction has proven to be the most effective means for a person to learn and advance in the carpentry craft. Developing job skills is only part of the solution; it is just as important to learn good work habits, convey a positive, cooperative attitude to those around you, and practice good safety habits every day.

Review Questions

1. The construction worker most likely to become knowledgeable about many trades is the _____.
   a. electrician
   b. carpenter
   c. plumber
   d. mason

2. The group that is responsible for the enforcement of apprenticeship standards and also ensures that proper apprentice training is conducted is the _____.
   a. contractor
   b. Occupational Safety and Health Administration
   c. Department of Commerce
   d. Apprenticeship Committee

3. The term used to describe the overall behavior and attitude expected in the workplace is _____.
   a. absenteeism
   b. responsibility
   c. tardiness
   d. professionalism

4. The *Code of Federal Regulations (CFR) 29:30* specifies requirements for _____.
   a. on-the-job safety classes
   b. supervisory training
   c. fall prevention procedures
   d. approved apprenticeship programs

5. The purpose of the Youth Apprenticeship Program is to _____.
   a. make sure all young people know how to use basic carpentry tools
   b. provide job opportunities for people who quit high school
   c. allow students to start in an apprenticeship program while still in high school
   d. make sure that people under 18 have proper supervision on the job

6. The work of a crew of craftworkers and laborers is usually directed by a(n) _____.
   a. lead carpenter
   b. apprentice supervisor
   c. general contractor
   d. carpenter/instructor

7. The foundation of an effective safety program is _____.
   a. holding frequent safety meetings
   b. using a process of hazard recognition, evaluation, and control
   c. rewarding workers who don't suffer injuries
   d. developing a master accident prevention campaign

8. Which of the follow statements about SkillsUSA is *not* true?
   a. It promotes understanding of the free enterprise system.
   b. It has more than 500,000 active members.
   c. It emphasizes high ethical standards and pride in the dignity of work.
   d. It consists of 54 state and territorial associations.

9. A combined total of 8,000 hours on-the-job and classroom training is needed for a carpentry apprentice to _____.
   a. become a master carpenter
   b. complete a degree in construction technology
   c. advance to journeyman
   d. receive a competency certificate

10. An important purpose of OSHA is to _____.
    a. catch people breaking safety regulations
    b. make rules and regulations governing all aspects of construction projects
    c. ensure that the employer provides and maintains a safe workplace
    d. assign a safety inspector to every project

11. Which of the following is *not* an advantage provided to students by SkillsUSA membership?
    a. Teamwork and leadership development
    b. Income-earning opportunities
    c. Community service opportunities
    d. Networking with potential employers

12. The Occupational Safety and Health Act was passed by the US Congress in _____.
    a. 1932
    b. 1956
    c. 1970
    d. 1998

13. Which of the following is *true* with respect to honesty?
    a. It is okay to borrow tools from the job site as long as you return them before anyone notices.
    b. You are doing your company a favor by using lower-grade materials than those listed in the specifications.
    c. It is okay to take materials or tools from your employer if you feel the company owes you for past efforts.
    d. Being late and not making up the time is the same as stealing from your employer.

14. If you must miss a day of work, you should _____.
    a. make up a good excuse
    b. call in early in the morning to tell your supervisor you won't be there
    c. ask a co-worker to let the supervisor know you'll be out
    d. deal with the problem tomorrow

15. When should an unsafe condition be reported to OSHA?
    a. As soon as it is discovered
    b. Never
    c. If company supervisors ignore your report
    d. After it causes an injury

16. A prospective carpenter apprentice must have a high school diploma or GED.
    a. True
    b. False

17. A safety program developed to meet the requirements of a specific situation and location is referred to as _____.
    a. locally oriented
    b. job-specific
    c. targeted
    d. site-specific

18. The activities and projects a SkillsUSA chapter will carry out during a school year is outlined in the _____.
    a. Master Schedule of Events
    b. National Program of Work
    c. Activities Planner and Guide
    d. National Action Plan

19. If one of your co-workers complains about your company, you should _____.
    a. contribute your own complaints to the conversation
    b. agree with the person to avoid conflict
    c. suggest that the person look for another job
    d. find some good things to say about the company

20. Since carpenters are usually on the job before other trades, their work must accommodate the other trades. This is called trade _____.
    a. etiquette
    b. coordination
    c. sequencing
    d. articulation

Trade Terms Quiz

Fill in the blank with the correct term that you learned from your study of this module.

1. The installation of roof trusses is a type of _______.
2. The building materials required for the completion of a project are listed on a _______.
3. The installation of kitchen cabinets is a type of _______.
4. When management and its craftworkers view safety as a benefit in the work environment, they create a _______.
5. A profession, such as carpentry, in which a person trains and receives personal satisfaction for a lifetime is referred to as a _______.
6. A man-made building material, such as plastic molding, is considered to be _______.
7. When preparing to move a load of 2 × 4s from the storage area to the second floor a building, a worker should first perform a _______.
8. To _______ is to reduce or minimize the number of accidents on a job site.
9. If an accident occurs on a job site, it may be investigated by the _______.
10. Safety goggles and a hard hat are two types of _______.
11. Information about specific hazards identified on a job site and potential solutions is found in a(n) _______.

Trade Terms

Abate
Career
Finish carpentry
Job hazard analysis (JHA)
Material takeoff
Occupational Safety and Health Administration (OSHA)
Personal protective equipment (PPE)
Rough carpentry
Safety culture
Site-specific safety program
Synthetic

Cornerstone of Craftsmanship

Hal Heintz

Director of Workforce Development
Associated Builders and Contractors
South Texas Chapter

How did you get started in the construction industry?
Being of an inquisitive nature, my DNA has prompted me to seek out why and how things are built and put together. As a young boy my parents knew I had great mechanical aptitude and nurtured me to be the best I could be at anything I aimed at.

Who or what inspired you to enter the industry?
My desire to enter the industry came about because of my upbringing, working on a family dairy farm and being able to operate farm equipment at a young age. My father was a competent carpenter/woodworker also.

What do you enjoy most about your career?
Satisfaction is what I enjoy the most. There is personal satisfaction in knowing I am helping to build a stronger America. Helping others as an instructor, mentor, and advocate for doing what is right has carried me to great heights.

Why do you think training and education are important in construction?
A person should be guided to aim at a goal or target in life. Training and education not only lend to aiming and guiding, but also to be proficient and morally responsible. Theory and knowledge and hard work and sacrifice are key ingredients for success. Aptitude from training and education, a positive, enforced attitude when combined bring altitude on the ladder of success,

Why do you think credentials are important in construction?
"Been there, done that, and got the T-shirt" is a common reply. It can be taken as hearsay; some say "show me the money." I say "show me your credentials." Credentials speak volumes for sacrifice, ability, and character.

How has training/construction impacted your life and career?
I was given the opportunity to teach carpentry in 1999. It has taken me years of hard work, night classes, and giving up of my private life to get through apprenticeship and eventually attain my BS in 1995. I was ready to give back to this great industry that had afforded me and my family great rewards. My career has opened many doors to recognition and job satisfaction.

Would you recommend construction as a career to others? Why?
A career is a progression through life. It can be of enormous value so one can achieve and excel in things that bring joy as a person. Not only is the profession rewarding and challenging, but it can bring out of an individual ideas and concepts that when used for good will positively impact and inspire others and move this nation forward.

What does craftsmanship mean to you?
It means using your hands to shape ideas, your mind to convey meaning and thought, and your heart to feel self-satisfaction knowing your skills are of value, your work is above reproach, and the outcome of your labor lends meaning to the words value and honesty.

Trade Terms Introduced in This Module

Abate: To reduce or minimize.

Career: A profession for which an individual trains and considers to be a lifelong calling.

Finish carpentry: The portion of the carpentry trade associated with interior and exterior trim, cabinetry, siding, wall finishes, and decorative work.

Job hazard analysis: An approach to hazard recognition in which the task to be performed is broken down into its individual parts or steps, and then each step is analyzed for its potential hazard(s). When the hazard is identified, certain actions or procedures are recommended that will correct the hazard before an accident occurs.

Material takeoff: A list of building materials obtained by analyzing the project drawings (also known as a takeoff).

Occupational Safety and Health Administration (OSHA): An agency of the US Department of Labor whose mission is to set occupational safety and health standards for all places of employment, enforce these standards, ensure that employers provide and maintain a safe workplace for all employees, and provide research and educational programs to support safe working practices.

Personal protective equipment (PPE): Equipment or clothing designed to prevent or reduce injuries.

Rough carpentry: The portion of the carpentry trade associated with framing and other work that will be covered with finish materials.

Safety culture: The culture created when the whole company sees the value of a safe work environment.

Site-specific safety program: A safety program developed for a job site that identifies and takes into account any specific potential hazards that may be encountered.

Synthetic: Man-made (not naturally occurring); typically a result of a chemical reaction.

Samples of NCCER Credentials

NCCER
The Standard for Developing Craft Professionals

This is to certify that

Steven Whitaker

has fulfilled the requirements for

Carpentry Level One

in NCCER's standardized training curriculum
on this Sixteenth day of September, 2012

Donald E. Whyte
President, NCCER

27101-13_A01.EPS

THE STANDARD FOR DEVELOPING CRAFT PROFESSIONALS

13614 Progress Boulevard, Alachua, Florida 32615 • p. 888.622.3720 f. 386.518.6255 • www.nccer.org

Official Transcript

January 17, 2012

NCCER Card #: 1720726
Trainee Name: John Q Smith
Sponsor: Austin Industrial Incorporated
Address: 2801 E 13th St
La Porte, TX 77571

Current Employer/School:
Solomon Plumbing Company

Module	Description	Instructor	Training Location	Date Completed
00101-04	Basic Safety	Kevin Jenkins	Solomon Plumbing Company	2/20/2008
00102-04	Introduction to Construction Math	Dave Buck	Building Trades Institute, LLC	8/8/2008
00103-04	Introduction to Hand Tools	Kevin Jenkins	Solomon Plumbing Company	1/1/2008
00104-04	Introduction to Power Tools	Dave Buck	Building Trades Institute, LLC	8/8/2008
00105-04	Introduction to Blueprints	Kevin Jenkins	Solomon Plumbing Company	3/20/2008
00106-04	Basic Rigging	Dave Buck	Building Trades Institute, LLC	8/8/2008
00108-04	Basic Employability Skills	Rod Blackburn	Utility Contractors, Inc.	3/15/2009
02101-05	Introduction to the Plumbing Profession	Kevin Jenkins	Solomon Plumbing Company	3/22/2008
26101-02	Electrical Safety	Don Whyte	National Center for Construction Education &	7/29/2002
26102-02	Hand Bending	Don Whyte	National Center for Construction Education &	7/29/2002
26103-02	Fasteners and Anchors	Don Whyte	National Center for Construction Education &	7/29/2002
26104-02	Electrical Theory One	Don Whyte	National Center for Construction Education &	7/29/2002
26105-02	Electrical Theory Two	Don Whyte	National Center for Construction Education &	7/29/2002
26106-02	Electrical Test Equipment	Don Whyte	National Center for Construction Education &	7/29/2002
26107-02	Introduction to the National Electrical Code	Don Whyte	National Center for Construction Education &	7/29/2002
26108-02	Raceways, Boxes, and Fittings	Don Whyte	National Center for Construction Education &	7/29/2002
26109-02	Conductors	Don Whyte	National Center for Construction Education &	7/29/2002

Page 1

Donald E. Whyte
President, NCCER

27101-13_A02.EPS

27101-13_A03.EPS

Figure Credits

SkillsUSA, Module Opener, Figures 4 and 5

The Carol M. Highsmith Archive, Library of Congress, Prints and Photographs Division, Figure SA03

APA—The Engineered Wood Association, Figure 2

Calculated Industries, Inc., Figure SA05

Section Review Answer Key

Answer	Section Reference	Objective Reference
Section One		
1. c	1.0.0	1a
Section Two		
1. c	2.0.0	2a
2. b	2.2.3	2b
3. d	2.3.1	2c
Section Three		
1. c	3.1.0	3a
2. d	3.2.0	3b
3. d	3.3.0	3c
Section Four		
1. b	4.0.0	4a
2. a	4.1.0	4a
3. c	4.2.1	4b

NCCER CURRICULA — USER UPDATE

NCCER makes every effort to keep its textbooks up-to-date and free of technical errors. We appreciate your help in this process. If you find an error, a typographical mistake, or an inaccuracy in NCCER's curricula, please fill out this form (or a photocopy), or complete the online form at **www.nccer.org/olf**. Be sure to include the exact module ID number, page number, a detailed description, and your recommended correction. Your input will be brought to the attention of the Authoring Team. Thank you for your assistance.

Instructors – If you have an idea for improving this textbook, or have found that additional materials were necessary to teach this module effectively, please let us know so that we may present your suggestions to the Authoring Team.

NCCER Product Development and Revision

13614 Progress Blvd., Alachua, FL 32615

Email: curriculum@nccer.org
Online: www.nccer.org/olf

❑ Trainee Guide ❑ Lesson Plans ❑ Exam ❑ PowerPoints Other ____________

Craft / Level: Copyright Date:

Module ID Number / Title:

Section Number(s):

Description:

Recommended Correction:

Your Name:

Address:

Email: Phone:

27102-13

Building Materials, Fasteners, and Adhesives

Overview

Carpenters are involved in just about every phase of construction, so they must be familiar with building materials and fasteners. A carpenter working in residential construction works primarily with wood products. In commercial construction, steel and concrete are the common structural materials. A high-rise building might be framed with structural steel, have cast concrete floors, and have glass and metal curtain wall panels attached to provide the exterior finish. A variety of nails, screws, bolts, and anchors are available for attaching materials to one another. In addition, adhesives may also be used.

Module Two

Trainees with successful module completions may be eligible for credentialing through the NCCER Registry. To learn more, go to **www.nccer.org** or contact us at **1.888.622.3720**. Our website has information on the latest product releases and training, as well as online versions of our *Cornerstone* magazine and Pearson's product catalog.

Your feedback is welcome. You may email your comments to **curriculum@nccer.org**, send general comments and inquiries to **info@nccer.org**, or fill in the User Update form at the back of this module.

This information is general in nature and intended for training purposes only. Actual performance of activities described in this manual requires compliance with all applicable operating, service, maintenance, and safety procedures under the direction of qualified personnel. References in this manual to patented or proprietary devices do not constitute a recommendation of their use.

27102-13
Building Materials, Fasteners, and Adhesives

Objectives

When you have completed this module, you will be able to do the following:

1. Identify various types of building materials and describe their uses.
   a. State the uses of various types of hardwoods and softwoods.
   b. Describe common lumber defects.
   c. Identify the different grades of lumber and describe uses for each.
   d. Explain how treated lumber differs from nontreated lumber.
   e. Describe how plywood is manufactured and cite common applications for plywood on a construction project.
   f. Identify uses of hardboard.
   g. Identify uses of particleboard.
   h. Identify uses of high- and medium-density overlay plywood.
   i. Describe how oriented strand board differs from particleboard and cite common applications for OSB.
   j. Cite common applications for mineral fiberboard.
   k. State the uses of various types of engineered lumber.
   l. Identify applications for wood I-beams.
   m. List advantages of glulam lumber over conventional solid lumber.
   n. Describe the composition of concrete and explain how hydration occurs.
   o. List uses of concrete masonry units for a construction project.
   p. Identify where metal framing members may be used in a structure.
2. Identify safety precautions associated with building materials.
   a. List general safety guidelines for working with building materials.
   b. Cite safety precautions for working with wood building materials.
   c. Cite safety precautions for working with concrete building materials.
   d. Cite safety precautions for working with metal building materials.
3. Describe the proper method of handling and storing building materials.
   a. List basic material-handling guidelines.
   b. Describe how to handle and store wood building materials.
   c. Describe how to handle and store concrete building materials.
   d. Describe how to handle and store metal building materials.
4. Explain how to calculate the quantities of lumber, panel, and concrete products using industry-standard methods.
   a. Calculate lumber quantities.
   b. Calculate panel quantities.
   c. Calculate the volume of concrete required for rectangular and cylindrical shapes.

5. Describe the fasteners, anchors, and adhesives used in construction work and explain their uses.
   a. Identify various types of nails and cite uses for each.
   b. Identify applications for staples.
   c. Identify various types of screws and cite uses for each.
   d. Describe uses for hammer-driven pins and studs.
   e. Identify various types of bolts and cite uses for each.
   f. Identify various types of mechanical anchors and cite uses for each.
   g. Identify various types of bolt anchors and explain how each is installed.
   h. Identify various types of screw anchors and cite uses for each.
   i. Identify various types of hollow-wall anchors and cite uses for each.
   j. List the types of glues and adhesives used in construction.

Performance Tasks

Under the supervision of the instructor, you should be able to do the following:

1. Given a selection of building materials, identify a particular material and state its use.
2. Calculate the quantities of lumber, panel, and concrete products using industry-standard methods.
3. Demonstrate safe and proper installation of drop-in anchors.

Trade Terms

Admixtures
Area
Board foot
Butt joint
Cantilever
Catalyst
Combustible
Cured concrete
Experience modification rate (EMR)
Galvanized
Green concrete
Gypsum
Hydration
Joist
Material safety data sheet (MSDS)
Millwork
Nail set
Penny
Plastic concrete
Rafter
Resins
Sheathing
Shiplap
Sill plate
Slag
Vaulted ceiling
Volume

Industry-Recognized Credentials

If you're training through an NCCER-accredited sponsor, you may be eligible for credentials from NCCER's Registry. The ID number for this module is 27102-13. Note that this module may have been used in other NCCER curricula and may apply to other level completions. Contact NCCER's Registry at 888.622.3720 or go to **nccer.org** for more information.

Contents

Topics to be presented in this module include:

Contents (*continued*)

Figures and Tables

Figures and Tables (*continued*)

SECTION ONE

1.0.0 BUILDING MATERIALS AND THEIR USES

Objective

Identify various types of building materials and describe their uses.

a. State the uses of various types of hardwoods and softwoods.
b. Describe common lumber defects.
c. Identify the different grades of lumber and describe uses for each.
d. Explain how treated lumber differs from nontreated lumber.
e. Describe how plywood is manufactured and cite common applications for plywood on a construction project.
f. Identify uses of hardboard.
g. Identify uses of particleboard.
h. Identify uses of high- and medium-density overlay plywood.
i. Describe how oriented strand board differs from particleboard and cite common applications for OSB.
j. Cite common applications for mineral fiberboard.
k. State the uses of various types of engineered lumber.
l. Identify applications for wood I-beams.
m. List advantages of glulam lumber over conventional solid lumber.
n. Describe the composition of concrete and explain how hydration occurs.
o. List uses of concrete masonry units for a construction project.
p. Identify where metal framing members may be used in a structure.

Performance Task 1

Given a selection of building materials, identify a particular material and state its use.

Trade Terms

Admixtures: Materials that are added to a concrete mix to change certain properties of the concrete such as retarding setting time, reducing water requirements, or making the concrete easier to work with.

Butt joint: The joint formed when one square-cut edge of a piece of material is placed against another material.

Cantilever: A beam, truss, or slab (floor) that extends past the last point of support.

Cured concrete: Concrete that has hardened and gained its structural strength.

Green concrete: Concrete that has hardened but has not yet gained its structural strength.

Gypsum: A chalky material that is a main ingredient in plaster and gypsum board.

Hydration: The catalytic action water has in transforming the chemicals in portland cement into a hard solid. The water interacts with the chemicals to form calcium silicate hydrate gel.

Joist: Generally, equally spaced framing members that support floors and ceilings.

Material safety data sheet (MSDS): Information that details any toxic, chemical, or potentially harmful substances that are contained in a product.

Millwork: Manufactured wood products such as doors, windows, and moldings.

Plastic concrete: Concrete when it is first mixed and is in a semiliquid and moldable state.

Rafter: A sloping structural member of a roof frame to which sheathing is secured.

Resins: Protective natural or synthetic coatings.

Sheathing: Panel material to which roofing material or siding is secured.

Shiplap: A method of cutting siding in which each board is tapered and grooved so that the upper piece fits tightly over the lower piece.

Sill plate: A horizontal member that supports the framework of a building on the bottom of a wall or box joist. It is also called a sole plate.

Slag: The ash produced during the reduction of iron ore to iron in a blast furnace.

Vaulted ceiling: A high, open ceiling that generally follows the roof pitch.

Volume: The amount of space occupied in three dimensions (length, width, and height).

Carpenters work with a wide variety of building materials. Wood framing materials are most common in residential construction, although residential foundations are usually built of concrete and/or concrete masonry units (blocks). Carpenters working on commercial construction projects are more likely to work with steel and concrete. This module introduces different types of building materials used in residential and commercial construction, as well as the fasteners and adhesives used to attach these materials.

Many kinds of lumber and other wood building products are used in construction. Different wood products suit different purposes, so it is important for a carpenter to know the types and grades of wood building materials, along with their uses and limitations. When selecting lumber, the carpenter must know what type best fits the application, and it is also important to be able to recognize defects in the material.

Although wood framing was the norm for many years, the use of steel framing has become increasingly popular in both residential and commercial construction. A carpenter must be able to identify steel framing members, steel beams, and steel joists.

Building a New House

If wood is used for a new house, expect to see a lot of it on the job site. Approximately 14,000 board feet of lumber is used to build an average-size house.

27102-13_SA01.EPS

1.1.0 Hardwoods and Softwoods

Throughout the ages, people have used wood for a variety of purposes: as fuel, as building materials, in weapons, and in transportation. In this module, it is important to distinguish lumber from wood. Lumber refers to the boards, timbers, etc., produced from sawmills, whereas wood refers to the material itself, which comes from many species of trees.

Wood has several advantages:

- It is easily worked.
- It has durability and beauty.
- It has great ability to absorb shocks from sudden loads.
- It is free from rust and corrosion, comparatively light in weight, and adaptable to a countless variety of purposes.

Lumber, as well as the trees it comes from, are classified as either hardwood or softwood. This classification can be somewhat confusing as some softwoods are actually harder than hardwoods, and certain hardwoods are softer than some softwoods. The general distinction is made that softwood trees have needles, are coniferous (cone bearing), and retain their green needles throughout the year. Pine and fir trees are two types of softwood trees. Hardwoods come from deciduous (leaf-bearing) trees, such as oak and maple, which lose their leaves each fall. Descriptions and uses of various hardwoods and softwoods are provided in *Appendix A*.

Wood is useful as a building material because of the manner in which a tree forms its fibers: growing by the addition of new material to the outer layer just under the bark and preserving its old fibers as it adds new ones.

Viewing a cross section of a tree, as shown in *Figure 1*, one can see that the trunk consists of a series of concentric rings covered by a layer of bark. Each annular ring represents one year of tree growth. This growth takes place just under the bark in the cambium layer. The cells that are formed there are long; their tubular fibers are composed mainly of cellulose. They are bound together by a substance known as lignin, which connects them into bundles. In wood, lignin is the material that acts as a binder between the cells, adding strength and stiffness to the cell walls. The bundles of fibers run the length of the tree and carry food from the roots to the branches and leaves.

Running at right angles to the annular rings are another group of cells known as medullary rays. These rays carry food from the inner bark to the cambium layer and act as storage tanks for food. The rays are more pronounced in some spe-

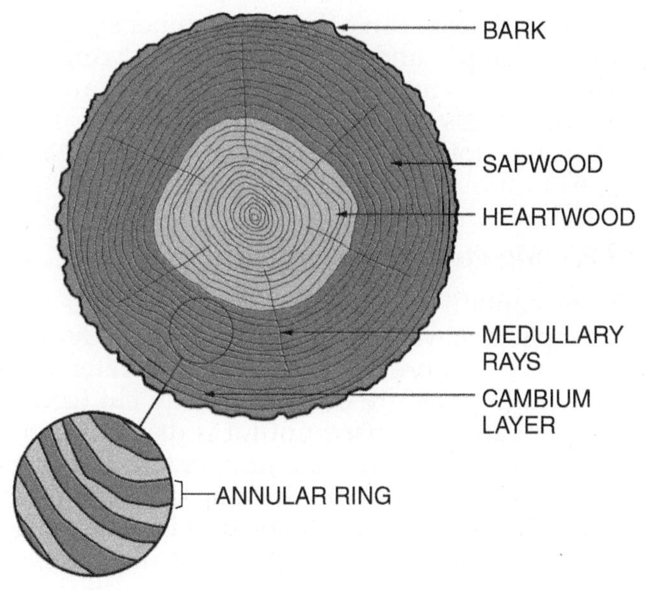

Figure 1 Cross section of a tree.

cies than others (such as oak) but are present in all trees. As one layer of wood succeeds another, the cells in the inner layer die (cease to function as food storage) and become useful only to give stiffness to the tree. This older wood, known as heartwood, is usually darker in color, drier, and harder than the living layer (sapwood).

Heartwood is more durable than sapwood. If wood is to be exposed to decay-producing conditions without the benefit of a preservative, a minimum percentage of heartwood will be specified. Sapwood takes preservative treatment more readily than heartwood, and it is equally durable when treated.

1.1.1 Lumber Cutting

Most trees are sawed so that the annular rings form an angle of less than 45 degrees with the surface of the boards produced. Such lumber is called flat grained in softwoods and plain sawed in hardwoods (*Figure 2*). This is the least expensive method and minimizes waste, but the lumber produced by this method is more likely to shrink or warp.

There is another method in which the wood is cut with the annular rings at an angle greater than 45 degrees with the surface of the boards. Lumber cut by this method is known as edge grained or vertical grained in softwoods and quarter sawed in hardwoods. The lumber produced by this method is usually more durable and less likely to warp or shrink. Quarter-sawed lumber is often used for hardwood floors.

In order to obtain the maximum amount of lumber from a log, most logs are cut using a combination of the two methods.

Ghost Wood

Longleaf pine and bald cypress are praised by carpenters for their outstanding beauty and durability. Unfortunately, clear-cutting the vast forests in the southern United States that occurred in the late 1800s significantly reduced the population of the two species. Today, lumber companies have resurrected this wood by retracing the rivers used to transport the logs to sawmills. Logs of original-growth pine and cypress have been lying along the bottom of these riverbeds for more than one hundred years. This wood is now being harvested by scuba divers, sawed, dried, and used for various construction purposes.

GOING GREEN

FSC Certified Wood

Look for sustainably harvested materials. The Forest Stewardship Council (FSC) is an organization that evaluates wood products to determine if they are sustainably harvested. More information on the FSC certification standards is available online at **http://us.fsc.org/**.

27102-13_SA02.EPS

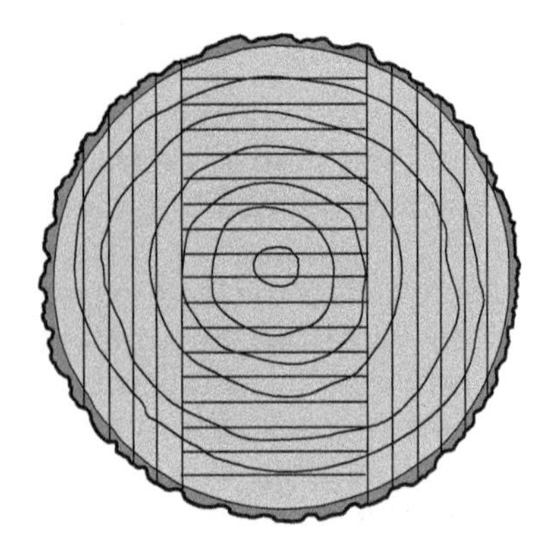

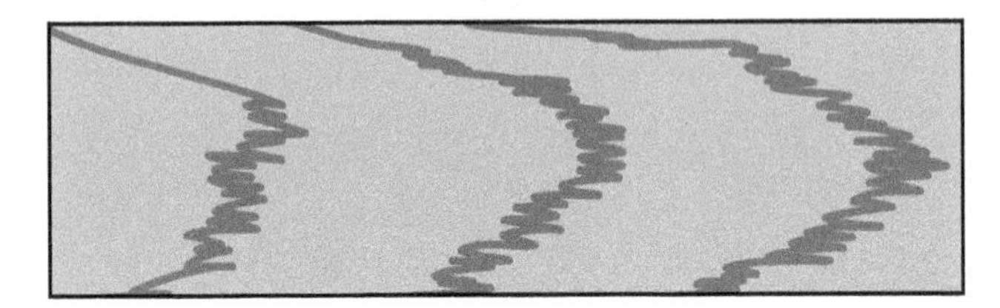

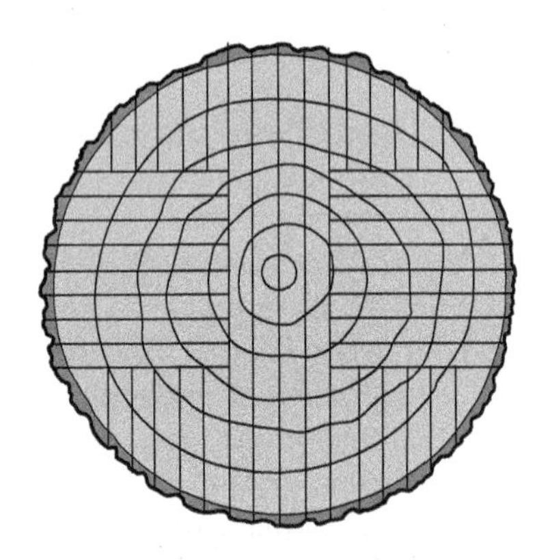

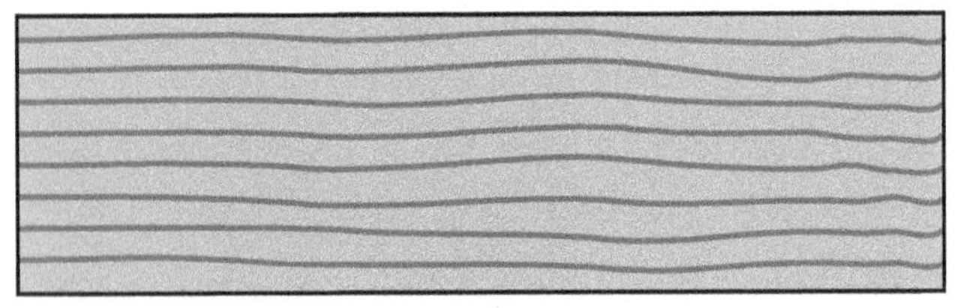

Figure 2 Common lumber-cutting methods.

1.2.0 Lumber Defects

Before covering lumber grading, which is to a large extent based on the amount and types of defects in the lumber, it is necessary to understand the various kinds of defects. Some lumber defects, such as knots, occur naturally. Defects can also occur during the cutting and planing processes. Further defects can occur while the lumber is drying, or because of improper storage or handling.

Lumber defects can affect the lumber's appearance, strength, and usability. For most common uses, a certain number and size of defects are permitted; unless lumber is severely damaged, some use can be found for just about any piece. *Figure 3* shows various types of lumber defects.

1.2.1 Moisture and Warping

Trees contain a large amount of moisture. For example, a newly cut 10' length of 2 × 10 lumber can contain more than 4 gallons of water. Due to its high moisture content, freshly cut (green) lumber cannot be used until it is dried. If green lumber is used, it may result in cracked ceilings or floors, squeaky floors, sticking doors, or other problems. Initially, some moisture is removed by heating lumber in kilns at the sawmill. However, the lumber continues to dry for a long time after that. As cut lumber dries, it will shrink. This can cause warping and splitting because the lumber dries unevenly. A carpenter must be able to tell when lumber contains too much moisture. Moisture meters can be used on site for that purpose. A simple way to check for moisture is to hit the lumber with a hammer. If moisture comes to the surface in the dent left by the hammer, the lumber is too wet and should not be used.

1.3.0 Lumber Grading

Lumber is graded by inspectors who examine the lumber after it is planed. Grading methods and standards are established by the American Lumber Standards Committee. Basically, the grade describes the type, size, and number of defects allowed in the worst board of that grade.

Inspection criteria are established by the US Department of Commerce and published as product standards. Inspections are performed by regional agencies and associations, such as the Southern Pine Inspection Bureau, Western Wood Products Association, California Redwood Association, etc. Hardwood grades are regulated by the National Hardwood Lumber Association. Different associations may have slightly different standards

Mahogany

Mahogany is used for the finest furniture and millwork. There are many species of mahogany; more than 300 species come from the Philippines alone. South America is a large supplier of mahogany, as are the Fiji Islands in the South Pacific. The most exotic mahogany comes from Africa.

for the numbers and types of defects permitted within a particular grade. Inspectors from accredited inspection agencies examine the lumber, determine the grade in accordance with the association's specifications, and apply the applicable grade stamp. Keep in mind that since grading is done right after the board is planed, it cannot account for warping that might occur if the board is improperly stored.

Carpenters must be familiar with the various lumber grades. The architect who designs the building will specify the lumber grades to be used on the job. It is absolutely essential that the carpenter or contractor who buys the lumber uses the specified grade. In order to do that, you must be able to read the grade stamp. *Figure 4* shows and explains a typical grade stamp.

Lumber is graded based on its strength, stiffness, and appearance. The highest grades have very few defects; the lowest grades may have knots, splits, and other problems. The primary agencies for lumber grading are the Western Wood Products Association and the Southern Forest Products Association. Although the end results of the grading processes are the same, their grade designations are somewhat different. It is important to interpret the grade stamps for the grading association in your area.

Softwood lumber is graded into three major categories: boards, dimension lumber, and timbers. These categories are further divided into specific types of lumber and then classified by appearance (*Table 1*) and strength (*Table 2*). The following is a breakdown of each category:

- *Boards* – Boards, sheathing, and form lumber
- *Dimension lumber* – Light framing, studs, structural light framing, and structural joists and planks
- *Timbers* – Beams and stringers

Hardwoods and Softwoods

Hardwoods and softwoods vary widely in color and grain pattern. Carpenters should be able to recognize common hardwood and softwood species and describe uses for them.

ASH

OAK

MAPLE

HARDWOODS

SOUTHERN YELLOW PINE

NORDIC PINE

FIR

SOFTWOODS

27102-13_SA03.EPS

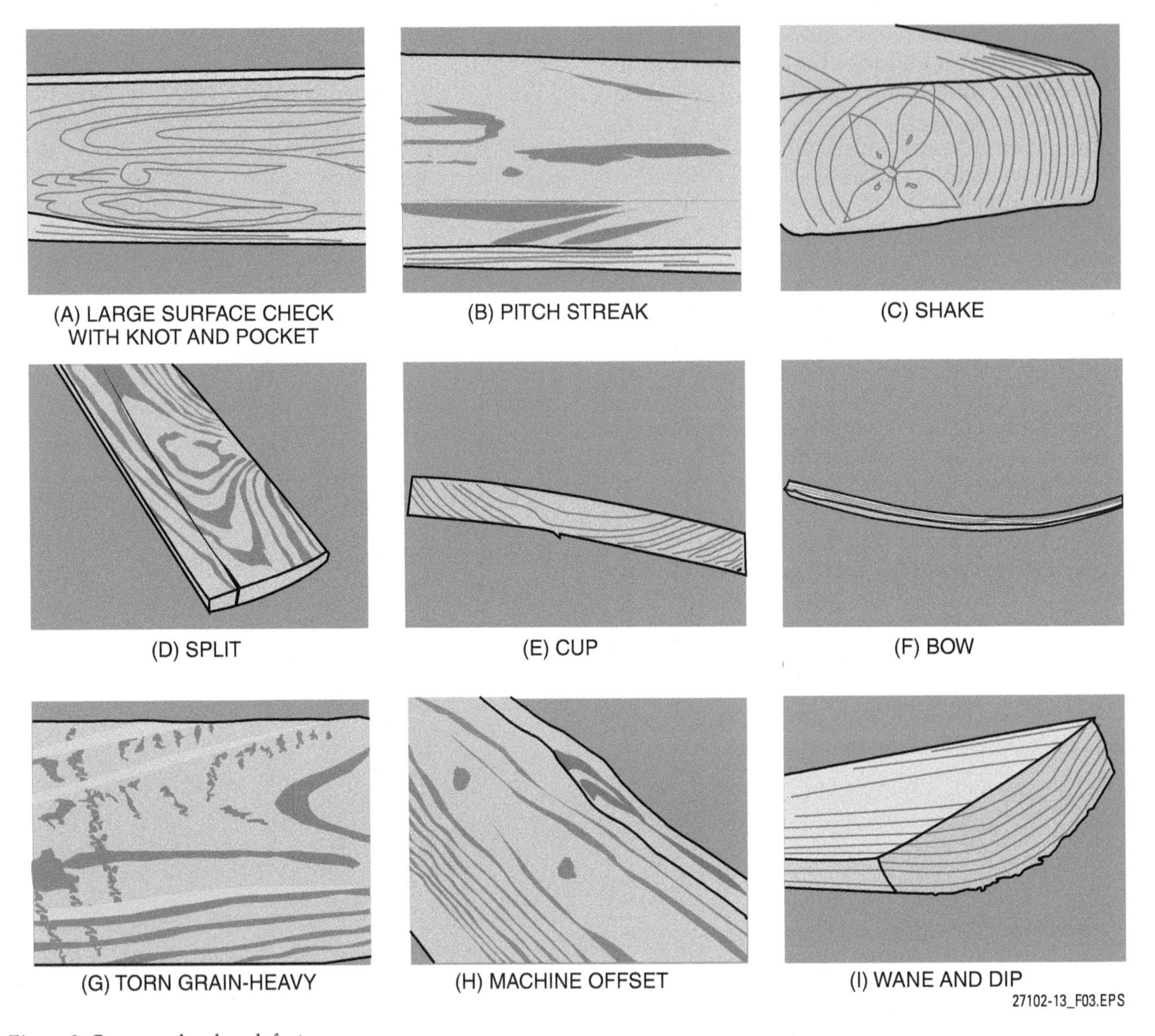

Figure 3 Common lumber defects.

Drying Lumber

There are two methods of drying lumber: kiln drying and air drying. In kiln drying, moisture is removed from lumber in large temperature-controlled buildings called kilns. In air drying, the lumber is stacked outdoors and allowed to dry for up to three months, as compared with only a few days for kiln drying. Even though the process is much longer, air-dried lumber is still likely to have a higher moisture content than kiln-dried lumber.

27102-13_SA04.EPS

(A) Inspection association trademark.

(B) Mill Identification – firm name, brand, or assigned mill number.

(C) Grade Designation – grade name, number, or abbreviation.

(D) Species Identification – indicates species individually or in combination.

(E) Condition of seasoning at time of surfacing:

S-Dry – 19% maximum moisture content

MC15 – 15% maximum moisture content

S-GRN – over 19% moisture content (unseasoned)

27102-13_F04.EPS

Figure 4 Typical lumber grade stamp.

Table 1 Board Appearance Grades

	Product	Grades	Uses
High-Quality Appearance Grades	Selects	B and Better Select C Select D Select	Selects are used for walls, ceilings, trim, and other areas where appearance is important but the use of hardwood is not desirable because of cost or other considerations.
	Finish	Superior Prime E	
	Paneling	Clear (any select or finish grade) No.2 Common (selected for knotty paneling) No. 3 Common (selected for knotty paneling)	
	Siding (bevel, bungalow)	Superior Prime	
General-Purpose Appearance Grades	Boards, Sheathing	No. 1 Common No. 2 Common No. 3 Common No. 4 Common No. 5 Common	Common boards are used for shelving, decking, window trim, and other applications where appearance is less critical, or where on-site selection and cutting is practical when appearance is a consideration.
	Alternate Board	Select merchantable Construction Standard Utility Economy	

Grading Lumber

An understanding of lumber grading is very important to proper building construction. For example, it allows you to communicate effectively about the details of the construction process with others on the job site. More importantly, the building drawings specify grades of lumber that should be used for various parts of the structure. You will have to interpret those drawings in order to construct the building properly.

Table 2 Grades of Dimension Lumber and Timbers

Product	Grades	Uses
Light Framing 2" to 4" thick 2" to 4" wide	Construction Standard Utility Economy	Light framing lumbar is used where great strength is not required such as for studs, plates, and sills.
Studs 2" to 4" thick	Stud Economy Stud	This optional grade is intended for vertical use, as in loadbearing walls.
Structural Light Framing 2" to 4" thick 2" to 4" wide	Select Structural No. 1 No. 2 No. 3 Economy	Structural lumber is used for light framing and forming where greater bending strength is required. Typical uses include trusses, concrete pier wall forms, etc.
Structural Joists and Planks 2" to 4" thick 5" to 18" wide	Select Structural No. 1 No. 2 No. 3 Economy	These grades are designed especially for lumber 5 inches and wider and are suitable for use as floor joists, rafters, headers, small beams, trusses, and general flooring applications.
Beams and Stringers 5" and thicker with this width more than 2" greater than the thickness.	Select Structural No. 1 No. 2 No. 3	These grades are used for stringers, beams, posts, and other support members.
Posts and Timbers 5" x 5" with the width not more than 2" greater than the thickness.	Select Structural No. 1 No. 2 No. 3	These grades are used for stringers, beams, posts, and other support members.

1.3.1 Grading Terms

Among the different agencies, criteria for grading may vary somewhat. Generally, all grading agencies use five basic size classifications as follows:

- *Boards (BD)* – Consists of members under 2" thick and 2" wide and wider.
- *Light framing (L.F.)* – Consists of members 2" to 4" thick and 2" to 4" wide.
- *Joists and planks (J&P)* – Consists of members 2" to 4" thick and 5" to 18" wide.
- *Beams and stringers (B&S)* – Consists of members 5" thick and thicker with the width more than 2" greater than the thickness.
- *Posts and timbers (P&T)* – Consists of members 5" × 5" with the width not more than 2" greater than the thickness, approximately square.

For each species and size classification, the grading agencies establish several stress-quality grades, for example, Select structural, No. 1, No. 2, No. 3, and assign allowable stresses for each.

Some other lumber terms used in grading are:

- *Nominal size* – The size by which it is known and sold in the market (e.g., 2 × 4), as opposed to the dressed size.
- *Dressed (actual) size* – The dimensions of lumber after surfacing with a planer. The dressed size is usually ½" to ¾" less than the nominal or rough size. A 2 × 4 stud, for example, measures 1½" by 3½". *Table 3* compares nominal lumber sizes to dressed sizes for dimension lumber. Additional data on softwood and hardwood lumber dimensions, including equivalent metric conversions, are provided in *Appendix C*.
- *Dressed lumber* – Lumber that is surfaced by a planer on one side (S1S), two sides (S2S), one edge (S1E), two edges (S2E), or any combination of sides and edges (S1S1E, S1S2E, or S4S). Dressed lumber may also be referred to as planed or surfaced.
- *Dimension lumber* – Lumber is supplied in nominal 2", 3", or 4" thicknesses with standard widths. Light framing, studs, joists, and planks are classified as dimension lumber.

Table 3 Nominal and Dressed (Actual) Sizes of Dimension Lumber (in inches)

Nominal	Dressed
2 × 2	1½ × 1½
2 × 4	1½ × 3½
2 × 6	1½ × 5½
2 × 8	1½ × 7¼
2 × 10	1½ × 9¼
2 × 12	1½ × 11¼

- *Matched lumber* – Lumber that is edge- or end-dressed and shaped to make a tongue-and-groove (T&G) joint at the edges or ends.
- *Patterned lumber* – Lumber that is shaped to a pattern in addition to being dressed, matched, or shiplapped, or any combination of these workings.
- *Rough lumber* – Lumber as it comes from the sawmill prior to any dressing or planing operation.
- *Stress-grade lumber* – Lumber grades having assigned working stress and elasticity values in accordance with basic accepted principles of strength grading. Stress is the force exerted on a unit area of lumber, usually expressed in terms of pounds per square inch. Allowable stress is the maximum stress established by the applicable building codes. Allowable stress is always less than the ultimate stress (the amount of stress that can be withstood before the material fails) for obvious structural and safety reasons.
- *Surfaced lumber* – Same as dressed lumber.
- *Framing lumber* – Lumber used for the structural members of a building, such as studs and joists.
- *Finish lumber* – Lumber suitable for millwork or for the completion of the interior of a building, chosen because of its appearance or ability to accept a high-quality finish.
- *Select lumber* – In softwoods, a general term for lumber of good appearance and finishing qualities.

Trimming is the act of crosscutting a piece of lumber to a given length.

- Double end-trimmed (DET) lumber is trimmed reasonably square by a saw on both ends.
- Precision-end-trimmed (PET) lumber is trimmed square and smooth on both ends to uniform lengths, with a manufacturing tolerance of 1⁄16" over or under (+/–) in length in 20 percent of the pieces.
- Square-end-trimmed lumber is trimmed square, permitting only a slight manufacturing tolerance of 1⁄64" for each nominal 2" of thickness or width.

1.3.2 Classification of Manufacturing Defects

Various types of defects that can occur while lumber is being cut, dressed, and dried. The following describes how manufacturing defects impact lumber grades:

- *Standard A* – Very light torn grain; occasional light chip marks; very slight knife marks.
- *Standard B* – Very light torn grain; very light raised grain; very light loosened grain; slight chip marks; average of one slight chip mark per lineal foot but not more than two in any lineal foot; very slight knife marks; slight mismatch.
- *Standard C* – Medium torn grain; light raised grain; light loosened grain; very light machine bite; very light machine gouge; very light machine offset; light chip marks if well scattered; occasional medium chip marks; very slight knife marks; slight mismatch.
- *Standard D* – Heavy torn grain; medium raised grain; very heavy loosened grain; light machine bite; light machine gouge; light machine offset; medium chip marks; slight knife marks; very light mismatch.
- *Standard E* – Very heavy torn grain; raised grain; very heavy loosened grain; medium machine bite; machine gouge; medium machine offset; chip marks; knife marks; light wavy dressing; light mismatch.
- *Standard F* – Very heavy torn grain; raised grain; very heavy loosened grain; heavy machine bite; machine gouge; heavy machine offset; chip marks; knife marks; medium wavy dressing; medium mismatch.
- *Standard G* – Loosened grain; raised grain; torn grain; machine bite; machine burn; machine gouge; machine offsets; chip marks; medium wavy dressing; mismatch.

1.3.3 Abbreviations

Most contract documents and construction drawings use abbreviations to convey information to a craftworker. *Appendix B* lists the common abbreviations used by the Southern Forest Products As-

Calculating Dressed Softwood Lumber Sizes

As can be seen in *Table 3*, nominal lumber sizes are different than the actual, or dressed, dimensions of the board or lumber. A general rule can be applied when calculating dressed sizes of softwood boards. Boards with nominal dimensions less than 1" have dressed dimensions that are ¼" smaller. Boards with a nominal width of 2" to 6" usually have dressed dimensions that are ½" smaller. For boards with widths greater than 6", subtract ¾" from the nominal dimensions to determine the dressed dimensions. Remember, this is a general rule and may not be accurate in every case. The only way to be certain is to perform an actual measurement.

sociation when referring to pine lumber. Many of them also hold for the other wood species.

1.4.0 Treated Lumber

Most types of wood will deteriorate when they have been exposed to excess moisture or humidity levels or are in contact with the ground for extended periods of time. Wood preservatives are commonly applied to the lumber to protect it against decay. Wood preservatives may be applied through pressure or nonpressure processes.

Pressure-treated lumber is softwood lumber protected by chemical preservatives forced deep into the wood through a vacuum-pressure process. Pressure-treated lumber has been used for many years for on-ground and belowground applications such as landscape timbers, sill plates, and foundations. In some parts of the country, pressure-treated (PT) lumber is also extensively used for decks, porches, docks, and other outdoor structures subject to decay from exposure to the elements. PT lumber is popular in areas where structures are exposed to insect or fungus attack. A major advantage of pressure-treated lumber is its relatively low price in comparison with redwood and cedar. When redwood or cedar is used in decay-prone areas, only the more expensive heartwood will resist decay and insects.

Pressure-treated lumber is available in three grades. The three grades are designated by their preservative retention (in pounds per cubic foot). Aboveground (0.25 pounds per cubic foot) grade is to be used only 18" or more above ground. Ground-contact grade (0.40 pounds per cubic foot) is used when there is contact with water or soil on or below ground. The third grade (0.60 pounds per cubic foot) is used when structural reliability is required, such as in wooden foundations and utility poles.

In the nonpressure-treating process, lumber is submerged in a vat of preservative or the preservative is sprayed on the lumber. This process is easier and less expensive than the pressure-treating process, but lumber treated in this manner cannot be substituted for pressure-treated lumber where specified by the building code.

Matched vs. Patterned Lumber

Matched lumber is end- or edge-dressed to create a smooth joint with another piece of lumber. Tongue-and-groove siding is an example of matched lumber that is edge dressed. Patterned lumber is shaped to a pattern in addition to being dressed or matched. Shiplap siding is a type of patterned lumber in which the two pieces are tapered opposite of each other so that when joined together, they form a smooth overlapping joint.

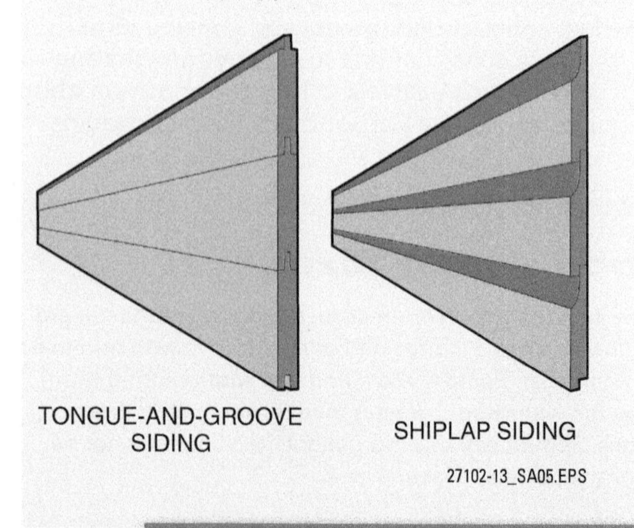

1.5.0 Plywood

Engineered wood products are a general classification of wood building materials that are manufactured from solid and reconstituted wood products and by-products. The wood products and by-products are combined with adhesives, resins, and other binders, and pressed under extreme heat and pressure conditions to form panel and lumber products. Engineered wood products are as strong and dimensionally stable as their solid wood counterparts.

Engineered wood products can be divided into two major categories: panel products and lumber products. Panel products include plywood, hardboard, particleboard, oriented strand board (OSB), and fiberboard. Lumber products can be divided into five general classifications: laminated veneer lumber (LVL), parallel strand lumber (PSL), laminated strand lumber (LSL), wood I-joists, and glued laminated lumber (glulam). Engineered wood products provide several benefits:

- They can be made from younger, more abundant trees.
- They can increase the yield from a tree by 30 percent to 50 percent.
- They are stronger than the same size of structural lumber.
- Greater strength allows the engineered lumber to span a greater distance.
- A length of engineered wood is lighter than the same length of solid lumber. It is therefore easier to handle.

- They are dimensionally accurate and do not warp, crown, or twist.

Plywood is made of layers (plies) of wood veneers. A layer may be 1⁄16" to 5⁄8" thick. The center layer is known as the core. As layers are added on either side of the core, they are placed alternately at right angles, which increases the strength of the panel. Layers with the grain at right angles to the core are called crossbands. The outer exposed layer is known as the face veneer.

Veneers are made by the rotary cutting of trees. First, logs are cut to a specific length, then softened with a hot water or steam bath. The bark is then removed and the log is locked into its center on a lathe. As the lathe turns the log, a long knife slices off a thin veneer. A steel roller at the rear of the knife assists in keeping the veneer intact and maintains a uniform thickness.

This continuous veneer is trimmed into smaller sheets that are fed into a hot oven or dryer to reduce the moisture content. The moisture content is reduced to a range of 3 to 8 percent. At this point, the veneers are separated into core and crossband materials. Veneers used for the panel faces now go through a patching machine to remove and correct defects in the veneer. The patching machine will also match face grains. The veneers go through one final step, called splicing, prior to becoming a sheet of plywood.

In the final manufacturing process, hot glue is applied by machine to the core, crossbands, faces, and plies. The rough plywood sheet will now go into a hot press. This hot press will apply a great amount of pressure to the pieces to squeeze out the excess glue and will compress the rough plywood to its approximate final thickness. This process takes from 2 to 20 minutes. Once the process is complete, the sheets are cut and sanded to their final thickness.

1.5.1 Plywood Panel Sizes

The standard size of a plywood panel is 4' × 8'. Other panel sizes, ranging from 6' to 8' widths and up to 16' lengths, are also available on special order. Sheathing-grade plywood is nominally sized by the manufacturer to allow for expansion. In other words, a 4'0" × 8'0" panel will actually measure 47¾" × 95¾".

Plywood panel thickness varies from 3⁄16" to 1¼". Common sizes are ¼", ½", and 5⁄8" for finish paneling and 3⁄8", ½", 5⁄8", and ¾" for some structural purposes. There are three types of edges on plywood:

- Butt joint (two standard pieces joined)
- A shiplap cut or edge
- Interlocking tongue-and-groove

Opposite edges or all four edges may be cut to match.

1.5.2 Grading for Softwood Construction Plywood

After the plywood panels have been manufactured, they are then graded. The grading procedure is similar to the grading of lumber. The grade stamp that appears on each panel lists the grade of the plywood. *Table 4* shows examples of plywood application data developed by APA–The

Wood Preservatives

Wood preservatives commonly used for construction lumber are waterborne preservatives such as chromated copper arsenate (CCA), alkaline copper quaternary (ACQ) and sodium borates (SBX). CCA-treated lumber is no longer permitted for use in residential and consumer-related applications. However, CCA-treated lumber can be used for industrial, highway, and agricultural applications such as poles, guardrail posts, and saltwater marine exposures. Quality stamps for treated wood, printed directly on the wood or attached to the ends of the lumber, provide information regarding the type of preservative and retention.

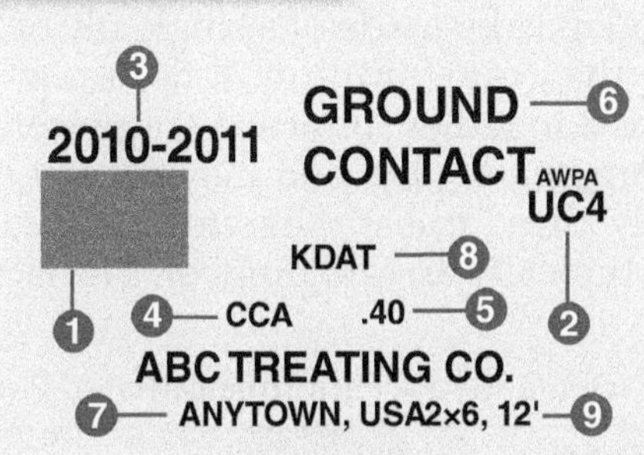

1 ALSC inspection agency mark
2 AWPA Use Category
3 Year of treatment, if required
4 Preservative used for treatment
5 Preservative retention
6 Exposure category
7 Treating company and location
8 DRY or KDAT, if applicable
9 Length and/or class

27102-13_SA06.EPS

Engineered Wood Association. Complete information is available from the APA website (**www.apawood.org**).

Plywood is rated for interior or exterior use. Exterior-rated plywood is used for sheathing, siding, and other applications where there may be exposure to moisture or wet weather conditions. Exterior-rated plywood panels are made of high-grade veneers bonded together with a waterproof glue that is as strong as the wood itself.

Interior plywood uses lower grades of veneer for the back and inner plies. Although the plies can be bonded with a water-resistant glue, waterproof glue is normally used. The lower-grade veneers reduce the bonding strength, however, which means that interior-rated panels are not suitable for exterior use.

Evolution of Wood Products

In the past, the primary source of structural beams, timbers, joists, and other loadbearing lumber was old-growth trees. These trees, which need many decades to mature, are tall and thick and can produce a large amount of high-quality, tight-grained lumber. Extensive logging of these trees to meet demand resulted in higher prices and conflict with forest conservation interests. The development of wood-laminating techniques by lumber producers has permitted the use of younger-growth trees in the production of structural building materials.

1.5.3 Plywood Glues

Although most plywood today is rated for interior use, most of the glue used on plywood is waterproof glue. The grade stamp will indicate whether the plywood is interior or exterior and if the glue is waterproof or not.

Plywood panels, designated as structural-grade panels, are manufactured for heavy-duty applications. Structural-grade panels are seldom used in residential construction, and may be either interior or exterior plywood.

1.5.4 Plywood Cores

Plywood for industrial use or rough-in work is often called softwood (soft) plywood. Softwood plywood has a veneer core that is manufactured using the same procedure as the plies. Softwood plywood may be made of a single ply or two plies having the grain running in the same direction and glued together to form a single core or layer (for example, ½" plywood may consist of veneer, but only three plies). Therefore, the number of plies is typically an odd number. The face veneers on sheets of plywood always run in the same direction.

The plywood used in the manufacture of cabinets, doors, furniture, and finished components is known as hardwood plywood. This type of plywood may have any of four different types of cores:

- Lumber core
- Particleboard core
- Veneer core (common in softwood plywood)
- Fiberboard core

The type of core will not be indicated on the grade stamp, but one can easily tell by looking at the edge of the plywood (*Figure 5*).

1.5.5 Face Veneers

Face veneers are the outermost plies of a plywood panel. Face veneer quality is indicated using the letters A, B, C, D, or N. Two letters separated by a hyphen are shown on a grade stamp to indicate the face veneer quality or grade. The quality of the front panel face is indicated by the first letter.

Table 4 Guide to APA Performance-Rated Plywood Panels

APA Plywood Grade	Description and Use
Rated Sheathing	Unsanded sheathing grade for wall, roof, subflooring, and industrial applications
Structural Rated Sheathing	Unsanded plywood grades to use where shear and cross-panel strength properties are of maximum importance
Rated Sturd-I-Floor	Combination subfloor-underlayment provides smooth surface for application of carpet and pad; available with tongue-and-groove edge
Underlayment	Used as underlayment under carpet and pad; available with tongue-and-groove edge
C-D Plugged	Used for built-ins as well as wall and ceiling tile backing
Appearance Grades	Sanded grade for use where a high-quality surface is required

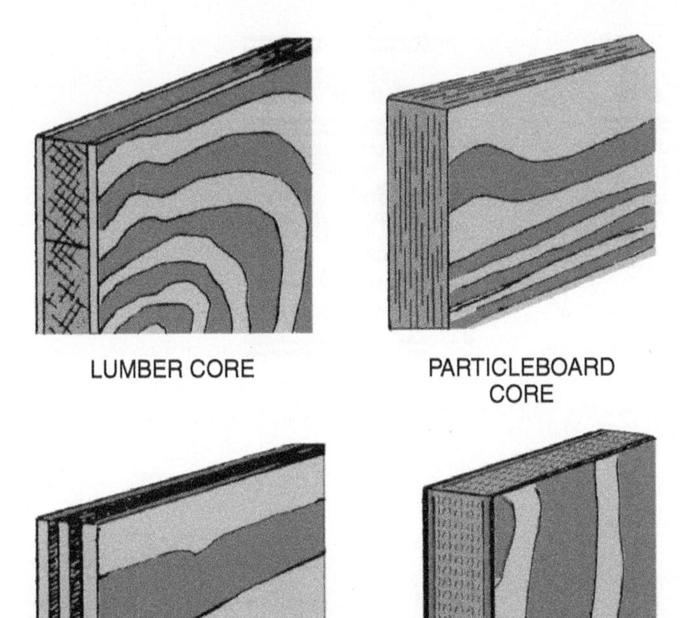

Figure 5 Types of plywood cores.

The quality of the back panel face is indicated by the second letter.

The best-quality veneer, which is the only grade free from defects, carries an N grade. This grade should be selected to receive a finish that is natural and exposed to view. N-grade veneers are typically used in high-quality applications where a natural finish is desired. N-grade veneers are not normally found in construction plywood.

A plywood face that is smooth, has no open defects, but could have some heat repairs is an A grade. This type of veneer accepts paint readily. The next grade is B. This veneer offers a surface that is solid with splits no larger than 1⁄32". Some defects may be larger than 1⁄32" and may be fixed or repaired with smooth plugs. In most cases, tight knots are allowed. Sanding may create minor flaws that are allowed. This product may be treated like the A grade, but it will not be as smooth. All A and B faces are sanded.

In the C grade, the veneer may have splits to a maximum of ½" and up to 1½" knot holes; however, they must not affect the required strength of the panel. On C-grade panels, there may be some sanding defects; these are permitted. The next grade is C Plugged. This veneer is an improved C. The veneer has tighter limits on splits and knot holes. The C Plugged grade has a fully sanded surface. C faces are sanded only if noted.

The poorest grade is the D grade. It has less strength and a poor appearance because of the defects. The D grade is typically used for the cores and backs of interior plywood.

Plywood that is designated for sheathing, subflooring, concrete forms, and panels used for special structural purposes is called performance-rated plywood. The veneers are either unsanded or lightly sanded. The plywood sheets carry slightly different grade markings. The face veneers are either C-C or C-D.

1.5.6 Plywood Grades

The APA–The Engineered Wood Association *Grades and Specifications* is the standard for plywood grading. Different grade stamps are used for various types of plywood (*Figure 6*). Sanded plywood panels are panels with B-grade or better face veneers, which are sanded smooth during manufacture. Performance-rated panels, also known as structural panels, include three grades: Rated Sheathing, Rated Sturd-I-Floor, and Rated Siding. Rather than listing the face veneer grade, grade marks for performance-rated panels include information on the panel grade, thickness, and span rating. Always be sure to check the project specifications for grades. Different associations can certify plywood. Some mills certify their own plywood, and this certification may not be accepted by the grading associations or agencies. Be sure to find out who is certifying plywood as a certain grade.

Plywood made from hardwoods such as birch and oak is used in making furniture and cabinets. Hardwood plywood uses a different grading system than the one described previously.

1.6.0 Hardboard

While plywood uses full sheets of veneers in manufacture, other engineered panel products, such as hardboard, particleboard, oriented strand

Selecting Plywood

Plywood that is expressly manufactured for either interior or exterior use may be used for other purposes in certain situations. Some local codes may require the use of pressure-treated plywood for exterior construction, in bathrooms, or in other high-moisture areas of a house. Pressure-treated plywood can withstand moisture better than interior plywood. Always check the local code(s) before beginning any construction project.

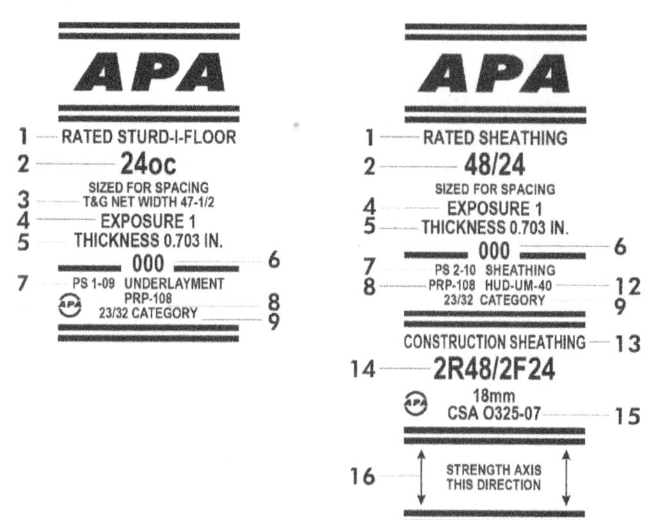

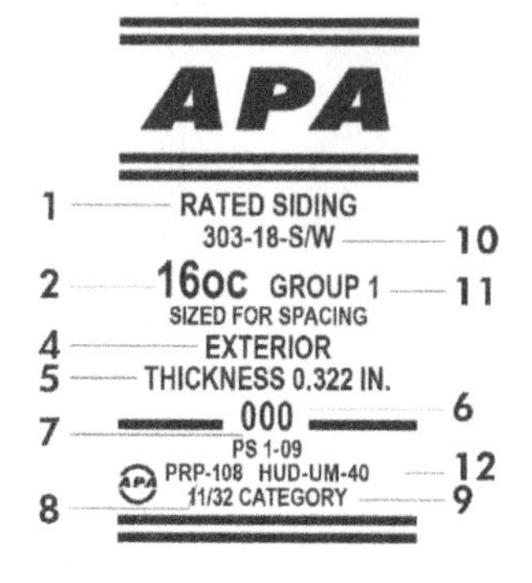

1 Panel grade
2 Span Rating
3 Tongue-and-groove
4 Bond classification
5 Mill thickness declaration
6 Mill number
7 Product Standard
8 APA's performance rated panel standard
9 Performance Category
10 Siding face grade
11 Species group number
12 HUD recognition
13 Panel grade, Canadian standard
14 Panel mark – Rating and end-use designation per the Canadian standard
15 Canadian performance rated panel standard
16 Panel face orientation indicator

27102-13_F06.EPS

Figure 6 Performance-rated panel grade stamps.

board, and fiberboard, use vegetable or mineral fibers, as well as waste sawdust, wood chips, and wood scraps, as their main ingredients. The ingredients are softened with heat and moisture, and a binder is added to the mixture. The mixture is then passed through a press, which uses heat and pressure to produce hardboard to the density and thickness desired by the manufacturer.

Hardboard is a manufactured building material, sometimes known as tempered board or pegboard. Hardboard is extremely dense. Common thicknesses for hardboard are 3/16", 1/4", and 5/16". The standard hardboard panel size is 4' × 8'; however, hardboard can be manufactured in widths up to 6' and lengths up to 16' for specialized uses.

GOING GREEN

Housewrap

Housewrap is a translucent plastic sheet material that is installed over exterior wall sheathing before windows and doors are installed. Housewrap prevents air and water infiltration, while allowing the building to breathe. Water vapor and gases can escape from the building's interior through the walls without trapping moisture in the walls.

Hardboard is water resistant, but is susceptible to the edges breaking if they are not properly supported. Holes must be predrilled for nailing; direct nailing into hardboard would cause it to fracture.

Three grades of hardboard are available: standard, tempered, and service grade. Standard hardboard is light brown in color, smooth, and can easily accept paint. Standard hardboard is suitable only for interior use such as cabinets.

Tempered hardboard, commonly known as Masonite™, is manufactured in a similar manner as standard hardboard, but the hardboard is coated with oils and resins, and heated or baked to a dark brown color. In this process, the tempered hardboard becomes denser, stronger, and more brittle. Tempered hardboard is suitable for either interior or exterior uses such as siding, wall paneling, and other decorating purposes. One type of tempered hardboard is perforated hardboard or pegboard in which holes are punched 1" apart. Special hooks are made to fit into the holes to support different types of items (shelves, brackets, and hangers for workshops and kitchens, for example).

Service-grade hardboard is not as dense, strong, or heavy as standard hardboard. The surfaces are not as smooth as standard hardboard, and it is less expensive than other hardboard. Service-grade hardboard can be used for most applications for which standard or tempered hardboard is used. Service-grade hardboard is manufactured for items such as cabinets, furniture components, and perforated hardboard.

1.7.0 Particleboard

Particleboard is composed of small particles of wood that are combined with a binder and formed into panels using heat and pressure. Some manufacturers may use southern pine materials, while others select the stock based on its species, fiber characteristics, moisture content, and color. There are two types of particleboard as follows:

- Type I is basically a mat-formed particleboard generally made with vera-formaldehyde resin, suitable for interior applications. Refer to the **material safety data sheet (MSDS)** for handling. Type I particleboard is available in three classes: A, B, and C. The classes are density grades. Class A is a high-density board weighing 50 pounds per cubic foot or over. Class B is a medium-density board of 37 to 50 pounds per cubic foot. Class C is a low-density board of 37 pounds per cubic foot and lower.
- Type II is a mat-formed particleboard made with durable moisture- and heat-resistant binders. Type II particleboard is suitable for interior and certain exterior applications (when so labeled). Type II particleboard is available in two classes: Class A is a high-density board weighing 50 pounds per cubic foot or over; Class B is a medium-density board weighing 37 to 50 pounds per cubic foot.

Each particleboard manufacturer has specification sheets that contain the technical information previously discussed. The information sheets are usually available from a local lumber dealer.

Particleboard panels range in size from ¼" to 1½" in thickness and from 3' to 8' in width. Panels with thicknesses of 3" and lengths ranging up to 24' are available for special purposes. Particleboard has no grain, is smoother than plywood, is more resilient, and is less likely to warp. Particleboard has many uses such as shelving and cabinets. In some situations, ¼" particleboard is used as a core for plywood panels.

Some types of particleboard can be used for underlayment if permitted by the local building code. If particleboard is used as underlayment, it should be placed with the long dimension across the joists and the edges staggered. Particleboard can be nailed, although some types of particleboard will crumble or crack when nailed close to the edges.

1.8.0 High-Density Overlay (HDO) and Medium-Density Overlay (MDO) Plywood

High-density overlay (HDO) plywood panels have a hard, resin-impregnated fiber overlay heat-bonded to both surfaces. HDO panels are abrasion- and moisture-resistant, and can be used for concrete forms, cabinets, countertops, and similar high-wear applications. HDO plywood also resists damage from chemicals and solvents. HDO plywood panels are available in five common thicknesses: ⅜", ½", ⅝", ¾", and 1".

Medium-density overlay (MDO) plywood panels are coated on one or both surfaces with a smooth, opaque overlay. MDO plywood accepts paint well and is very suitable for use as structural siding, exterior decorative panels, and soffits. MDO plywood panels are available in eight common thicknesses ranging from 11⁄32" to 23⁄32".

Both HDO and MDO panels are manufactured with waterproof adhesive and are suitable for exterior use. If MDO panels are to be used outdoors, however, the panels should be edge sealed with one or two coats of a good-quality exterior house-paint primer. An easy way to efficiently seal the edges is to stack panels and paint the edges of several panels at one time.

1.9.0 Oriented Strand Board (OSB)

Oriented strand board (OSB) is a manufactured structural panel used for wall and roof sheathing (*Figure 7*) and single-layer floor construction. OSB consists of compressed wood strands arranged in five or more crossbanded layers and bonded with phenolic resin under intense heat and pressure. OSB offers dimensional stability, stiffness, fastener holding capacity, and no core voids. OSB panels may be sprayed with a surface treatment to provide additional qualities. For example, OSB panels used for roof sheathing may have a reflective coating applied to the underside to provide a higher fire rating, or a nonskid surface may be applied to the top side for better traction for workers on the roof.

Fire-rated OSB provides greater burn-through resistance than common OSB panels, and is commonly used as a roof or wall sheathing. Fire-rated panels are coated with a noncombustible, inert material (typically magnesium oxide) and water mixture that is chemically bonded to the OSB. When the panels are exposed to intense heat, the water is released, resisting the burn-through and slowing the spread of flames. Fire-rated panels have a special grade stamp.

27102-13_F07.EPS

Figure 7 OSB panel.

Before cutting OSB, be sure to check the applicable MSDS for safety hazards. The MSDS is the most reliable source for safety information.

> **CAUTION**
>
> Whenever working with older materials that may contain asbestos, contact your supervisor for the company's policies on safe handling of the material. State and federal regulations require specific procedures to be followed prior to and during removal, cutting, or disturbing of any suspect materials. Also, some materials emit a harmful dust when cut. Check the MSDS before cutting.

1.10.0 Mineral Fiberboards

Glass and gypsum rock are the most common materials used in the manufacture of mineral fiberboards. Glass fibers or gypsum powder are mixed with a binder and pressed or sandwiched between two layers of asphalt-impregnated paper, producing a rigid insulation board. Mineral fiberboard will not support combustion and will not burn.

Certain types of mineral fiberboard have chemical foam mixed with the glass fibers to make a good, rigid insulation. However, this mineral insulation will crush and should not be used when it must support a heavy load.

1.11.0 Engineered Lumber Products

Engineered lumber, such as laminated veneer lumber (LVL), parallel strand lumber (PSL), and laminated strand lumber (LSL), make efficient use of natural resources (trees) and provide exceptional structural qualities. *Figure 8* shows three common types of engineered lumber.

27102-13_F08.EPS

Figure 8 LVL, PSL, and LSL.

1.11.1 Laminated Veneer Lumber (LVL)

Like plywood, LVL is made from laminated wood veneers. Douglas fir and southern pine are the primary sources. Thin (1⁄10" to 3⁄16") sheets are peeled from the tree in widths of 27" or 54". The veneers are laid-up in a staggered pattern with the veneers overlapping to increase strength. Unlike plywood, the grain of each layer runs in the same direction as the other layers. The veneers are bonded with an exterior-grade adhesive, then pressed together and heated under pressure.

LVL is used for floor and roof beams and for the support members called headers that are used over window and door openings. It is also used in scaffolding and concrete forms. No special cutting tools or fasteners are required.

1.11.2 Parallel Strand Lumber (PSL)

PSL is made from long strands of Douglas fir and southern pine. The strands are about 1⁄8" or 1⁄10" thick and are bonded together with waterproof adhesive in a special heating process. PSL is used for beams, posts, studs, rafters, and columns. It is manufactured in thicknesses up to 7". Columns can be up to 7" wide; beams range in size up to 18" in width.

1.11.3 Laminated Strand Lumber (LSL)

LSL is manufactured from small logs of almost any species of wood. Aspen, red maple, and poplar that cannot be used for standard lumber are commonly used. In the manufacturing process, the logs are cut into short strands, which are bonded together and pressed into long blocks (billets) up to 5½" thick, 8' wide, and 40' long.

LSL is used for millwork such as doors and windows and any other product that requires high-grade lumber. However, LSL will not support as much of a load as a comparable size of PSL because PSL is made from stronger wood.

1.12.0 Wood I-Beams

Wood I-beams, also known as wood I-joists (*Figure 9*), consist of a web with flanges bonded to the top and bottom. This arrangement, which mimics a steel I-beam, provides exceptional strength. The webs are made of OSB or plywood and the flanges are grooved to fit over the web. Wood I-beams are most commonly used for floor and ceiling joists and for rafters in residential and commercial construction. Because of their strength, wood I-beams can be used in greater spans than a comparable length of dimension lumber. Lengths of up to 60' are available.

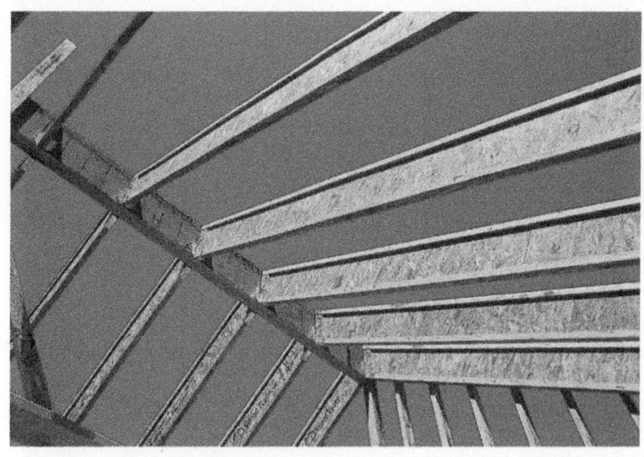

27102-13_F09.EPS

Figure 9 Wood I-beams.

1.13.0 Glued Laminated Lumber (Glulam)

Glulam is manufactured from lengths of solid, kiln-dried lumber that are glued together. Glulam is popular in architectural applications where exposed beams are used (*Figure 10*). Because of its exceptional strength and flexibility, glulam can be used in areas subject to high winds or earthquakes. Glulam is available in three appearance grades: industrial, architectural, and premium. Industrial grade is used in open buildings such as warehouses and garages where appearance is not a priority or where beams are not exposed. Architectural grade is used where beams are exposed and appearance is important. Premium, the highest grade, is used where the highest-quality appearance is needed (*Figure 11*).

Glulam beams used in residential construction are available in standard widths of 3⅛", 3½", 5⅛", 5½", and 6¾". Depths range from 5½" to 28½". They are available in very long lengths. Glulam members are used for many applications, including ridge beams; basement beams; window and door headers; stair treads, supports, and stringers; and cantilever and vaulted ceilings. Since glulam members are laminated, they can be formed into arches and a variety of curved configurations. Glulam beams are especially popular for use in churches.

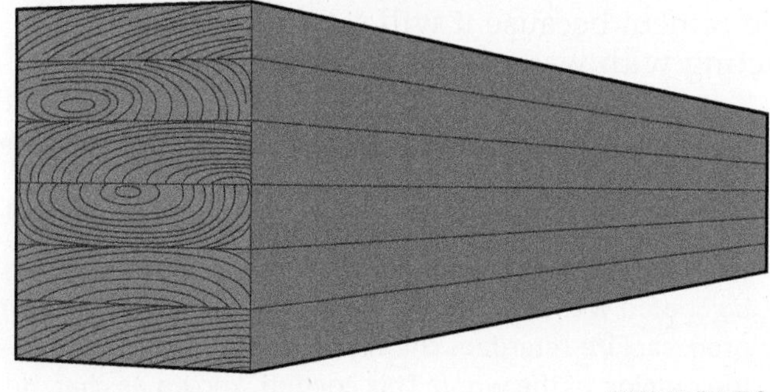

27102-13_F10.EPS

Figure 10 Glulam beam construction.

27102-13_F11.EPS

Figure 11 Glulam beam application.

1.14.0 Concrete and Concrete Materials

Concrete is a mixture of three basic materials: portland cement, aggregates, and water. When first mixed, concrete is in a semiliquid state and is referred to as plastic concrete. When the concrete hardens but has not yet gained structural strength, it is called green concrete. After concrete has hardened and gained its structural strength, it is called cured concrete. Various types of concrete can be obtained by varying the basic materials and/or by adding other materials, called admixtures, to the mix.

1.14.1 Portland Cement

Portland cement is a finely ground powder consisting of varying amounts of lime, silica, alumina, iron, and other trace components. While dry, it may be moved in bulk or can be bagged in moisture-resistant sacks and stored for relatively long periods of time. Portland cement is a hydraulic cement because it will set and harden by reacting with water with or without the presence of air. This chemical reaction is called hydration, and it can occur even when the concrete is submerged in water. The reaction creates a calcium silicate hydrate gel and releases heat. Hydration begins when water is mixed with the cement and continues as the mixture hardens and cures. The reaction occurs rapidly at first, depending on how finely the cement is ground and what admixtures are present. Then, after its initial cure and strength are achieved, the concrete mixture continues to slowly cure over a longer period of time until its ultimate strength is attained.

Today, portland cement is manufactured by heating lime mixed with clay, shale, or slag to about 3,000°F. The material that is produced is pulverized, and gypsum is added to regulate the hydration process. Manufacturers use their own brand names, but nearly all portland cements are manufactured to meet American Society for Testing and Materials (ASTM) or American National Standards Institute (ANSI) specifications.

1.14.2 Aggregates for Concrete

Natural sand, gravel, crushed stone, blast furnace slag, and manufactured sand (from crushed stone, gravel, or slag) are the most commonly used aggregates in concrete. In some cases, recycled crushed concrete is used for low-strength concrete applications. There are other types of special aggregates for concrete.

Use of Engineered Wood Products

Engineered wood products are used in a wide array of applications that were once exclusively served by solid lumber. For example, PSL is used for columns, ridge beams, girders, and headers. LVL is also used for form headers and beams. Wood I-beams are used to frame roofs as well as floors. An especially noteworthy application is the use of LSL studs, top plates, and sill plates in place of lumber to frame walls.

Fire-Retardant Building Materials

Lumber and panel materials are sometimes treated with fire-retardant chemicals. The lumber or panels can either be coated with the chemical in a nonpressure process or impregnated with the chemical in a pressure-treatment process. Fire-retardant chemicals react to extreme heat, releasing vapors that form a protective coating around the outside of the wood. This coating, known as char, delays ignition and inhibits the release of smoke and toxic fumes.

Glulam Beams vs. Steel Beams

When exposed to fire, glulam beams outperform steel beams for a longer period of time. Metal that is unprotected will quickly lose its strength and collapse suddenly. On the other hand, wood will lose its strength slowly.

Aggregates make up 60 percent to 80 percent of the volume of concrete and function not only as filler material, but also provide rigidity to greatly restrain volume shrinkage as the concrete mass cures. At the usual aggregate content of about 75 percent by volume, shrinkage of concrete is only one-tenth that of pure cement paste. The presence of aggregates in concrete provides an enormous contact area for an intimate bond between the paste and aggregate surfaces.

Since aggregates are such a large portion of the concrete volume, they have a substantial effect on the quality of the finished product. In addition to concrete volume stability, they influence concrete weight, strength, and resistance to environmental destruction. Therefore, aggregates must be strong, clean, free of chemicals and coatings, and of the proper size and weight.

1.14.3 Water for Concrete

Generally, any drinkable water (unless it is extremely hard or contains too many sulfates) may be used to make concrete. If the quality of the water is unknown or questionable, the water should be analyzed, or mortar cubes should be made and tested against control cubes made with drinkable water. If the water is satisfactory, the test cubes should show the same compressive strength as the control cubes after a 28-day cure time.

High concentrations of chlorides, sulfates, alkalis, salts, and other contaminants have corrosive effects on concrete and/or steel reinforcing bars, mesh, and cables. Sulfates can cause disintegration of concrete, while alkalis, sodium carbonate, and bicarbonates may affect the hardening times as well as the strength of concrete.

1.14.4 Admixtures for Concrete

Admixtures are materials added to a concrete mixture before or during mixing to modify the characteristics of the final concrete. They may be used to improve workability during placement, increase strength, retard or accelerate strength development, increase frost resistance, and impart color. Usually, an admixture will affect more than one characteristic of concrete. Therefore, its effect on all the properties of the concrete must be considered. Admixtures may increase or decrease the cost of concrete work by lowering cement requirements, changing the volume of the mixture, or lowering the cost of handling and placement.

Think About It

Concrete, Mortar, and Grout

What is the difference between concrete, mortar, and grout?

The History of Portland Cement

Portland cement is a successor to a hydraulic lime that was first developed by John Smeaton in 1756 when he built a lighthouse in the English Channel. He used a burned mixture of limestone and clay capable of setting and hardening under water as well as in air. The next developments took place around 1800 in England and France where a cement material was made by burning nodules of clayey limestone. About the same time in the United States, a similar material was obtained by burning a naturally occurring substance called cement rock. These materials belong to a class known as natural cement, similar to portland cement but more lightly burned and not of a controlled composition.

In 1824, Joseph Aspdin, another Englishman, patented a manufacturing process for making a hydraulic cement out of limestone and clay that he called portland cement because when set it resembled portland stone, a type of limestone from the Isle of Portland, England. In 1850, this was followed by a more heavily burned portland cement product developed by Isaac Charles Johnson in southeastern England. Since then, the manufacture of portland cement has spread rapidly throughout the world accompanied by numerous new developments and improvements in its uses and manufacture.

Strength of Concrete

The compression strength required for a specific concrete structure, such as a footing or slab, is normally specified on the construction drawings and/or specifications. Compression strength is expressed in pounds per square inch (psi).

Control of concrete setting time may reduce costs by decreasing waiting time for floor finishing and form removal. Extending the setting time may reduce costs by keeping the concrete plastic, thereby eliminating bulkheads and construction joints. In practice, it is desirable to fully pretest all admixtures with the specific concrete mix to be used since the mix may affect the admixture efficiency and, ultimately, the final concrete.

There are a number of other admixtures available that are not in general use, such as gas forming, air entraining, grouting, expansion producing or expansion reducing, bonding, corrosion-inhibiting, fungicidal, germicidal, and insecticidal. These admixtures are for very specialized work, generally under the supervision of the manufacturer or a specialist.

1.15.0 Concrete Masonry Unit Construction

Concrete masonry units (CMUs), also referred to as concrete block, are commonly used to build foundations and basement walls. Concrete block was once called cinder block because it contained the cinders left after burning coal.

Today, CMUs are manufactured under a more controlled environment using materials such as portland cement, sand, and gravel. After water is added to the portland cement, aggregates such as

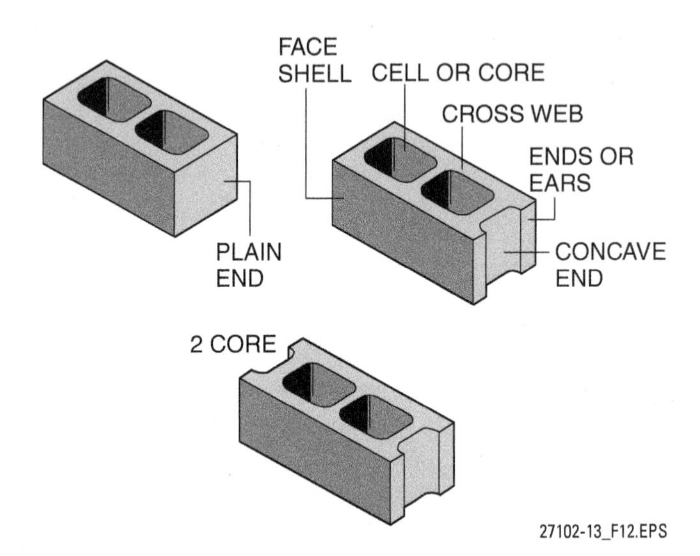

Figure 12 Concrete masonry units.

sand and gravel, and admixtures are added to the mixture.

CMUs are machine-molded into shape. They are compacted in the molds and cured, typically using live steam. After curing, the CMUS are dried and aged. The moisture content is checked; it must be a specified minimum amount before the blocks can be shipped for use.

CMUs are available in a variety of sizes, with 8" CMUs being one of the most common sizes used for residential construction. The nominal dimensions of 8" CMUs are 8" × 8" × 16" (actual dimensions are 7⅝" × 7⅝" × 15⅝"). CMUs are available in modular sizes and colors determined by their ingredients. A variety of surface and admixtures can give CMUs varied and attractive surfaces. Certain finishing techniques give CMUs the appearance of brick, rough stone, or cut stone. Like clay masonry units, CMUs can be placed in structural pattern bonds. *Figure 12* shows the names of the parts of a CMU.

CMUs take up more space than other building units (such as brick), so fewer are needed. CMU

CMU Construction

CMUs are often used in commercial construction. In some parts of the country, they are also used for the walls of residential construction in place of wood framing. The CMUs can be painted on the outside or faced with brick, stucco, or other finish material.

27102-13_SA07.EPS

Sand

Desert sand is the most plentiful commodity in Middle Eastern countries such as Egypt and Saudi Arabia. However, grains of desert sand are round and smooth, whereas concrete requires irregularly shaped grains. If desert sand is used, the concrete might disintegrate. The irony is that countries with plentiful supplies of sand must import sand for use in making concrete.

bed joints usually need mortar only on the shells and webs, so there is less mortaring as well.

CMUs are available in three weights: normal, lightweight, and aerated. Normal-weight CMUs can be manufactured with regular-, high-, and extra-high-strength concrete. CMUs using high- and extra-high-strength concrete are made with different aggregates and curing times, and are used to limit wall thickness in buildings over 10 floors high. Lightweight block is made with lightweight aggregates. The loadbearing and appearance qualities of the first two weights are similar; the major difference is that lightweight block is easier and faster to lay. Aerated block is made with an admixture that generates gas bubbles inside the concrete for a lighter block.

Common CMUs have two hollow cores, making it easy to reinforce CMU walls. Grout alone, or steel reinforcing bars combined with grout, is commonly used to fill the hollow cores. Reinforcement increases loadbearing strength, rigidity, and wind resistance.

Loadbearing CMUs are used as backing for veneer walls, bearing walls, and other structural uses. Both regular and specially shaped CMUs are used for paving, retaining walls, and slope protection. Nonstructural CMUs are used for screening, partition walls, and as a veneer wall for wood, steel, or other backing. Both kinds of CMUs are available in a variety of shapes and modular sizes.

Spacing

The spacing between steel reinforcing bars must be taken into consideration when selecting the size of the coarse aggregate to use in a concrete mixture. If too large an aggregate is used when rebar is spaced closely, the aggregate may become jammed between the rebar and create a void in the concrete.

GOING GREEN

Autoclaved Aerated Concrete

Autoclaved aerated concrete (AAC) is a lightweight precast building material cured under high pressure inside special ovens called autoclaves. AAC contains millions of very small bubbles that are generated during the manufacturing process, making the concrete extremely lightweight. AAC is nontoxic material whose primary material is fly ash. Fly ash is a recycled material that is a generated by the electric utility industry. More than 50 million tons of fly ash are generated annually, making it plentiful for the production of AAC.

1.16.0 Steel Framing Materials

Steel framing materials include light- and heavy-gauge framing members, such as steel joists and steel trusses (*Figure 13*). Steel framing offers many advantages including: noncombustibility, dimensional uniformity, light weight, freedom from rot and moisture, and ease of construction. The steel framing components are manufactured to fit together easily. There are a variety of steel framing systems, both loadbearing and nonbearing types. Some nonbearing partitions are designed to be demountable or moveable and still meet the requirements of sound insulation and fire resistance when covered with the proper gypsum system.

Light-gauge steel framing members are manufactured from 25-gauge sheet steel (the higher the gauge, the lighter the steel). Light-gauge steel framing members include C-shapes, tracks, U-channels, and furring channels. C-shapes are

27102-13_F13.EPS

Figure 13 Steel trusses.

used as studs; tracks are used as top and bottom plates. These framing members may have a knurled or dimpled pattern along the exterior surface for positive screw settings. Light-gauge steel studs are available in a wide range of standard web widths ranging from 1⅝" to 6". The flanges are 1¼" deep, and the lip size varies by stud size. (See *Figure 14*.) Lengths of 6' to 16' are commercially available; other lengths are available by special order.

Heavy-gauge steel framing members are available in the same shapes as light-gauge framing members (*Figure 15*). These framing members are made from 20-gauge sheet steel and have knurled or dimpled sides for positive screw setting. Web and flange dimensions of heavy-gauge members are similar to those of light-gauge steel framing members. They also have cutouts and utility knockout holes 12" from each end and at the midpoint of the stud. Heavy-gauge studs can be ordered to the lengths that are needed.

Steel framing members are also available in thicknesses greater than 20 gauge. Steel studs are available in 18- 16-, 14-, 12-, and 10-gauge thicknesses, and are classified as structural steel studs. They are available in web widths of 2 ½" to 8" and flange widths of 2" or 2 ½". They can be ordered in whatever length is needed.

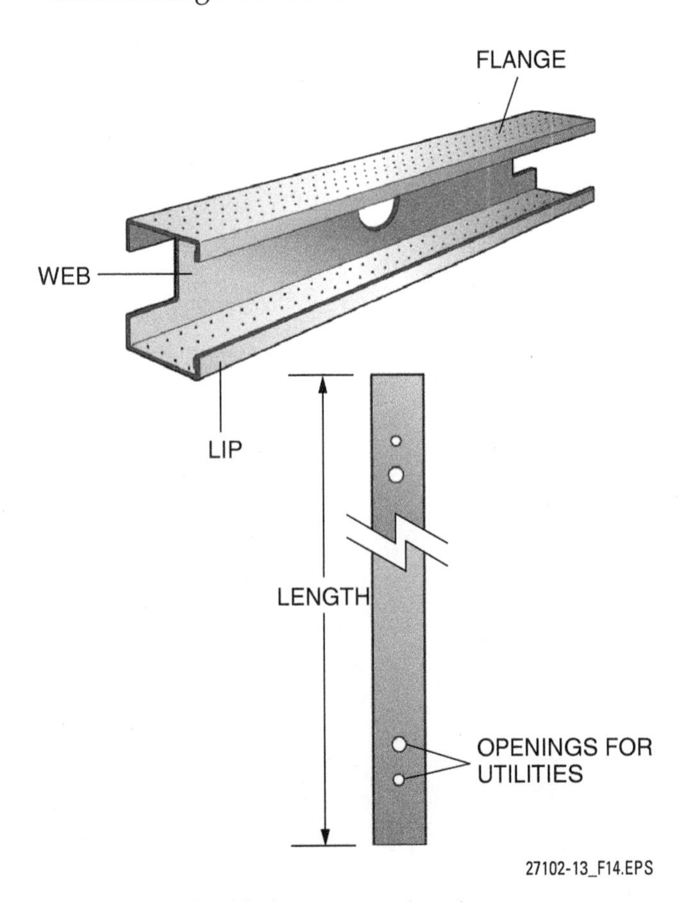

Figure 14 Standard light-gauge steel stud.

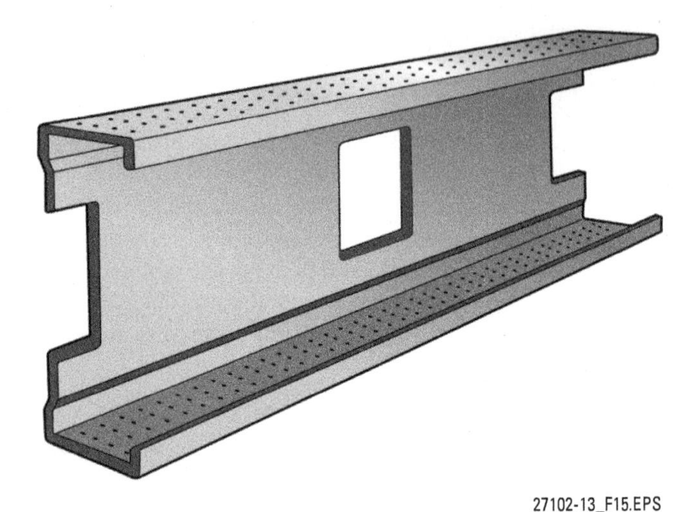

Figure 15 Heavy-gauge steel stud.

Although different materials are used, the general approach to framing with steel studs is the same as that used for wood studs. In fact, steel studs can be used with either wood plates or steel tracks.

Like wood framing, steel framing is installed 12", 16", or 24" on center (OC), openings are framed with headers and cripples, and special framing is needed for corners and partition Ts.

Depending on the load, reinforcement may be needed when framing openings. Bracing of walls to keep them square and plumb until sheathing is applied is also required. *Table 5* shows the framing spacing for various gypsum board applications.

The erection of steel studs typically starts by laying steel tracks in position on the floor and ceiling and securing them (*Figure 16*). If the tracks are being fastened to concrete, a low-velocity powder-actuated fastener is generally used. If the tracks are being fastened to wood, such as in a residence, screws can be driven with a screw gun.

WARNING!

The use of a powder-actuated fastener requires special training and certification. The use of these tools may be prohibited by local codes because of safety concerns.

Once the tracks are in place, the studs and openings are laid out in the same manner as a wood frame wall. The studs may be secured to the tracks with screws or they may be welded. In some cases, the entire wall will be laid out on the floor, then raised and secured. When heavy-gauge walls are used, they may be assembled and welded in a fabrication shop and brought to the site.

Table 5 Maximum Framing Spacing

	Single-Ply Gypsum Board Thicknesses		Application to Framing		Maximum OC Spacing of Framing
Ceilings	⅜"		Perpendicular		16"
	½"		Perpendicular or Parallel		16"
	½"		Perpendicular		24"
	⅝"		Perpendicular		24"
Sidewalls	⅜"		Perpendicular or parallel		16"
	½" or ⅝"		Perpendicular or Parallel		24"
Fasteners Only - No Adhesive Between Piles					
	Multi-Ply Gypsum Board Thicknesses		Application to Framing		Maximum OC Spacing of Framing
	Base	Face	Base	Face	
Ceilings	⅜"	⅜"	Perpendicular	Perpendicular	16"
	½"	⅜"	Parallel	Perpendicular	16"
		½"	Parallel	Perpendicular	16"
	½"	½"	Perpendicular	Perpendicular	24"
	⅝"	½"	Perpendicular	Perpendicular	24"
		⅝"	Perpendicular	Perpendicular	24"

Sidewalls*

* For two-layer applications with no adhesive between piles, ⅜", ½", or ⅝" gypsum board may be applied perpendicularly (horizontally) or parallel (vertically) on framing spaced a maximum of 24" OC. Maximum spacing should be 16" OC when ⅜" board is used as the face layer.

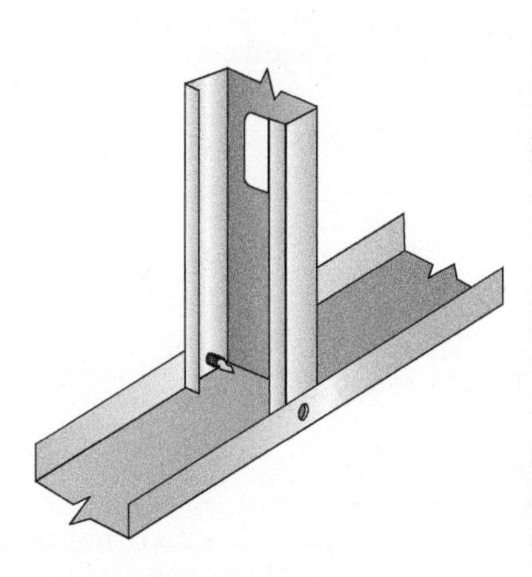

27102-13_F16.EPS

Figure 16 Steel framing.

Identifying Light-Gauge Framing Members

The Steel Studs Manufacturers Association has developed an identification system that identifies the shape, web width, flange width, and steel thickness. This system, known as *The Right S-T-U-F,* provides the needed information in one location. S-T-U-F refers to the common light-gauge shapes: S = studs or joists (C shapes), T = tracks, U = U-channels or channel studs (without lips), and F = furring channels. The web and flange widths are expressed in hundredths of an inch. The steel thickness is expressed in mils. For example, the following identifier on a steel framing member means what? **400S125-33**

The web width is 400, or 4" (400 ÷ 100 = 4"); the shape is S (stud); the flange width is 125, or 1¼" (125 ÷ 100); and the steel thickness is 33 mils, or 20 gauge.

1.0.0 Section Review

1. In the cross section of a tree, one year of tree growth is represented by one _____.
   a. cambium layer
   b. medullary ray
   c. annular ring
   d. lignin

2. Which of the following defects do *not* occur naturally in wood?
   a. Shakes
   b. Pitch streaks
   c. Knots
   d. Torn grain

3. The softwood lumber grading category that includes light framing, studs, structural light framing, and structural joists and planks is _____.
   a. dimension lumber
   b. boards
   c. timbers
   d. beams

4. Pressure-treated lumber grades are designated by their preservative retention. How many grades of pressure-treated lumber are available?
   a. Two
   b. Three
   c. Four
   d. Eight

5. The standard size of a plywood panel is _____.
   a. 2' × 4'
   b. 4' × 8'
   c. 4' × 6'
   d. 6' × 10'

6. Hardboard is only available in thicknesses greater than ½".
   a. True
   b. False

7. The class of Type I particleboard that weighs 37–50 pounds per cubic foot is Class _____.
   a. A
   b. B
   c. C
   d. D

8. The type of plywood with a hard, resin-impregnated fiber overlay heat-bonded to both surfaces is _____.
   a. hardwood plywood
   b. medium-density overlay plywood
   c. high-density overlay plywood
   d. hardboard

9. The wood strands in OSB are bonded together with _____.
   a. mastic
   b. phenolic resin
   c. epoxy
   d. interior adhesive

10. The panel product that will *not* support combustion and will *not* burn is _____.
    a. plywood
    b. particleboard
    c. OSB
    d. mineral fiberboard

11. The engineered lumber product made of laminated wood veneers that are bonded together with exterior-grade adhesive is _____.
    a. PSL
    b. LVL
    c. OSB
    d. LSL

12. The flanges of wood I-beams are made of _____.
    a. particleboard
    b. MDO plywood
    c. OSB
    d. solid lumber

13. The grade of glulam member that is required where the highest-quality appearance is needed is _____.
    a. industrial
    b. architectural
    c. premium
    d. N

14. The process through which concrete hardens by the reaction between water and cement is known as _____.
    a. hydration
    b. calculation
    c. regeneration
    d. formation

15. The openings in CMUs that are filled to reinforce CMU walls are referred to as _____.
    a. voids
    b. cores
    c. pitch pockets
    d. bed joints

16. The light-gauge steel framing member typically used for top and bottom plates is called a _____.
    a. C-shape
    b. track
    c. U-channel
    d. furring channel

SECTION TWO

2.0.0 SAFETY PRECAUTIONS WITH BUILDING MATERIALS

Objective

List safety precautions associated with building materials.

a. List general safety guidelines for working with building materials.
b. Cite safety precautions for working with wood building materials.
c. Cite safety precautions for working with concrete building materials.
d. Cite safety precautions for working with metal building materials.

Trade Terms

Combustible: Capable of easily igniting and rapidly burning; used to describe a fuel with a flash point at or above 100°F.

Experience modification rate (EMR): A rate computation to determine surcharge or credit to workers' compensation premium based on a company's previous accident experience.

When taking a job, you have a safety obligation to the employer, co-workers, family, and yourself. In exchange for your wages and benefits, you agree to work safely. You are also obligated to make sure anyone you work with is working safely. Your employer is likewise obligated to maintain a safe workplace for all employees. The ultimate responsibility for on-the-job safety, however, rests with you. Whether it is installing window units or unloading and storing building materials and supplies, safety is part of your job!

A safety culture is created when the whole company sees the value of a safe work environment. Creating and maintaining a safety culture is an ongoing process that includes a sound safety structure and attitude. Everyone in the company, from management to laborers, must be responsible for safety every day they come to work. A strong safety culture can also lower your **experience modification rate (EMR)**, which leads to winning more bids and keeping workers employed.

The mission of the Occupational Safety and Health Administration (OSHA) is to save lives, prevent injuries, and protect the health of America's workers. *OSHA 29 CFR 1926* is the regulatory source pertaining to construction site safety.

2.1.0 General Safety

You are responsible for wearing appropriate personal protective equipment (PPE) on the job. When worn correctly, PPE is designed to protect you from injury. Keep PPE in good condition and know when to use PPE for a given task. Many workers are injured on the job because they are not using proper PPE. Not all potentially dangerous conditions can be seen just by looking around a job site. It's important to stop and consider what type of accidents could occur for any task you are about to undertake. Using common sense and knowing how to use PPE will greatly reduce your chance of getting injured.

While the bulk of the building materials on a construction project are put to use, there is a certain amount of scrap and other debris that is produced during typical construction processes. A certain amount of housekeeping will be required to help prevent accidents. Some good housekeeping rules include:

- Remove from work areas all scrap material and lumber with nails protruding.
- Clean up spills to prevent falls.
- Remove all **combustible** scrap materials regularly.

- Make sure there are containers for the collection and separation of refuse. Containers for flammable or harmful refuse must be covered.
- Dispose of wastes often.
- Store all tools and equipment when you're finished using them.

Depending on the size of the construction project, many types of building materials are delivered to the job site in bulk (large quantities). Always use appropriate material-handling equipment when lifting and/or handling heavy objects or large quantities of building materials. When lifting heavy materials, bend your knees and keep your back straight. A large number of construction injuries can be attributed to improper lifting techniques.

2.2.0 Wood Product Safety

One of the most common injuries associated with wood products is embedded splinters. Many of these injuries can be avoided if the proper gloves are worn when handling building materials. As well as preventing embedded splinters, gloves help to prevent cuts or scrapes when handling sharp objects. However, never wear gloves around rotating or moving equipment as they can easily get caught up in the equipment.

When stacking materials, such as lumber, ensure that they will not fall or slide. For lumber, place the pieces flat on the ground, and do not stand them on end. Depending on the area where the lumber is stored, scrap lumber or short 2 × 4s may be placed under the lumber at a 90-degree angle to prevent the lumber from contacting the ground. Do not pile lumber more than 6' high if moving it manually.

Because the chemicals used in treated lumber present a hazard to people and the environment, special precautions apply:

- When cutting treated lumber, always wear eye protection and a dust mask.
- Wash any skin that is exposed while cutting or handling the lumber.
- Wash clothing that is exposed to sawdust separately from other clothing.

Plywood Safety

Plywood is awkward to carry. Remember, carry only one sheet of plywood at a time, and do not hold it over your head. In strong winds, use caution as the plywood could act like a sail and injure you and others on the job site.

- Do not burn treated lumber, because the ash poses a health hazard. Check local regulations for proper disposal.
- Be sure to read and follow the manufacturer's safety instructions.

2.3.0 Concrete Safety

Those working with dry cement or wet concrete should be aware that it is harmful. Dry cement dust can enter open wounds and cause blood poisoning. The cement dust, when it comes in contact with body fluids, can cause chemical burns to the membranes of the eyes, nose, mouth, throat, or lungs. It can also cause a fatal lung disease known as silicosis. Wet cement or concrete can also cause chemical burns to the eyes and skin. Make sure that appropriate personal protective equipment is worn when working with dry cement or wet concrete. If wet concrete enters boots from the top, remove the boots and rinse your legs, feet, boots, and clothing with clear water as soon as possible. Repeated contact with cement or wet concrete can also cause an allergic skin reaction known as cement dermatitis.

2.4.0 Steel Product Safety

Many of the general safety guidelines previously discussed also apply to steel product safety. Additional steel product safety guidelines are as follows:

- Wear gloves when handling steel framing members. The gloves should be thick enough to prevent penetration by sharp edges of the framing members.

Construction Fatal Four

In calendar year 2010, 751 workers were killed in construction-related accidents. The leading cause for these accidents were falls, struck by object, caught–in/between, and electrocution. These Fatal Four were responsible for nearly three out of five (57%) construction worker deaths in 2010. Eliminating the Fatal Four would save 431 workers' lives in the United States every year.

- Since metal does not absorb moisture (like wood), steel framing members may become slippery when wet. Use caution when handling wet steel framing members.
- Ensure that proper PPE is used, including hearing protection and goggles, when cutting steel framing members with a cutoff saw. When metal is cut with a cutoff saw, a loud noise may be emitted and flying metal fragments may be produced.
- The edges of steel framing members may be sharp. Avoid dropping members or placing heavy loads of steel framing members on electrical cords as they may cut through the cord and create an electrical hazard.

CAUTION

Steel framing members are galvanized with a zinc coating to prevent corrosion. When zinc reaches an elevated temperature (over 900°F), such as when welding or cutting the framing members, the zinc burns and emits a zinc oxide vapor. The vapor irritates the lungs, which leads to difficulty breathing. Craftworkers exposed to high concentrations of zinc oxide over prolonged periods of time can develop a condition known as the "zinc chills." Zinc chills can produce fevers and tremors.

Job Hazard Analysis

A hazard is the potential for harm, and is often associated with a condition or activity that if left uncontrolled can result in an injury or illness. A job hazard analysis is a technique used to identify hazards in a specific task before they occur, and focuses on the relationship between the worker, the task, the tools to be used, and the work environment. Before starting any task, consider hazards that may be encountered and take corrective action to minimize the opportunity for injury.

2.0.0 Section Review

1. When moving lumber manually, do *not* pile lumber more than _____.

   a. two feet high
   b. three feet high
   c. four feet high
   d. six feet high

2. A fatal lung disease associated with cement is _____.

   a. bronchial pneumonia
   b. cement dermatitis
   c. silicosis
   d. asthma

3. A potential for harm, often associated with a condition or activity that, if left uncontrolled, could result in an injury or illness, is referred to as a(n) _____.

   a. experience modification rate
   b. hazard
   c. accident
   d. job hazard analysis

SECTION THREE

3.0.0 HANDLING AND STORING BUILDING MATERIALS

Objective

Describe the proper method of handling and storing building materials.

a. List basic material-handling guidelines.
b. Describe how to handle and store wood building materials.
c. Describe how to handle and store concrete building materials.
d. Describe how to handle and store metal building materials.

Trade Terms

Galvanized: Protected from corrosion (rust) by a zinc coating.

Building materials may be delivered to a job site weeks before they are used. Unless the lumber, plywood, and other building materials are properly handled and stored, they can be seriously damaged. Although a contractor allows for a certain amount of waste and spoilage in an estimate, such an allowance does not account for damage due to improper storage. It is up to the work crew to take care of the building materials.

3.1.0 Material-Handling Basics

To reduce the risk of injury when manually handling material, plan your task before doing it, wear the appropriate personal protective equipment (PPE), and follow proper lifting procedures. Be aware of hazards when working from heights or working near suspended loads.

3.1.1 Pretask Planning

Most material-handling accidents occur when workers are new and relatively inexperienced at a job, or when experienced workers think that accidents cannot happen to them. Before handling any material, think about what you are doing. Always assess the situation before attempting to lift any material by doing the following:

- Check to make sure the load is not too big, too heavy, or hard to grasp.
- Make sure the load does not have protruding nails, wires, or sharp edges.
- Make sure the material is something you are able to lift by yourself. If not, ask a co-worker for assistance.
- Inspect your path of travel. Look out for hazards that could result in a slip, trip, or fall. If there are hazards in your path of travel, move them or go around them.
- Always read the warning labels or instructions on materials before they are moved, and be aware of the potential dangers if you mishandle a particular product.

3.1.2 Personal Protective Equipment

It's important to wear the proper clothing and PPE when moving or handling materials. Remember the following safety guidelines when dressing to perform materials-handling operations:

- Do not wear loose clothing that can get caught in moving parts.
- Be sure to button shirt sleeves and tuck in shirt tails.
- Remove all rings and jewelry.
- Tie back and secure long hair underneath your hard hat.
- If a wristwatch is worn, wear one that will easily break away if it gets caught in machinery.
- Wear gloves whenever cuts, splinters, blisters, or other hand injuries are possible. Select gloves that fit properly. Tight gloves may increase hand fatigue. Loose gloves reduce grip strength and may get caught on moving objects or machinery.
- Gloves should be removed when working with rotating machinery and equipment with exposed moving parts.

3.1.3 Proper Lifting Procedures

Use proper lifting techniques to reduce the risk of back injuries. Determine the weight of the load prior to lifting and plan your lift. Know where it is to be unloaded and remove any slipping or tripping hazards from your path of travel. When lifting objects, it is important to avoid unnecessary physical stress and strain. Know your limits and what you are able to physically handle when lifting a load. If the load is too heavy, ask a co-worker for help.

When lifting an object, make sure you have firm footing. Bend your knees and get a good grip. Be sure to lift with your legs, keep your back straight, and keep your head up. Keep the load close to your body (*Figure 17*). Face in the planned direction of travel and then lift. To avoid musculoskeletal injury, avoid twisting until you are

standing straight, and then proceed in the direction of travel.

When unloading materials, use your leg muscles to set them down. Space your feet far enough apart to maintain good balance and control of the object. When placing the load down, move your fingers out of the way to avoid pinching or crushing them.

3.2.0 Handling and Storing Wood Products

When a complete framing package for a residence is delivered to the job site, the framing package looks great. All the material is neatly packed. Some material is strapped and some neatly tied. However, when the load is dumped, it will shift. What was a neat load suddenly becomes a pile of lumber spread out in different directions. Some lumber may be split or broken. Other material may be partially buried in sand, mud, or snow. If the framing package is left the way it is, any of the following may result:

- The lumber is exposed to the elements. Sun, wind, rain, and snow can create serious problems.
- What would have been good lumber may now be crooked, warped, or cupped.
- Some lumber will crack when strain is placed against it; there is a good possibility of splits.
- It will cost the contractor money to replace the materials.
- The time involved in getting additional material delivered to the job site could affect the schedule.
- There is a safety hazard associated with using damaged materials.
- Additional time required in working with defective materials is costly.

A simple way to avoid these problems is to have knowledgeable workers at the site when the materials are delivered. It takes a few hours to stack the lumber and protect it from the elements, saving your employer the cost of replacing the damaged materials. Also, delivery people are more likely to be careful if an observer is present.

There are several factors to consider when unloading and stacking building materials at the job site. These may include the following:

- The order in which the material will be used
- The weight of the material
- The weather conditions

Get instructions from your supervisor before deciding how and where to stack the materials. Here are some general guidelines.

Be sure to consider the order in which the material will be used. For example, do not put the siding on top of the sheathing.

All material should be stacked on a level surface or as close to level as possible. If stacking material indoors, the amount of material stacked

BEND YOUR KNEES AND GRASP THE OBJECT FIRMLY.

LIFT THE OBJECT BY STRAIGHTENING YOUR LEGS.

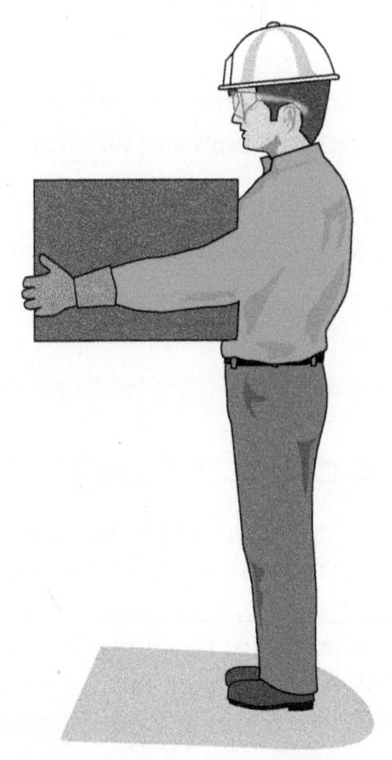

MOVE FORWARD ONCE YOU ARE IN FULL VERTICAL POSITION.

27102-13_F17.EPS

Figure 17 Proper lifting technique.

on the subfloor should not exceed its weight-bearing capacity. If stacking outdoors, try not to block the foundation or slab. The next step is to find some scrap lumber and place it flat on the ground. Two or three 2 × 4s spaced approximately 4' apart from each other will work nicely. Stack the widest lumber on top of this scrap lumber. Be sure the good lumber does not come in contact with the ground. Ensure that there is enough support under the lumber so that it will not sag.

Separate the lumber, keeping the treated material from the regular material.

Keep your materials banded until it is time to use them. Try to avoid using lumber that has been exposed to heavy rain.

Protect the lumber from the elements by covering the materials with a waterproof material (e.g., roofing paper or polyethylene). Cover the lumber to protect it from the elements, but not to prevent air from circulating around the stack. Preventing air circulation could cause twisting or warping, depending on the moisture content of the air.

Special care is needed for interior finish materials. If you are using knock-down (KD) door frames (either wood or metal) for the doors, these usually come tied together. Always ensure that the door frames (also known as jambs) are stacked flat and off the floor or concrete. Some moldings will come wrapped or tied. Take care to lay them flat and off the floor or concrete. Wall paneling should be treated the same way. Extreme caution should be taken not to damage the edges or mar the face of the panel.

Stacking interior finish materials on top of scrap lumber serves two purposes:

1. It prevents moisture in the concrete floor from coming in contact with the bottom piece.
2. Although the finish material is under cover, there is always a possibility of a pipe bursting and water flowing on the floor.

3.2.1 Plywood Storage

Remember, when stacking plywood only two methods may be used. Sheets may be stacked flat on a solid surface and off the floor or ground (*Figure 18*), or the sheets may be stacked in a full vertical position between two posts. Keep the plywood off the floor or ground. Storing plywood at an angle will cause the material to warp or twist.

3.3.0 Handling and Storing Concrete Products

Portland cement is a moisture-sensitive material. For smaller quantities, portland cement is commonly available in paper bags that hold one cubic foot of cement by volume and weigh 94 pounds.

27102-13_F18.EPS

Figure 18 Storing plywood panels.

Wood Defects

When lumber is properly stored at a job site, there is a good chance that most of it will be usable. Situations may be encountered in which the lumber may be too damaged to use, at least for its intended purpose. Experienced carpenters can find other uses for damaged and defective lumber. Wood with cosmetic defects, like knots, may be desirable for roughing-in. Some warped lumber can be used in joists and rafters if the proper precautions are taken to account for the distortion. It is a good idea, however, to consult with an experienced carpenter or supervisor when you encounter these types of situations.

Portland cement can be stored indefinitely as long as it is kept in dry conditions. Bags of cement should always be stored off the ground and in a dry location to prevent the cement from absorbing moisture that causes lumps to form in the cement powder. The bags should be stacked closely together to reduce air circulation. If bagged portland cement is stored outdoors, it should be stacked on pallets and covered with a waterproof covering. For larger concrete quantities, portland cement is provided in bulk and should be stored in watertight bins or silos.

3.4.0 Handling and Storing Steel Products

Steel framing members should be handled carefully since the edges of the members may be sharp. The members should be stored in a dry location and/or covered to ensure moisture does not affect them. Light-gauge framing members typically receive a galvanized coating to prevent corrosion. However, if the coating is deeply scratched, corrosion may occur, which will weaken the framing member.

3.0.0 Section Review

1. When loading and unloading materials, to assist in your task use your _____.
   a. upper back muscles
   b. arm muscles
   c. leg muscles
   d. lower back muscles

2. When covering lumber to protect it from the elements, wrap it tightly with waterproof material to prevent air from circulating around it.
   a. True
   b. False

3. Bags of cement contain one cubic foot of cement by volume and weigh approximately _____.
   a. 25 pounds
   b. 68 pounds
   c. 86 pounds
   d. 94 pounds

4. To prevent corrosion, light-gauge steel framing members are typically _____.
   a. painted
   b. galvanized
   c. lacquered
   d. recycled

Section Four

4.0.0 Calculate Lumber, Panel, and Concrete Quantities

Objective

Explain how to calculate the quantities of lumber, panel, and concrete products using industry-standard methods.

a. Calculate lumber quantities.
b. Calculate panel quantities.
c. Calculate the volume of concrete required for rectangular and cylindrical shapes.

Performance Task 2

Calculate the quantities of lumber, panel, and concrete products using industry-standard methods.

Trade Terms

Area: The surface or amount of space occupied by a two-dimensional object such as a rectangle, circle, or square.

Board foot: Lumber quantity measure equivalent to a piece of lumber that is 1" thick, 12" wide, and 1' long.

For commercial construction projects, the material quantities are typically calculated by an estimator and ordered to be delivered to the job site. For residential construction projects, a carpenter may need to order building materials and supplies. Therefore, an understanding of basic material quantities is needed to communicate the proper information to the lumber dealer or supply house.

4.1.0 Calculate Lumber Quantities

Large quantities of lumber are normally ordered by the board foot. A board foot is equivalent to a piece of lumber that is 1" thick, 12" wide, and 1' long. Each of the boards in *Figure 19* represents one board foot. When calculating board feet, always make sure nominal lumber dimensions are used. The following formula is used to calculate board feet:

$$\text{Board feet} = \text{number of pieces} \times \text{thickness (in inches)} \times \text{width (in inches)} \times \text{length (in feet)} \div 12$$

For example, 20 pieces of 2 × 6 lumber that are each 8' long equals 160 board feet.

$$\frac{20 \times 2 \times 6 \times 8}{12} = \frac{1{,}920}{12} = 160 \text{ board feet}$$

Trim and moldings are priced by the lineal foot and are ordered by the dimension of the piece (for example, 150' of 1" quarter round molding). *Appendix C* contains lumber conversion tables.

4.2.0 Calculate Panel Quantities

Panel products, such as plywood, particleboard, and OSB, are calculated by first determining the area to be covered. Area is calculated by multiplying the length and width (length × width = area). The area is then divided by the panel area

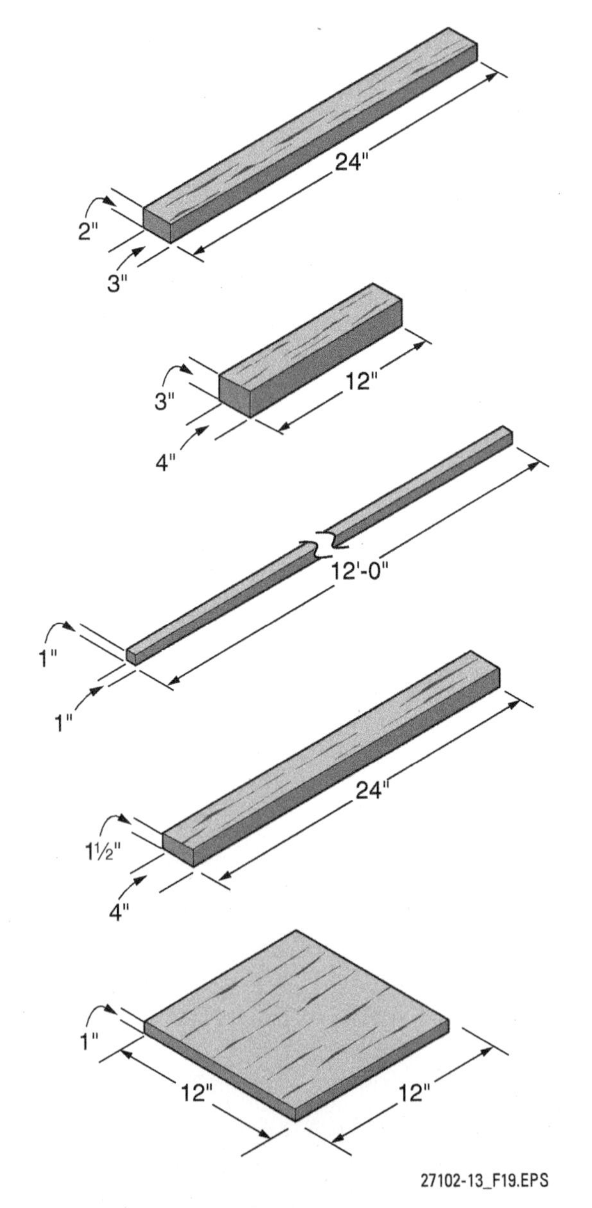

Figure 19 Examples of a board foot.

to determine the number of panels that will be required. The following formula is used to calculate board feet:

No. of panels =
roof area (length × width [in feet])
÷ panel area (length × width [in feet])

For example, to determine the number of 4' × 8' OSB panels needed for a 20' by 41' roof, use the following procedure:

Roof area = 20 × 41 = 820 square feet

Panel area = 4 × 8 = 32 square feet

No. of panels = 820 ÷ 32 = 25⅝ sheets, which is then rounded to 26 sheets.

4.3.0 Estimating Concrete Volume

Accurate measuring and estimating of concrete quantities is required for concrete work. Fortunately, most concrete structures can be divided into rectangular or circular shapes, individually estimated, and the results added together to obtain the required volume of concrete. For instance, a floor is a rectangular horizontal slab. A footing and a foundation wall can be divided into a long rectangular shape representing the footing and a vertical slab representing the wall. The volume of each can be calculated and the results added together for the total volume. Construction drawings for a project will provide the dimensions for the various portions of a concrete structure, and these dimensions are used to calculate the volume of concrete required for the structure. For reference purposes, *Appendix D* provides area or volume formulas for various geometric shapes and a table for conversion of inches to fractions of a foot or a decimal equivalent.

4.3.1 Rectangular Volume Calculations

A number of methods can be used to determine the volume of concrete required for a rectangular object. One method uses the following formula:

Cubic yards of concrete (rounded up to next 1/4 yard) =

$$\frac{\text{thickness (in feet)} \times \text{width or height (in feet)} \times \text{length (in feet)}}{27 \text{ (cubic feet/cubic yard)}}$$

To use the formula, all dimensions in inches must be converted to feet and/or fractions of a foot and then into a decimal equivalent. For example:

7" = (7⁄12)' or (7 ÷ 12)' = 0.58'

8" = (8⁄12)' or (8 ÷ 12)' = 0.66'

23" = (23⁄12)' or (23 ÷ 12)' = 1.92'

The other two methods involve knowing the width or height and length of an area in feet, along with the thickness in inches, and then using a construction calculator (*Figure 20*) or a concrete table (*Figure 21*) and a simple formula to determine the volume.

4.3.2 Example Calculation Using the Formula

Using the calculation formula, determine the amount of concrete required for the partial wall, footing, and floor slab plan shown in *Figure 22*.

Step 1 The entire footing and wall length must be determined. Since the wall is centered on the footing, the wall length is the same as the footing length:

Footing length = wall length = 20' + (15' – 2') = 33'

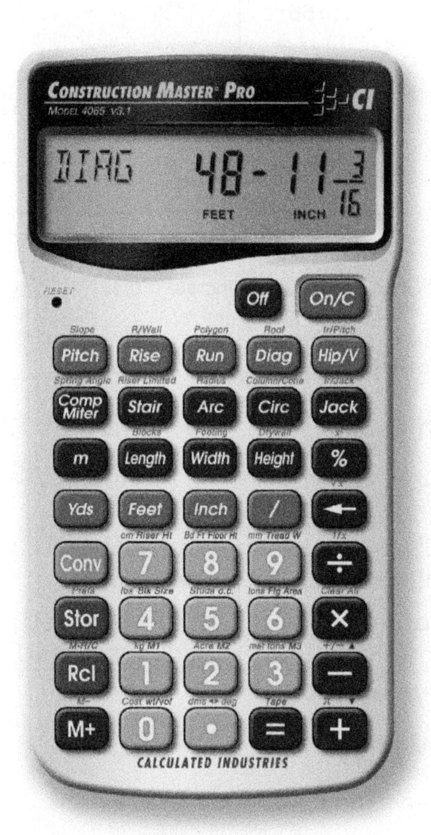

Figure 20 Typical construction calculator.

ONE CUBIC YARD OF CONCRETE WILL PLACE:			
THICKNESS	SQ FT	THICKNESS	SQ FT
1"	324	7"	46
1¼"	259	7¼"	45
1½"	216	7½"	43
1¾"	185	7¾"	42
2"	162	8"	40
2¼"	144	8¼"	39
2½"	130	8½"	38
2¾"	118	8¾"	37
3"	108	9"	36
3¼"	100	9¼"	35
3½"	93	9½"	34
3¾"	86	9¾"	33
4"	81	10"	32.5
4¼"	76	10¼"	31.5
4½"	72	10½"	31
4¾"	68	10¾"	30
5"	65	11"	29.5
5¼"	62	11¼"	29
5½"	59	11½"	28
5¾"	56	11¾"	27.5
6"	54	12"	27
6¼"	52	15"	21.5
6½"	50	18"	18
6¾"	48	24"	13.5

27102-13_F21.EPS

Figure 21 Portion of typical concrete table.

Step 2 Determine the floor slab length and width:

$$\text{Length} = 20' - 16" = 20' - 1.33' = 18.67'$$

$$\text{Width} = 15' - 16" = 15' - 1.33' = 13.67'$$

Step 3 Using the formula, determine the volume of the wall, footing, and slab:

$$\text{Volume} = \frac{\text{width or height (feet)} \times \text{length (feet)} \times \text{thickness (feet)}}{27 \text{ (cubic feet/cubic yard)}}$$

$$\text{Wall} = \frac{3 \times 33 \times 0.67}{27} = \frac{66.33}{27}$$

$$= 2.46 \text{ cubic yards}$$

$$\text{Footing} = \frac{2 \times 33 \times 0.67}{27} = \frac{44.22}{27}$$

$$= 1.64 \text{ cubic yards}$$

$$\text{Slab} = \frac{13.67 \times 18.67 \times 0.5}{27} = \frac{127.6}{27}$$

$$= 4.73 \text{ cubic yards}$$

Step 4 Add the wall, footing, and slab volumes:

$$2.46 + 1.64 + 4.73 =$$
8.83 rounded up to 9 cubic yards

For the plan shown, 9 cubic yards of concrete will be required.

4.3.3 Cylindrical Volume Calculations

The volume of concrete required for a circular column or pier can be calculated by using the following formula:

$$2.46 + 1.64 + 4.73 =$$
8.83 rounded up to 9 cubic yards

$$\text{Cubic yards of concrete} = \frac{\pi \times \text{radius}^2 \text{ (in square feet) height (in feet)}}{27 \text{ (cubic feet/cubic yard)}}$$

Where:

- $\pi = 3.14$
- Radius = diameter (in feet) of column ÷ 2
- Measurements in inches are converted to feet

As with rectangular volume calculations, a certain percentage of waste must be added to the circular volume estimate based on the job site conditions.

Calculating the Volume of Concrete

When using any of the three methods for calculating the volume of concrete needed for a job, a waste factor must be taken into consideration. Typically, about 5 percent of the calculated volume is added to account for waste. For larger jobs, experienced carpenters order the required number of full truckloads (typically 12 cubic yards per load), then specify the required amount needed for the last truckload once the total volume of concrete needed is known.

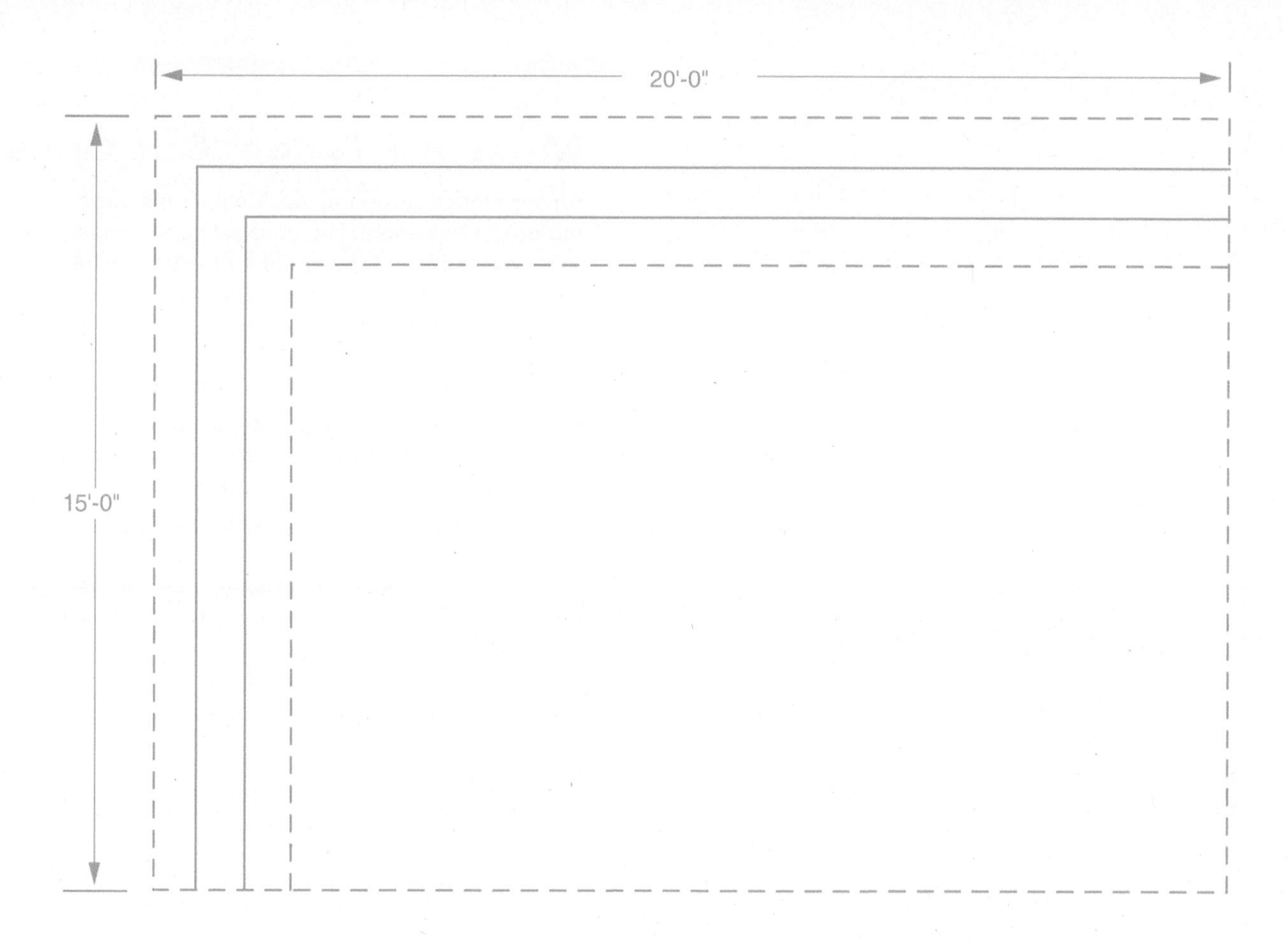

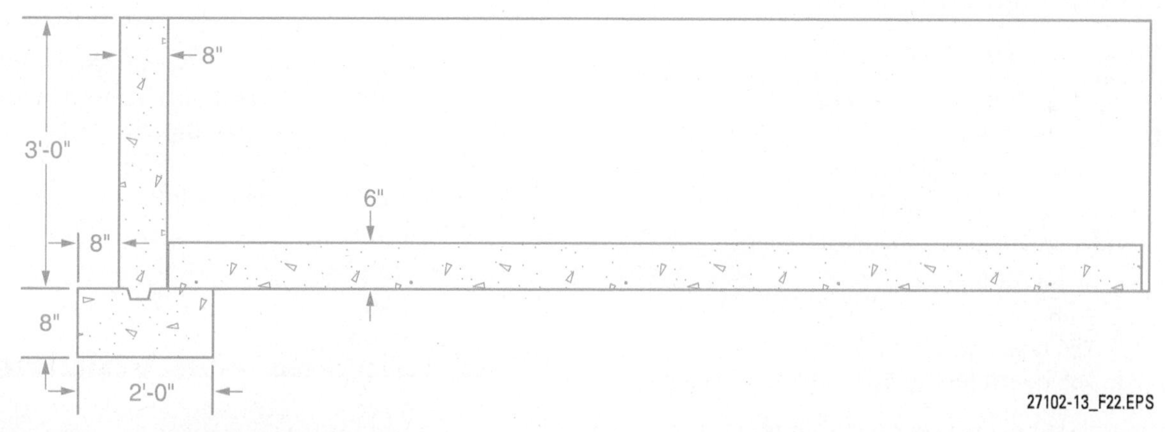

Figure 22 Partial wall, footing, and floor slab plan.

4.3.4 Example Calculation of Cylindrical Volume

Using the circular volume formula, determine the volume of concrete required for the circular column shown in *Figure 23*.

Step 1 Determine the radius of the column:

$$3' \div 2 = 1.5'$$

Step 2 Calculate the volume using the formula:

$$\text{Cubic yards of concrete} = \frac{\pi \times \text{radius}^2 \text{ (in square feet) height (in feet)}}{27 \text{ (cubic feet/cubic yard)}}$$

$$= \frac{3.14 \times (1.5)^2 \times 20}{27} = \frac{141.3}{27}$$

$$= 5.23 \text{ cubic yards}$$

rounded up to $5\frac{1}{4}$ cubic yards

For the plan shown, $5\frac{1}{4}$ cubic yards of concrete will be required.

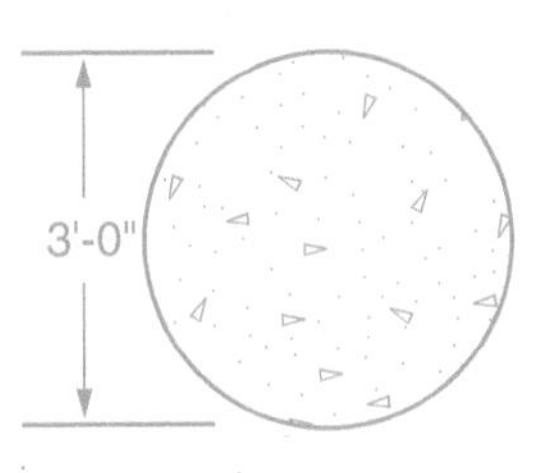

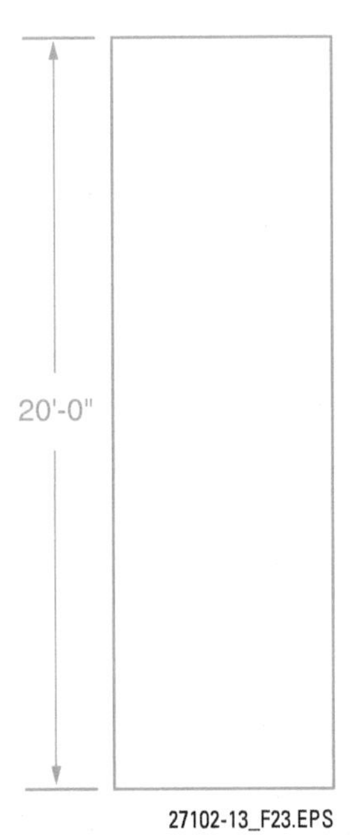

Figure 23 Typical circular column plan.

Measure Twice, Cut Once

An apprentice carpenter was helping his boss remodel a basement. The boss asked the apprentice to measure and cut the studs for the walls. The apprentice, looking to impress the boss, attacked the job with gusto. Instead of cutting and checking the fit of the first stud, he cut all of them at once, based on a quick measurement. When all the studs were cut, they started to place them, only to find out that each stud was 3¼" short. The apprentice had forgotten to add the measuring tape case to the tape reading.

4.0.0 Section Review

1. Four pieces of lumber measuring 2" thick, 12" wide, and 8' long contains a total of _____.
   a. 16 board feet
   b. 64 board feet
   c. 256 board feet
   d. 768 board feet

2. Multiply the length by the width to determine the _____.
   a. area
   b. volume
   c. diameter
   d. circumference

3. Multiply the thickness, width, and length to determine the _____.
   a. area
   b. volume
   c. diameter
   d. circumference

SECTION FIVE

5.0.0 FASTENERS, ANCHORS, AND ADHESIVES

Objective

Describe the fasteners, anchors, and adhesives used in construction and explain their uses.

a. Identify various types of nails and cite uses for each.
b. Identify applications for staples.
c. Identify various types of screws and cite uses for each.
d. Describe uses for hammer-driven pins and studs.
e. Identify various types of bolts and cite uses for each.
f. Identify various types of mechanical anchors and cite uses for each.
g. Identify various types of bolt anchors and explain how each is installed.
h. Identify various types of screw anchors and cite uses for each.
i. Identify various types of hollow-wall anchors and cite uses for each.
j. List the types of glues and adhesives used in construction.

Performance Task 3

Demonstrate safe and proper installation of drop-in anchors.

Trade Terms

Catalyst: A substance that causes a reaction in another substance.

Nail set: A punch-like tool used to recess finish nails.

Penny: A measure of nail length; commonly abbreviated with the letter *d*.

Various types of fasteners, anchors, and adhesives are used to temporarily or permanently fasten materials together. The project specifications and construction drawings contain information regarding the type of fastener, anchor, or adhesive to use in a given situation.

5.1.0 Nails

Nailing is the most common method used for attaching two pieces of lumber. Some nails are made specifically to be driven with a hammer; others are made for use with pneumatic or cordless nailers. Nails are often referred to as 8-**penny**, 10-penny, etc. This method of designating nail length dates back many years. In written form, the penny designation appears as a lowercase d (*Figure 24*); an 8-penny nail would read 8d, and so forth. Nails longer than 16d are commonly referred to as spikes. Spikes range from 20d (4") to 60d (6"). The thickness (gauge) of the nail shank increases as the length increases.

5.1.1 Types of Nails

A variety of nails are available (*Figure 25*), each with specific applications as follows:

- *Common nails* – Common nails are the most frequently used of all nails. They have a flat head, smooth and round shank (shaft), and diamond point. They are available from 2d to 60d in length. Common nails are used when appearance is not of prime importance, such as in rough framing or building concrete forms.
- *Box nails* – Box nails are very similar to common nails in appearance, except they have a thinner head and shank and have less of a tendency to split the wood. They often have a resin coating to resist rust and create more holding power. Box nails are available from 2d to 40d.
- *Finish nails* – Finish nails have a small barrel-shaped head with a slight indentation in the top to receive a **nail set**, so the nail head can be driven slightly below the surface of the wood. The void above the head is then filled with a putty or filler before paint or finish is applied to conceal the nail. Finish nails are used for installing millwork and trim where the final appearance is important. Their small heads reduce their holding power considerably. If holding power is important, finish nails can be obtained with galvanized coatings.
- *Casing nails* – Casing nails look a lot like finish nails; however, they have a conical-shaped head and the shank is larger. In addition, casing nails are available in lengths from 2d to 40d (as opposed to the 20d maximum length of a finish nail). Casing nails have more holding power because of the larger shank and the shape of the head.

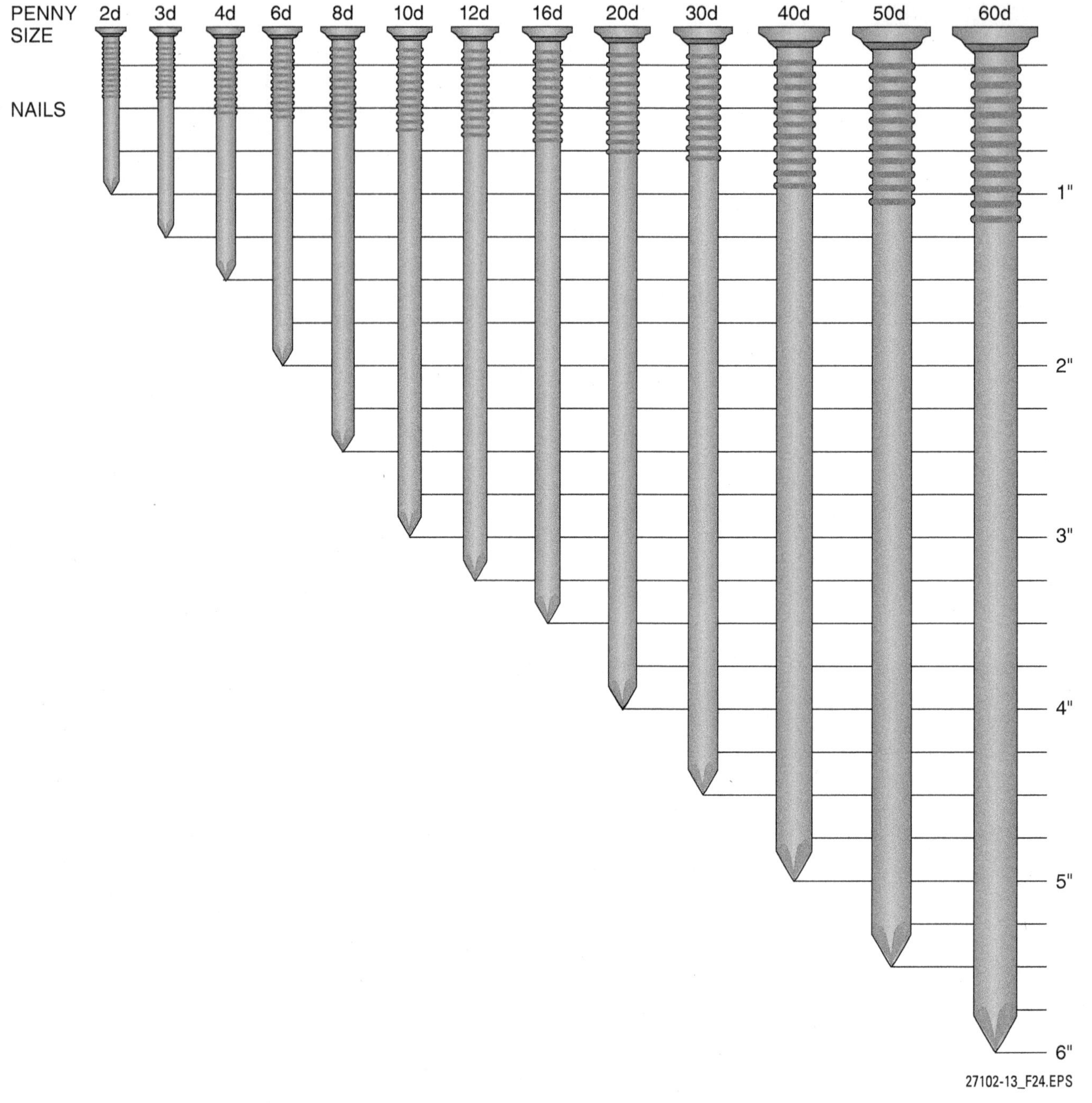

Figure 24 Nail sizes.

- *Duplex or double-headed nail* – Duplex nails are designed for building concrete forms and other temporary work that will later be disassembled. The advantage of duplex nails is that they can be driven to the first head, which gives them sufficient holding power. They can be easily removed at a later time with a claw hammer or wrecking bar.
- *T-nails* – T-nails are specially designed nails used in pneumatic nailers. They are coated with resin and available in strips for insertion into the nailer. T-nails are available in lengths of 1¼", 1½", 1¾", 2", 2⅜", and 2½". These nails are also available with galvanized coatings. The use of a pneumatic nailer reduces installation time as it only takes a pull of the trigger to drive them home.
- *Cement-coated nails* – Cement-coated nails are coated with an adhesive (cement) to provide greater holding power. As the nails are driven, the coating melts due to the friction created when driving the nails. The coating then hardens when it cools, creating an excellent bond with the base material.
- *Drywall nails* – Drywall nails have an annular or ring shank. Their principle use is to fasten gyp-

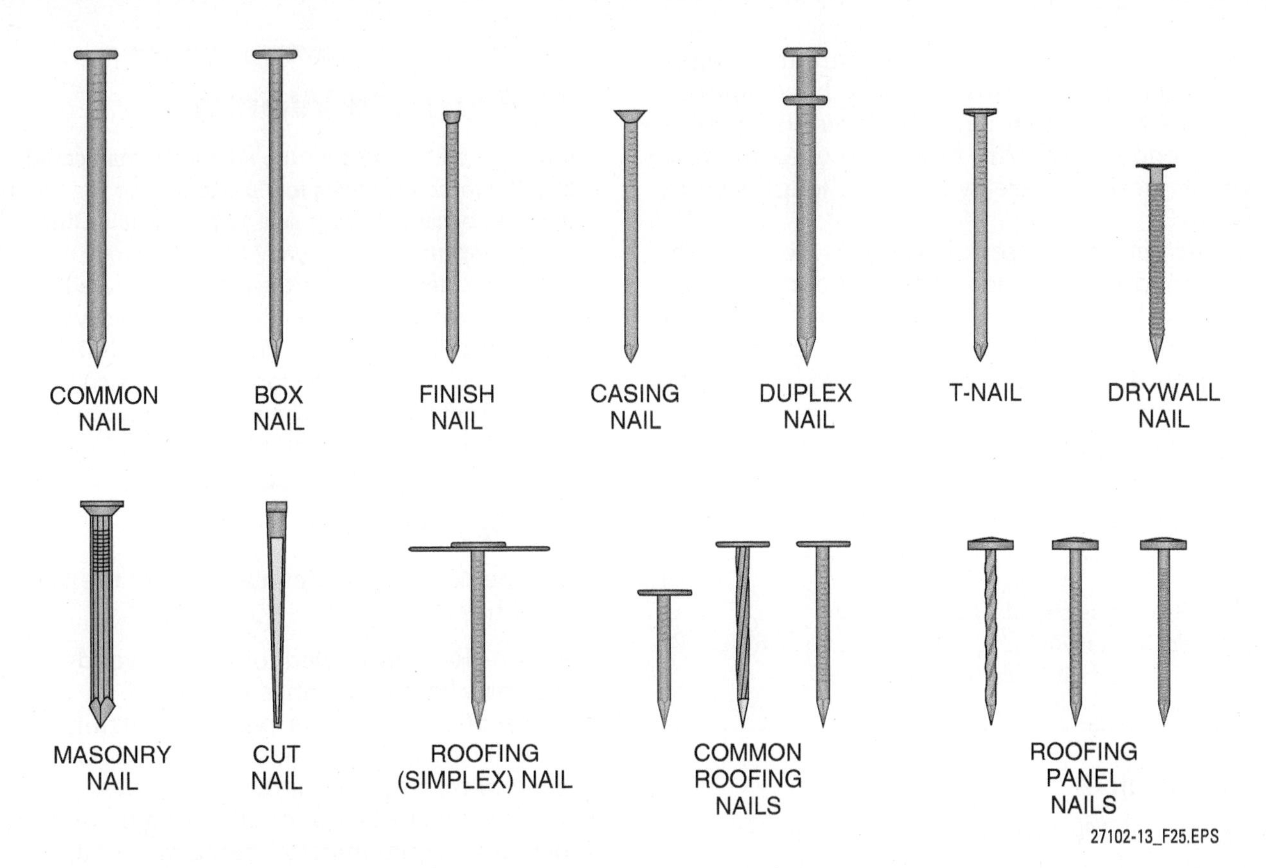

Figure 25 Kinds of nails.

sum board (drywall) to wood studs. The ring shanks increase holding power and prevent the nail from backing out of the stud. Drywall nails are available in lengths ranging from 1" to 2½".

- *Masonry nails* – Masonry nails are available in lengths of ½" to 4". The shanks on longer masonry nails are grooved, and the nails have a sinker head. Masonry nails are case-hardened and are used to fasten metal or wood to masonry or concrete.
- *Cut nails* – Cut nails could also be grouped in with masonry nails, because in today's construction they are seldom used for other than masonry purposes. Cut nails are rectangular in cross section, have a blunt point, and have little tendency to bend. They are available in lengths up to about 5" long.
- *Roofing nails* – Most roofing nails are galvanized to resist rusting. There are several types of roofing nails:
  - Common roofing nails are available in lengths of ¾" to 2". These roofing nails have a thin head but different types of shanks. Common roofing nails are primarily used to fasten asphalt or fiberglass shingles to roof sheathing. They are also used to fasten insulation board to studs. Longer roofing nails are used for reroofing jobs, while shorter ones are used on new roofs.
  - Roofing panel nails have a neoprene washer. They range in length from 1¾" to 2½", with either a helical (spiral) or plain shank. The helical shank has greater holding power. Roofing panel nails are used on fiberglass, steel, or aluminum roofing panels. The neoprene washer serves as a weatherproofing seal.

The following are some additional types of nails that may be encountered:

- *Hardboard siding nails* – These nails are designed for installing aluminum, vinyl, and steel siding.
- *Insulated siding nails* – These nails have a zinc coating and are available in various colors to match the color of the siding being installed.
- *Panel nails* – These nails are also available in various colors and have a plain- or ring-shank design. The head is small and blends in with the paneling for a good appearance.

Panel Nails

Panel nails are available in a variety of colors to match the paneling. These nails have annular rings to provide more holding capacity. Nails used to fasten paneling should be long enough to penetrate ¾" into the framing.

- *Wire brads* – Wire brads look similar to regular finish nails. However, they have smaller shanks than finish nails, thus reducing splitting tendencies. Wire brads are used for fine finish work and are available in lengths from 3⁄16" to 3".
- *Escutcheon pins* – Escutcheon pins are specially designed nails, which generally come as part of a hardware kit. They have an oval head and vary in length from 3⁄16" to 2". They are used for such things as metal trim on store cabinets or fastening house numbers to the structure. Escutcheon pins add a nice appearance to the finish trim and generally match the color of the hardware being installed.

WARNING!

Wear a face shield when driving cut nails, because they are subject to breakage, which could do serious damage to an eye.

5.2.0 Staples

There are a wide variety of staples manufactured to do specific jobs. Refer to the manufacturers' catalogs for specific information regarding uses. The factors that should be considered when selecting staples are:

- The type of point
- The crown (or top) width
- The length
- What type of metal they are made of
- The various coatings that are available

A Penny's Worth

Back in the fifteenth century, an Englishman could buy 100 three-inch nails for 10 pennies, so the three-inch nail became known as a 10-penny nail. Other nails were known as 8-penny nails, 12-penny nails, and so on, depending on the cost per hundred. Although a penny isn't worth much these days, the designation is still used to identify the nails. The d that is used in the written form (8d, 10d, etc.) is said to represent the denarius, a small Roman coin used in Britain that was equal to a penny.

5.2.1 Types of Staples

The type of staple is determined by the type of point it has (*Figure 26*).

- *Chisel* – Recommended for grainy woods; keeps legs parallel to the entire leg length.
- *Crosscut chisel* – Legs penetrate straight and parallel through cross-grain wood. Good for general tacking and nailing purposes.
- *Outside chisel* – Good for clinching inward after penetrating the material being fastened.
- *Inside chisel* – Good for clinching outward after penetrating the material being fastened.
- *Divergent* – After the start of penetration, legs diverge to allow use of longer legs in thin material; very good for wallboard insulation.
- *Outside chisel divergent* – Legs diverge, then cross, locking the staple; provides excellent penetration ability.

Selecting the Right Nail

As a rule of thumb, select a nail that is three times longer than the thickness of the material being fastened. However, the nail length should not exceed the total thickness of the two pieces being fastened together. For example, if you are nailing two 2 × 8 joists where they overlap on a girder, three times the thickness of one joist would be 4½". Since the total thickness of the two joists is only 3", however, it would not make sense to use a 4½" nail.

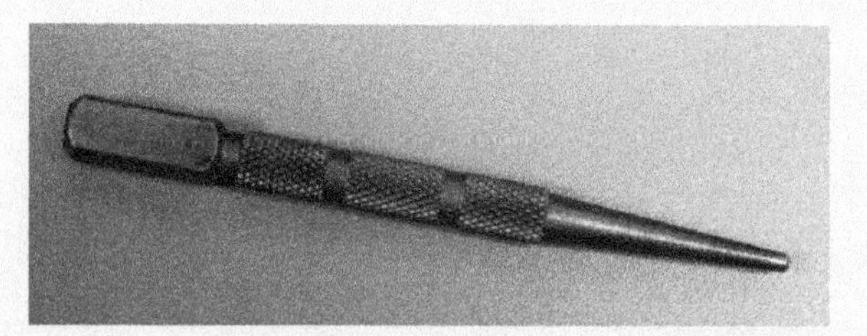

27102-13_SA08.EPS

Common nails are used primarily for rough framing. Box nails, which are thinner and have slightly smaller heads than common nails, are used when working near the edge of the lumber (toenailing studs, for example) because they are less likely to split the wood. Box nails are also used for fastening exterior insulation board and siding.

Three types of nails may be used for trim. Finish nails are used primarily for fastening interior trim. Casing nails, which are thicker and have slightly larger heads than finish nails, are used to fasten heavy trimwork and exterior finish material. Wire brads are thinner than finish and casing nails and are used for very light trimwork such as small moldings. These trim nails all have a small head with a slight indentation, which makes it easy to drive the nail head below the surface of the wood using a nail set.

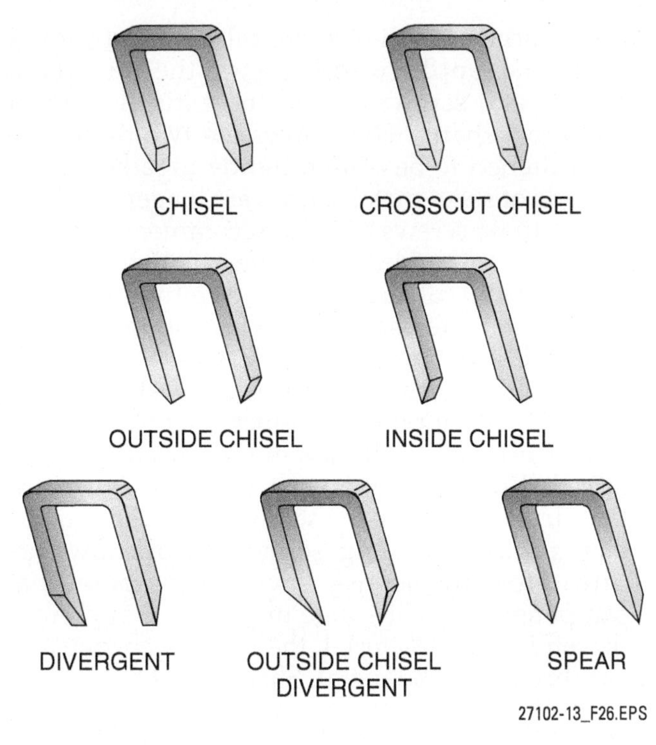

Figure 26 Staples.

- *Spear* – Good for penetrating dense materials; will deflect easily when hitting an obstruction.

The most common crown widths vary from ⅜" to 1". Wider crowns are used mostly for millwork, crating, and the application of asphalt shingles. The most common wire gauge (thickness) is 14 to 16 and the length is normally ½" to 2".

Staples are glued together in strips for insertion into a stapler. In addition, the adhesive provides greater holding power. The adhesive melts due to friction when the staple is driven. When it cools, the adhesive bonds well to the base material. Some staplers are manually operated; others are pneumatically or electrically powered.

5.3.0 Screws

Screws have greater holding power than nails and can easily be removed without damaging the materials involved. However, when selecting the appropriate fastener for a job, screw cost and the time involved in installation must be considered.

5.3.1 Wood Screws

Wood screws are commonly available in more than 20 different gauges (diameters) and lengths. The screw gauge ranges from 0 to 24 (the higher the number, the larger the diameter). See the wood screw chart in *Figure 27.* Wood screws range in length from ¼" to 5".

Screws are available with several head shapes, including round, flat, and pan (*Figure 28*). The most common types of screw head recesses are slotted, Phillips, and Robertson (square).

When using wood screws, it is usually necessary to drill pilot holes, particularly in hardwoods, to receive the screw. Pilot holes perform four important functions:

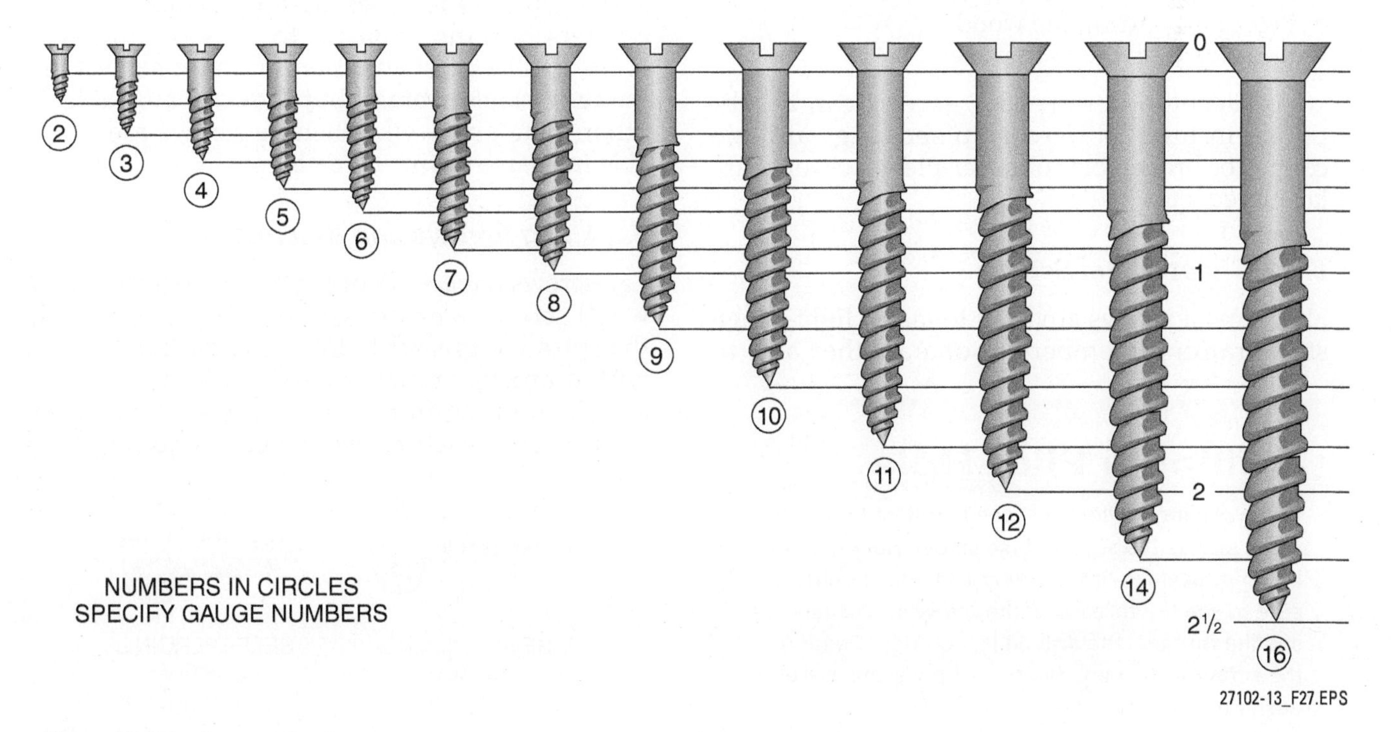

Figure 27 Wood screw chart.

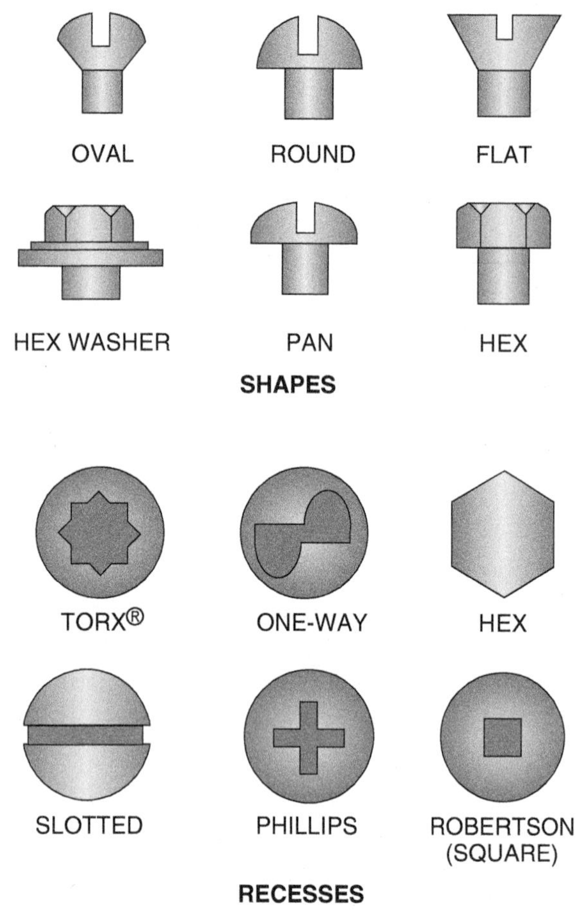

Figure 28 Common screw head shapes and recesses.

- Ensures pulling the materials tightly together
- Makes the screw easier to drive
- Prevents splitting the wood
- Prevents damage to the screw

For decorative purposes, screws can be obtained in many different finishes (e.g., antique copper or brass, gold- or silver-plated, chromium, lacquered, etc.).

5.3.2 Sheet-Metal Screws

Sheet-metal screws are used to fasten light-gauge steel framing members to one another and to fasten other building materials to the framing members. Unlike wood screws, the threads of sheet-metal screws extend the entire length of the screw shank. This allows the two pieces being fastened to be drawn tightly together. Since threads of sheet-metal screws are longer and hold better, these screws are also recommended when attaching softer building materials like particleboard and hardboard to steel framing members. Sheet-metal screws are available in approximately the same size range as wood screws.

Pilot holes are not required for sheet-metal screws. Self-tapping sheet-metal screws include self-drilling and self-piercing screws (*Figure 29*). Self-drilling screws have a drill bit–like point that drills a hole for the screw before the threads engage. Self-piercing screws, also known as thread-forming screws, have a sharp point that can penetrate light-gauge metal. As the screw is driven into the material, the threads engage and cut threads in the metal. Sheet-metal screws are typically driven with a power screwdriver.

Drilling a Pilot Hole

When drilling a pilot hole, use a drill bit that is the same size as the shank of the screw. Hold the drill bit lengthwise over the screw and you should be able to see the threads of the screw. If you can't see the threads, the drill bit is too large. Coating the screw with soap, paraffin, or beeswax makes it easier to drive.

5.3.3 Machine Screws

Machine screws are used to fasten butt hinges to metal jambs or door closers to their brackets, and to install lock sets. It may be necessary to drill and tap holes in metal to receive machine screws as fasteners for wood or various kinds of trim. Machine screws have four basic head designs, as shown in *Figure 30*.

Machine screws are available with a straight slot or Phillips recesses, and are available in diameters ranging from $\frac{1}{16}$" to $\frac{3}{8}$" and lengths of $\frac{1}{8}$" to 3". The most common machine screws are made of steel or brass, but they are available in other metals as well. Machine screws are available in coarse or fine threads.

5.3.4 Lag Screws and Shields

Lag screws (*Figure 31*), or lag bolts, are heavy-duty wood screws with square- or hex-shaped heads that provide greater holding power. Lag screws with diameters ranging between $\frac{1}{4}$" and $\frac{1}{2}$" and lengths ranging from 1" to 6" are common. They are typically used to fasten heavy equipment to

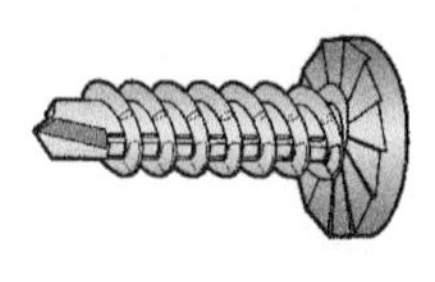

Figure 29 Sheet-metal screws.

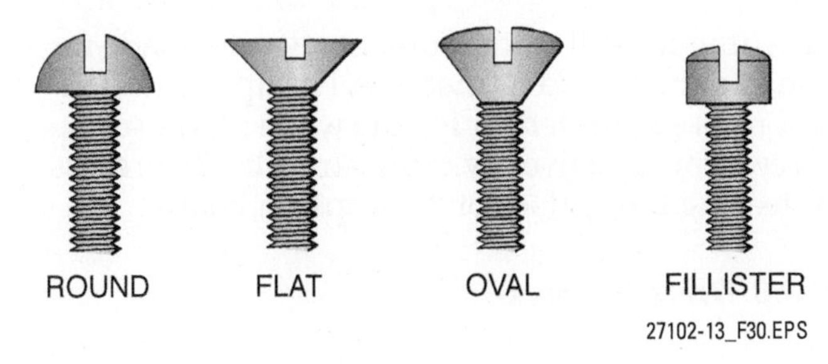

Figure 30 Machine screws.

wood, but can also be used to fasten equipment to concrete when a lag shield is used.

A lag shield is a lead tube that is split lengthwise but remains joined at one end. It is placed in a predrilled hole in the concrete or CMUs. When a lag screw is screwed into the lag shield, the shield expands in the hole, firmly securing the lag screw. In hard masonry, short lag shields (typically 1" to 2" long) may be used to minimize drilling time. In soft or weak masonry, long lag shields (typically 1½" to 3" long) should be used to achieve maximum holding strength.

Make sure to use the correct length lag screw to achieve proper shield expansion. The length of a lag screw should be equal to the thickness of the component being fastened plus the length of the lag shield. Also, drill the hole in the masonry to a depth approximately ½" longer than the shield being used. If the head of a lag screw rests directly on wood when installed, a flat washer should be placed under the head to prevent the head from digging into the wood as the lag screw is tightened. Be sure to take the thickness of washers into account when selecting the length of the screw.

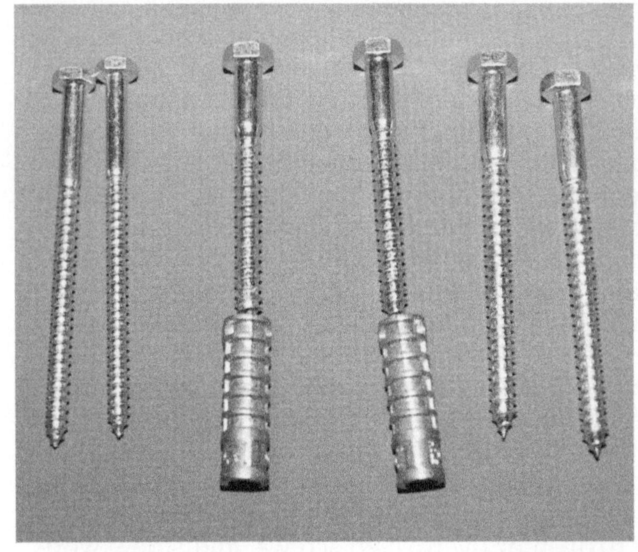

Figure 31 Lag screws and shields.

5.3.5 Concrete/Masonry Screws

Concrete/masonry screws (*Figure 32*), commonly called self-threading anchors, are used to fasten a device or fixture to concrete, CMUs, or brick without using an anchor. To provide a close tolerance between the pilot hole and the screw threads, a specially designed carbide drill bit and installation tool are used when installing these screws. The drill bit and installation tool are typically used with a rotary hammer or hammer-drill. The pilot hole must be drilled to the diameter and depth specified by the screw manufacturer. Once the pilot hole is drilled, the installation tool, along with an appropriate drive socket or bit, are used to drive the screws. When being driven into the concrete, the widely spaced threads on the screws cut into the walls of the hole to provide a tight friction fit. Most types of concrete/masonry screws can be removed and reinstalled to allow for shimming and leveling of the fastened device.

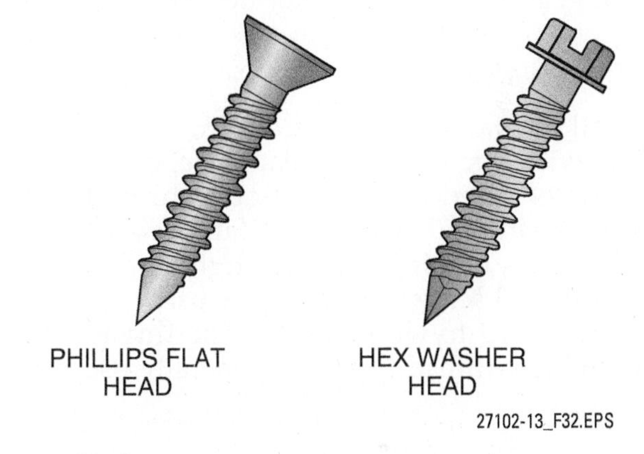

Figure 32 Concrete screws.

Threaded vs. Nonthreaded Fasteners

For most applications, either threaded or nonthreaded fasteners could be used. However, threaded fasteners are sometimes preferred because they can usually be tightened or removed without damaging the surrounding material. To maintain holding power, ensure you are using the proper-diameter drill bit when drilling the pilot hole. In addition, drill the pilot hole to approximately two-thirds the length of the threaded portion of the screw.

5.3.6 Deck Screws

Deck screws (*Figure 33*) are available in a wide variety of shapes and sizes for different indoor and outdoor applications. Some deck screws are used to fasten treated lumber and other types of wood decking to wood framing. Self-drilling deck screws are used to fasten wood decking to different gauges of metal support structures. Similarly, other self-drilling deck screws are used to fasten metal decking and sheeting to different gauges and types of metal structural support members. Due to their diversity, ensure that manufacturer's recommendations are followed when selecting the proper screw for a particular application. A self-feeding power screwdriver may be used to drive deck screws, eliminating angle driving, underdriven or overdriven screws, and screw wobble. The self-feeding power screwdriver, which allows screws to be driven while the operator is standing, also reduces operator fatigue.

5.3.7 Drywall Screws

Drywall screws (*Figure 34*) are thin, self-drilling screws with bugle-shaped heads. Depending on the type of drywall screw, it cuts through the gypsum board and anchors itself into wood and/or steel studs, holding the panel tight to the stud. Coarse-thread screws are typically used to fasten gypsum board to wood studs. Fine-thread and high-and-low-thread types are generally used for fastening to steel studs. Some screws are made for use in either wood or metal. A Phillips or Robertson recess is preferred for drywall screws so the screws can be driven without stripping the recess or tearing the surface of the gypsum board.

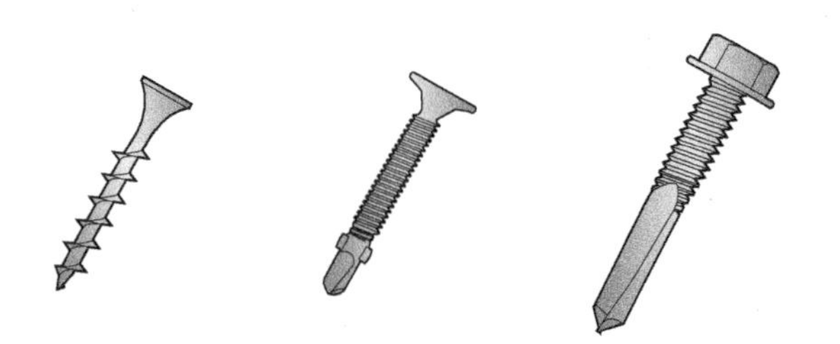

Figure 33 Typical deck screws.

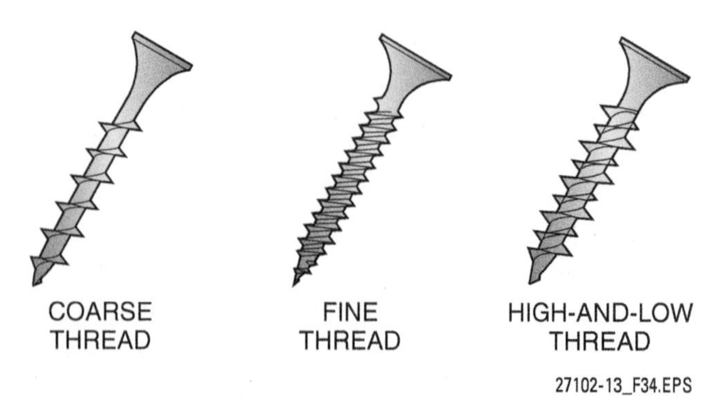

Figure 34 Drywall screws.

5.3.8 Drive Screws

Drive screws are installed by hammering the screw into a drilled or punched hole of the proper size; they do not require the hole to be tapped. Drive screws are mostly used to fasten parts that will not be exposed to much pressure. A typical use of drive screws is to attach permanent name plates on electric motors and other types of equipment. *Figure 35* shows typical drive screws.

5.4.0 Hammer-Driven Pins and Studs

Hammer-driven pins or threaded studs (*Figure 36*) use a special tool to fasten wood or steel to concrete or CMUs without pilot holes. To install these fasteners, insert the pin or threaded stud into the installation tool with the washer seated in the recess. Position the pin or stud against the

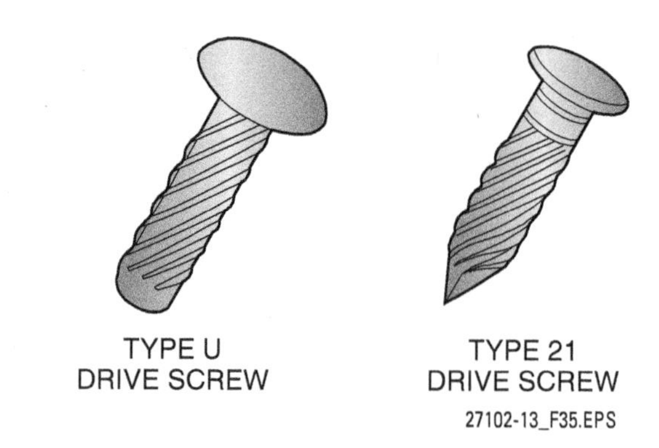

Figure 35 Drive screws.

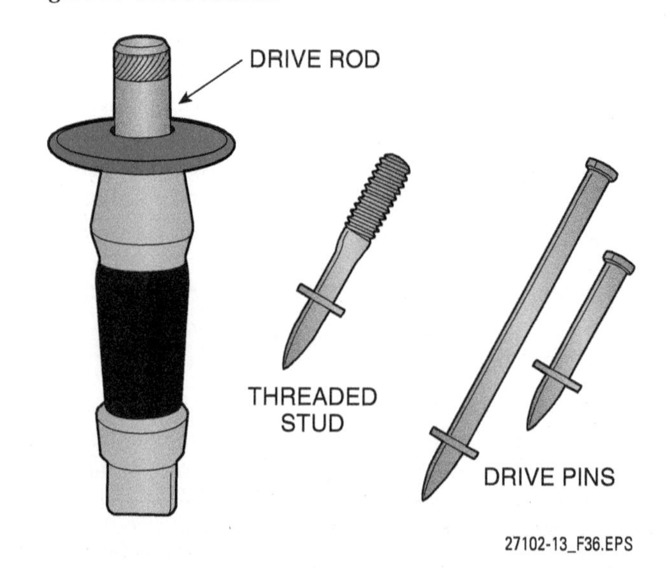

Figure 36 Hammer-driven pins and installation tool.

concrete or CMUs and tap the drive rod of the tool lightly until the striker pin contacts the pin or stud. Strike the drive rod using heavy blows with about a two-pound hammer. The force of the hammer blows transmits through the tool directly to the head of the fastener, causing it to be driven into the concrete or CMU. For best results, the drive pin or stud should be embedded a minimum of ½" in hard concrete to 1¼" in softer CMUs.

5.5.0 Bolts

Bolts are often used by the carpenter to attach one unit or member to another. Bolts differ from screws in that bolts require a nut. The most common bolts used by a carpenter are stove bolts, machine bolts, and carriage bolts. However, there are many different designs and types for special jobs (*Figure 37*).

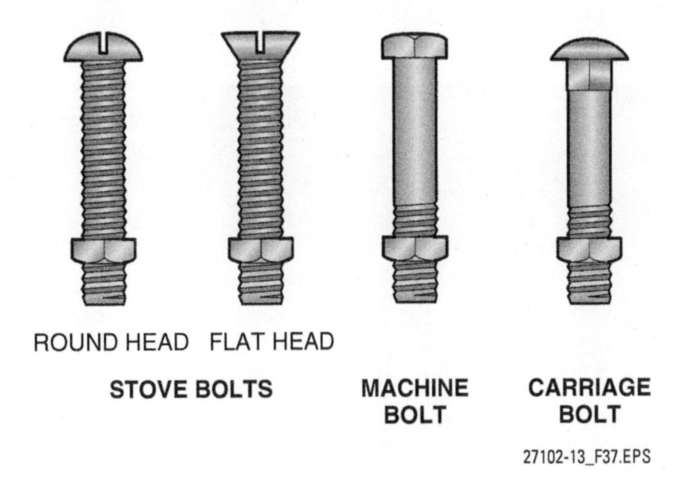

Figure 37 Bolts.

5.5.1 Stove Bolts

Stove bolts are used in lighter types of construction assemblies. They are available in either round or flat heads and lengths from ⅜" to 6". The shanks are threaded all the way to the head for stove bolts up to 2" long. If they are longer than 2", they are threaded to a maximum of 2" from the end. They can be obtained in several different materials such as steel, brass, and copper.

5.5.2 Machine Bolts

Machine bolts are used to fasten wood or metal pieces together. The most common diameters range from ¼" to 2". They are available in lengths from ¾" to 30", and can have either regular or fine threads. The bolts typically have square or hexagon heads with nuts to match the head shape. Specially designed nuts, such as self-locking nuts or cap nuts, are available for specific purposes. For a decorative appearance, machine bolts can be ordered with special finishes such as antique or chrome.

5.5.3 Carriage Bolts

Carriage bolts are similar to machine bolts except for the head design. The head of a carriage bolt is oval and the shank immediately below the head is square. This design allows the square shank to be driven or drawn into the wood, and the nut tightened without the bolt turning. Due to the head shape, other framing members can be placed over the heads with little or no interference. Carriage bolts are available in lengths of ¾" to 20" with either regular or fine threads. Common carriage bolt diameters range from ¼" to ¾". They are also available in different materials and finishes, such as brass or galvanized.

5.6.0 Mechanical Anchors

Mechanical anchors are devices that give fasteners a firm grip in a variety of materials, where the fasteners by themselves would otherwise have a tendency to pull out. Anchors are classified in many ways by different manufacturers. In this module, mechanical anchors are divided into five broad categories:

- Anchor bolts
- One-step anchors
- Stud bolt anchors
- Screw anchors
- Self-drilling anchors
- Hollow-wall anchors

In addition to the mechanical anchors discussed here, a wide variety of specialty anchors are also available from suppliers.

5.6.1 Anchor Bolts

One of the most common anchors in frame construction is the anchor bolt. Anchor bolts are commonly used to anchor the sill plate to a concrete or CMU foundation. In addition, larger anchors bolts (ranging from 2 ½" to 5" in diameter and 10' to 12' long) may be used to attach equipment to its foundation (*Figure 38*). Anchor bolts must be accurately positioned and affixed prior to concrete placement. The anchor bolt position must once again be checked to ensure the bolt did not become displaced during concrete placement. The sill plate is then placed over the anchor, and a washer and nut are used to secure the sill plate to the foundation. The sill plate is the base for the floor framing.

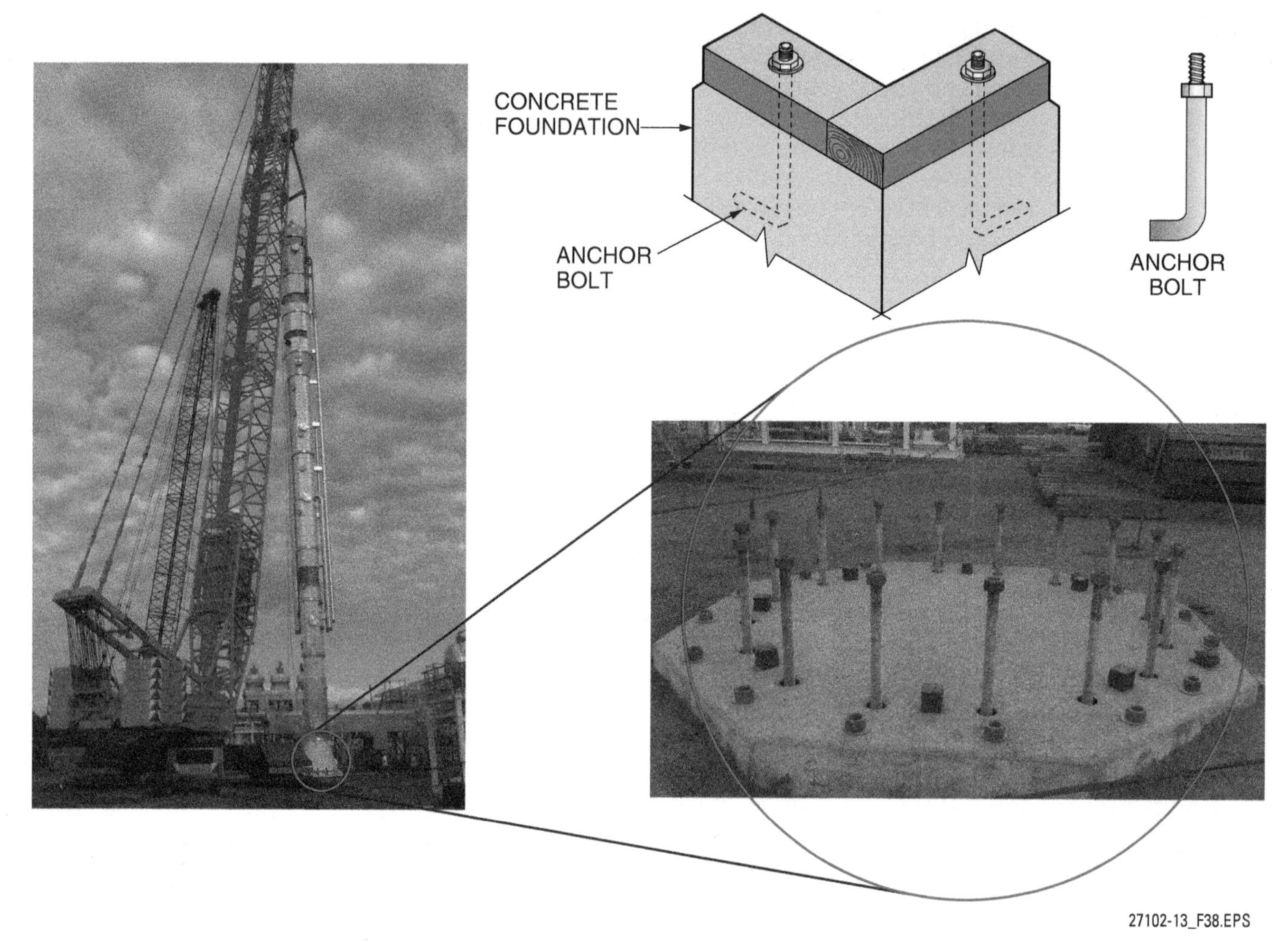

Figure 38 Anchor bolts.

5.6.2 One-Step Anchors

One-step anchors are designed to be installed through the mounting holes in the component to be fastened. Initially, the anchor and the hole into which the anchor is installed have the same diameter. As the bolt or screw is tightened, the anchor expands to create a tight fit in the hole. One-step anchors are available in various diameters ranging from ¼" to 1¼" and lengths ranging from 1 ¾" to 12". Wedge, stud bolt, sleeve, one-piece, and screw and nail hammer-set anchors (*Figure 39*) are common types of one-step anchors.

5.6.3 Wedge Anchors

Wedge anchors are heavy-duty anchors supplied with nuts and washers. The drill bit size used for the hole is the same diameter as the anchor. The depth of the hole is not critical as long as it meets the minimum manufacturer recommendations. Blow the hole clean of dust and other material, insert the anchor into the hole, and drive it with a hammer far enough so that at least six threads are below the top surface of the component. Then, tighten the anchor nut to expand the anchor and secure it in the hole.

5.6.4 Stud Bolt Anchors

Stud bolt anchors are heavy-duty threaded anchors commonly used to level components or equipment fastened to them. These anchors are designed to fit to the bottom in its mounting hole; therefore, it is critical that the hole depth specified by the manufacturer is achieved to ensure proper expansion. Blow the hole clean of dust and other material, then insert the anchor in the hole with the expander plug end down. Drive the anchor into the hole with a hammer or setting tool to expand the anchor and tighten it in the hole. The anchor is fully set when it can no longer be driven into the hole. Fasten the component or equipment to the anchor using a bolt with the correct size and thread type.

One type of stud bolt anchor, called a TZ® expansion anchor, is a medium-duty anchor especially suited for seismic and safety-relevant applications. TZ® expansion anchors can be used

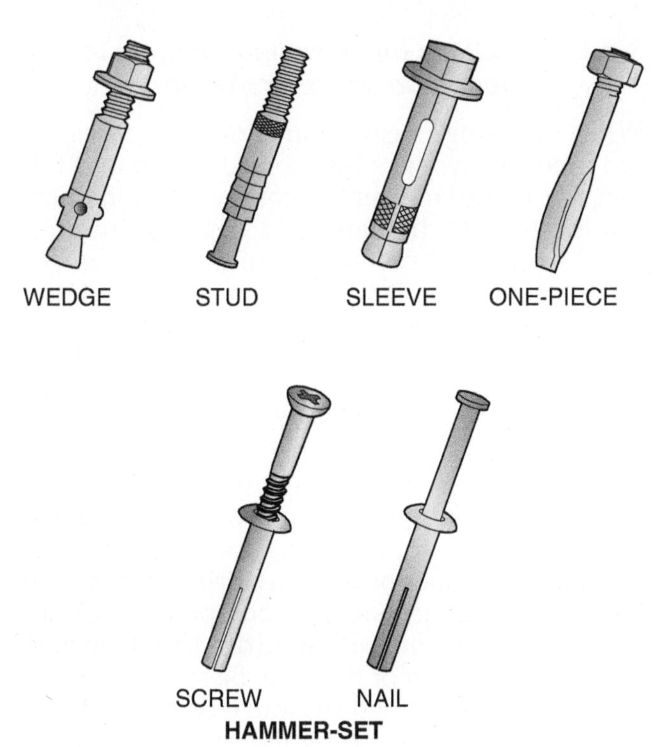

Figure 39 One-step anchors.

in cracked and uncracked concrete, and are available in ⅜" to ¾" diameters and in lengths ranging from 3" to 10".

5.6.5 Sleeve Anchors

Sleeve anchors are multipurpose anchors. The depth of the anchor hole is not critical as long as the minimum depth recommended by the manufacturer is drilled. Blow the hole clean of dust and other material, insert the anchor into the hole, and tap it until it is flush with the component. Tighten the anchor nut or bolt to expand the anchor and tighten it in the hole.

5.6.6 One-Piece Anchors

One-piece anchors are multipurpose anchors. They work on the principle that as the anchor is driven into the hole, the spring force of the expansion mechanism is compressed and flexes to fit the size of the hole. Once set, it tries to regain its original shape. The depth of the hole drilled in the masonry must be at least one-half inch deeper than the required embedment. The proper depth is critical. Overdrilling is as bad as underdrilling. Blow the hole clean of dust and other material, insert the anchor through the component, and drive it into the hole with a hammer until the head is firmly seated against the component. It is important to make sure that the anchor is driven to the proper embedment depth. Some manufacturers make specially designed drivers and installation tools. These tools allow anchors to be installed in confined spaces and help prevent damage to the component from stray hammer blows.

5.6.7 Hammer-Set Anchors

Hammer-set anchors are made for use in concrete and masonry. There are two types of hammer-set anchors: screw and nail. An advantage of the screw-type anchors is that they are removable. Both types have a diameter the same size as the anchoring hole. For both types, the anchor hole must be drilled to the diameter of the anchor and to a depth at least a quarter-inch deeper than that required for embedment. Blow the hole clean of dust and other material, then insert the anchor into the hole through the mounting holes in the component to be fastened. Drive the screw or nail into the anchor body to expand it. It is important to make sure that the head is seated firmly against the component and is at the proper embedment depth.

5.7.0 Bolt-Type Anchors

Bolt-type anchors are designed to secure two or more pieces of material together. Threaded rod may be used with some types of bolt-type anchors. Drop-in and expansion anchors (*Figure 40*) are common types of bolt-type anchors. Self-drilling anchors are another type of bolt-type anchor.

5.7.1 Drop-In Anchors

Drop-in anchors are typically used for heavy-duty applications. There are two types of drop-in anchors. The first type, made for use in solid concrete and filled masonry, has an internally threaded expansion anchor with a preassembled internal expander plug. A pilot hole must be drilled to the diameter and depth specified by the manufacturer. Clean the hole of dust and other material, insert the anchor into the hole, and tap it until it is flush with the surface. Drive the setting

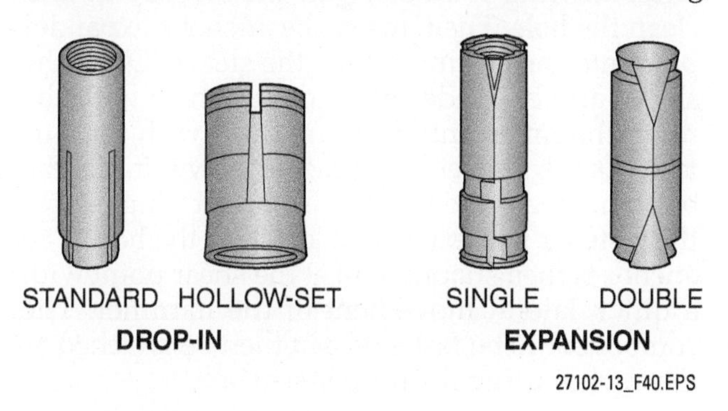

Figure 40 Bolt-type anchors.

tool supplied with the anchor into the anchor to expand it. Position the component to be fastened in place, and thread and tighten the correct-size machine bolt into the anchor.

The second type of drop-in anchor, called a hollow-set anchor, is made for use in hollow concrete and masonry base materials. They can also be used in solid concrete and filled masonry. Hollow-set drop-in anchors have a slotted, tapered expansion sleeve and a serrated expansion cone. The anchor hole must be drilled to the diameter specified by the manufacturer. When installed in hollow base materials, drill the hole into the cell or void. Blow the hole clean of dust and other material, insert the anchor into the hole, and tap it until it is flush with the surface. Position the component to be fastened in place, then thread the proper-size machine bolt or screw into the anchor and tighten it to expand the anchor in the hole.

5.7.2 Expansion Anchors

Single- and double-expansion anchors are made for use in concrete and other masonry. Double-expansion anchors are used mainly when fastening into concrete or masonry of questionable strength. For both types, the anchor hole must be drilled to the diameter and depth specified by the manufacturer. Blow the hole clean of dust and other material, then insert the anchor into the hole, threaded cone end first. Tap it until it is flush with the surface. Position the component to be fastened in place, then thread the proper-size machine bolt into the anchor and tighten it to expand the anchor in the hole.

5.7.3 Self-Drilling Anchors

Self-drilling anchors are used in concrete and masonry (*Figure 41*). Some types of self-drilling anchors have a cutting sleeve that is first used as a drill bit and later becomes the expandable anchor itself. A rotary hammer is used to drill the hole in the concrete or masonry using the cutting sleeve. After drilling the hole, pull the sleeve out and clean the hole. Then, insert the anchor's expander plug into the cutting end of the sleeve. Drive the sleeve and expander plug into the hole with the rotary hammer until they are flush with the surface. As the sleeve is hammered down, it hits the bottom, where the tapered expander plug causes the fastener to expand and lock into the hole. The anchor is then snapped off at the shear point with a quick lateral movement of the hammer. The component to be fastened can then be attached to the anchor using the proper-size bolt.

5.7.4 Guidelines for Installing Anchors in Hardened Concrete or Masonry

When selecting masonry anchors, regardless of the type, always take into consideration and follow the manufacturer's recommendations pertaining to hole diameter and depth, minimum embedment in concrete, maximum thickness of the material to be fastened, and the pullout and shear load capacities.

When installing anchors and/or stud bolt anchors in hardened concrete, make sure the area where the equipment or component is to be fastened is smooth so it will have solid footing. An uneven footing might cause the equipment to twist, warp, not tighten properly, or vibrate when in operation. Before starting, carefully inspect the rotary hammer or hammer-drill and the drill bit(s) to ensure they are in good operating condition. Set the drill or hammer-drill depth gauge to the required depth of the hole. Be sure to use the type of carbide-tipped masonry or rotary hammer bits recommended by the drill/hammer or anchor manufacturer because these bits are made to take the higher impact of the masonry mate-

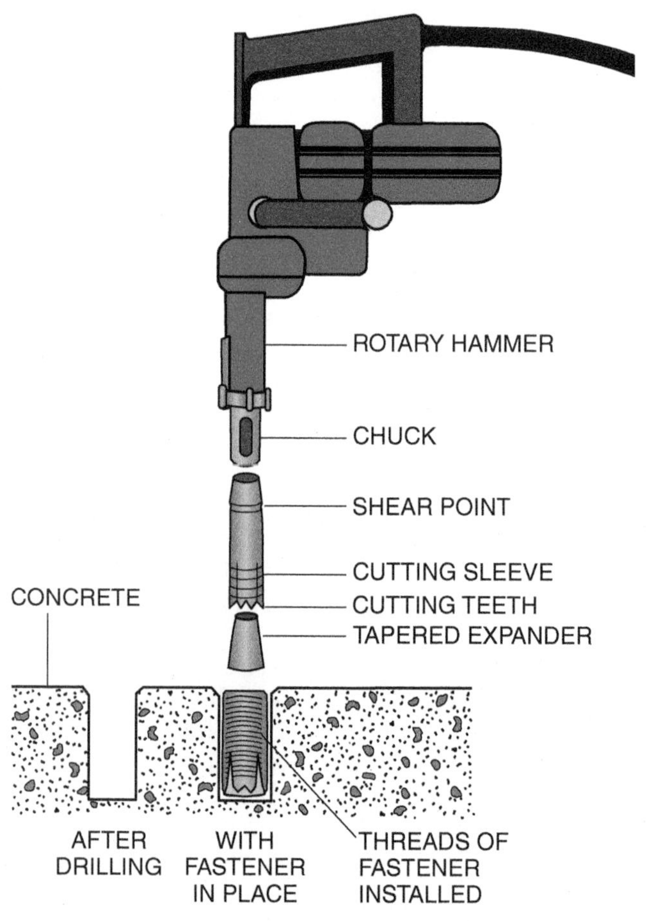

Figure 41 Self-drilling anchor.

rials. When drilling into concrete or masonry, do not force the bit into the material by pushing down hard on the drill. Use a little pressure and let the drill do the work. For large holes, start with a smaller bit, then change to a larger bit.

The methods for installing the different types of anchors in hardened concrete or masonry were briefly described in this section. Always install the selected anchors according to the manufacturer's specifications. Here is an example of a typical procedure used to install many types of expansion anchors in hardened concrete and masonry. Refer to *Figure 42* as you study the procedure.

> **WARNING!**
> Drilling in concrete generates noise, dust, and flying particles. Always wear safety glasses, ear protection, a dust mask, and gloves. Ensure that other workers in the area are also wearing protective equipment before you begin drilling.

Step 1 Drill the anchor bolt hole the same diameter as the anchor bolt. The hole must be deep enough for six threads of the bolt to be below the surface of the concrete. Clean out the hole using a squeeze bulb.

Step 2 Drive the anchor bolt into the hole using a hammer. Protect the threads of the bolt with a nut that does not allow any threads to be exposed.

Step 3 Put a washer and nut on the bolt, and tighten the nut with a wrench until the anchor is secure in the concrete.

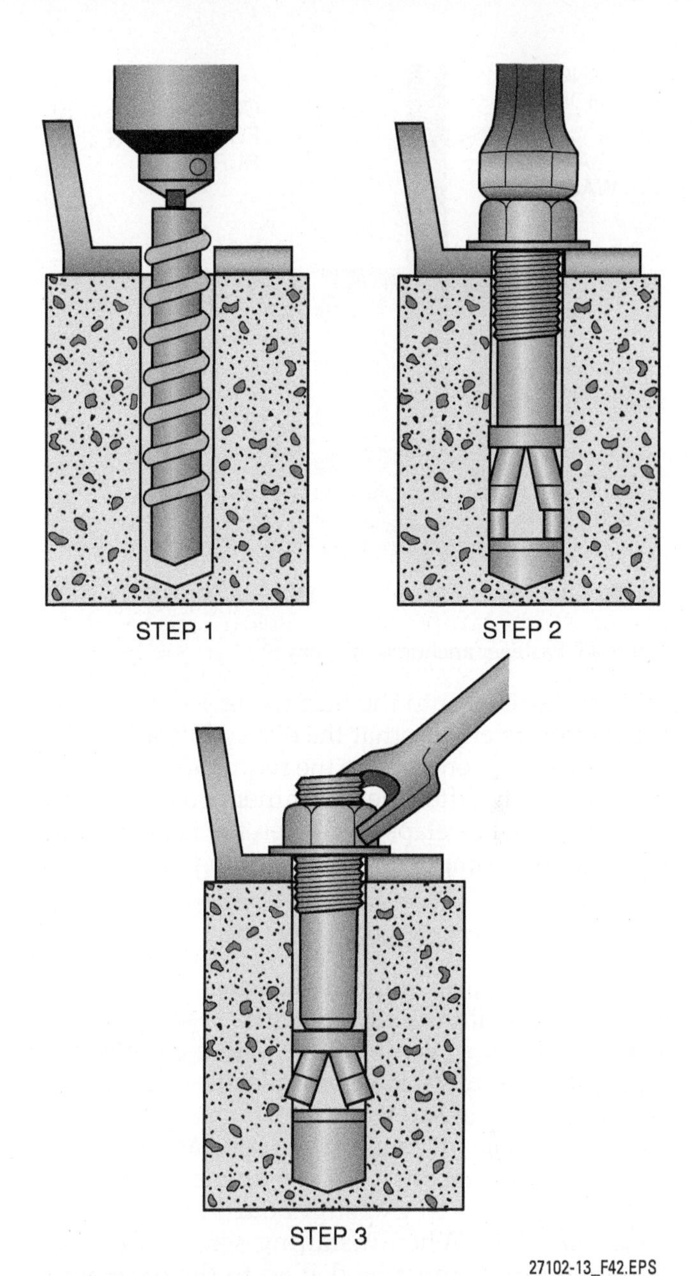

Figure 42 Installing an anchor bolt in hardened concrete.

5.7.5 Epoxy Anchoring Systems

Epoxy anchoring systems can be used to anchor threaded rods, dowels, and other fasteners in solid concrete, CMUs, and brick. For one type of epoxy anchoring system, a two-part epoxy is packaged in a two-chamber cartridge that keeps the catalyst and hardener ingredients separated until use. This cartridge is placed into a special tool similar to a caulking gun. When the gun handle is pumped, the epoxy catalyst and hardener are mixed within the gun; then the epoxy is ejected from the gun nozzle.

To use epoxy to install an anchor in solid concrete (*Figure 43*), drill a hole of the proper size in the concrete and clean it using a nylon (not metal) brush. Dispense a small amount of epoxy from the gun to make sure that the catalyst and hardener have mixed properly; proper mixing is indicated by the epoxy being of a uniform color. Place the gun nozzle into the hole, and inject the epoxy into the hole until half the hole is filled. Push the

Eye Injuries

The average direct cost of an eye injury is $5,700. That includes both the direct and indirect cost of accidents, not to mention the long-term effects on the health of the worker.

Source: The Center to Protect Workers' Rights

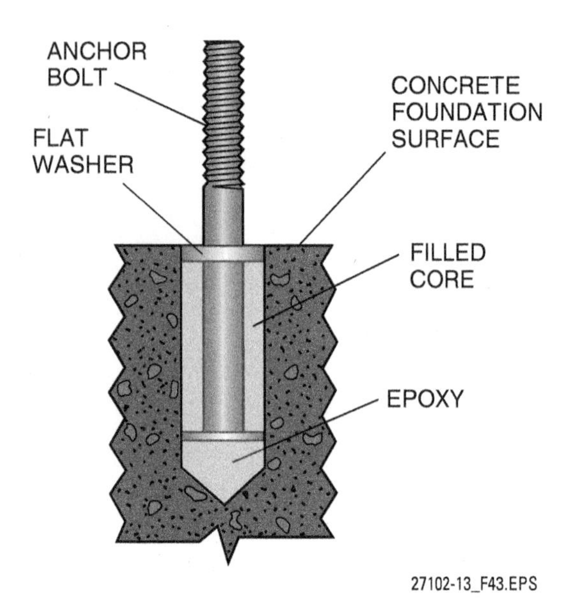

Figure 43 Fastener anchored in epoxy.

selected fastener into the hole using a slow twisting motion to ensure that the epoxy fills all voids and crevices. Then, set it to the required plumb (or level) position. After the recommended cure time for the epoxy has elapsed, tighten the fastener nut to secure the component or fixture in place.

5.8.0 Screw Anchors

Screw anchors are light-duty anchors that are installed flush with the surface of the base material. They are used in conjunction with sheet-metal, wood, or lag screws. Fiber and plastic anchors are common types of screw anchors (*Figure 44*).

Fiber and plastic anchors are typically used in concrete and masonry. Plastic anchors are also commonly used in gypsum board and similar base materials. When installing screw anchors, the anchor hole must be drilled to the diameter specified by the manufacturer. The minimum depth of the hole is equal to the anchor length. Clean the hole of dust and other material, insert the anchor into the hole, and tap it until it is flush with the surface. Position the component to be fastened in place, then drive the proper type and size of screw through the component mounting hole and into the anchor to expand the anchor.

5.9.0 Hollow-Wall Anchors

Hollow-wall anchors are used for hollow materials and thin materials such as concrete plank, CMUs, structural steel, gypsum board, and plaster. Some types of hollow-wall anchors can also be used in solid materials. Toggle bolts, sleeve-type wall anchors, wallboard anchors, and metal

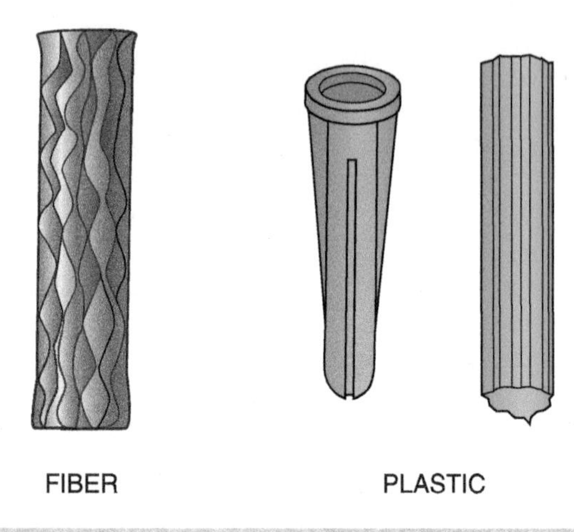

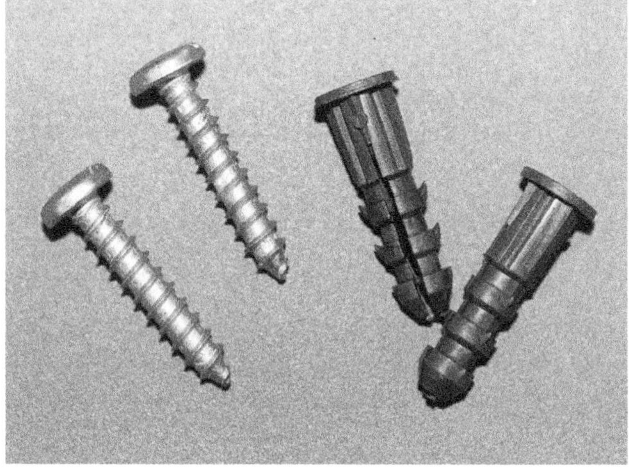

Figure 44 Screw anchors and screws.

drive-in anchors are common anchors used when fastening to hollow materials.

When installing anchors in hollow walls or ceilings, regardless of the type, always follow the manufacturer's recommendations pertaining to use, hole diameter, wall thickness, grip range (thickness of the anchoring material), and the pullout and shear load capacities.

5.9.1 Toggle Bolts

Toggle bolts (*Figure 45*) are used to fasten equipment, hangers, supports, and similar items to hollow surfaces such as walls and ceilings. Toggle bolts consist of a slotted bolt or screw and spring-loaded wings. When inserted through the item to be fastened, then through a predrilled hole in the wall or ceiling, the wings spring apart and provide a firm hold on the inside of the hollow wall or ceiling as the bolt is tightened. Note that the hole drilled in the wall or ceiling should be just large enough for the compressed wings to pass through. Once the toggle bolt is installed, be careful not to completely unscrew the bolt because the wings will fall off, making the fastener useless.

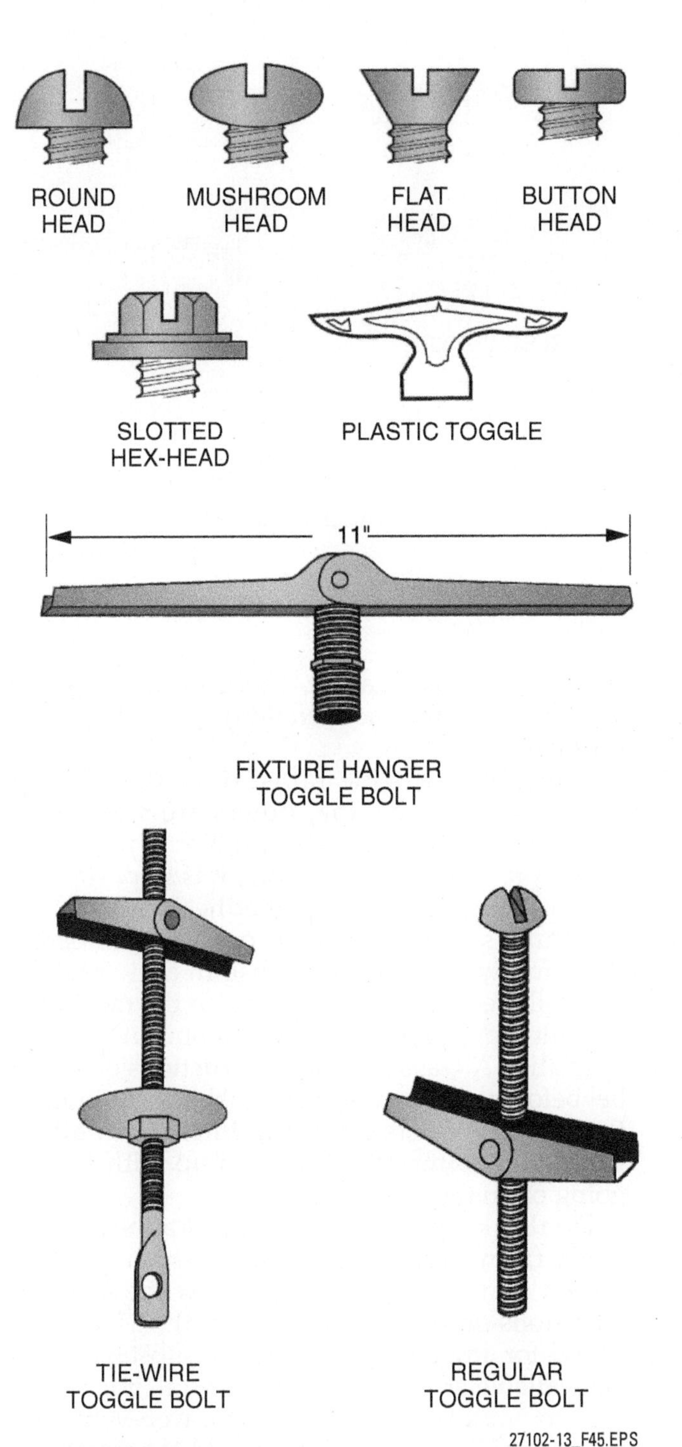

Figure 45 Toggle bolts.

Screw-actuated plastic toggle bolts are also available. The plastic toggle bolts are similar to metal toggle bolts, but they come with a pointed screw and do not require as large a hole. Unlike the metal version, the plastic wings remain in place if the screw is removed.

Toggle bolts are used to fasten a part to hollow block, gypsum board, plaster, panel, or tile. The following procedure can be used to install toggle bolts:

WARNING!

Follow all safety precautions to avoid injury.

Step 1 Select the proper-size toggle bolt and drill bit for the job.

Step 2 Check the toggle bolt for damaged or dirty threads or a malfunctioning wing mechanism.

Step 3 Drill a hole completely through the surface to which the part is to be fastened.

Step 4 Insert the toggle bolt through the opening in the item to be fastened.

Step 5 Screw the wings onto the end of the toggle bolt, ensuring that the flat side of the wing is facing the bolt head.

Step 6 Fold the wings completely back and push them through the drilled hole until the wings spring open.

Step 7 Pull back on the item to be fastened to hold the wings firmly against the inside surface to which the item is being attached.

Step 8 Tighten the toggle bolt with a screwdriver until it is snug.

5.9.2 Wall Anchors

Sleeve-type wall anchors (*Figure 46*) are suitable for use in plywood, gypsum board, and similar materials. The two types of sleeve-type wall anchors are standard and drive. Standard sleeve-type wall anchors are commonly used in walls and ceilings and are installed by drilling a mounting hole to the required diameter. Insert the anchor into the hole and tap it until the gripper prongs embed in the base material. Tighten the anchor screw to draw the anchor tight against the inside of the wall or ceiling.

Drive-in wall anchors are driven into the base material without the need for a mounting hole.

Toggle Bolts

What will happen if a toggle bolt is fastened too tightly?

After the anchor is installed, remove the anchor screw. Position the component being fastened in place, then reinstall the screw through the mounting hole in the component and into the anchor. Tighten the screw into the anchor to secure the component.

Wallboard anchors are self-drilling light- and medium-duty anchors used for fastening in gypsum board. Using a Phillips screwdriver, drive the anchor into the wall until the head of the anchor is flush with the wall or ceiling surface. Position the component being fastened over the anchor, then secure it by driving the proper-size sheet-metal screw into the anchor.

Metal drive-in anchors are used to fasten light to medium loads to gypsum board. They have two pointed legs that stay together when the anchor is hammered into a wall and spread out against the inside of the wall when a No. 6 or No. 8 sheet-metal screw is driven in.

The procedure for installing a fastener in a hollow CMU wall or brick when using epoxy is basically the same as the one just described. The difference is that the epoxy is first injected into an anchor screen to fill the screen, then the anchor screen is installed into the drilled hole. Use of the anchor screen is necessary to hold the epoxy intact in the hole until the anchor is inserted into the epoxy.

5.10.0 Adhesives

The term *adhesives* describes a variety of products that are used to attach one material to another. Construction adhesives are used to attach paneling and gypsum board to framing, install subflooring, and attach ceiling tiles. Special types of adhesives are used to install ceramic tile, floorcovering, and carpet. There are also a variety of glues used in a household or workshop environment.

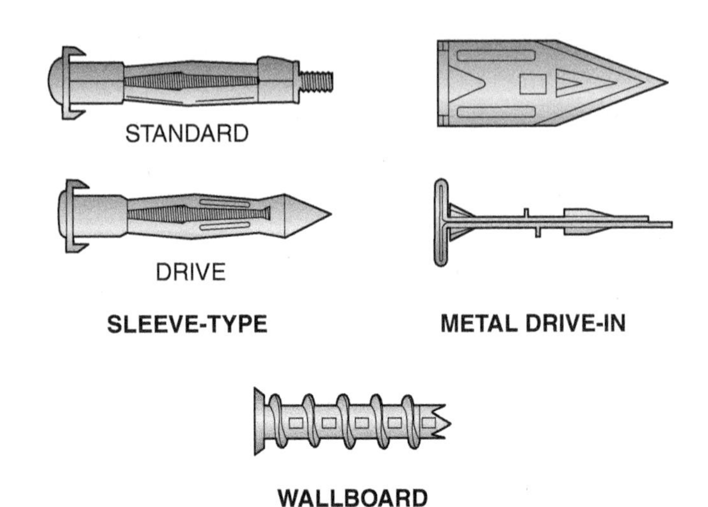

Figure 46 Sleeve-type, metal drive-in, and wallboard anchors.

Case History

Installation Requirements

In a college residence hall, battery-powered emergency lights were anchored to gypsum board hallway ceilings with drywall screws, without using any anchors and with no additional support. These fixtures weigh 8–10 pounds each and might easily have fallen from the ceiling, causing severe injury. When the situation was discovered, the contractor was responsible for removing and replacing dozens of fixtures.

The Bottom Line: Incorrect or inadequate anchoring methods can be both costly and dangerous.

Glues have a wide range of characteristics. Some are quick drying; others are slow. Some are waterproof; some are not. Some must cure under pressure, etc. Therefore, it is important for the carpenter to know what adhesive to use in a particular situation. At some point in your career, knowing what adhesive to use in a given situation will become second nature. In the meantime, consult a professional such as a building supply specialist. Always read the instructions on the label before using any adhesive. Also keep in mind that new products are being introduced all the time, so it is important to keep up with what is going on in the industry.

Finally, keep in mind that most adhesives are made from chemicals, some of which are toxic and/or flammable. Read and follow the safety instructions on the label and check the MSDS. The MSDS for any toxic or flammable substance must be available at the job site. In general, it is a good idea to work with adhesives in a well-ventilated area away from open flames, and to wear a face mask and eye protection.

5.10.1 Glues

Glues are used in laminated construction such as arches, curved members, and beams, and in the construction of cabinets and millwork. Carpenters may carry a plastic bottle of liquid glue in their toolboxes for making interior trim joints on the job.

Glues are available in two forms: liquid or dry. The dry forms are mixed into water on the job. Some glues are provided in two or more liquid

forms that when mixed together in proper proportions create a very strong adhesive. In cases where mixing is involved, the manufacturer's MSDS instructions must be followed, both for the quality desired and safety.

Some of the most common types of glue a carpenter might use are:

- *Animal or hide glue* – Made from animal hides, hooves, and bones. It has a long setting time, which makes it suitable when extra assembly time is needed. It is typically used for indoor applications (furniture and other wood products) since it is not waterproof.
- *Polyvinyl or white glue* – Sets quickly, becomes transparent when dry, and is very good for furniture, cabinets, and other interior woodwork. It is a liquid and is available in containers up to a gallon in size, or by special order in larger quantities.
- *Casein glue* – Available in a powder form and is mixed with water. Casein glue can be used in temperatures down to 35°F. It is water resistant and should only be used on oily woods. Under normal circumstances, casein glue powder will last for years; however, the powder tends to deteriorate with age.
- *Urea formaldehyde or plastic resin glue* – Available in powder form and is mixed with water. It is slow setting and has good resistance to heat and water. It is not recommended for oily woods. It is often used for laminated timbers and general woodworking where moisture is present.
- *Resorcinol resin or waterproof glue* – An expensive glue that is provided in two liquid chemical parts. One part acts as a catalyst on the hardener when the two parts are mixed together. It is excellent for use in exterior woodwork or laminated timbers that are exposed to extremely cold temperatures. Eye goggles and a respirator are required when mixing and working with this glue.
- *Contact cement* – A type of rubber cement. It is a liquid and comes ready-mixed in containers ranging in size from a pint to a 55-gallon drum. Contact cement is not a high-strength glue. It is commonly used when applying plastic laminates to the base material for cabinets and countertops. Contact cement is applied to both surfaces and allowed to dry to the touch before contact is made between the two materials. Upon contact, the bonding is instantaneous and movement or sliding the material to a different position is impossible.

WARNING!

Observe all MSDS precautions for this material. Extreme caution must be taken when using contact cement, as it is highly flammable. It should be used in a well-ventilated area, as the fumes can overcome a worker.

5.10.2 Construction Adhesives

Construction adhesives are available for many types of heavy-duty construction applications such as installing paneling, structural decking, and gypsum board. In addition to providing a strong bond between materials, construction adhesives offer such advantages as sound deadening and the bonding of dissimilar materials such as wood, gypsum, glass, metal, and concrete.

Construction adhesives are typically applied from a cartridge with a manual ratchet-type caulking gun or pneumatic caulking gun (*Figure 47*). Cartridges range in capacity from about 10 ounces to 30 ounces.

Adhesive Safety

Some types of adhesives emit noxious and harmful vapors before they cure (dry). Here are some general safety precautions to follow when using adhesives:

- Ensure good ventilation is provided.
- Keep all flames and lit tobacco products out of the area.
- Make sure switches, electric tools, or other sparking devices are not used in the area.
- Close off any area where people that are smoking could wander through.
- Do not breathe vapors for any length of time.
- Immediately remove any adhesive that comes in contact with your skin.
- Wear protective goggles and clothing when required.
- Follow the manufacturer's MSDS instructions and check the label on the container for specific safety precautions.
- When in doubt, ask questions.

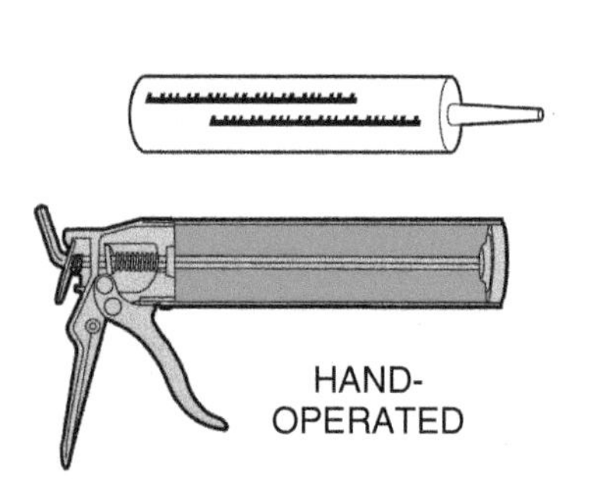

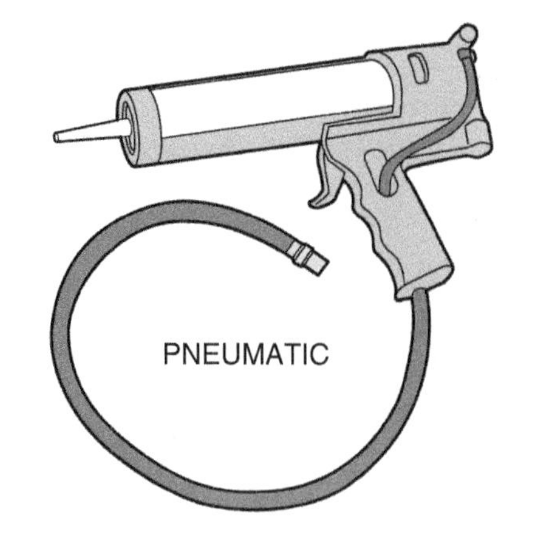

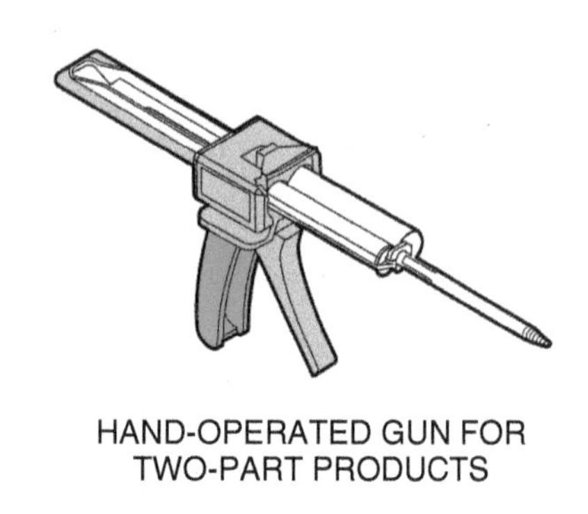

27102-13_F47.EPS

Figure 47 Adhesive applicators.

There are several types of adhesives for fastening sheet materials to framing members. They include the following:

- *Construction adhesive* – Used to apply wood or gypsum board to framing members. It has a solvent base. A caulking gun is used to apply construction adhesive.
- *Neoprene adhesive* – Can be used to attach wood to wood and gypsum to wood or metal. Neoprene adhesive will also bond to concrete. Like construction adhesive, neoprene adhesive is also applied with a caulking gun. Neoprene adhesive has a solvent base.
- *Contact cement* – Contact cement will bond dissimilar materials together, such as gypsum board to framing members and gypsum board to paneling. It is commonly used to bond plastic laminates to their base material. Application is by brush or roller. Contact cement should be dry to the touch before joining the two materials together.
- *Drywall adhesive* – Although suitable for lamination, it is used primarily for applying gypsum board to framing members. A caulking gun is used for application. It is available as a water-based product; however, solvent-based adhesive is more widely used.
- *Instant-bond glue* – Known by trade names such as Krazy Glue® and Super Glue®. They work well on nonporous materials such as glass, ceramics, metal, and many plastics. Some types of instant-bond glue are designed for porous materials such as wood and paper. When using this glue, do not touch your fingers to each other or to anything else. Acetone or nail polish remover will dissolve the glue, but be sure to keep it away from your eyes.
- *Epoxy* – Made by mixing a catalyst and a hardening agent at the time of use. Epoxies are good for bonding dissimilar materials. Fast-acting epoxies set almost instantly. Epoxies are typically provided in special syringe-type applicators that keep the two components separate until needed. The catalyst and hardener blend together when the plungers are depressed. Slower-setting epoxies are mixed in a container at the job site.

WARNING!

Many adhesives and cements are flammable and/or explosive. Ensure they are stored in a well-ventilated area and always read all warning labels on the adhesive packaging. Take the proper precautions to avoid an explosion or fire.

5.10.3 Mastics

Mastics are generally used to apply floor coverings, roofing materials, ceramic tile, or wall paneling. Most mastics have a synthetic rubber or latex base. Some are thick pastes and can be used where moisture is not present. Others are waterproof and can be used in bathrooms, kitchens, and laundries, or in direct contact with concrete. Some mastics require hot application with special equipment to keep them fluid while being applied. Others are applied with brushes, trowels, or rollers. Some mastics are available in tubes and are applied with manual or pneumatic caulking guns.

Keep in mind that good coverage with no voids is necessary for satisfactory bonding. Also, if applying floor tiles or paneling, use caution to prevent mastic from squeezing up between the joints or cracks and thus staining or marring the finish surface. To avoid this, place each piece without sliding after contact is made. When applying the mastic, make sure it is evenly spread. For satisfactory bonding, smooth, clean, dry surfaces are a must. Flaking plaster or old paint must be pre-

treated with special chemicals or removed. Leveling cements or compounds may be obtained and applied to overcome unevenness in walls or concrete floors. If moisture is likely to be encountered (e.g., concrete on basement walls or floors), a waterproofing material must be applied before the mastic. Only waterproof mastic should be used in these circumstances. Under adverse conditions, check the manufacturer's recommendations for secure installation. Many manufacturers recommend the type of adhesive to be used with their product.

5.10.4 Shelf Life

All adhesives have a limited shelf life (life span) that ranges from 12 to 24 months. If they are not used within that time, the adhesives will begin to lose their bonding ability. Most manufacturers claim a shelf life of at least two years. Keep in mind, however, that three to six months or more may have elapsed from the time the adhesive was manufactured until it was purchased. When purchasing adhesives, write the date on the container and do not use it beyond the shelf life listed on the label.

5.0.0 Section Review

1. Spikes are nails that are longer than _____.
   a. 8d
   b. 16d
   c. 48d
   d. 60d

2. A small barrel-shaped head with a slight indentation in the top to receive a nail set is known as a _____.
   a. common nail
   b. box nail
   c. finish nail
   d. roofing nail

3. To penetrate dense materials, use staples with _____.
   a. outside chisel points
   b. divergent points
   c. spear points
   d. inside chisel points

4. When fastening light-gauge steel framing members to one another and when fastening other building materials to the framing members use _____.
   a. wood screws
   b. lag bolts
   c. sheet-metal screws
   d. machine screws

5. Drive pins or studs should be embedded in hard concrete a minimum of _____.
   a. ⅛"
   b. ¼"
   c. ½"
   d. 1"

6. The type of bolt with a square shank, which allows the nut to be tightened without the bolt turning, is a _____.
   a. stove bolt
   b. machine bolt
   c. lag bolt
   d. carriage bolt

7. To tie the sill plate to a concrete or CMU foundation use a(n) _____.
   a. machine bolt
   b. stove bolt
   c. carriage bolt
   d. anchor bolt

8. The pilot hole for a drop-in anchor must be drilled to _____.
   a. ⅛" smaller than the bolt size
   b. an oversized diameter
   c. whatever diameter is convenient
   d. the diameter specified by the manufacturer

9. The type of anchor that has a cutting sleeve, which is used first as a drill bit and later becomes the expandable anchor itself is a(n) _____.
   a. self-drilling anchor
   b. expansion anchor
   c. drop-in anchor
   d. self-threading anchor

10. Screw anchors are considered to be _____.
    a. light-duty anchors
    b. medium-duty anchors
    c. heavy-duty anchors
    d. extra-heavy-duty anchors

11. The type of bolt that consists of a slotted screw or bolt and spring-loaded wings is a(n) _____.
    a. lag bolt
    b. anchor bolt
    c. toggle bolt
    d. carriage bolt

12. An epoxy anchoring system consists of two parts: a catalyst and a(n) _____.
    a. resin
    b. hardener
    c. adhesive
    d. glue

13. A type of water-resistant glue that should only be used on oily woods is _____.
    a. polyvinyl glue
    b. plastic resin
    c. casein glue
    d. hide glue

SUMMARY

Lumber and other wood building materials are commonly used in framing and sheathing of single-family residences. Steel framing members are being used more and more as the price grows and the supply of lumber shrinks.

In commercial work, concrete and steel are much more common as building materials. In many instances, entire structures are made of cast concrete. In others, a steel structural frame is combined with concrete floors and curtain wall panels.

It is important to know how to select the right material for any given situation. There are many types and grades of lumber and panel materials; each has specific uses. More importantly, each has certain limitations that the carpenter must know in order to avoid harmful or costly errors.

Likewise, there is a correct size and type of fastener for every fastening situation and an adhesive that is just right for occasions when an adhesive is needed. This module provided an introduction to the many types of fasteners and adhesives a carpenter uses on the job. As your training progresses, you will learn more about how and when to use them.

Review Questions

1. In a tree, growth takes place in the _____.
   a. heartwood
   b. medullary rays
   c. cambium layer
   d. bark

2. In the grading stamp shown in *Review Question Figure 1*, the term *S-GRN* means that the lumber _____.
   a. is made from short-grain wood
   b. has a maximum 19 percent moisture content
   c. has a maximum 15 percent moisture content
   d. has moisture content over 19 percent

12 STAND

ABC S-GRN D FIR

27102-13_RQ01.EPS

Figure 1

3. A piece of lumber 3" in thickness and 8" wide would be described by which of these grading classifications?
   a. boards
   b. joists and planks
   c. beams and stringers
   d. posts and timbers

4. If lumber is graded as double-end-trimmed, its ends are trimmed _____.
   a. square within a tolerance of $\frac{1}{64}$" for every 2"
   b. twice at each end
   c. reasonably square on both ends
   d. square and smooth on both ends within a tolerance of $\frac{1}{16}$" for the entire length

5. Ground-contact-grade pressure-treated lumber has a preservative retention of _____.
   a. 0.25 pounds per cubic foot
   b. 0.40 pounds per cubic foot
   c. 0.45 pounds per cubic foot
   d. 0.60 pounds per cubic foot

6. Plywood layers (plies) with grain at right-angles to the core are called _____.
   a. spacers
   b. crossbands
   c. stiffeners
   d. lateral bands

7. Plywood that is typically used to manufacture doors, furniture, and cabinets is known as _____.
   a. softwood plywood
   b. finished plywood
   c. hardwood plywood
   d. premium plywood

8. The type of hardboard that is coated with oils and resins during its manufacture is _____.
   a. tempered hardboard
   b. service-grade hardboard
   c. standard-grade hardboard
   d. exterior hardboard

9. Which of the following types of sheet material is used in high-wear applications such as concrete forms, cabinets, and countertops?
   a. HDO
   b. MDO
   c. OSB
   d. Mineral fiberboard

10. Which of the following is *not* a characteristic of oriented strand board (OSB)?
    a. Dimensional stability
    b. Stiffness
    c. Fastener holding capacity
    d. Attractive face grain

11. An engineered wood product commonly used to make doors and other millwork is _____.
    a. glulam
    b. LVL
    c. PSL
    d. LSL

12. A glulam beam consists of _____.
    a. laminated veneers bonded together under pressure
    b. solid, kiln-dried lumber glued together
    c. veneers bonded to a core with glue
    d. long strands of veneer bonded together with adhesive

13. The typical aggregate content of concrete is approximately _____.
    a. 45%
    b. 60%
    c. 75%
    d. 85%

14. The nominal dimensions of a typical concrete masonry unit are _____.
    a. 8 × 8 × 16 inches
    b. 4 × 4 × 8 inches
    c. 6 × 12 × 8 inches
    d. 2 × 4 × 12 inches

15. Ten-gauge steel studs are classified as _____.
    a. light-gauge studs
    b. extra-heavy studs
    c. structural steel studs
    d. reinforcing studs

16. Which of the following is *not* a mission of the Occupational Safety and Health Administration (OSHA)?
    a. Saving lives
    b. Establishing prevailing wages
    c. Preventing injuries
    d. Protecting the health of America's workers

17. An action that will *not* reduce your chance of getting injured on the job site is to_____.
    a. consider what type of accidents might occur
    b. wear appropriate personal protective equipment
    c. use common sense
    d. disregard safety rules

18. Which of these statements about pressure-treated lumber is *not* true?
    a. It is often used to make landscape timbers.
    b. Scrap material should be burned.
    c. Eye protection and a dust mask must be worn when cutting it.
    d. It is less expensive than redwood.

19. Material-handling accidents are most likely to happen when _____.
    a. the weather is bad
    b. equipment is defective
    c. the job is behind schedule
    d. workers are new or inexperienced at a job

20. When lifting an object, _____.
    a. bend over as far as possible, then straighten up
    b. lift with your legs, not your back
    c. use mostly your arm muscles
    d. raise the load as quickly as possible

21. Lumber that is stored outdoors should be _____.
    a. laid flat on the ground
    b. stood on end
    c. stacked on lengths of scrap lumber
    d. on the south side of a building

22. Portland cement is usually packaged _____.
    a. in 55-gallon drums
    b. in paper bags holding one cubic foot
    c. in five-gallon pails
    d. in plastic bags holding 50 pounds of material

23. A piece of lumber that is 1" thick, 12" wide, and 1' long is equivalent to _____.
    a. one board foot
    b. one square foot
    c. one cubic foot
    d. one metric foot

24. How many board feet are there in thirty 2" × 10" × 15' joists?
    a. 480
    b. 640
    c. 750
    d. 1,250

25. How many 4' × 8' sheets of plywood would be needed to provide subfloor for a 25' by 30' room?

   a. 24
   b. 25
   c. 28
   d. 30

26. When calculating the amount of concrete needed, the unit of measure is _____.

   a. short tons
   b. cubic yards
   c. cubic meters
   d. foot-pounds

27. In construction, the most common fasteners used for attaching two pieces of lumber are _____.

   a. lag bolts
   b. rivets
   c. nails
   d. wood screws

28. Nail lengths are expressed in penny designations; the symbol used to represent penny is _____.

   a. p
   b. d
   c. #
   d. c

29. Use the chart in *Review Question Figure 2* to find the length of an 8-penny nail.

   a. 1"
   b. 2½"
   c. 4"
   d. 6"

30. A fastener that is very good for wallboard insulation is the _____.

   a. crosscut chisel staple
   b. spear staple
   c. outside chisel staple
   d. divergent staple

31. Which of the following types of nail would normally not be used to secure interior trim?

   a. Box nail
   c. Finish nail
   c. Casing nail
   d. Wire brad

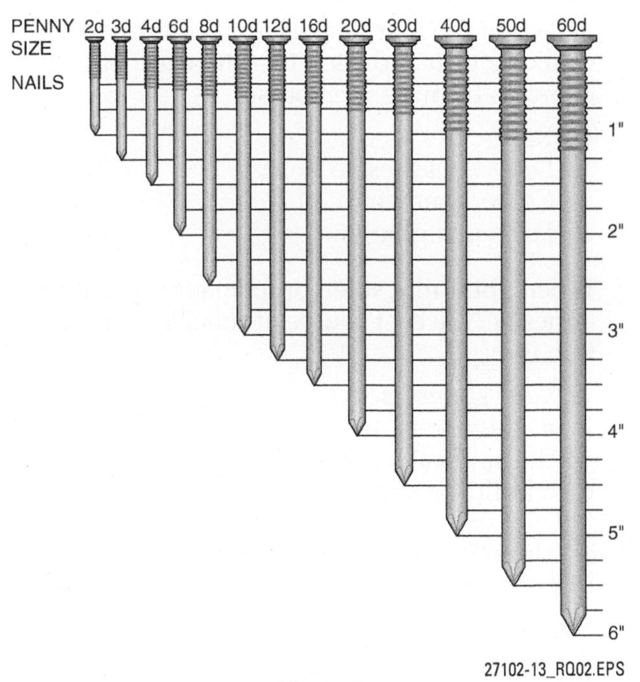

Figure 2

32. The type of staple point that clinches outward after it penetrates the material is the _____.

   a. crosscut chisel
   b. outside chisel
   c. spear
   d. inside chisel

33. If a lag screw's head will rest directly on wood when installed, use a _____.

   a. lag shield
   b. flat washer
   c. split washer
   d. sleeve anchor

34. The type of bolt that is used to attach the sill plate to the foundation is the _____.

   a. machine bolt
   b. carriage bolt
   c. anchor bolt
   d. stove bolt

35. Which of the anchoring devices listed below is *not* used in solid masonry?

   a. Toggle bolt
   b. Wedge anchor
   c. Expansion anchor
   d. Drop-in anchor

36. Which of the following is applied with a caulking gun and used to adhere gypsum board to framing members?

    a. Resorcinol glue
    b. Drywall adhesive
    c. Contact cement
    d. Polyvinyl glue

37. When obtaining safety information about an adhesive, the best source of reliable information is _____.

    a. your supervisor
    b. a co-worker
    c. the applicable MSDS
    d. a building supply dealer

38. The adhesive that requires two components that are mixed together at the time of use is _____.

    a. epoxy
    b. polyvinyl glue
    c. contact cement
    d. mastic

39. Many adhesives and cements are flammable and/or explosive. Store them in an area that is _____.

    a. enclosed
    b. well ventilated
    c. pressurized
    d. secure

40. Wood products that are sustainably harvested are certified by the _____.

    a. Forest Stewardship Council
    b. Wood Products Institute
    c. National Forestry Council
    d. Federal Standards Agency

Trade Terms Quiz

Fill in the blank with the correct term that you learned from your study of this module.

1. The main ingredient in plaster and gypsum board is _______.
2. After installing finishing nails, use a(n) _______ to recess the nails, then fill the holes with putty.
3. Crown molding is a type of _______.
4. The _______ describes safe handling methods for a particular substance.
5. The roof sheathing is attached to the _______.
6. Use _______ nails to prevent rusting.
7. A floor is supported by equally spaced _______.
8. When two square-cut pieces of material are placed end-to-end, it is called a(n) _______.
9. A(n) _______ floor is built out to extend beyond the last point of support.
10. A(n) _______is a popular homebuilding option because the higher ceiling makes the room feel more spacious.
11. _______ are materials added to a concrete mix to alter certain properties.
12. A lumber quantity measure equivalent to a board measuring 1" thick, 12" wide, and 1' long is a _______.
13. _______ has hardened and gained its structural strength.
14. The amount of space occupied by a two-dimensional object is referred to as the _______.
15. When referring to nail length, the letter *d* is an abbreviation for _______.
16. When mixing epoxy, the _______ is the substance that causes the reaction in the hardener.
17. _______ is the ash produced during the reduction of iron ore to iron in a blast furnace.
18. To determine the _______, multiply the length, width, and height together.
19. _______ are protective natural or synthetic coatings.
20. Concrete that is in a semiliquid and moldable state is referred to as _______.
21. When applying finish roofing or siding material to a structure, it is fastened to the _______.
22. The horizontal members that supports the framework of a building at the bottom of a wall is the _______.
23. The chemicals in portland cement react with water through a process known as _______.
24. A(n) _______ material is capable of easily igniting and rapidly burning.
25. _______ is concrete that is hardened but not achieved is structural strength.
26. The _______ is used to calculate the workers' compensation premium based on the company's previous accident experience.
27. In _______ siding, each board is tapered and grooved so the upper piece fits tightly over the lower piece.

Trade Terms

Admixtures
Area
Board foot
Butt joint
Cantilever
Catalyst
Combustible
Cured concrete
Experience modification rate (EMR)
Galvanized
Green concrete
Gypsum
Hydration
Joist
Material safety data sheet (MSDS)
Millwork
Nail set
Penny
Plastic concrete
Rafter
Resins
Sheathing
Shiplap
Sill plate
Slag
Vaulted ceiling
Volume

Cornerstone of Craftsmanship

Curtis Haskins

Superintendent/Site Manager
Bowden Contracting Company, Inc.

How did you get started in the construction industry?

In 1978, I was a 17-year-old high school dropout with a pregnant girlfriend. With a child on the way, I needed to work full time to support the baby and girlfriend. An older friend of mine told me about his job building houses and told me they were looking for more workers. When I attended high school, I enjoyed wood-shop classes and knew I liked working with my hands. My first rough-in job paid almost twice what my other friends were making ($4.10/hour vs. $2.65/hour–minimum wage back then). I continue to bring home a larger paycheck than many family members and friends–even those who have attended and graduated from college with an advanced degree. Staying current with the trade by acquiring advanced training has contributed largely to my success.

Who or what inspired you to enter the industry?

I've always enjoyed being outdoors. I was involved in the Boy Scouts, and enjoyed camping with my family when we went on vacation…what better job to have than one where I could be outside a lot. After I was in the trade for a short time, I realized construction was something I enjoyed. I began to seek out information on carpentry and building construction, looking at large commercial construction sites and wanting to be a part of the construction industry.

What do you enjoy most about your career?

My career has spanned a number of different positions–from being a laborer on a rough-in crew, to carpenter, to lead carpenter, to foreman, and on to my current position being a superintendent/site manager. I am currently responsible for the operations of multimillion-dollar commercial construction projects. I enjoy starting a new project as a bare piece of ground and leaving there seeing a new building I am proud of. I like visiting old projects I have been involved in (even the first houses I worked on) and remembering the time there. It is as if you are building your legacy. I enjoy the fact I am not in an office cubicle for eight hours a day, and instead can be outside a majority of the time, moving around (read "exercise"). I enjoy that my workplace is constantly in a state of transformation. It never looks the same from day to day. Every day brings a new set of challenges and rewards.

Why do you think training and education are important in construction?

I believe all craftspeople, from laborers through managers, should consider themselves professionals. We are responsible for the buildings, bridges, roads, etc., that millions of people use daily. We should all take pride in our craft and strive to learn as much as possible. I feel privileged to be able to pass on some of what I have learned through the years, as well as having the opportunity to instruct upcoming craft professionals through the Associated Builders and Contractors Apprenticeship Training Program.

Why do you think credentials are important in construction?

For someone to have a document that states they are a formally trained craftsman gives them a sense of pride. It also shows employers the person has taken a proactive step in their career, and is serious about their trade. It's a win-win situation.

How has training/construction impacted your life and career?
I have enjoyed a satisfying career as a construction professional. Some of the opportunities which have been presented to me include: working on a multitude of different projects, meeting hundreds of different people; being invited to judge craft competitions at the local and national levels; being involved with trade organizations, such as Associated Builders and Contractors (ABC) and SkillsUSA, which allows you to network with your peers. (My current and two previous places of employment came about directly because of my involvement with ABC.)

Would you recommend construction as a career to others? Why?
While construction may not be the best job for many people, it *is* the best job for some (me included). I believe this type of career is best suited for individuals who enjoy learning new things and working with their hands, like changing work environments, can work well with a team, and are not afraid to take on increasing leadership roles.

What does craftsmanship mean to you?
Dedication. Ingenuity. Accuracy. Pride.

Trade Terms Introduced in This Module

Admixtures: Materials that are added to a concrete mix to change certain properties of the concrete such as retarding setting time, reducing water requirements, or making the concrete easier to work with.

Area: The surface or amount of space occupied by a two-dimensional object such as a rectangle, circle, or square.

Board foot: Lumber quantity measure equivalent to a piece of lumber that is 1" thick, 12" wide, and 1' long.

Butt joint: The joint formed when one square-cut edge of a piece of material is placed against another material.

Cantilever: A beam, truss, or slab (floor) that extends past the last point of support.

Catalyst: A substance that causes a reaction in another substance.

Combustible: Capable of easily igniting and rapidly burning; used to describe a fuel with a flash point at or above 100°F.

Cured concrete: Concrete that has hardened and gained its structural strength.

Experience modification rate (EMR): A rate computation to determine surcharge or credit to workers' compensation premium based on a company's previous accident experience.

Galvanized: Protected from corrosion (rust) by a zinc coating.

Green concrete: Concrete that has hardened but has not yet gained its structural strength.

Gypsum: A chalky material that is a main ingredient in plaster and gypsum board.

Hydration: The catalytic action water has in transforming the chemicals in portland cement into a hard solid. The water interacts with the chemicals to form calcium silicate hydrate gel.

Joist: Generally, equally spaced framing members that support floors and ceilings.

Material safety data sheet (MSDS): Information that details any toxic, chemical, or potentially harmful substances that are contained in a product.

Millwork: Manufactured wood products such as doors, windows, and moldings.

Nail set: A punch-like tool used to recess finishing nails.

Penny: A measure of nail length; commonly abbreviated with the letter *d*.

Plastic concrete: Concrete when it is first mixed and is in a semiliquid and moldable state.

Rafter: A sloping structural member of a roof frame to which sheathing is secured.

Resins: Protective natural or synthetic coatings.

Sheathing: Panel material to which roofing material or siding is secured.

Shiplap: A method of cutting siding in which each board is tapered and grooved so that the upper piece fits tightly over the lower piece.

Sill plate: A horizontal member that supports the framework of a building on the bottom of a wall or box joist. It is also called a sole plate.

Slag: The ash produced during the reduction of iron ore to iron in a blast furnace.

Vaulted ceiling: A high, open ceiling that generally follows the roof pitch.

Volume: The amount of space occupied in three dimensions (length, width, and height).

Appendix A

COMMON HARDWOODS AND SOFTWOODS

Tree	Use	Class (H—Hard, S-Soft)
Alder	Used in cabinets and furniture. The color is reddish-brown. Alder is a less expensive substitute for cherry. It is also know as red alder and fruitwood.	H
Ash	Coarse-grained, hard, and heavy. The heartwood is brown; the sapwood is almost white. Used for furniture, cabinets, trim, tool handies, and plywood.	H
Basswood	One of the softest hardwoods. It has an open, coarse grain. The color is creamy white. Used for veneer core and moldings.	H
Beech	Strong, heavy, hard. Takes a good polish, works easily. The color is white to brown. Also used for tool handles, furniture, and plywood.	H
Birch	Very strong, heavy, and fine-grained. It takes an excellent finish. Can be beautifully hand polished. The sapwood is a soft yellow; the heartwood is brown, sometimes slightly colored with red. Due to the climate, birch has a wide variety of color changes. Used for furniture, cabinets, flooring, and plywood.	H
Western red cedar	The thin sapwood layer is a pale yellow. The heartwood varies from a deep dark brown to a very light pink. The wood is light and soft, but lacks strength. It is straight-grained and has high resistance to changes in moisture. After seasoning, it retains its shape and size exceptionally well. Used for shingles, shakes, siding, planking, building logs, and yard lumber.	S
Tennessee cedar	This product is very light, soft, and close-grained. The sapwood is white in color. The heartwood is reddish brown. This cedar has a very strong odor that acts as an insect repellent. Used for trunk linings and closets.	S
Cherry	Cherry is a close-grained and light wood. It takes a good finish and works easily. The sapwood is an off-white. The heartwood is reddish brown. This product can be used for cabinets, interior trim, and furniture.	H
Chestnut	This beautiful, open-grained wood is worked very easily. The sapwood is light brown in color. The heartwood is a dark tan. Can be used for cabinets or plywood and paneling.	H
Cypress	Cypress is a very durable, light product, and is very easily worked. The heartwood is a brilliant yellow; the sapwood is light brown in color. It is used for interior trim and can also be used for cabinet work and exterior trim.	S
Elm	This strong, magestic tree is very hard and the lumber is very heavy. The color of this tree is light to medium brown, with slight variations of red and gray. Used primarily for tool handles and heavy construction purposes.	H
Douglas Fir	Fir is very strong and durable. The sapwood is slightly off-white and the heartwood varies in color from a light red to a deep yellow. Fir is used for all types of construction and in the manufacturing of plywood.	S
Western hemlock	This lumber is straight-gained and fine-textured, and it works easily. It has very good qualities for gluing. It is almost totally free from pitch and is often interchanged with fir. Hemlock is used for general construction.	S
Gum	This soft-textured, close-grained tree is very durable. The color of the heartwood varies from red to brown. The sapwood is yellowish white. It closely resembles walnut in color. This wood works easily, but has a tendency to warp. It is sometimes substituted for walnut in the manufacture of furniture.	H
Larch	This product is strong, fairly hard, and resembles Douglas fir more than any other softwood. This lumber is also suitable for general construction purposes.	H
Mahogany	This lumber is an open-pored, strong, durable product. Mahogany can be worked easily. The color is basically a reddish brown. Poor-grade mahogany has small gray spots or flecks. Much mahogany is from overseas. Can be used for paneling, furniture, plywood, and interior trim.	H
Sugar maple	This lumber is very difficult to work with, but it is very strong and can take an excellent finish. The product is very close-grained and light brown to dark brown in color. Furniture and flooring are some of its uses.	H

Tree	Use	Class (H—Hard, S-Soft)
Silver maple	This lumber is not very strong or durable. It is soft and light, and it works easily. This lumber is frequently used for turning and interior trim.	H
White oak	White oak is very hard and heavy. It has a close grain with open pores. The product is very difficult to work with, but it takes an excellent finish. The heartwood is tan in color, with a light sapwood. Can be used for flooring, plywood, interior trim, and furniture.	H
Red oak	Red oak has a coarser grain than white oak and is also a little softer. The variation in color is from a light tan to a medium brown. The uses for red oak are furniture, plywood, bearing posts, interior trim, and paneling. It could also be used for cabinets.	H
White pine	This lumber is very easily worked. It is soft, light, and very durable. The heartwood is a very slightly yellowish white. Can be used for construction (code permitting) and is also used in furniture and millwork.	S
Sugar pine	This is a light and very soft lumber. It is uniform in texture and works easily. The sapwood is yellow. The heartwood is light red to medium brown. Sugar pine is mostly quarter-sawed and has specks of reddish brown, which makes it easy to recognize. Can be used for inexpensive furniture and interior millwork.	S
Ponderosa pine	This pine is fairly soft and uniform in texture. Besides being strong, it works smoothly without splintering. The sapwood is a light pale yellow with a darker heartwood. Ponderosa pine is ideal for woodworking in the shop and is common in door and window frames and moldings.	S
Lodgepole pine	This pine lumber seasons very easily and is of moderate durability. This product is soft and straight-grained with a fine, uniform texture. The sapwood is almost white. This product is used largely in making propping timbers for miners.	S
Southern yellow pine	This lumber is also called southern pine or yellow pine. It is a very difficult product to work with. It is strong, hard, and tough. The heartwood is orange-red in color, with the sapwood being lighter in color. This product is used for heavy structural purposes such as floor planks, beams, and timbers.	S
Poplar	This product is uniform in texture, light, and soft. The color varies from white to yellow. This lumber is easy to work with. Can be used for inexpensive furniture, crates, plywood, and moldings.	H
Redwood	The redwood is the largest tree in North America. Redwood is easily worked, light, and coarse-grained. The color is dull red. Although very beautiful, redwood lacks strength. It is used primarily for interior and exterior trim, planking for patio decks, and shingles.	S
White spruce	This lumber is soft, works very easily, and is very lightweight. There is very little contrast betwen the heartwood and sapwood. The color is a pale yellow. It is not very durable, but among the softwoods it is one of medium strength. In certain situations, it is vulnerable to decay. White spruce has value as pulp wood and is used in light construction.	S
Engelmann spruce	This type of spruce takes a good finish. The lumber is very light and straight-grained. This species is usually larger than white spruce and has a larger percentage of clear lumber with little or no defects. It is used for oars, paddles, sounding boards for musical instruments, and construction work.	S
Sitka spruce	These trees have a straight grain and grow very tall. This species is a very tough, strong wood. The color is a very creamy white with a pinkish or light red tinge. This species resists shattering and splintering. It is used primarily in constructing masts, spars, scaffolding, and general construction (code permitting).	S
Walnut	This product is durable, very hard, strong and it can easily be finished to a beautiful luster. The sapwood is a light brown to a medium brown. The heartwood is a reddish brown to a dark brown. This product is used in the manufacture of fine cabinets, furniture, flooring, plywood, and interior trim.	H

Appendix B

Common Abbreviations Used in the Lumber Industry

AD	Air Dried
ADF	After Deducting Freight
ALS	American Lumber Standard
AST	Anti-Stain Treated
AVG	Average
B & B	Band Better
B & S	Beams and Stringers
Bd.	Board
BF or Bd. Ft.	Board Feet
Bdl.	Bundle or Bundled
B/L	Bill of Lading
Bev.	Bevel or Beveled
CB1S	Center Bead on One Side
CB2S	Center Bead on Two Sides
CC	Cubical Content
Cft. or Cu. Ft.	Cubic Feet or Foot
CG2E	Center Groove on Two Edges
CIF	Cost, Insurance, and Freight
CIFE	Cost, Insurance, Freight, and Exchange
Clg.	Ceiling
Com.	Common
Csg.	Casing
Ctr.	Center
CV1S	Center V on One Side
CV2S	Center V on Two Sides
D &M	Dressed and Matched
DB Clg.	Double-Beaded Ceiling
DB Part	Double-Beaded Partition
DET	Double-End-Trimmed
Dim.	Dimension
Dkg.	Decking
D/sdg	Drop Siding
EB1S	Edge Bead on One Side
EB2S	Edge Bead on Two Sides
EE	Eased Edges
EG	Edge Vertical or Rift Grain
EM	End Matched
EV1S	Edge V on One Side
EV2S	Edge V on Two Sides
E & CB1S	Edge and Center Bead on One Side
Part.	Partition
Pat.	Pattern
Pcs.	Pieces
PE	Plain End
P & T	Post and Timbers
P1S and P2S	See S1S and S2S
Reg.	Regular
Rfg.	Roofing
Rgh.	Rough
E & CB2S	Edge and Center Bead on Two Sides
E & CV1S	Edge and Center V on One Side
E & CV2S	Edge and Center V on Two Sides
FA	Facial Area
Fac.	Factory
FBM	Feet Board Measure
FG	Flat (slash) Grain
Flg.	Flooring
FOB	Free on Board
FOHC	Free of Heart Centers
FOK	Free of Knots
Frt.	Freight
Ft.	Foot or Feet
G	Girth
GM	Grade Marked
G/R	Grooved Roofing
HB	Hollow Back
Hrt.	Heart
H & M	Hit and Miss
in.	Inch or Inches
Ind.	Industrial
J & P	Joints and Planks
KD	Kiln Dried
Lbr.	Lumber
LCL	Less than Carload
Lft.	Lineal Foot or Feet
Lin.	Lineal
LL	Longleaf
Lng.	Lining
M	Thousand
MBM	Thousand (feet) Board Measure
M.C.	Moisture Content
Merch.	Merchantable
MG	Medium Grain
MLDG.	Molding
Mft.	Thousand Feet
No.	Number
N1E or N2E	Nosed on One or Two Edges
Ord.	Order
Par.	Paragraph
STR	Structural
SYP	Southern Yellow Pine
S & E	Side and Edge (surfaced on)
S1E	Surfaced on One Edge
S2E	Surfaced on Two Edges
S1S	Surfaced on One Side
S2S	Surfaced on Two Sides
S1S1E	Surfaced on One Side and One Edge

27102-13_A02.EPS

R/L	Random Length	S2S2E	Surfaced on Two Sides and Two Edges
RES	Resawn	S2S1E	Surfaced on Two Sides and One Edge
Sdg.	Siding	S2S & CM	Surfaced on Two Sides and Center Matched
S E	Square Edge	S2S - SL	Surfaced on Two Sides and Shiplapped
Sel.	Select	S2S & SM	Surfaced on Two Sides and Standard Matched
S E & S	Square Edge and Sound	S4S	Surfaced on Four Sides
SL	Shiplap	T & G	Tongued and Grooved
SM	Surface Measure	Wdr.	Wider
Specs.	Specifications	Wt.	Weight
SR	Stress Rated	YP	Yellow Pine
Std.	Standard		
Std. Lgths.	Standard Lengths		
STDM	Standard Matched		
SSND	Sap Stained No Defects (stained)		

27102-13_A02A.EPS

Appendix C

LUMBER CONVERSION TABLES

Nominal and Actual Softwood Lumber Sizes		
Nominal	Actual (in inches)	Actual (in millimeters)
1 × 2	¾ × 1-½	19 × 38
1 × 4	¾ × 3-½	19 × 89
1 × 6	¾ × 5-½	19 × 140
1 × 8	¾ × 7-¼	19 × 184
1 × 10	¾ × 9-¼	19 × 235
1 × 12	¾ × 11-¼	19 × 286
5/4 × 4	1 × 3-½	25 × 89
5/4 × 6	1 × 5-½	25 × 140
2 × 2	1-½ × 1-½	38 × 38
2 × 4	1-½ × 3-½	38 × 89
2 × 6	1-½ × 5-½	38 × 140
2 × 8	1-½ × 7-¼	38 × 184
2 × 10	1-½ × 9-¼	38 × 235
2 × 12	1-½ × 11-¼	38 × 286
4 × 4	3-½ × 3-½	89 × 89
4 × 6	3-½ × 5-½	89 × 140
6 × 6	5-½ × 5-½	140 × 140
6 × 8	5-½ × 7-½	140 × 190
8 × 8	7-½ × 7-½	190 × 190

Nominal and Actual Hardwood Lumber Sizes		
Nominal (Rough)	Actual (Plain)	Actual (S4S)
½"	5/16"	5/16"
⅝"	7/16"	7/16"
¾"	9/16"	9/16"
1"	25/32"	25/32"
1¼"	1 1/16"	1 1/32"
1½"	1 5/16"	1 9/32"
2"	1¾"	1 11/16"
2½"	2¼"	2¼
3"	2¾"	2¾"
4"	3¾"	3¾"

Rapid Calculation of Board Measure		
Width	**Thickness**	**Board Feet**
3"	1" or less	¼ of the length
4"	1" or less	⅓ of the length
6"	1" or less	½ of the length
9"	1" or less	¾ of the length
12"	1" or less	Same as the length
15"	1" or less	1¼ of the length

Nominal Size (Inches)	Board Feet								
	Actual Length (in Feet)								
	8	10	12	14	16	18	20	22	24
1 × 2		1⅔	2	2⅓	2⅔	3	3½	3⅔	4
1 × 3		2½	3	3½	4	4½	5	5½	6
1 × 4	2¾	3⅓	4	4⅔	5⅓	6	6⅔	7⅓	8
1 × 5		4⅙	5	5⅚	6⅔	7½	8⅓	9⅙	10
1 × 6	4	5	6	7	8	9	10	11	12
1 × 7		5⅝	7	8⅙	9⅓	10½	11⅔	12⅚	14
1 × 8	5⅓	6⅔	8	9⅓	10⅔	12	13⅓	14⅔	16
1 × 10	6⅔	8⅓	10	11⅔	13⅓	15	16⅔	18⅓	20
1 × 12	8	10	12	14	16	18	20	22	24
1¼ × 4		4⅙	5	5⅚	6⅔	7½	8⅓	9⅙	10
1¼ × 6		6¼	7½	8¾	10	11¼	12½	13¾	15
1¼ × 8		8⅓	10	11⅔	13⅓	15	16⅔	18⅓	20
1¼ × 10		10 5/12	12½	14 7/12	16⅔	18¾	20⅚	22 11/12	25
1¼ × 12		12½	15	17½	20	22½	25	27½	30
1½ × 4	4	5	6	7	8	9	10	11	12
1½ × 6	6	7½	9	10½	12	13½	15	16½	18
1½ × 8	8	10	12	14	16	18	20	22	24
1½ × 10	10	12½	15	17½	20	22½	25	27½	30
1½ × 12	12	15	18	21	24	27	30	33	36
2 × 4	5⅓	6⅔	8	9⅓	10⅓	12	13⅓	14⅔	16
2 × 6	8	10	12	14	16	18	20	22	24
2 × 8	10⅔	13⅓	16	18⅔	21⅓	24	26⅔	29⅓	32
2 × 10	13⅓	16⅔	20	23⅓	26⅔	30	33⅓	36⅔	40
2 × 12	16	20	24	28	32	36	40	44	48
3 × 6	12	15	18	21	24	27	30	33	36
3 × 8	16	20	24	28	32	36	40	44	48
3 × 10	20	25	30	35	40	45	50	55	60
3 × 12	24	30	36	42	48	54	60	66	72
4 × 4	10⅔	13⅓	16	18⅔	21⅓	24	26⅔	29⅓	32
4 × 6	16	20	24	28	32	36	40	44	48
4 × 8	21⅓	26⅔	32	37⅓	42⅔	48	53⅓	58⅔	64
4 × 10	26⅔	33⅓	40	46⅔	53⅓	60	66⅔	73⅓	80

Appendix D

FORMULAS FOR GEOMETRIC SHAPES AND CONVERSION TABLES

Conversion Table for Changing Measurements (in inches) to Fractions and Decimal Parts of a Foot		
Inches	**Fractional Part of Foot**	**Decimal Part of Foot**
1	1/12	0.08
2	1/16	0.17
3	1/4	0.25
4	1/3	0.33
5	5/12	0.42
6	1/2	0.50
7	7/12	0.58
8	2/3	0.67
9	3/4	0.75
10	5/6	0.83
11	11/12	0.92
12	1	1.00

Metric Conversion of Pounds to Grams or Kilograms
1 pound (16 ounces) = 453.6 grams or 0.4536 kilogram (kg)

AREAS OF PLANE FIGURES

NAME FORMULA	SHAPE
(A = Area) Parallelogram $A = B \times h$	
Trapezoid $A = \frac{B + C}{2} \times h$	
Triangle $A = \frac{B \times h}{2}$	
Trapezium (Divide into 2 triangles) A = Sum of the 2 triangles (See above)	
Regular Polygon $A = \frac{aP}{2}$ Where a is the length of the *apothem* (perpindicular distance from center to a side) P is the *perimeter* (sum of the sides, s)	
Circle $\pi = 3.14$ (1) πR^2 A = (2) $.7854 \times D^2$	
Sector (1) $\frac{a^\circ}{360^\circ} \times \pi R^2$ A = (2) Length of arc $\times \frac{R}{2}$ ($\pi = 3.14$, a = angle of sector)	
Segment A = Area of sector minus triangle (see above)	
Ellipse $A = M \times m \times .7854$	
Parabola $A = B \times \frac{2h}{3}$	

VOLUMES OF SOLID FIGURES

NAME FORMULA	SHAPE
(V - volume) Cube $V = a^3$ (in cubic units)	
Rectangular Solids $V = L \times W \times h$	
Prisms $V(1) = \frac{B \times A}{2} \times h$ $V(2) = \frac{s \times R}{2} \times n \times h$ V = Area of end x h n = Number of sides.	
Cylinder $V = \pi R^2 \times h$ ($\pi = 3.14$)	
Cone $V = \frac{\pi R^2 \times h}{3}$ ($\pi = 3.14$)	
Pyramids $V(1) = L \times W \times \frac{h}{3}$ $V(2) = \frac{B \times A}{2} \times \frac{h}{3}$ V = Area of Base $\times \frac{h}{3}$	
Sphere $V(1) = \frac{1}{6} \pi D^3$ $V(2) = \frac{4}{3} \pi R^3$	
Circular Ring (Torus) $V = 2\pi^2 \times Rr^2$ V = Area of section $\times 2\pi R$ R = radius of the ring at its center, i.e., (OD – ID)/2 + ½ID; r = radius of the cross section of the ring at R.	

27102-13_A09.EPS

METRIC CONVERSION CHART

inches		mm	inches		mm	inches		mm	inches		mm
fractions	decimals		fractions	decimals		fractions	decimals		fractions	decimals	
–	.0004	.01	25/32	.781	19.844	–	2.165	55.	3-11/16	3.6875	93.663
–	.004	.10	–	.7874	20.	2-3/16	2.1875	55.563	–	3.7008	94.
–	.01	.25	51/64	.797	20.241	–	2.2047	56.	3-23/32	3.719	94.456
1/64	.0156	.397	13/16	.8125	20.638	2-7/32	2.219	56.356	–	3.7401	95.
–	.0197	.50	–	.8268	21.	–	2.244	57.	3-3/4	3.750	95.250
–	.0295	.75	53/64	.828	21.034	2-1/4	2.250	57.150	–	3.7795	96.
1/32	.03125	.794	27/32	.844	21.431	2-9/32	2.281	57.944	3-25/32	3.781	96.044
–	.0394	1.	55/64	.859	21.828	–	2.2835	58.	3-13/16	3.8125	96.838
3/64	.0469	1.191	–	.8661	22.	2-5/16	2.312	58.738	–	3.8189	97.
–	.059	1.5	7/8	.875	22.225	–	2.3228	59.	3-27/32	3.844	97.631
1/16	.062	1.588	57/64	.8906	22.622	2-11/32	2.344	59.531	–	3.8583	98.
5/64	.0781	1.984	–	.9055	23.	–	2.3622	60.	3-7/8	3.875	98.425
–	.0787	2.	29/32	.9062	23.019	2-3/8	2.375	60.325	–	3.8976	99.
3/32	.094	2.381	59/64	.922	23.416	–	2.4016	61.	3-29/32	3.9062	99.219
–	.0984	2.5	15/16	.9375	23.813	2-13/32	2.406	61.119	–	3.9370	100.
7/64	.109	2.778	–	.9449	24.	2-7/16	2.438	61.913	3-15/16	3.9375	100.013
–	.1181	3.	61/64	.953	24.209	–	2.4409	62.	3-31/32	3.969	100.806
1/8	.125	3.175	31/32	.969	24.606	2-15/32	2.469	62.706	–	3.9764	101.
–	.1378	3.5	–	.9843	25.	–	2.4803	63.	4	4.000	101.600
9/64	.141	3.572	63/64	.9844	25.003	2-1/2	2.500	63.500	4-1/16	4.062	103.188
5/32	.156	3.969	1	1.000	25.400	–	2.5197	64.	4-1/8	4.125	104.775
–	.1575	4.	–	1.0236	26.	2-17/32	2.531	64.294	–	4.1338	105.
11/64	.172	4.366	1-1/32	1.0312	26.194	–	2.559	65.	4-3/16	4.1875	106.363
–	.177	4.5	1-1/16	1.062	26.988	2-9/16	2.562	65.088	4-1/4	4.250	107.950
3/16	.1875	4.763	–	1.063	27.	2-19/32	2.594	65.881	4-5/16	4.312	109.538
–	.1969	5.	1-3/32	1.094	27.781	–	2.5984	66.	–	4.3307	110.
13/64	.203	5.159	–	1.1024	28.	2-5/8	2.625	66.675	4-3/8	4.375	111.125
–	.2165	5.5	1-1/8	1.125	28.575	–	2.638	67.	4-7/16	4.438	112.713
7/32	.219	5.556	–	1.1417	29.	2-21/32	2.656	67.469	4-1/2	4.500	114.300
15/64	.234	5.953	1-5/32	1.156	29.369	–	2.6772	68.	–	4.5275	115.
–	.2362	6.	–	1.1811	30.	2-11/16	2.6875	68.263	4-9/16	4.562	115.888
1/4	.250	6.350	1-3/16	1.1875	30.163	–	2.7165	69.	4-5/8	4.625	117.475
–	.2559	6.5	1-7/32	1.219	30.956	2-23/32	2.719	69.056	4-11/16	4.6875	119.063
17/64	.2656	6.747	–	1.2205	31.	2-3/4	2.750	69.850	–	4.7244	120.
–	.2756	7.	1-1/4	1.250	31.750	–	2.7559	70.	4-3/4	4.750	120.650
9/32	.281	7.144	–	1.2598	32.	2-25/32	2.781	70.6439	4-13/16	4.8125	122.238
–	.2953	7.5	1-9/32	1.281	32.544	–	2.7953	71.	4-7/8	4.875	123.825
19/64	.297	7.541	–	1.2992	33.	2-13/16	2.8125	71.4376	–	4.9212	125.
5/16	.312	7.938	1-5/16	1.312	33.338	–	2.8346	72.	4-15/16	4.9375	125.413
–	.315	8.	–	1.3386	34.	2-27/32	2.844	72.2314	5	5.000	127.000
21/64	.328	8.334	1-11/32	1.344	34.131	–	2.8740	73.	–	5.1181	130.
–	.335	8.5	1-3/8	1.375	34.925	2-7/8	2.875	73.025	5-1/4	5.250	133.350
11/32	.344	8.731	–	1.3779	35.	2-29/32	2.9062	73.819	5-1/2	5.500	139.700
–	.3543	9.	1-13/32	1.406	35.719	–	2.9134	74.	–	5.5118	140.
23/64	.359	9.128	–	1.4173	36.	2-15/16	2.9375	74.613	5-3/4	5.750	146.050
–	.374	9.5	1-7/16	1.438	36.513	–	2.9527	75.	–	5.9055	150.
3/8	.375	9.525	–	1.4567	37.	2-31/32	2.969	75.406	6	6.000	152.400
25/64	.391	9.922	1-15/32	1.469	37.306	–	2.9921	76.	6-1/4	6.250	158.750
–	.3937	10.	–	1.4961	38.	3	3.000	76.200	–	6.2992	160.
13/32	.406	10.319	1-1/2	1.500	38.100	3-1/32	3.0312	76.994	6-1/2	6.500	165.100
–	.413	10.5	1-17/32	1.531	38.894	–	3.0315	77.	–	6.6929	170.
27/64	.422	10.716	–	1.5354	39.	3-1/16	3.062	77.788	6-3/4	6.750	171.450
–	.4331	11.	1-9/16	1.562	39.688	–	3.0709	78.	7	7.000	177.800
7/16	.438	11.113	–	1.5748	40.	3-3/32	3.094	78.581	–	7.0866	180.
29/64	.453	11.509	1-19/32	1.594	40.481	–	3.1102	79.	–	7.4803	190.
15/32	.469	11.906	–	1.6142	41.	3-1/8	3.125	79.375	7-1/2	7.500	190.500
–	.4724	12.	1-5/8	1.625	41.275	–	3.1496	80.	–	7.8740	200.
31/64	.484	12.303	–	1.6535	42.	3-5/32	3.156	80.169	8	8.000	203.200
–	.492	12.5	1-21/32	1.6562	42.069	3-3/16	3.1875	80.963	–	8.2677	210.
1/2	.500	12.700	1-11/16	1.6875	42.863	–	3.1890	81.	8-1/2	8.500	215.900
–	.5118	13.	–	1.6929	43.	3-7/32	3.219	81.756	–	8.6614	220.
33/64	.5156	13.097	1-23/32	1.719	43.656	–	3.2283	82.	9	9.000	228.600
17/32	.531	13.494	–	1.7323	44.	3-1/4	3.250	82.550	–	9.0551	230.
35/64	.547	13.891	1-3/4	1.750	44.450	–	3.2677	83.	–	9.4488	240.
–	.5512	14.	–	1.7717	45.	3-9/32	3.281	83.344	9-1/2	9.500	241.300
9/16	.563	14.288	1-25/32	1.781	45.244	–	3.3071	84.	–	9.8425	250.
–	.571	14.5	–	1.8110	46.	3-5/16	3.312	84.1377	10	10.000	254.001
37/64	.578	14.684	1-13/16	1.8125	46.038	3-11/32	3.344	84.9314	–	10.2362	260.
–	.5906	15.	1-27/32	1.844	46.831	–	3.3464	85.	–	10.6299	270.
19/32	.594	15.081	–	1.8504	47.	3-3/8	3.375	85.725	11	11.000	279.401
39/64	.609	15.478	1-7/8	1.875	47.625	–	3.3858	86.	–	11.0236	280.
5/8	.625	15.875	–	1.8898	48.	3-13/32	3.406	86.519	–	11.4173	290.
–	.6299	16.	1-29/32	1.9062	48.419	–	3.4252	87.	–	11.8110	300.
41/64	.6406	16.272	–	1.9291	49.	3-7/16	3.438	87.313	12	12.000	304.801
–	.6496	16.5	1-15/16	1.9375	49.213	–	3.4646	88.	13	13.000	330.201
21/32	.656	16.669	–	1.9685	50.	3-15/32	3.469	88.106	–	13.7795	350.
–	.6693	17.	1-31/32	1.969	50.006	3-1/2	3.500	88.900	14	14.000	355.601
43/64	.672	17.066	2	2.000	50.800	–	3.5039	89.	15	15.000	381.001
11/16	.6875	17.463	–	2.0079	51.	3-17/32	3.531	89.694	–	15.7480	400.
45/64	.703	17.859	2-1/32	2.03125	51.594	–	3.5433	90.	16	16.000	406.401
–	.7087	18.	–	2.0472	52.	3-9/16	3.562	90.4877	17	17.000	431.801
23/32	.719	18.256	2-1/16	2.062	52.388	–	3.5827	91.	–	17.7165	450.
–	.7283	18.5	–	2.0866	53.	3-19/32	3.594	91.281	18	18.000	457.201
47/64	.734	18.653	2-3/32	2.094	53.181	–	3.622	92.	19	19.000	482.601
–	.7480	19.	2-1/8	2.125	53.975	3-5/8	3.625	92.075	–	19.6850	500.
3/4	.750	19.050	–	2.126	54.	3-21/32	3.656	92.869	20	20.000	508.001
49/64	.7656	19.447	2-5/32	2.156	54.769	–	3.6614	93.			

27102-13_A10.EPS

Additional Resources

This module presents thorough resources for task training. The following reference material is suggested for further study.

Basic Construction Materials. Upper Saddle River, NJ: Prentice Hall.

Principles and Practices of Light Construction. Upper Saddle River, NJ: Prentice Hall.

Principles and Practices of Commercial Construction. Upper Saddle River, NJ: Prentice Hall.

Buildings in Wood: The History and Traditions of Architecture's Oldest Building Material. New York: Rizzoli/Universe International Publications.

Concrete Masonry Handbook for Architects, Engineers, and Builders, Fifth Edition. W.C. Panarese, S.H. Kosmatka, and F.A. Randall, Jr. Portland Cement Association.

The Homeowner's Guide to Building with Concrete, Brick, and Stone. Portland Cement Association.

Figure Credits

SkillsUSA, Module opener

Southern Forest Products Association, Figures SA01, SA04, SA06, and 18

APA–The Engineered Wood Association, Figures 6, 7, 11

The Haskell Company, Figures SA07, 13

Calculated Industries, Inc., Figure 20

PCL Industrial, Figure 38 (photos)

Section Review Answer Key

Answer	Section Reference	Objective Reference
Section One		
1. c	1.1.0	1a
2. d	1.2.0	1b
3. a	1.3.0	1c
4. b	1.4.0	1d
5. b	1.5.1	1e
6. b	1.6.0	1f
7. b	1.7.0	1g
8. c	1.8.0	1h
9. b	1.9.0	1i
10. d	1.10.0	1j
11. b	1.11.1	1k
12. d	1.12.0	1l
13. c	1.13.0	1m
14. a	1.14.1	1n
15. b	1.15.0	1o
16. b	1.16.0	1p
Section Two		
1. d	2.1.0	2b
2. c	2.3.0	2c
3. b	2.3.0	2a
Section Three		
1. c	3.1.3	3a
2. b	3.2.0	3b
3. d	3.3.0	3c
4. b	3.4.0	3d
Section Four		
1. b	4.1.0	4a
2. a	4.2.0	4b
3. b	4.3.1	4c
Section Five		
1. b	5.1.0	5a
2. c	5.1.1	5
3. c	5.2.1	5b
4. c	5.3.2	5c
5. c	5.4.0	5d
6. d	5.5.3	5e
7. d	5.6.1	5f
8. d	5.7.1	5g
9. a	5.7.3	5g
10. a	5.8.0	5h
11. c	5.9.1	5i
12. b	5.7.5	5j
13. c	5.10.1	5j

NCCER CURRICULA — USER UPDATE

NCCER makes every effort to keep its textbooks up-to-date and free of technical errors. We appreciate your help in this process. If you find an error, a typographical mistake, or an inaccuracy in NCCER's curricula, please fill out this form (or a photocopy), or complete the online form at **www.nccer.org/olf**. Be sure to include the exact module ID number, page number, a detailed description, and your recommended correction. Your input will be brought to the attention of the Authoring Team. Thank you for your assistance.

Instructors – If you have an idea for improving this textbook, or have found that additional materials were necessary to teach this module effectively, please let us know so that we may present your suggestions to the Authoring Team.

NCCER Product Development and Revision

13614 Progress Blvd., Alachua, FL 32615

Email: curriculum@nccer.org
Online: www.nccer.org/olf

❑ Trainee Guide ❑ Lesson Plans ❑ Exam ❑ PowerPoints Other ______________

Craft / Level: Copyright Date:

Module ID Number / Title:

Section Number(s):

Description:

Recommended Correction:

Your Name:

Address:

Email: Phone:

27103-13

Hand and Power Tools

Overview

A carpenter is only as good as his or her tools. You must have a good set of tools, know how to use them safely and properly, and take good care of them. You must also know the proper tool to use for each task. Choosing the wrong tool can damage the material being worked on. A carpenter uses a variety of power tools including power saws and pneumatic nailers. These tools are dangerous to the user and others if they are not handled correctly. It is of utmost importance that you are properly trained in using these tools and that you remain conscious of the related safety concerns every time they are used.

Module Three

Trainees with successful module completions may be eligible for credentialing through the NCCER Registry. To learn more, go to **www.nccer.org** or contact us at **1.888.622.3720.** Our website has information on the latest product releases and training, as well as online versions of our *Cornerstone* magazine and Pearson's product catalog.

Your feedback is welcome. You may email your comments to **curriculum@nccer.org,** send general comments and inquiries to **info@nccer.org,** or fill in the User Update form at the back of this module.

This information is general in nature and intended for training purposes only. Actual performance of activities described in this manual requires compliance with all applicable operating, service, maintenance, and safety procedures under the direction of qualified personnel. References in this manual to patented or proprietary devices do not constitute a recommendation of their use.

27103-13
Hand and Power Tools

Objectives

When you have completed this module, you will be able to do the following:

1. Identify the hand tools commonly used by carpenters.
   a. Describe the safe use and maintenance of levels.
   b. Describe the safe use and maintenance of squares.
   c. Describe the safe use and maintenance of planes.
   d. Describe the safe use and maintenance of clamps.
   e. Describe the safe use and maintenance of hand saws.
2. Identify the power tools commonly used by carpenters.
   a. Describe the general safe use and maintenance of power tools.
   b. Describe the safe use of power saws.
   c. Describe the safe use of drill presses.
   d. Describe the safe use of routers and laminate trimmers.
   e. Describe the safe use of portable power planes.
   f. Describe the safe use of power metal shears.
   g. Describe the safe use of pneumatic and cordless nailers and staplers.

Performance Tasks

Under the supervision of the instructor, you should be able to do the following:

1. Demonstrate the safe and proper use of the following hand tools:
   - Level
   - Square
   - Clamp
   - Saw
2. Demonstrate or describe the safe and proper use of five of the following power tools:
   - Circular saw
   - Portable table saw
   - Compound miter saw
   - Drill press
   - Router/laminate trimmer
   - Portable power plane
   - Power metal shears
   - Pneumatic nailer/stapler

Trade Terms

Bevel cut
Compound cut
Crosscut
Dados
Door jack
Dovetail joints
Kerf
Kickback
Miter box
Miter cut
Mortise and tenon joints
Plane iron
Pocket (plunge) cuts
Rabbet cuts
True

Industry-Recognized Credentials

If you're training through an NCCER-accredited sponsor you may be eligible for credentials from NCCER's Registry. The ID number for this module is 27103-13. Note that this module may have been used in other NCCER curricula and may apply to other level completions. Contact NCCER's Registry at 888.622.3720 or go to **nccer.org** for more information.

Contents

Topics to be presented in this module include:

Figures

SECTION ONE

1.0.0 HAND TOOLS

Objective

Identify the hand tools commonly used by carpenters.

a. Describe the safe use and maintenance of levels.
b. Describe the safe use and maintenance of squares.
c. Describe the safe use and maintenance of planes.
d. Describe the safe use and maintenance of clamps.
e. Describe the safe use and maintenance of hand saws.

Performance Task 1

Demonstrate the safe and proper use of the following hand tools:

- Level
- Square
- Clamp
- Saw

Trade Terms

Bevel cut: A cut made across the sloping edge or side of a workpiece at an angle of less than 90 degrees.

Dovetail joints: Interlocking wood joints with a triangular shape like that of a dove's tail.

Kerf: The width of the cut made by a saw blade. It is also the amount of material removed by the blade in a through (complete) cut or the slot that is produced by the blade in a partial cut.

Miter box: A device used to cut lumber at precise angles.

Miter cut: A cut made at the end of a piece of lumber at any angle other than 90 degrees.

Mortise and tenon joints: Wood joints in which a rectangular cutout or opening receive the tongue that is cut on the end of an adjoining member.

Plane iron: Flat metal blade that is sharpened on one end to provide a cutting edge for a hand plane.

True: To accurately shape adjoining members so they fit well together.

In the *Core Curriculum,* hand and power tools that are commonly used by all trades were introduced. This module builds on that information and introduces some new tools widely used by carpenters on a construction site. The focus of this module is to familiarize you with each of these tools. While under the supervision of your instructor and/or supervisor, you will learn how to properly operate and use each of the tools described in this module. Keep in mind that this material is introductory; you will receive specific training in the use of these tools as your training progresses.

Every profession has its tools. A surgeon uses a scalpel, a teacher uses a whiteboard, and an accountant uses a calculator. Carpenters use a wide variety of hand tools, such as levels, squares, planes, clamps, and saws. Even if some of these tools are familiar, you need to learn to use them safely and maintain them properly. The better that your tools are used and maintained, the better carpenter you will be. This section describes how to safely use and maintain some common hand tools used by carpenters.

1.1.0 Safely Using and Maintaining Levels

A level is used to determine both how level a horizontal surface is and how plumb a vertical surface is. In addition to the spirit level covered in the *Core Curriculum,* the leveling tools and instruments shown in *Figure 1* are used:

- *Carpenter's level* – A level that is used to establish level and plumb lines, often consisting of glass tubes mounted in a lightweight rigid frame. The frame is placed against the surface to be leveled or plumbed and moved until the bubble is centered in the tube. Carpenter's levels are available in lengths ranging from 24" to 78".
- *Line level* – A line level is used to roughly level a long span. It consists of a glass tube mounted in a sleeve, which has hooks on both ends. The hooks are attached at the center of a tightly stretched line, which is moved up or down until the bubble is centered.
- *Water level* – A water level is a simple, but very accurate, tool consisting of a length of clear plastic tubing. It works on the principle that water seeks its own level. The water level is limited only by the length of the tubing. It can be used effectively for checking the level

CARPENTER'S LEVEL

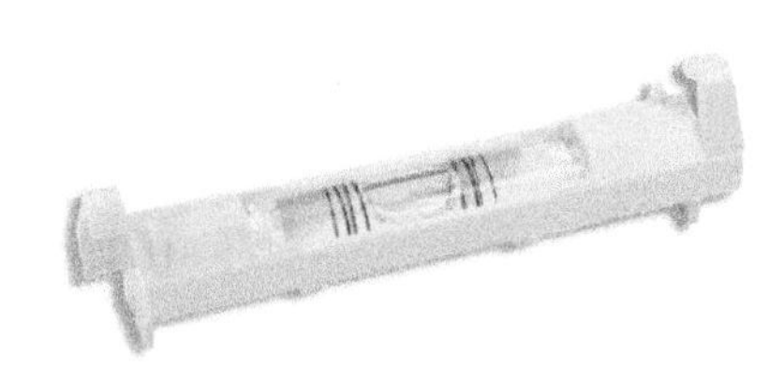

LINE LEVEL

WATER LEVEL

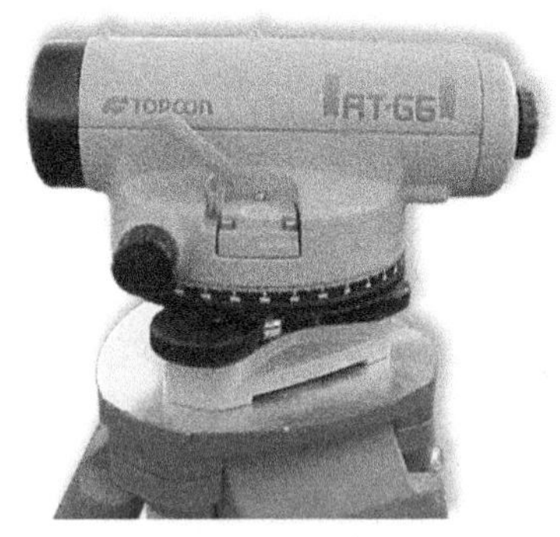

AUTOMATIC
BUILDER'S LEVEL

LASER LEVEL

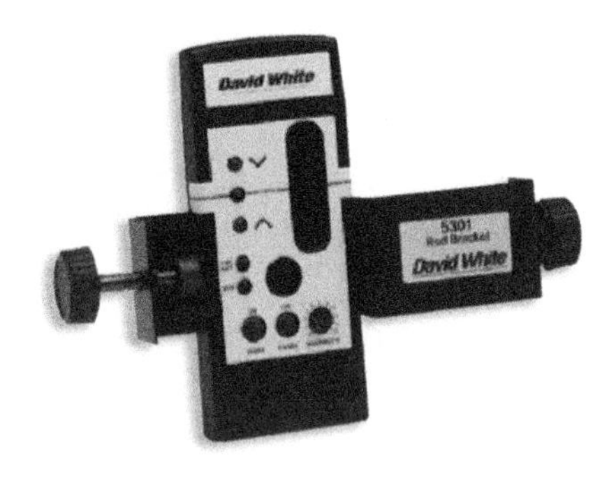

LASER DETECTOR

27103-13_F01.EPS

Figure 1 Levels.

around obstructions; for example, checking the level from one room to the next when a wall is in the way.

- *Builder's level* – The builder's level is basically a telescope with a spirit level mounted on top. It can be rotated horizontally 360 degrees but cannot be tilted up and down. It is used to check grades and elevations (also known as differential leveling) and to set up level points over long distances.
- *Auto level* – An auto (automatic) level is a leveling instrument that includes an internal compensator mechanism that, when set close to level, automatically eliminates any remaining variation from level. This feature reduces the need for an auto level to be set up perfectly level. Auto levels are one of the most commonly used leveling instruments used for building layout, construction, and surveying due to their rapid setup time and ease of use.
- *Transit level* – The telescope of a transit level can be moved up and down 45 degrees, making more operations possible than with a builder's level. When the telescope is locked in a level position, the transit level can be used to perform all the functions of a builder's level. Moving the telescope up and down allows the transit level to be used to plumb columns, building corners, or any other vertical members. Transit levels are most commonly used for building layout.

A leveling rod is used with a builder's level or transit level. The rod is held vertically in the area where the grade or elevation is being established by a second person. The horizontal crosshair of

Water Level

A water level is very effective for distances where a 4- or 6-foot level is not long enough, like decks and siding. It is also ideal for checking levels around obstructions, such as a wall that separates two rooms. The water level may be filled with colored water for increased visibility. Water levels are the most accurate leveling device (unless the water is frozen) as they are not affected by the distance between the two points being leveled. Water levels have been used in construction since ancient times.

the level is then aligned with a mark or measurement on the leveling rod.

Using a spirit level is very simple. It just requires a careful eye to ensure that it is being read correctly. Follow these steps to use a spirit level properly:

Step 1 Ensure the surface is free of debris.

Step 2 Place the spirit level on the object being leveled.

Step 3 Look at the air bubble. If the bubble is centered between the lines, the object is level (or plumb).

Levels are precision instruments that must be handled with care. Although there is little risk of personal injury when working with most levels, there is a chance of damaging or breaking a level if it is handled improperly. Remember these maintenance guidelines when working with levels:

- Replace a level if a crack or break appears in any of the vials.
- Keep levels clean and dry.
- Don't bend or apply too much pressure on a level.
- Don't drop or bump a level.

1.1.1 Safely Using and Maintaining Laser Levels

Laser levels are precise electronic devices with a rotating beacon that projects an intense laser beam. Therefore, laser levels must be handled carefully and should not be subjected to adverse weather conditions such as rain or snow. Low-power laser levels project a laser beam that is not as intense as high-power laser levels. The laser beam projected from high-power laser levels can cause serious eye and skin damage if not used properly. A label on the body of the laser level lists the laser level classification and maximum output. The laser beam is detected by an electronic laser detector at distances up to 1,000'.

Laser levels enable one person, instead of two, to perform layout operations. When a laser level is operated in the sweep mode, the head rotates through 360 degrees, allowing the laser beam to sweep multiple sensors placed at different locations. Rotating-beam laser levels only require one worker to set up and operate (*Figure 2*).

WARNING!

Lasers can cause serious damage to your eyes. Never point a laser at anyone or look directly at a laser beam.

Safety precautions for use of a laser level include, but are not limited to:

- Workers in the area of an operating laser level with output greater than 5mW (milliwatts) should wear antilaser eye protection.
- Standard laser warning placards should be posted in areas of operating lasers (*Figure 3*).
- Laser beams should not be directed at other workers.
- When possible, set up laser levels above or below the eye level of workers.
- Only qualified and trained workers are permitted to operate laser levels. Proof of worker qualification should be in the worker's possession at all times when using a laser level.

Spirit Levels with Lasers

A wide variety of spirit levels with lasers are available, allowing you to perform traditional spirit-level operations as well as projecting a fixed laser beam to level or plumb across long distances. The model shown here extends the level's reference line to 100 feet with an accuracy of ±¼".

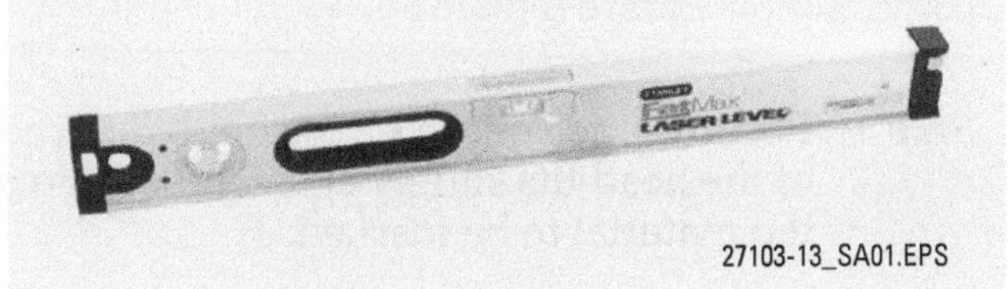

27103-13_SA01.EPS

27103-13_F02.EPS

Figure 2 Operating a laser level.

27103-13_F03.EPS

Figure 3 Laser-level safety placards.

1.2.0 Safely Using and Maintaining Squares

Squares are used for marking, checking, and measuring. Establishing squareness is an essential part of carpentry; therefore, several types of squares are needed. In the *Core Curriculum,* the combination square was introduced. *Figure 4* shows a few more types of squares. Squares are made from steel, aluminum, and plastic and are available in both standard and metric measurements.

- *Combination square* – A combination square is used to lay out lines at 45- and 90-degree angles. The square can be adjusted by sliding the handle along the length of the blade.
- *Miter square* – A miter square is used to lay out 45- and 90-degree angles. When the handle is placed flush against the surface, it forms a 90-degree angle. When the beveled corner of the handle is placed against the wood, it forms a 45-degree angle for laying out and marking lines such as for a miter cut.
- *Try square* – A try square is similar to a miter square except a try square does not have a beveled corner on the handle. Because of the lack of a beveled corner, a try square can be used to lay out lines only at a right angle (90 degrees) to an edge or surface and to check the squareness of adjoining surfaces of planed lumber.
- *Sliding T-bevel* – A sliding T-bevel is an adjustable gauge for setting, testing, and transferring angles, such as when making a miter or bevel cut. The metal blade pivots and can be locked at any angle.
- *Drywall square* – A drywall square, also called a T-square, is used for laying out square and angled lines on large surfaces such as plywood, paneling, or gypsum board. The blade on drywall squares is approximately 4 feet long to extend across a 4-foot panel.
- *Framing square* – A framing square is used for squaring lines and marking angles. Additional information about the use of the various types of squares is included in later modules.
- *Rafter square* – A rafter square is a type of framing square that has useful tables and formulas imprinted on each side. Rafter tables (*Figure 5*) and an Essex board measure table are typically included.
- *Speed Square™* – A Speed Square™, also called a rafter angle square, is used to lay out and check 45- to 90-degree angles. It is especially useful when laying out angle cuts for roof rafters and stairs. Standard Speed Squares™ are 7" triangular tools. The outside edges have a 7" scale on one edge, a full 90-degree scale on another edge, and a T-bar on the third edge. There are also 12" Speed Squares™ available for larger projects and stair layout.

The type of square used depends on the type of job being performed and your preference. Follow these steps to mark a 90-degree angle using a combination square (*Figure 6*):

Step 1 Position the square so the angled corner of the head fits snugly against the edge of the material to be marked.

> **Keeping It Level**
>
> Use care when handling and using a level. Levels are precision instruments and should be treated as such. In order to provide accurate readings, levels need to be calibrated properly. If they aren't, what appears to be a level reading will actually be slightly off. These errors are magnified over long distances, so a small miscalculation can wind up being a very large problem. Avoid this by taking proper care of your instruments.

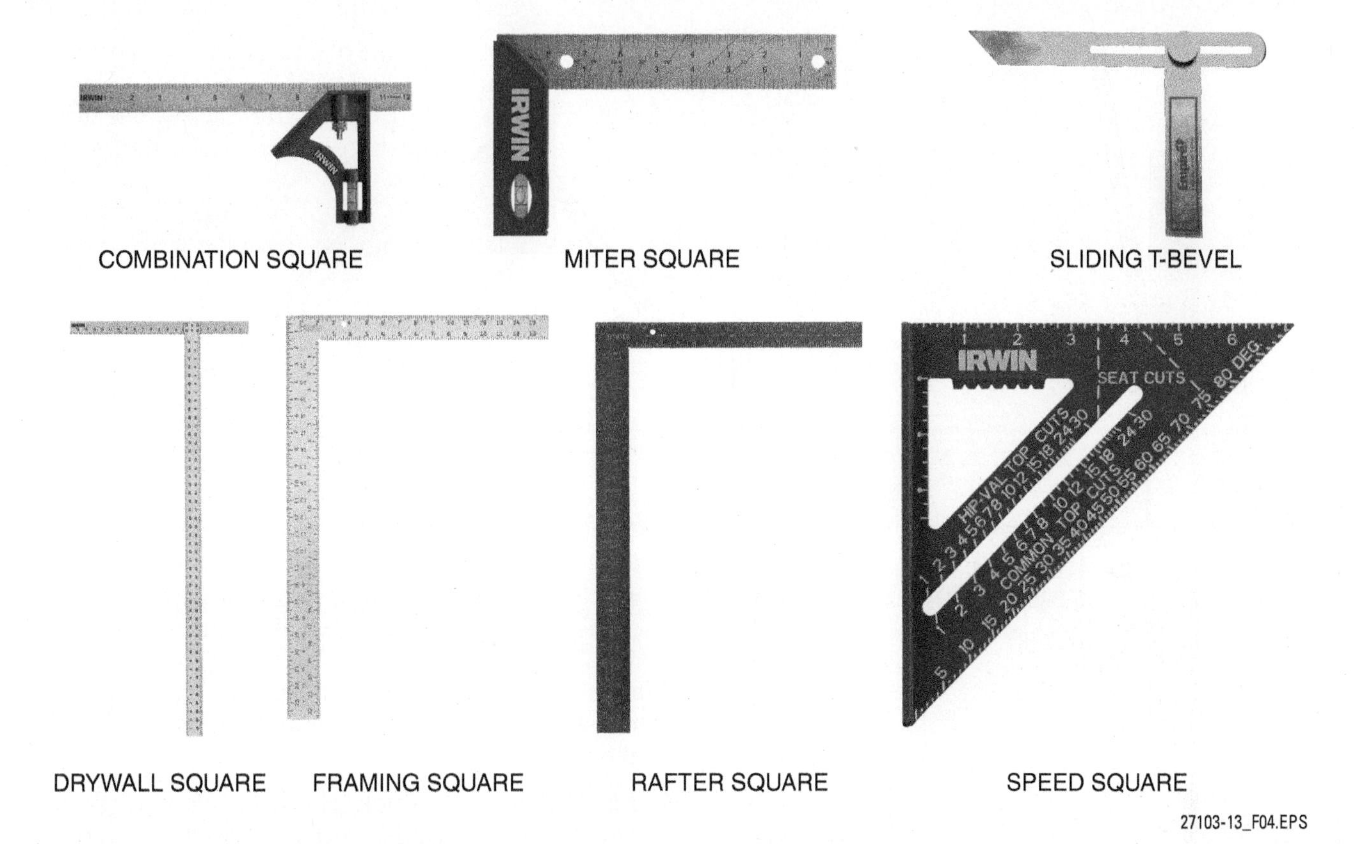

Figure 4 Squares.

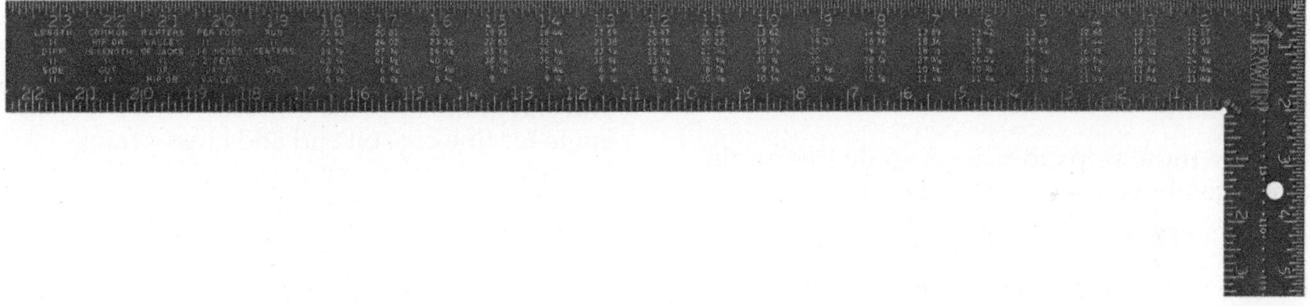

Figure 5 Rafter tables.

Laser Levels

The term *laser* is an abbreviation for Light Amplification by Stimulated Emission of Radiation. In other words, the narrow highly focused beam of light projected from a laser level is produced by amplifying the beam using stimulated radiation. Lasers are common in CD and DVD players, laser printers, and bar-code scanners, as well as in laser levels used in construction.

There are four major classifications and two subclassifications of lasers, all which are based on the hazards they pose. Classes I, II, and IIa pose little threat to a user. These are the types of lasers used in laser printers, CD/DVD players, and bar-code scanners. Classes IIIa, IIIb, and IV are more powerful lasers typically used in construction and industrial applications; these lasers have higher output power than Class I, II, and IIa lasers. Lasers with higher output are brighter than ones with a lower output.

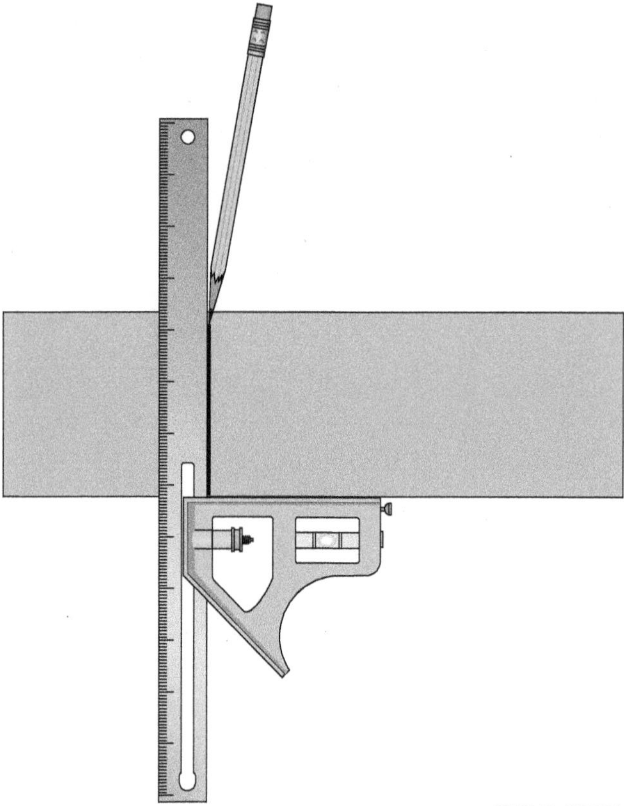

27103-13_F06.EPS

Figure 6 Using a combination square to make a 90-degree angle.

Step 2 Starting at the edge of the material, use the blade as a straightedge to guide the mark.

Follow these steps to mark a 45-degree angle using a combination square (*Figure 7*):

Step 1 Set the blade at a 45-degree angle.

Step 2 Position the square so that the head fits snugly against the edge of the material to be marked.

Step 3 Starting at the edge of the material, use the blade as a straightedge to guide the mark.

The safety and maintenance guidelines to remember when using a steel square are:

- Keep the square dry to prevent it from rusting.
- Use a light coat of oil on the blade, and occasionally clean the blade's grooves and the setscrew.
- Don't use a square for anything it wasn't designed for, especially prying or hammering.
- Don't bend a square or use one for any kind of horseplay.
- Don't drop or strike the square hard enough to change the angle between the blade and the head.

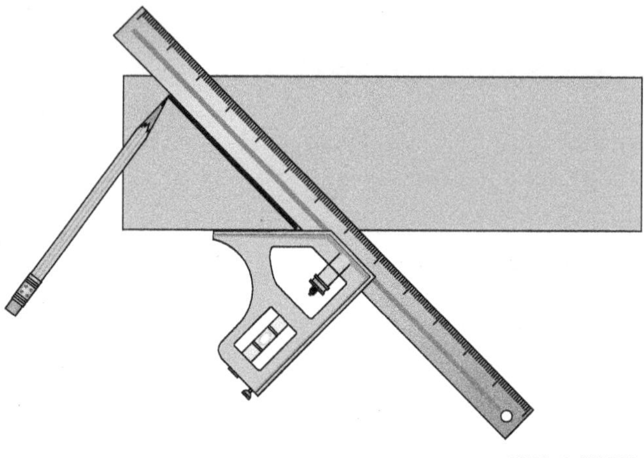

27103-13_F07.EPS

Figure 7 Using a combination square to mark a 45-degree angle.

1.3.0 Safely Using and Maintaining Planes

Planes (*Figure 8*) are used to remove excess wood from surfaces and edges. Some types of planes that are available are:

- *Block plane* – A block plane is used to plane small pieces, edges, and joint surfaces when a small amount of change is needed. Block planes are available in 4" to 7" lengths and 1⅜" to 1⅝" widths. **Plane irons** are usually adjustable from 12 degrees to 21 degrees. A low-angle block plane is used for shaping fine trimwork and end grain. Its plane iron rests at a 12-degree angle for fine cuts on end and cross-grains.
- *Jack plane* – A jack plane is a type of bench plane used to plane rough work. Jack planes are 14" long and 2" wide and have an adjustable blade.
- *Smoothing plane* – A smoothing plane is another type of bench plane. Smoothing planes are used to plane small work to a smooth and even surface without removing too much material. Smoothing planes are available in lengths of 7" to 10" and widths of 1⅝" to 2".
- *Jointer plane* – A jointer plane is used to plane and **true** long boards such as the edges of doors. They are available in 18" to 30" lengths. The cutting edge is ground straight and is set for a fine cut.

Square Accuracy

Using tools that don't provide accurate measurements or readings can result in serious errors on a job site. Squares may be damaged accidentally from time to time. Periodically check squares with a known straightedge to ensure the accuracy of your measurements, marks, and cuts.

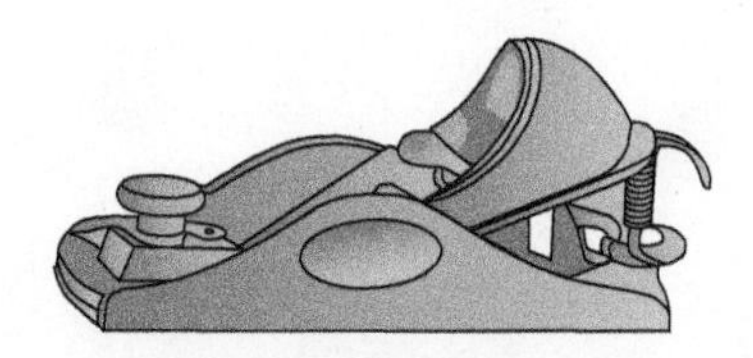

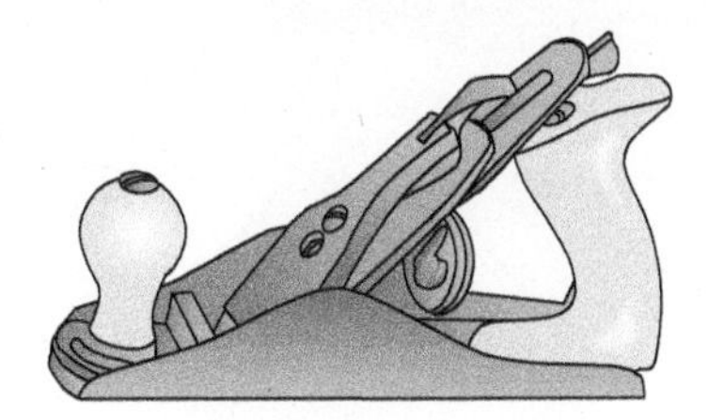

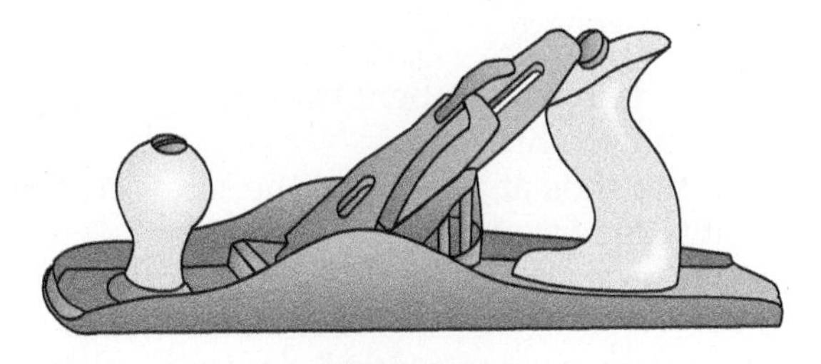

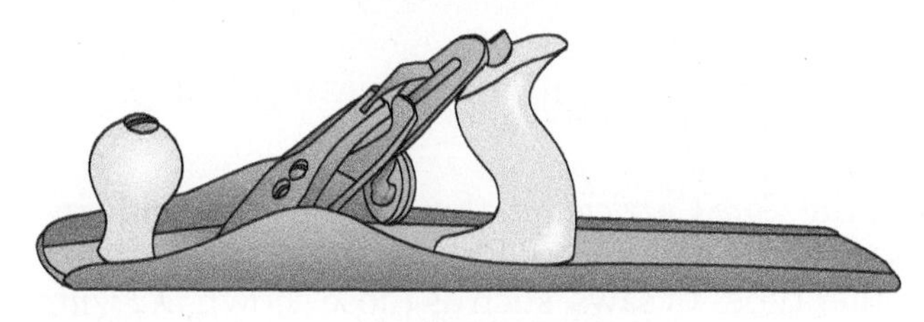

Figure 8 Planes.

Although several types of planes are available, most can be used for the same purpose: to joint and true boards and lumber. The plane selected for a particular task is typically based on personal preference. The general procedure for using a plane is as follows:

Step 1 Push the plane in the direction of the grain. At the beginning of the stroke, press down on the knob. At the end of the stroke, press down on the handle. If the grain appears torn after the initial stroke, reverse the direction of planing.

Step 2 When planing face grain, hold the plane square with the surface of the wood.

Step 3 When planing end grain, plane from one edge to the middle of the work. Then, reverse direction and plane from the other edge to the middle of the work. This helps to prevent the corners of the workpiece from tearing out.

Step 4 When completed working with a plane, lay it on its side so the cutting edge does not contact a surface, thereby dulling the cutting edge.

A plane iron must be sharp for it to cut effectively. Some maintenance guidelines to consider when using planes are:

WARNING!

Ensure all power tools are unplugged when performing maintenance. Never change the adjustment of tool rests when the grinder is on or when the grinding wheels are spinning. Doing so may damage the work or cause injury to you or another worker.

Step 1 If a plane iron is nicked or otherwise damaged, use a bench grinder with a tool rest to grind the cutting edge (*Figure 9*). A grinder attachment may be used to keep the plane iron square with the grinding wheel.

Step 2 When grinding a plane iron, frequently dip it in water to cool the cutting edge and prevent it from burning.

Step 3 After grinding the cutting edge of the plane iron, the edge must be dressed manually to remove the burr that is produced when grinding. Move the cutting edge back and forth on an oilstone (bevel on the cutting edge facing downward). Then turn the plane iron over and make a few strokes with the flat side of the plane iron flush with the oilstone surface.

GOING GREEN

Take Pride in Being a Green Carpenter

As a carpenter, you contribute to the efficiency of a building in many ways. For example, the method of framing that is used can improve the energy efficiency of a building. The materials installed on the exterior can improve the durability of the building. Resources can be conserved by reducing the amount of material used to build a structure. Waste materials can also be recycled. For example, cutoffs from studs can be used as blocking when assembling corner posts. Your daily activities on a job site can affect the efficiency of a building and impact the environment.

Step 4 Lightly oil the bottom of the plane to prevent rusting.

1.4.0 Safely Using and Maintaining Clamps

In addition to the C-clamp, which was introduced in the *Core Curriculum,* several other types of clamps may be used by a carpenter. Some of the most commonly used clamps are shown in *Figure 10.*

- *Hand-screw clamp* – A hand-screw clamp can be used to hold pieces together while adhesive dries. The screws can turn independently of one another to adjust the angle of the jaws, making it useful for clamping up beveled or tapered pieces.
- *Locking C-clamp* – Locking C-clamps are available in a variety of sizes up to 24". They are based on the same design as clamping pliers; when the handles are squeezed together, the jaws lock into place. The jaw opening is adjustable. Locking C-clamps are commonly used when assembling steel framing members.
- *Spring clamp* – This simple clamp is spring loaded so that the jaws will lock down on the work when the handles are released.
- *Quick clamp* – Quick clamps are heavy-duty clamps that are designed to apply extreme pressure. The jaws are tightened by pulling on the trigger and loosened by pulling on the release lever.
- *Web (strap) clamp* – Web clamps are designed to secure round, oval, and oddly shaped work. A ratchet assembly is used to tighten the strap.
- *Pipe clamp* – This heavy-duty clamp can also be used as a spreader. Pipe clamps are made to fit ½" or ¾" pipe. The length of the clamp is determined by the length of the pipe. A sliding jaw operates with a spring-locking device, and a middle jaw is controlled by a screw set in a third, fixed jaw. A reversing pipe clamp has a sliding jaw that can also be used in the reverse direction for spreading by applying pressure away from the clamp instead of between the clamp jaws.

When clamping wood or other soft material, place pads or thin blocks of wood between the work piece and the clamp to protect the work. Tighten the clamp's pressure mechanism; don't force it. Some guidelines to remember when using and storing clamps are:

- Store clamps by clamping them to a rack.
- Use pads or thin wood blocks when clamping wood or other soft materials.

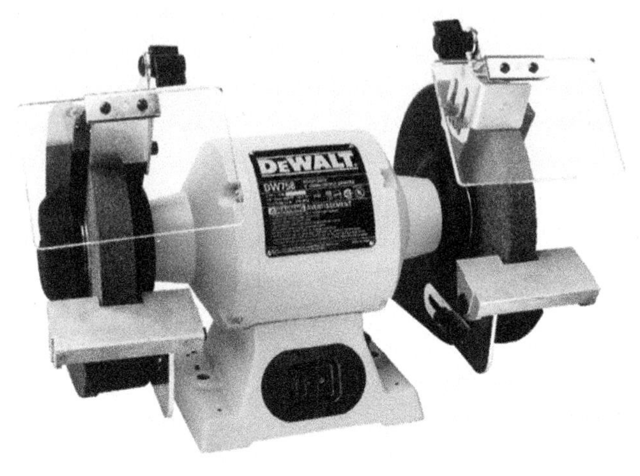

27103-13_F09.EPS

Figure 9 Bench grinder.

- Discard clamps with bent frames.
- Clean and oil threads.
- Check the shoe at the end of the screw to make sure it turns freely.
- Don't use a clamp for hoisting (pulling up) work.
- To maintain proper control, don't use pliers or pipe on the handle of a clamp.

1.5.0 Safely Using and Maintaining Handsaws

In the *Core Curriculum,* the crosscut saw and ripsaw were introduced. Carpenters also use many other types of saws, such as those shown in *Figure 11.* Keep your saws sharpened and properly protected when not in use.

- *Hacksaw* – A hacksaw is used to cut metal. It has a replaceable blade.
- *Backsaw (miter saw)* – This saw has a strong back and very fine teeth. It is commonly used with a **miter box** to cut trim.
- *Dovetail saw* – A dovetail saw, with its very fine teeth, is used to cut very smooth joints, such as **mortise and tenon joints** and **dovetail joints**. The blade of a dovetail saw is extremely thin, but is reinforced across the top edge to stiffen it.

Planes

Planes must be handled carefully and inspected regularly. Damage to the plane blade will result in unsightly grooves in your work. A nail, for example, can cause a sharp burr or take a chunk out of the blade. Always keep your blades sharp and replace them when necessary.

Figure 10 Clamps.

- *Compass (keyhole) saw* – This saw is used for making rough cuts in gypsum board, plywood, hardboard, and paneling. It is often used to cut access holes for piping and electrical boxes.
- *Coping saw* – A coping saw has a thin, fine-toothed blade and is ideal for making curved and scroll cuts in moldings and trimwork.
- *Drywall saw* – The drywall saw has a stiff blade with coarse teeth, and has a tapered nose for easy penetration through gypsum board. Drywall saws are commonly used to complete cuts for stringers and rafter birdsmouths where a circular saw cannot reach.

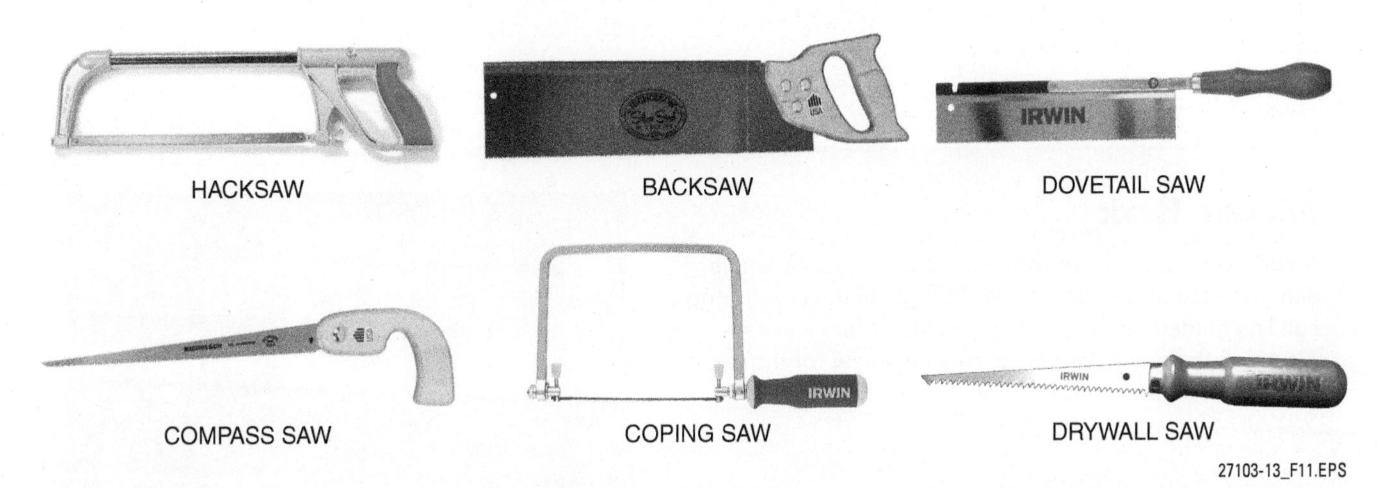

Figure 11 Special-purpose handsaws.

Using the right saw for the job makes cutting easy. The crosscut saw cuts across the grain of wood. Because it has 8 to 14 teeth per inch (tpi), it will cut slowly but smoothly. Follow these steps to use a crosscut saw properly:

Step 1 Mark the cut to be made with a square or other measuring tool.

Step 2 Make sure the piece to be cut is well supported (on a sawhorse, jack, or other support). Support the scrap end as well as the main part of the wood to keep it from splitting as the kerf nears the edge. With short pieces of wood, support the scrap end of the piece with your free hand. With longer pieces, additional support will be needed.

Step 3 Place the saw teeth on the edge of the wood farthest from you, just at the outside edge of the mark.

Step 4 Start the cut with the part of the blade closest to the handle end of the saw, because you will pull the first stroke toward your body.

Step 5 Use the thumb of the hand that is not sawing to guide the saw so it stays vertical to the work. Ensure the blade stays in contact with the wood.

Step 6 Place the saw at about a 45-degree angle to the wood, then pull the saw to make a small groove.

Step 7 Start sawing slowly, increasing the length of the stroke as the kerf deepens.

Clamps

As you gain experience, various uses for different types of clamps will be learned. For example, a quick clamp is helpful in tight spots when only one hand is available. A hand-screw clamp can be used to hold pieces together when laminating.

NOTE

Don't push or ride the saw into the wood. Let the weight of the saw set the cutting rate. It's easier to control the saw and less tiring that way.

Step 8 Continue to saw with the blade at a 45-degree angle to the wood.

NOTE

If the saw starts to wander from the line, angle the blade toward the line. If the saw blade sticks in the kerf, wedge a thin piece of wood into the cut to hold it open.

The ripsaw cuts along the grain of wood. Because it has fewer points (5 to 9 tpi) than the crosscut saw, it will make a coarser, but faster, cut. To use a ripsaw properly, mark and start a ripping cut the same way as when starting a cut with a crosscut saw. Once the kerf has been started, saw with the blade at a steeper angle to the wood—about 60 degrees. Also, it is very important to focus on your work when sawing—saws can be dangerous if used incorrectly or if you are not paying attention.

Miter Box

A backsaw is often used with a miter box. A miter box is either slotted or has saw guides that guide the saw during a cut. The guided saw movement results in very accurate cuts. Although the miter box was once a necessity for trim work, most carpenters now use a compound miter saw.

27103-13_SA02.EPS

Maintain your saws for them to work properly. Some guidelines for properly maintaining handsaws are:

- Clean the saw blade with a fine emery cloth and apply a coat of light machine oil if it starts to rust—rust ruins the saw blade.
- Apply a light coat of paraffin wax to a cleaned saw blade. The wax will minimize rusting and also allow the blade to move freely within the kerf.
- Always lay a saw down gently.
- Have the saw sharpened by an experienced sharpener.
- Brace yourself when sawing so you are not thrown off balance on the last stroke.
- Don't let saw teeth come in contact with stone, concrete, or metal.

1.0.0 Section Review

1. The tool that includes an internal compensator mechanism, which allows for rapid setup time is the _____.
   a. router
   b. builder's level
   c. leveling rod
   d. auto level

2. Which of the following tools is especially useful when laying out angle cuts for roof rafters?
   a. Try square
   b. Combination square
   c. Drywall square
   d. Speed Square™

3. The type of plane used to smooth rough lumber is the _____.
   a. block plane
   b. jack plane
   c. smoothing plane
   d. jointer plane

4. Which of the following clamps might be used when assembling an irregular-shaped object such as an octagonal table?
   a. Hand screw clamp
   b. Quick clamp
   c. Web clamp
   d. C-clamp

5. The type of saw commonly used with a miter box is the _____.
   a. backsaw
   b. coping saw
   c. drywall saw
   d. compass saw

Section Two

2.0.0 Power Tools

Objective

Identify the power tools commonly used by carpenters.

a. Describe the general safe use and maintenance of power tools.
b. Describe the safe use of power saws.
c. Describe the safe use of drill presses.
d. Describe the safe use of routers and laminate trimmers.
e. Describe the safe use of portable power planes.
f. Describe the safe use of power metal shears.
g. Describe the safe use of pneumatic and cordless nailers and staplers.

Performance Task 2

Demonstrate or describe the safe and proper use of five of the following power tools:

- Circular saw
- Portable table saw
- Compound miter saw
- Drill press
- Router/laminate trimmer
- Portable power plane
- Power metal shears
- Pneumatic nailer/stapler

Trade Terms

Compound cut: A simultaneous bevel and miter cut.

Crosscut: A cut made across the grain in lumber.

Dados: Rectangular grooves that are made part of the way through and across the grain of lumber.

Door jack: A holder or stand used to hold a door on edge while planing, routing, or installing hinges.

Kickback: A sharp, uncontrolled grabbing and throwing of the workpiece by a tool as it rejects material being forced into it.

Pocket (plunge) cuts: A cut made to remove an interior section of a workpiece or stock (such as a sink cutout for a countertop) or to make square or rectangular openings in floors or walls.

Rabbet cuts: Rectangular cuts made along the edge or end of a board to receive another board.

In this section, some of the power tools studied earlier in the *Core Curriculum* will be reviewed and several new ones will be introduced. The intent of this section is to familiarize you with each of the power tools and their associated safety rules. Before being allowed to operate a specific power tool, you must be able to show that you understand the safety rules for the tool. As training progresses, you will learn how to operate each of the power tools under the supervision of your instructor. Specific operating procedures and safety rules for using a power tool are provided in the operator's manual supplied by the manufacturer. Before operating any power tool for the first time, always read the operator's manual to familiarize yourself with the tool. If the manual is missing, contact the manufacturer (or ask your supervisor to contact the manufacturer) for a replacement manual. Power tools that were covered in the *Core Curriculum* include: drills, saws, grinders, pneumatic nailers, and powder-actuated tools.

2.1.0 Safely Using and Maintaining Power Tools

Before describing the power tools, first it is important to review the general safety rules that apply when using all power tools, regardless of type. It is also important to review the general guidelines that should be followed in order to properly care for power tools.

> **NOTE**
> Power tools may be operated by electricity (AC or DC), air, combustion engines, or explosive powder.

2.1.1 Safety Rules Pertaining to All Power Tools

Rules for the safe use of all power tools include, but are not limited to, the following:

- Keep all tools in good condition with regular maintenance.
- Do not attempt to operate any power tool before being cleared by the instructor on that particular tool.
- Use only equipment that is approved to meet Occupational Safety and Health Administration (OSHA) standards.
- Examine each tool for damage before use and do not use damaged tools.
- Always wear eye protection and other appropriate personal protective equipment (PPE) when operating power tools.

- Wear face and hearing protection when required.
- Wear proper respirator equipment when necessary.
- Wear the appropriate clothing for the job being done. Always wear tight-fitting clothing that cannot become caught in the moving tool. Roll up or button long sleeves, tuck in shirttails, and tie back long hair. Do not wear any jewelry or watches.
- Do not distract others or let anyone be a distraction while operating a power tool.
- Do not engage in horseplay.
- Do not throw objects or point tools at others.
- Consider the safety of others, as well as yourself.
- Do not leave a power tool unattended while it is running.
- Assume a safe and comfortable position before using a power tool.
- Do not remove ground plugs from electrical equipment or extension cords.
- Be sure that a power tool is properly grounded and connected to a ground fault circuit interrupter (GFCI) circuit before using it.
- Be sure that portable or stationary power tools are unplugged at the power source or disabled before performing maintenance or changing accessories.
- Do not use a dull or broken tool or accessory.
- Use a power tool only for its intended purpose.
- Keep your feet, fingers, and hair away from the blade and/or other moving parts of a power tool.
- Do not use a power tool with guards or safety devices removed or disabled.
- Do not operate a power tool if your hands or feet are wet.
- Keep the work area clean at all times.
- Become familiar with the correct operation and adjustments of a power tool before attempting to use it. Always follow the manufacturer's instructions as it pertains to its intended use.
- Keep a firm grip on the power tool at all times.
- Use electric extension cords of sufficient size to service the particular power tool being used.
- Do not use worn or frayed extension cords.
- If a tool or extension cord is defective, it must be brought to the attention of the supervisor so it can be tagged and immediately removed from service.
- Extension cords should not be hung by nails or wire, or fastened with staples.
- Report unsafe conditions to your instructor or supervisor.

2.1.2 Guidelines Pertaining to the Care of All Power Tools

The following are some guidelines for the proper care of power tools:

- Keep all tools clean and in good working order.
- Keep all machined surfaces clean and waxed or lightly oiled.
- Follow the manufacturer's maintenance procedures.
- Protect cutting edges.
- Keep all tool accessories (such as blades and bits) sharp.
- Always use the appropriate blade for the arbor size.
- Report any unusual noises, sounds, or vibrations to your instructor or supervisor.
- Regularly inspect all tools and accessories.
- Keep all tools in the proper place when not in use.
- Use the proper blade for the job being done.

GOING GREEN

Wind Farms

The US Department of Energy states that wind farms may be a major source of power as we approach the year 2030. Their analysis shows that up to 20 percent of US power needs could be handled by wind power, which would reduce pollution to the same extent as taking 140 million cars off the road.

27103-13_SA03.EPS

2.2.0 Safely Using and Maintaining Power Saws

Power saws, especially the circular saw, are among the carpenter's most commonly used tools. In addition to the reciprocating saw covered in the *Core Curriculum,* there are several types of power saws a carpenter might use on a job site, including the following:

- Circular saws
- Portable table saws
- Power miter saws/compound miter saws
- Abrasive saws

2.2.1 Safely Using and Maintaining Circular Saws

Circular saws (*Figure 12*) are versatile, portable saws used to perform a variety of tasks, including the following:

- Ripping (rip cut)
- Crosscutting (crosscut)
- Mitering
- Pocket (plunge) cuts
- Bevel cuts

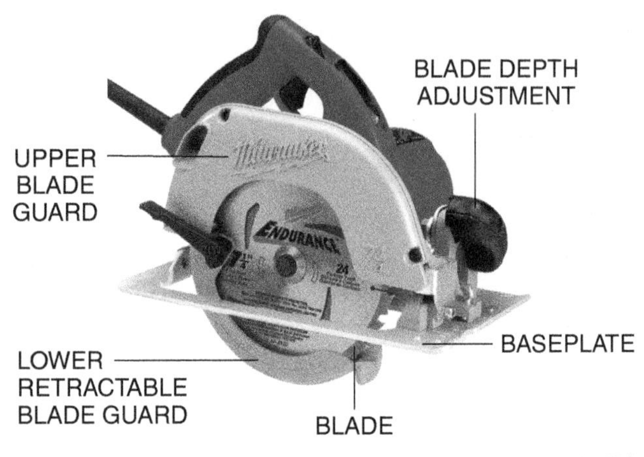

Figure 12 Circular saws.

The size of a circular saw is determined by the diameter of the largest-size blade that can be used with the saw. The size of the saw blade determines the thickness of material that can be cut. Saws using a 7¼" blade are popular. Circular saws have upper and lower guards that surround the blade. The upper guard is fixed; the lower guard is spring loaded and retracts as the saw cuts into

Extension Cords

Electrical cords are frequently used on job sites, but their care and maintenance is often overlooked. Use the following safety guidelines to ensure your safety and the safety of other workers.

- Every electrical cord should have an Underwriters Laboratory (UL) label affixed.
- Check the UL label for specific amperage. Use an extension cord that will supply the correct amperage per the manufacturer's instructions. See chart below.

Maximum Amperage for Extension Cords

Copper Wire Size (AWG)	2-Wire and 3-Wire Cords (Single Phase)
10	30A
12	25A
14	18A
16	13A

- A cord set not marked for outdoor use is to be used indoors only. Check the UL label on the cord for an outdoor marking.
- Do not remove, bend, or modify any metal prongs or pins of an electrical cord.
- Extension cords used with portable tools and equipment must be three-wire type and designated for hard or extra-hard use. Check the UL label for the cord's use designation.
- Avoid overheating an electrical cord. Make sure the cord is uncoiled, and that is does not run under any covering materials, such as tarps, insulation rolls, or lumber.
- Do not run a cord through doorways or through holes in ceilings, walls, and floors, which might pinch the cord. Also, check to see that there are no sharp corners along the cord's path. Any of these situations could lead to cord damage.
- Extension cords are a tripping hazard. They should never be left unattended and should always be put away when not in use.

the stock. The saw has a baseplate that rests on the material being cut and can be adjusted to change the depth of the cut or to make bevel cuts ranging from 0 to 45 degrees. The saw is started by pressing the trigger in the handle and stopped by releasing it. It will run only while the trigger is pressed. To make ripping of narrower boards easier, most saws can be equipped with an adjustable rip fence that is attached to the saw baseplate in slots in front of the blade.

> **WARNING!**
> When using a circular saw to cut on an angle, the blade guard may have the tendency to bind. Do not reach below the saw to release the guard. Severe injury, such as lost fingers, could result. If you cannot complete the cut, find an alternative saw to finish the job. Never jeopardize your safety for any reason.

Rules for the safe use of a portable circular saw include, but are not limited to the following:

- Always wear safety glasses and other appropriate PPE.
- Hold the saw firmly.
- Always start the saw before making contact with the stock.
- Keep the electric cord clear of the blade to avoid cutting it.
- Check the condition of the blade and be sure it is secure before starting the saw. Never use a dull saw blade; dull blades are more dangerous than sharp ones.
- Be sure the blade guards are in place and working properly.
- Set the blade only deep enough to cut through the stock.
- Check the stock for nails and any other metal before cutting it.
- Allow the saw to reach maximum speed before starting the cut.
- Keep hands clear of the blade.
- If the blade binds in a cut, stop the saw immediately and hold it in position until the blade stops. Then, back the blade out of the cut with the saw stopped.
- Stop the saw and lay it on its side after finishing the cut.
- Do not hold stock in your hands while ripping. Secure the material to sawhorses.
- Use saw blades only for their intended application.
- Ensure the saw blade is placed on the arbor correctly and that the teeth are facing in the proper direction.

> **Circular Saw Brakes**
>
> Most new models of circular saws are equipped with electric brakes. An electric brake stops the circular saw motor in approximately two seconds after the trigger is released. Saws without electric brakes take about 10 to 15 seconds to stop. An electric brake works by the principle of reversing the flow of electricity to the saw motor.

> **WARNING!**
> Always unplug any power saw before changing the blade.
>
> Circular saws require frequent blade changes, either to replace a worn-out blade or to change the blade for a different kind of work. Always refer to the operator's manual for information on the safe and proper procedure for changing a saw blade.

2.2.2 Safely Using and Maintaining Portable Table Saws

Portable table saws (*Figure 13*) are used on job sites to do the following sawing and related woodworking tasks:

- Ripping
- Crosscutting
- Mitering
- Rabbeting (making rabbet cuts)
- Dadoing (making dados)
- Cutting molding

The table saw consists of a machined flat metal table surface with a slotted table insert in the center through which a circular saw blade extends. The table provides support for the stock as it is fed into the rotating blade. The size of a table saw is determined by the diameter of the largest-size circular saw blade it can use, with 10" and 12" diameters being common. The depth of the cut is adjusted by moving the blade vertically up or down. The blade can also be tilted to make beveled cuts up to 45 degrees, as indicated on a related degree scale.

For crosscutting, the table is grooved to accept a guide called a miter gauge. The stock is held against the vertical face of the miter gauge to provide support and prevent the stock from twisting when being fed through the blade. Pushing the miter gauge forward moves the stock through and past the blade so that each cut is flat and square. Typically, the miter gauge can be turned to any angle up to 60 degrees, with positive stops at 0 and 45 degrees right and left for mitering.

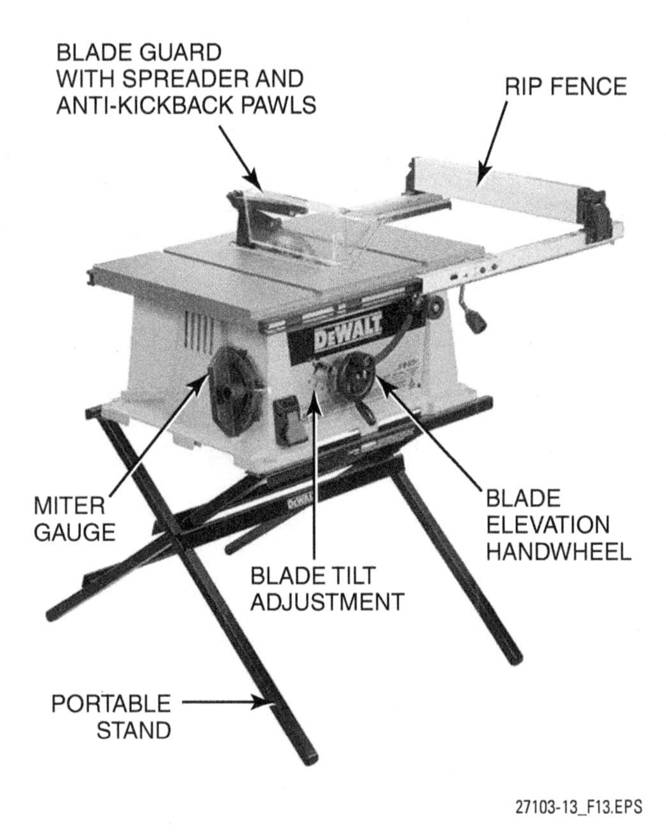

Figure 13 Portable table saw.

For ripping, a rip guide or fence must be used to support the stock. The fence or guide clamps to the front and rear edges of the table, and has rails that allow it to be moved toward or away from the blade while remaining parallel. The fence or guide is positioned to the width desired for the work, then locked in place. The stock is pushed against the fence while making the ripping cut as it is being pushed through the blade.

A blade guard protects the user's hands from the rotating blade and prevents flying chips. As stock is being cut, a kerf is created. At the rear of the blade guard assembly are a kerf spreader and an antikickback device. The kerf spreader keeps the kerf in the stock open behind the blade. If the kerf is allowed to close, the blade can become pinched, causing the stock to abruptly move backward while cutting and resulting in possible injury to the operator. If kickback starts to occur, sharp pawls on the antikickback device grip the stock and hold it in place. Rules for the safe use of a table saw include, but are not limited to, the following:

- Keep the guard over the blade while the saw is being used.
- Do not stand directly in line with the blade.
- Make sure that the blade does not project more than ⅛" above the stock being cut.
- Never reach across the saw blade.
- Use a pushstick for ripping all stock less than 4" wide.
- Disconnect power when changing blades or performing maintenance.
- Never adjust the fence or other accessories until the saw has stopped.
- Enlist a helper or use a work support when cutting long or wide stock. The weight of the stock may cause the table to lean.
- Before cutting stock to length using a miter gauge and rip fence, clamp on a step block for clearance.
- Do not rip without a fence, or crosscut without a miter gauge.
- When the blade is tilted, be sure that the blade will clear the stock and the blade cover plate before turning on the machine.
- Be sure that the stock has a straight edge before ripping.
- When using a dado or molding head, take extra care to hold the stock firmly.
- Never remove scraps from the saw table with your hands while the saw blade is running.
- Use the proper blade for the job being done.

2.2.3 Safely Using and Maintaining Power Miter Saws/Compound Miter Saws

Power miter saws and compound miter saws (*Figure 14*) are also commonly called power miter boxes. These saws combine a miter box or table with a circular saw, allowing it to make accurate 90-degree and miter cuts. The saw blade pivots horizontally from the rear of the table and locks in position to cut angles from 0 to 45 degrees right and left. Stops are set for common angles.

The difference between the power miter saw and compound miter saw is that the blade on a compound miter saw can be tilted vertically, allowing the saw to be used to make a compound cut (combined bevel and miter cut). Some compound miter saws also have a sliding pivot, which

Table Saw Kickback

When making cuts on a table saw, particularly crosscuts and dado cuts, the saw blade may get pinched in the piece being cut. This could cause the stock to kick back, resulting in an injury. The antikickback device (discussed elsewhere) is designed to prevent a kickback if one should occur. When ripping stock, it's a good idea to use a featherboard to guide the stock when feeding it into the saw blade. The fingers of a featherboard also assist in preventing kickback.

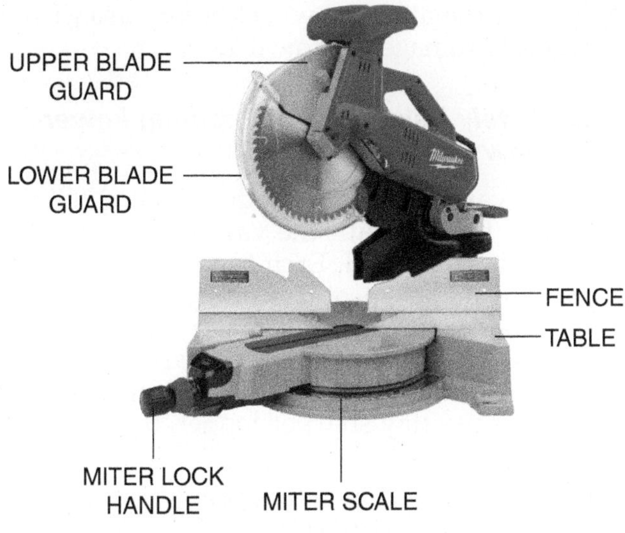

Figure 14 Compound miter saw.

allows wider stock to be cut. Sliding compound miter saws combine the features of a radial arm saw and a compound miter saw. The table rotates with the track arm, and the saw head can be adjusted to make crosscuts, straight miter cuts, bevel cuts, and compound miter cuts. This type of saw is ideal for cutting crown molding.

Rules for the safe use of a power miter saw/ compound miter saw are:

- Always check the condition of the blade and be sure the blade is secure before starting the saw.
- Keep your fingers clear of the blade.
- Never make adjustments while the saw is running.
- Never reach around or across the blade.
- Never leave a saw until the blade stops.
- Ensure the saw is sitting on a level and firm base and is properly fastened to the base.
- Be sure the saw is locked securely at the correct angle before turning on the saw.
- Do not attempt to cut oversized material.
- Turn off the saw immediately after making the cut and use the brake to stop the blade.
- Enlist a helper to support the end of long stock.
- Be sure the blade guards are in place and working properly.
- Allow the saw blade to reach its maximum speed before starting the cut.
- Hold the stock firmly against the fence when making the cut.

2.2.4 Safely Using and Maintaining Abrasive Saws

Abrasive saws use a special wheel that can slice through either metal or masonry. The most common types of abrasive saws are the demolition saw and the chop saw. The main difference between the two is that the demolition saw is not mounted on a base.

Demolition saws (*Figure 15*) operate on either electricity or gasoline. These saws are effective for cutting through most materials found at a construction site. Although all saws are potentially hazardous, the demolition saw is a particularly dangerous tool that requires the full attention and concentration of the operator.

CAUTION

When using an abrasive saw (demolition saw or chop saw), be sure to use the appropriate eye, ear, face, and hand personal protective equipment. When cut with these saws, metal will throw hot sparks or small pieces of molten metal. Ensure that ear muffs, a full-face shield, and gloves are worn when using abrasive saws. In addition, ensure that sparks or hot metal will not come into contact with flammable or combustible materials. Also, if you have long hair, be sure to tie it back or cover it properly.

WARNING!

Gas-powered demolition saws must be handled with caution. Using a gas-powered saw where there is little ventilation can cause carbon monoxide poisoning. Carbon monoxide gas is hard to detect, and it is deadly! Use fans to circulate air, and have a trained person monitor air quality.

SawStop®

The SawStop® safety system is an advanced table saw safety device that immediately stops a rotating blade when a person contacts the blade. The SawStop® system applies a small electrical signal to the blade and then monitors the signal for changes. The human body has relatively good conduction properties. When a person contacts the moving blade, the electrical signal drops quickly. (Wood has relatively low conduction properties and does not affect the electrical signal.) When contact with the blade occurs, a fast-acting brake stops the blade by forcing a block of aluminum into the path of the blade. All of this occurs within 1⁄200th of a second.

> **WARNING!**
> Because a demolition saw is handheld, there is a tendency to use it in positions that can easily compromise your balance and footing. Be very careful not to use the saw in any awkward positions.

A chop saw is a versatile and accurate tool that is similar to a power miter saw (*Figure 15*), and is commonly used to cut steel framing members. A chop saw uses an abrasive wheel to cut materials similar to demolition saws. A chop saw is a better choice than a demolition saw to obtain an exact square or angled cut. Metal pipe, channel, tubing, conduit, or metal materials can be cut with a chop saw using an abrasive wheel. Chop saws can also be fitted with blades to cut other materials such as polyvinyl chloride (PVC) or wood. A vise on the base of the chop saw holds the material securely and pivots to allow miter cuts. Some can be pivoted past 45 degrees in either direction.

Chop saws are sized according to the diameter of the largest abrasive wheel they accept. Two common sizes are 12-inch and 14-inch wheels. Each wheel has a maximum safe speed. Never exceed that speed. Typically the maximum safe speeds are 5,000 revolutions per minute (rpm) for 12-inch wheels and 4,350 rpm for 14-inch wheels.

Abrasive saws require extreme care during use. Follow these guidelines when using an abrasive saw:

- Wear personal protective equipment to protect your eyes, ears, face, and hands.
- Use abrasive wheels that are rated for a higher rpm than the tool can produce. This way, the maximum safe speed for the wheel cannot be exceeded. The blades are matched to a safety rating based on their rpm.
- Be sure the abrasive wheel is properly secured on the arbor before using a demolition saw. Point the wheel of a demolition saw away from yourself and others before starting the saw.
- Use two hands when operating a demolition saw.
- Secure the materials being cut in a vise. The jaws of the vise hold the pipe, tube, or other material firmly and prevent it from turning while being cut.
- Be sure the guard is in place and the adjustable shoe is secured before using a demolition saw.
- Keep the saw and the blades clean.
- Inspect the saw before use. Never operate a damaged saw. Ask your supervisor if you have a question about the condition of a saw.
- Inspect the abrasive wheel before use. If the wheel is damaged, throw it away.

2.2.5 Safely Using and Maintaining Power Saw Blades

A wide variety of saw blades (*Figure 16*) are available for use with the various power saws previously discussed. Each type of blade is designed to make an optimum cut in different types and/or densities of material. Generally, standard high-speed steel or carbide-tipped blades are used with most power saws. Carbide-tipped blades stay sharper longer, but they are

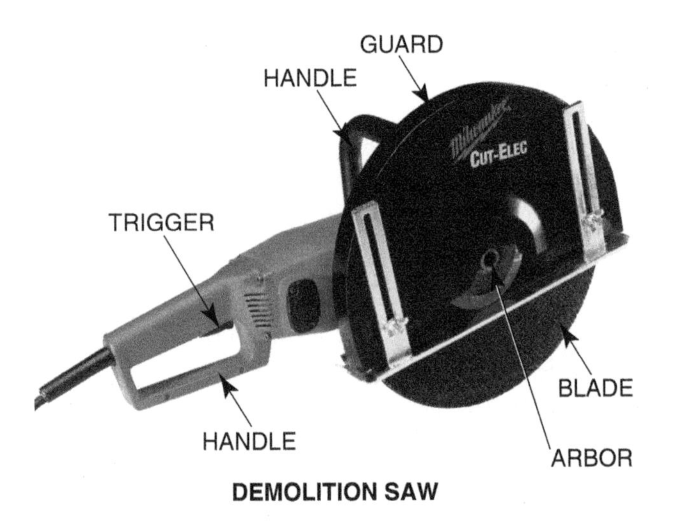

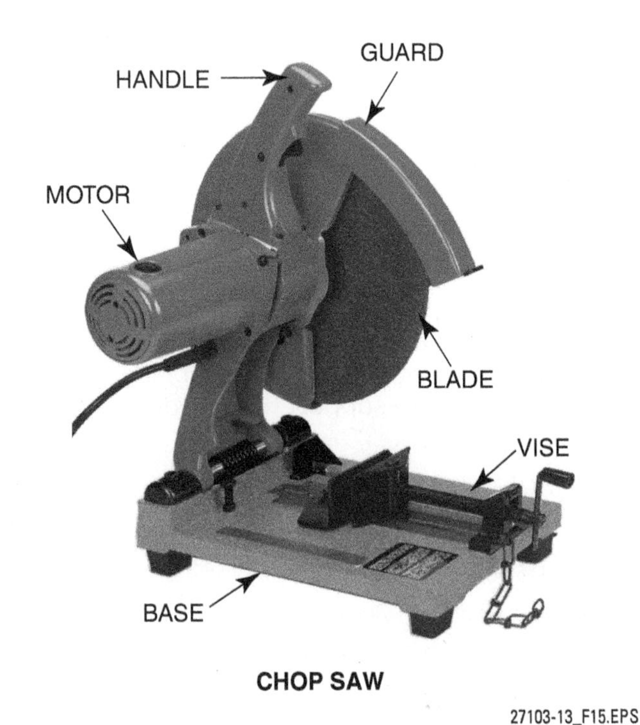

Figure 15 Abrasive saws.

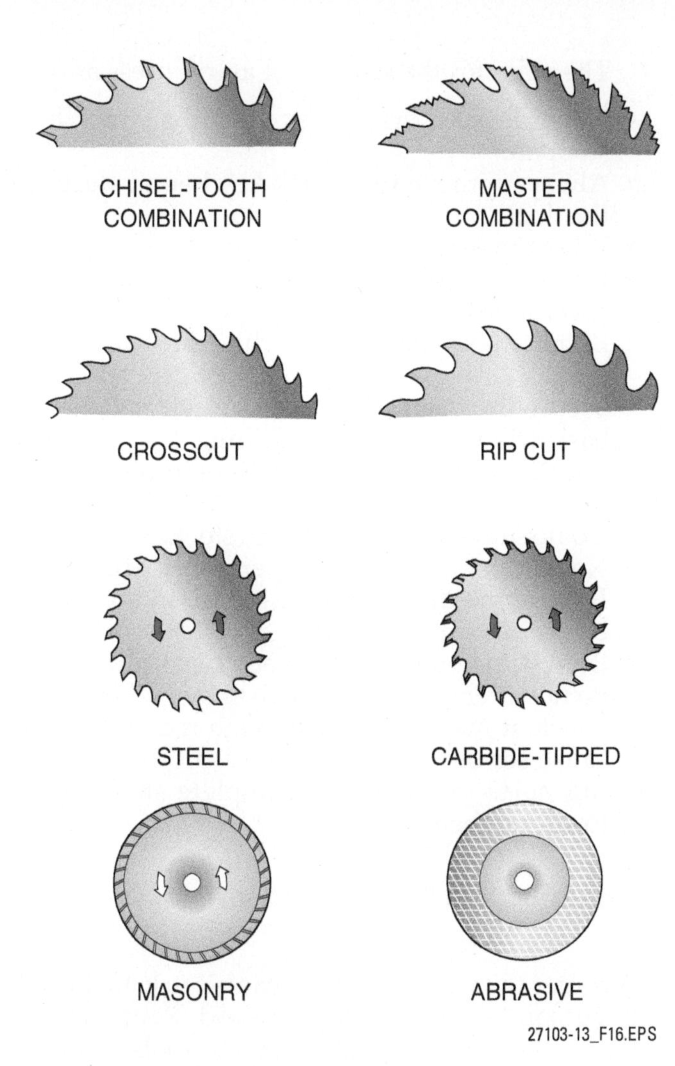

Figure 16 Circular saw blades.

more brittle and can be damaged if improperly handled. The number of teeth on a blade, the grind of each tooth, and the space between the teeth (gullet depth) determines the smoothness and speed of the cut.

Crosscutting requires the severing of wood fibers, whereas during ripping, the teeth must gouge out lengths of wood fiber. Blades made specifically for crosscutting have small teeth that are alternately beveled across their front edges. The gullets are small because only fine sawdust is produced. Blades made specifically for ripping have large, scraper-like teeth with deep, curved gullets.

When both types of teeth are combined in a combination blade, the scraper, called a raker, has a much deeper gullet than the crosscut teeth. In chisel-tooth combination blades, the teeth and gullets are a compromise between rip and crosscut designs. They can be used for both types of cutting, but they cut less smoothly than blades designed to do just one job.

Blades designed for cutting plywood and other easily splintered or chipped materials have small, scissor-like teeth. The smoothness of the cut is determined by the number of teeth and their sharpness.

Use a blade recommended by the blade manufacturer for the type of material being cut. Always make sure that the blade diameter, arbor hole size, and maximum rotation speed are compatible with the saw on which it is to be used. Manufacturers have different names for similar circular saw blades. Some types of circular saw blades widely used in carpentry include the following:

- *Chisel-tooth combination blade (24 to 30 teeth)* – This is an all-purpose blade that is free cutting in any direction. It can be used for crosscutting, ripping, mitering, and general-purpose work.
- *Master combination blade (48 to 60 teeth)* – This is an all-purpose blade used for smooth, fast cutting. It can be used for crosscutting, ripping, mitering, and general-purpose work.
- *Combination rip blade (36 to 44 teeth)* – This is an all-purpose blade used for crosscutting, ripping, mitering, and general-purpose work.
- *Carbide-tipped blade* – Most of the combination blades previously described are available with carbide-tipped teeth. These blades are popular because of their economy and quality.
- *Hollow-ground plane (miter) blade (48 to 60 teeth)* – Hollow-ground plane blades are used to make smooth finish cuts on all solid woods and are ideal for cabinet work. They can be used for crosscutting, ripping, mitering, and finish work.

2.3.0 Safely Using and Maintaining Drill Presses

Drill presses consist of an electric drill and work table combined into one unit (*Figure 17*). Floor and benchtop drill presses are common for shop work,

How to Tell When Saw Blade Is Dull

Keep your saw blades sharp in order to maintain clean, consistent cuts. There are several things that will indicate when a circular saw blade starts to dull. First, a subtle change in the pitch and sound of the saw motor may be noticed because the motor has to work harder to compensate for the dullness. Second, crosscuts will become increasingly ragged (feathered). Third, smoke may come from the cuts as a result of the increased friction and heat from the saw blade. Sharpen or replace blades as often as needed.

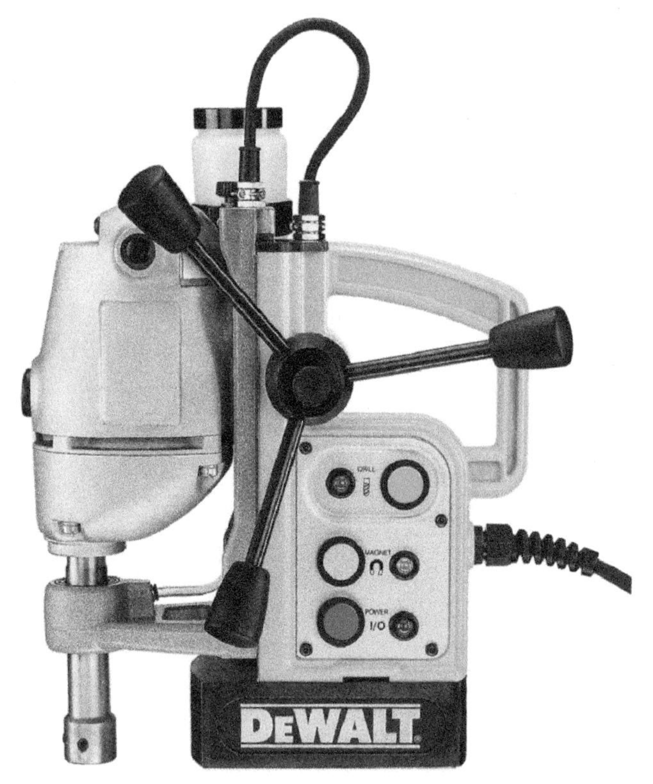

Figure 17 Magnetic drill press.

while electromagnetic-base drill presses are frequently used on a job site. The throat capacity of a drill press (the distance between its rear post and the center of the bit) determines the maximum size of the stock it can accept.

On floor and benchtop models, the drill table can be raised and lowered. On some models, the table can be tilted for drilling angled holes. The motor mounts on an adjustable bracket and is controlled by a belt-and-pulley drive. The belt is shifted between different grooves on the pulley to vary the speed of drilling. A chart is usually mounted on the drill press that gives guidelines for the approximate speed to use for different drilling operations. On electromagnetic-base drill presses, the base is placed firmly against a steel or iron surface and the magnet is engaged. The surface can be either vertical or horizontal.

For many models of drill presses, drill bits are tightened into or loosened from the chuck using a chuck key. Some newer models have keyless chucks, meaning that a chuck key is not required for tightening the drill chucks. For drilling, the drill is lowered into the stock by means of a feed handle, which returns to its original position when released. Operations such as sanding and mortising with a drill press require the use of special accessories made for that purpose.

The rules for the safe use of a drill press are:

- Never leave a drill press until the chuck has stopped.
- Always remove the chuck key before starting the drill.
- Wear proper PPE.
- If you have long hair, be sure to tie it back or cover it properly.
- Do not make adjustments while the drill is running.
- Secure the magnetic lock (for electromagnetic-base drill presses) before drilling.
- Use the correct speed for the job being performed.
- Do not force the drill into the stock.
- Never wear gloves when operating a drill press.
- Be sure the drill or tool is secure in the chuck before starting the drill.
- Even if the power has already been turned off, do not attempt to stop the drill by grabbing the chuck; it may still have enough momentum to cause serious injury. Always wait for all moving parts to come to a complete stop before touching them.

2.4.0 Safely Using and Maintaining Routers/Laminate Trimmers

A router (*Figure 18*) is a portable electric tool used to cut joints and patterns in wood. Routers have high-speed motors that allow the tools to make clean, smooth cuts. The motor shaft powers a chuck (collet) in which a wide variety of specially designed router bits and cutters can be installed. The motor's vertical position adjusts up or down in order to change the depth of the cut.

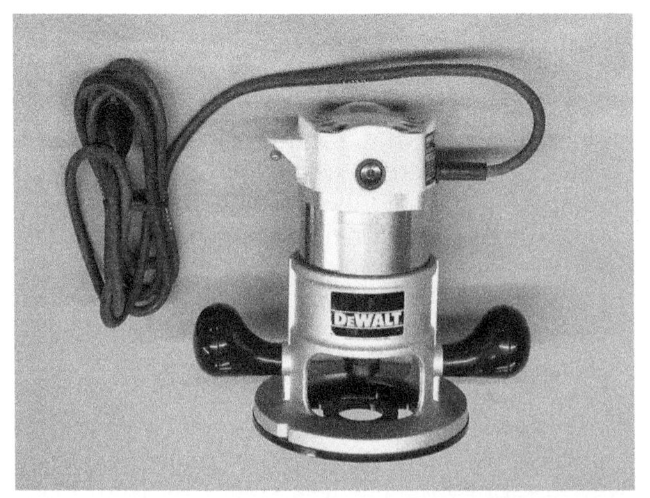

Figure 18 Router.

Router bits and cutters are made of carbide-tipped steel or high-speed steel. Carbide-tipped bits are preferred for most router applications because of durability. With different bits and available accessories, the router can be used to make many different types of cuts, including the following:

- Shaping edges
- Rabbeting
- Beveling
- Dovetailing
- Dadoing, fluting, and reeding
- Mortising

A laminate trimmer (*Figure 19*) is a specialized type of router used to trim and shape the edges of plastic laminate materials, such as those used for countertops. It can be used for bevel trimming, flush trimming, and cutting obtuse and acute angles.

The following are rules for the safe use of a router or laminate trimmer:

- Always install bits and cutters with the router or trimmer unplugged.
- Keep your hands clear of router bits.
- Hold the tool firmly.
- Be sure that the material is secured properly before routing.
- Allow the bit to contact the stock gently.
- Whenever possible, use jigs and guides with the router to ensure straight cuts.
- Turn off the tool immediately after making the cut, then wait for the bit to stop before putting the tool down. Place the router or laminate trimmer on its side so as not to dull the bit or cutter.

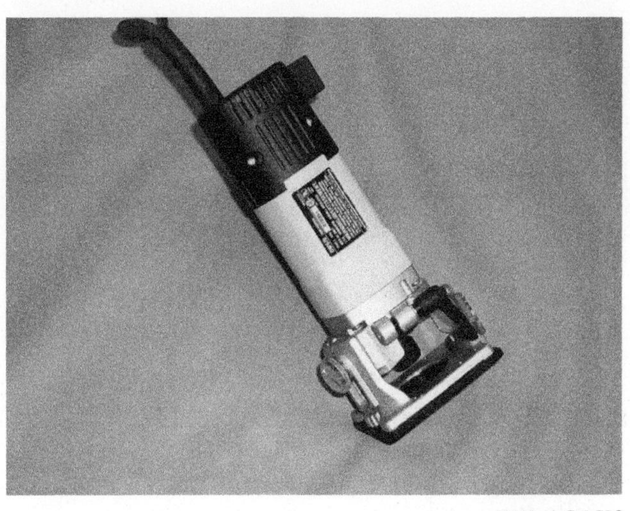

27103-13_F19.EPS

Figure 19 Laminate trimmer.

- Avoid making cuts too deep. Note that a deep cut can cause kickbacks.

Routing with a Straightedge

There are several ways to limit the amount of lateral motion while using a router. One very effective way is to use a straightedge. A clamp holds the straightedge to the stock and provides a solid guide for the carpenter. When making the cut, keep the base of the router firmly in contact with the straightedge. The result will be a cut that is accurate and straight.

Router Bits

Using a router on certain materials will cause excessive bit wear if the proper technique is not used. Particleboard, for example, will quickly wear out a router bit if a deep cut is attempted in one pass. To avoid this type of wear, make a deep cut through a series of shallower cuts. Problems may also be encountered if the correct bit isn't selected for the material being cut. Ensure the correct bit is being used for the application. The router shown here is being used to make a rabbet cut.

27103-13_SA05.EPS

Router Table

A router table consists of a flat table with a router attached to the underside of the table. The table holds the router securely and allows a carpenter to keep both hands on the workpiece.

2.5.0 Safely Using and Maintaining Portable Power Planes

Portable power planes (*Figure 20*) are excellent for fitting trim and framing members that will be fastened together. They are commonly used for the following tasks:

- Straightening
- Edge planing
- Chamfering
- Beveling

Smaller-block power planes are used for finish work and jointer planes are used for heavier work. Jointer planes can be equipped with adjustable fences to cut bevels or chamfers.

Rules for the safe use of power planes are:

- Hold the plane firmly with both hands.
- Be sure the material is secure before planing.
- Always use a door jack or other means to secure the door when planing door edges.

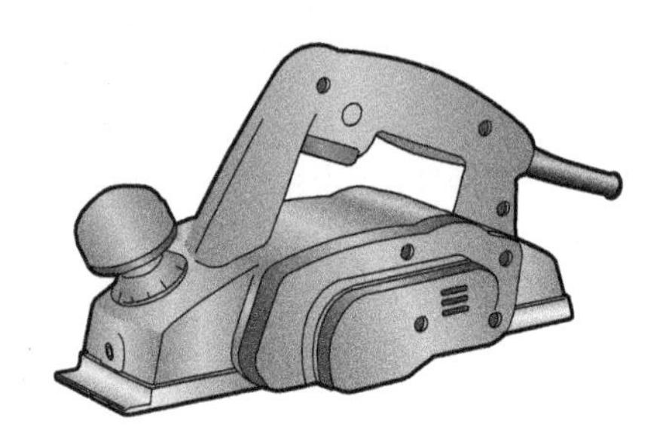

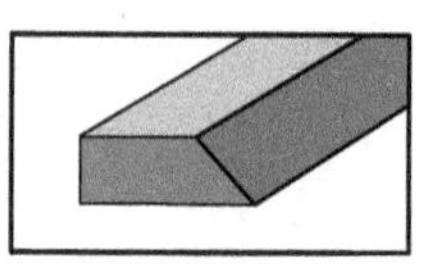

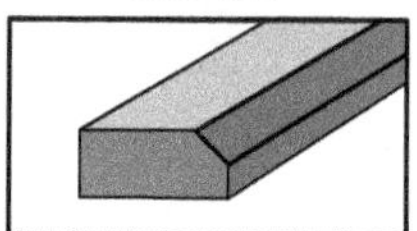

Figure 20 Power plane.

- Remove only small amounts at one time; make multiple passes for deeper cuts.

2.6.0 Safely Using and Maintaining Power Metal Shears

Power metal shears (*Figure 21*) are used to make burr-free cuts in sheet metal, metal strips, special lightweight materials, and steel studs. They can cut straight lines, tight right and left curves, and round, square, and irregularly shaped holes. A trigger-operated control allows variable-speed operation. Some models have a head that can swivel 360 degrees. Common applications for power metal shears are light-gauge metal material such as chimney and roof flashing, making openings in ductwork, trimming metal roof panels, and clipping strapping and stud hangers.

Cordless Power Tools

Today's cordless tools offer both substantial power capability and convenience. Once limited to only a few types, they are now available to serve a wide variety of applications. The battery life must be considered when selecting a cordless tool. Some materials, such as pressure-treated lumber and hardwoods, are hard to cut and drill, thereby reducing the battery life for a given charge.

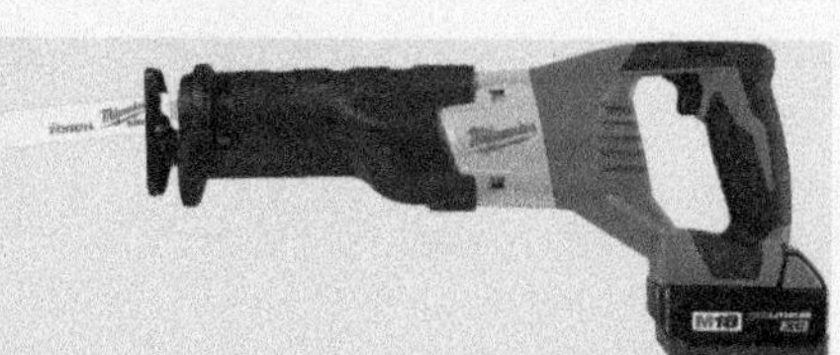

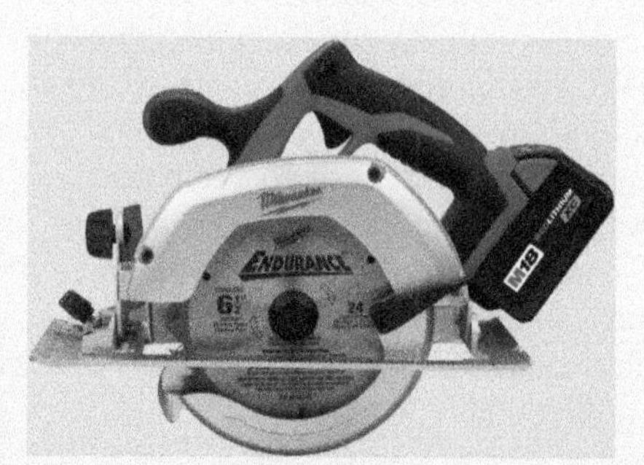

27103-13_F21.EPS

Figure 21 Power metal shears.

Rules for the safe use of power metal shears are:

- Keep your fingers away from the sharp edges of the tool and stock being cut.
- Never force the tool into the material.
- Do not attempt to cut material that is too thick.
- Keep the electric cord clear of the blade.

2.7.0 Safely Using and Maintaining Pneumatic/Cordless Nailers and Staplers

Pneumatic nailers and staplers are fastening tools (*Figure 22*) powered by compressed air, which is fed to the tool through an air hose connected to an air compressor. These tools, known as nail or staple guns, are widely used for quick, efficient fastening of framing, subflooring, or sheathing. Nailers and staplers are available in a variety of sizes to serve different purposes.

Under some conditions, staples have advantages over nails. For example, staples do not split wood as easily as a nail when driven near the end of a board. Staples are also excellent for fastening sheathing, shingles, building paper, and other materials because their two-legged design covers more surface area. However, both tools are sometimes used to accomplish the same fastening jobs. Nailers are typically used for the following:

- Applying sheathing
- Applying decking
- Applying roofing
- Installing framing
- Installing finish work
- Constructing cabinets

Staplers are typically used for the following:

- Applying sheathing
- Applying decking
- Applying roofing
- Installing insulation
- Installing ceiling tile
- Installing paneling

For most models of pneumatic nailers or staplers, the nails or staples are available in strips and are loaded into a magazine, which typically holds 100 or more fasteners. Some tools have an angled magazine, which makes it easier for the tool to fit into tight places. Coil-fed models typically use a coil of 300 nails loaded into a circular magazine. Lightweight nailers can handle tiny finishing nails. Larger framing nailers can shoot smooth-shank nails of up to 3¼" in length.

WARNING!

Pneumatic nailers and staplers are extremely dangerous. They are just as dangerous and deadly as a firearm. Never point these tools at anyone or in anyone's direction. When nailing or stapling, keep your hands away from the contact area.

Rules for the safe use of pneumatic nailers and staplers are:

- Be sure all safety devices are in good condition and are functioning properly.
- Use the pressure specified by the manufacturer.
- Never disengage the firing safety mechanism.
- Always assume that the fasteners are loaded into the tool.
- Never point a pneumatic fastener at yourself or anyone else.
- If someone is using these tools improperly, report it to your instructor or supervisor.
- Be sure the tool is disconnected from the power source before making adjustments or repairs.
- Use caution when attaching the nailer or stapler to the air supply because the fastener may discharge.
- Never leave a pneumatic tool unattended while it is still connected to the power source.
- Use nailers and staplers only for the type of work for which they were intended.
- Use only nails and staples designed for the tool being used.
- Never use fasteners on soft or thin materials that nails may completely penetrate.

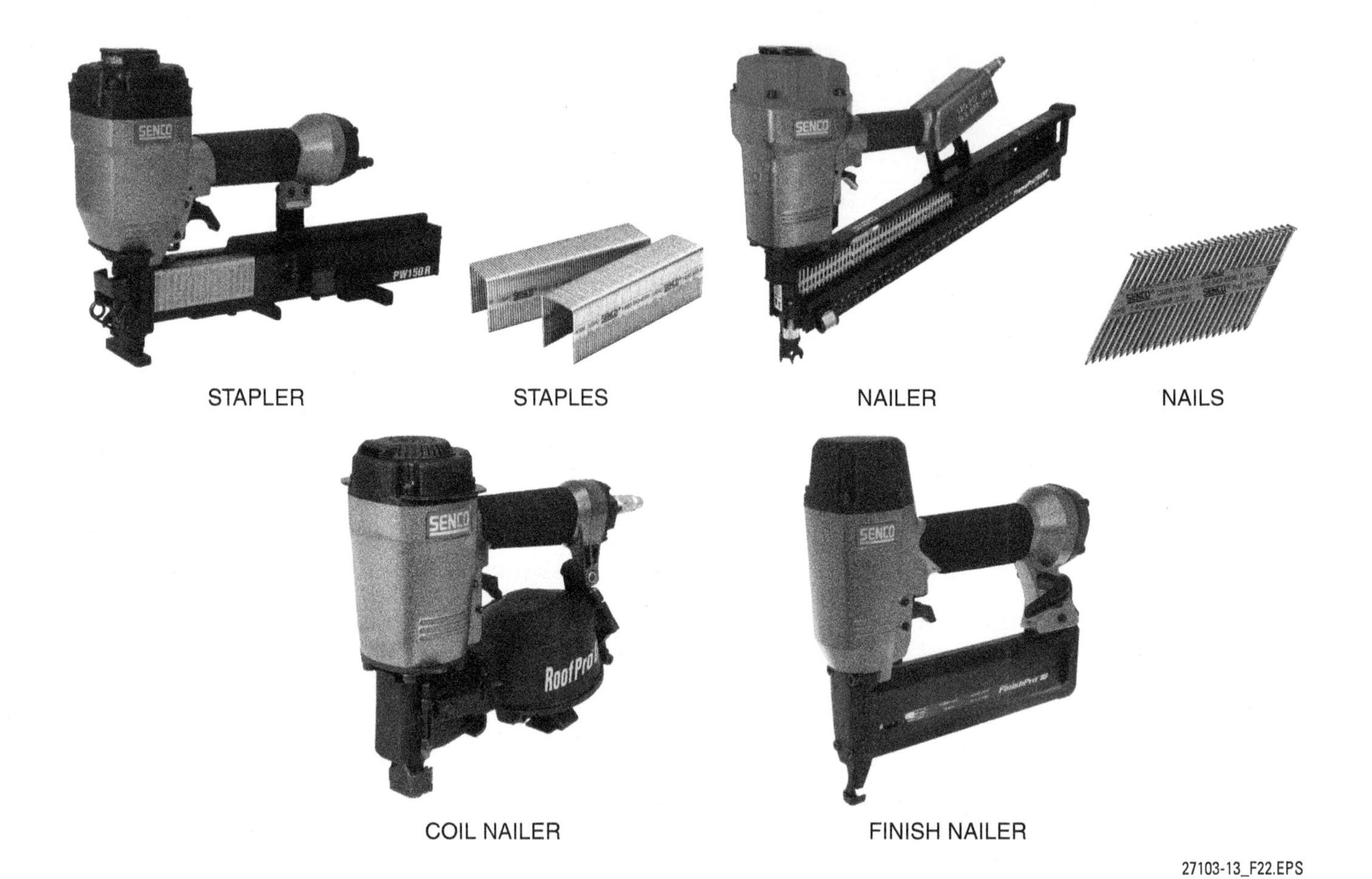

Figure 22 Pneumatic nailers and stapler.

Nailers and staplers are also available in cordless models (*Figure 23*). Cordless nailers and staplers use a tiny internal combustion engine powered by a disposable fuel cell and/or a rechargeable battery. The action of the piston drives the fastener into the stock. A cordless stapler can drive about 2,500 staples with one fuel cell. A cordless framing nailer can drive about 1,200 nails on one fuel cell. The battery on a cordless tool must be periodically recharged. It pays to have a spare battery to use while one is being charged. Rules for the safe operation of a cordless nailer or stapler are basically the same as those described previously for pneumatic nailers and staplers.

27103-13_F23.EPS

Figure 23 Cordless finish nailer.

2.0.0 Section Review

1. A power tool should be disconnected before performing maintenance on it.
   a. True
   b. False

2. The size of a circular saw is determined by the _____.
   a. length of the baseplate
   b. diameter of the largest blade that can be used
   c. diameter of the electrical cord
   d. throat capacity

3. The maximum size of stock that a drill press can accept is determined by the _____.
   a. length of the baseplate
   b. diameter of the largest blade that can be used
   c. diameter of the electrical cord
   d. throat capacity

4. When routing plastic laminates and plywood, it is recommended that you use a(n) _____.
   a. high-speed steel bit
   b. aluminum bit
   c. carbide-tipped bit
   d. drill bit

5. When cutting bevels or chamfers with a jointer plane, use a(n) _____.
   a. T-bevel
   b. adjustable fence
   c. miter gauge
   d. featherboard

6. Which of the following materials should *not* be cut with power metal shears?
   a. Sheet metal
   b. Steel studs
   c. Metal strips
   d. Steel I-beams

7. Pneumatic nailers and staplers are powered by _____.
   a. electricity
   b. propane
   c. compressed air
   d. hydraulics

SUMMARY

Most of the hand and power tools a carpenter will use on the job were covered in the *Core Curriculum*. This module introduced some additional tools and reviewed some of the main tools used by carpenters, such as the circular saw.

Knowing how to choose the proper tool for the job, use it skillfully, and keep it in good condition are essential carpentry skills. The use of power tools allows more work to be done in a shorter period of time. However, these tools can cause serious injury if operated incorrectly and/or unsafely. The carpenter must always be alert to the dangers of operating power tools and must follow applicable safety rules without fail when using power tools. In addition to the safety rules covered in this module, be sure to check the manufacturer's literature for the safety precautions that apply to the specific piece of equipment.

Review Questions

1. A tool that consists of a glass tube in a sleeve with hooks on each end is called a _____.
   a. builder's level
   b. line level
   c. transit level
   d. water level

2. Which type of square can be used to check only right angles?
   a. Framing square
   b. Combination square
   c. Speed Square™
   d. Try square

3. A large hand plane used for rough work is the _____.
   a. jack plane
   b. block plane
   c. smoothing plane
   d. jointer plane

4. After using a grinding wheel to sharpen a plane iron, the edge must be dressed using a(n) _____.
   a. leather strop
   b. oilstone
   c. mill file
   d. soapstone

5. The clamp whose jaws can be adjusted independently of one another is the _____.
   a. quick clamp
   b. pipe clamp
   c. hand-screw clamp
   d. locking C-clamp

6. The handsaw that would be used to make a scroll cut on a trim piece is the _____.
   a. dovetail saw
   b. back saw
   c. coping saw
   d. ripsaw

7. The saw blade should be held at about a 60-degree angle to the wood when cutting with a _____.
   a. hacksaw
   b. ripsaw
   c. miter saw
   d. crosscut saw

8. Before performing maintenance or changing accessories on any power tool, be sure that it is _____.
   a. not running
   b. OSHA-approved
   c. checked by your supervisor
   d. disconnected

9. A small extension cord can be used for any type of power tool.
   a. True
   b. False

10. If a power tool is making unusual noises, sounds, or vibrations, _____.
    a. clean and oil it
    b. ask another student to trade tools with you
    c. report the problem to your supervisor or instructor
    d. ignore it and keep working

11. Which of the following operations is not typically done with a portable table saw?
    a. Dadoing
    b. Ripping
    c. Mitering
    d. Mortising

12. When ripping stock with a portable table saw, the blade should not project above the material being cut by more than _____
    a. 1⁄16"
    b. 1⁄8"
    c. 1⁄4"
    d. 3⁄8"

13. Before starting a drill press, always _____.
    a. remove the chuck key
    b. lower the feed handle
    c. insert the chuck key
    d. remove the belt from the pulley

14. A laminate trimmer is a specialized type of _____.
    a. portable table saw
    b. power plane
    c. router
    d. shaper

15. A pneumatic nailer or stapler is powered by _____.
    a. compressed air
    b. batteries
    c. a small fuel cell
    d. kinetic energy

Trade Terms Quiz

Fill in the blank with the correct term that you learned from your study of this module.

1. One of the hazards associated with a table saw is the possibility of ________ as the stock gets pinched as it travels through the saw blade.
2. The ________ should be sharpened on a bench grinder to ensure smooth cutting.
3. When cutting trim molding around a door frame, you would most likely be performing ________.
4. If a long board has an imperfect edge, a jointer plane can be used to make it ________.
5. ________ are formed when the tongue of a workpiece is fit into a rectangular cutout or opening on an adjoining piece.
6. Use a(n) ________ to hold a door safely on edge when planing it.
7. A cut that is made across the sloping edge or side of a workpiece at an angle of less than 90 degrees is known as a(n) ________.
8. The installation of a strike plate and hinges in a door jamb would require ________.
9. When measuring, account for the amount of material removed by the saw blade, which is known as the ________.
10. A(n) ________ is a triangular joint often used to attach drawer fronts to the sides in antique furniture.
11. A cut made across the grain in lumber is known as a(n) ________.
12. A rectangular cut made along the edge or end of a board to receive another board is known as a(n) ________.
13. A(n) ________ can be used to cut lumber at precise angles.
14. The installation of crown molding often requires that wood be cut with both a bevel and a miter, which is known as a(n) ________.
15. For a cabinet drawer, ________ should be cut in the sides, front, and back to receive the drawer bottom.

Trade Terms

Bevel cut
Compound cut
Crosscut
Dados
Door jack
Dovetail joints
Kerf
Kickback
Miter box
Miter cut
Mortise and tenon joints
Plane iron
Pocket (plunge) cuts
Rabbet cuts
True

Cornerstone of Craftsmanship

Keith Bennett

Adjunct Professor
Santa Fe College
Gainesville, Florida

How did you get started in the construction industry?
My love for construction (building things) dates back to when I was young, when my brother and I would build forts in the woods. In high school, I enrolled in shop class and realized that I loved working with wood and tools. I took as many years in woodworking class as the school allowed; I even worked as a shop assistant for one semester. I started gaining a sense of accomplishment when looking at whatever I just created.

Who or what inspired you to enter the Industry?
No one really ever suggested that I work in the construction industry. I believe my inspiration was my love for building…creating and knowing that what I just accomplished and no one can take it from me. It was a real sense of pride. Later, I did find out that one of my uncles was a carpenter and that I was somehow following a family tradition.

What do you enjoy most about your career?
I enjoy passing on the knowledge that I have obtained to young men and women who are up-and-coming in the construction field. I also still love to drive around my hometown and see the houses that I helped to build.

Why do you think training and education are important in construction?
Let me put it like this, if you were going in for an operation, do you want the doctor performing the procedure to be well-educated and who has trained until he/she feels comfortable doing the procedure? The same idea holds for construction. A good carpenter will only become a great carpenter if they are willing to obtain training in the latest and greatest techniques and train until they fully understand them.

Why do you think credentials are important in construction?
As an employer, many people will tell you anything you want to hear when interviewing for a job. However, if someone provides the employer with credentials from an accredited training program, there's no denying the dedication and commitment involved in receiving the credentials. And someone can't take the credentials away from you.

How has training/construction impacted your life and career?
Construction has opened a number of doors for me… many doors that I would have never guessed possible. Back in high school, if someone were to tell me that I would be teaching at a college, I would have probably laughed at them. Taking part in the revision of the NCCER Carpentry curriculum is just one example of the types of doors opened to me.

Would you recommend construction as a career to others? Why?
Yes! The opportunities and the possibilities in construction are endless. The construction industry is one of the biggest career fields that exist. Career opportunities exist in a wide range of disciplines—from carpentry to plumbing. Construction is still one of the only career areas where you can be the owner of your own company.

What does craftsmanship mean to you?
To me, craftsmanship means being the best at what you do. A lot of people can call themselves a carpenter or electrician, but the number of true craftsmen is very limited.

Trade Terms Introduced in This Module

Bevel cut: A cut made across the sloping edge or side of a workpiece at an angle of less than 90 degrees.

Compound cut: A simultaneous bevel and miter cut.

Crosscut: A cut made across the grain in lumber.

Dados: Rectangular grooves that are made part of the way through and across the grain of lumber.

Door jack: A holder or stand used to hold a door on edge while planing, routing, or installing hinges.

Dovetail joints: Interlocking wood joints with a triangular shape like that of a dove's tail.

Kerf: The width of the cut made by a saw blade. It is also the amount of material removed by the blade in a through (complete) cut or the slot that is produced by the blade in a partial cut.

Kickback: A sharp, uncontrolled grabbing and throwing of the workpiece by a tool as it rejects material being forced into it.

Miter box: A device used to cut lumber at precise angles.

Miter cut: A cut made at the end of a piece of lumber at any angle other than 90 degrees.

Mortise and tenon joints: Wood joints in which a rectangular cutout or opening receive the tongue that is cut on the end of an adjoining member.

Plane iron: Flat metal blade that is sharpened on one end to provide a cutting edge for a hand plane.

Pocket (plunge) cuts: A cut made to remove an interior section of a workpiece or stock (such as a sink cutout for a countertop) or to make square or rectangular openings in floors or walls.

Rabbet cuts: Rectangular cuts made along the edge or end of a board to receive another board.

True: To accurately shape adjoining members so they fit well together.

Additional Resources

This module presents thorough resources for task training. The following reference material is suggested for further study.

The Art of Fine Tools. Newtown, CT: Taunton Press, Inc.

Field Guide to Tools. Philadelphia, PA: Quirk Publishing.

Measure Twice, Cut Once. Boston, MA: Little, Brown & Company.

Power Tools. Newtown, CT: Taunton Press, Inc.

Selecting and Using Hand Tools. Newtown, CT: Taunton Press, Inc.

Tools Rare and Ingenious: Celebrating the World's Most Amazing Tools. Newtown, CT: Taunton Press, Inc.

Tricks of the Trade: Jigs, Tools and Other Labor-Saving Devices. Newtown, CT: Taunton Press, Inc.

Black & Decker, **www.blackanddecker.com**

Bosch Tool Corporation, **www.boschtools.com**

Delta Machinery, **www.deltamachinery.com**

DeWalt Industrial Tool Company, **www.dewalt.com**

Irwin Tools, **www.irwin.com**

Makita Tools USA, **www.makita.com**

Milwaukee Electric Tool Company, **www.milwaukeetool.com**

Porter-Cable Corporation, **www.portercable.com**

Ridge Tool Company, **www.ridgid.com**

The Stanley Works, **www.stanleytools.com**

L.S. Starrett Company, **www.starrett.com**

Figure Credits

SkillsUSA, Module opener

Irwin Tools, Figures 1 (carpenter's level), 4 (combination square, miter square, drywall square, framing square, rafter square, speed square), 10 (quick clamp, spring clamp, hand-screw clamp, locking C-clamp, pipe clamp), and 11 (dovetail saw, coping saw)

Bowden Contracting Company, Figures 1 (laser level) and 2

Milwaukee Electric Tool Co., Figures 12, 14, 15 (chop saw), and SA04

Senco Brands Inc., Figure 22

Section Review Answer Key

Answer	Section Reference	Objective Reference
Section One		
1. d	1.1.0	1a
2. d	1.2.0	1b
3. b	1.3.0	1c
4. c	1.4.0	1d
5. a	1.5.0	1e
Section Two		
1. a	2.2.1	2a
2. b	2.2.1	2b
3. d	2.3.0	2c
4. c	2.4.0	2d
5. b	2.5.0	2e
6. d	2.6.0	2f
7. c	2.7.0	2g

NCCER CURRICULA — USER UPDATE

NCCER makes every effort to keep its textbooks up-to-date and free of technical errors. We appreciate your help in this process. If you find an error, a typographical mistake, or an inaccuracy in NCCER's curricula, please fill out this form (or a photocopy), or complete the online form at **www.nccer.org/olf**. Be sure to include the exact module ID number, page number, a detailed description, and your recommended correction. Your input will be brought to the attention of the Authoring Team. Thank you for your assistance.

Instructors – If you have an idea for improving this textbook, or have found that additional materials were necessary to teach this module effectively, please let us know so that we may present your suggestions to the Authoring Team.

NCCER Product Development and Revision

13614 Progress Blvd., Alachua, FL 32615

Email: curriculum@nccer.org
Online: www.nccer.org/olf

❑ Trainee Guide ❑ Lesson Plans ❑ Exam ❑ PowerPoints Other ______________

Craft / Level: Copyright Date:

Module ID Number / Title:

Section Number(s):

Description:

Recommended Correction:

Your Name:

Address:

Email: Phone:

27104-13

Introduction to Construction Drawings, Specifications, and Layout

Overview

Design and construction information for a structure is documented in its drawings and specifications. These documents provide a wide range of information—from where to place the structure on the site to what types of materials to use. Reading and interpreting construction drawings plays a large role in becoming a carpenter. A carpenter's work precedes that of other trades such as electricians, plumbers, and HVAC installers, all of whom need spaces set aside for their equipment, wiring, piping, and ductwork. Construction drawings show how and where to construct these spaces, and the carpenter must make sure it is done right.

Module Four

Trainees with successful module completions may be eligible for credentialing through the NCCER Registry. To learn more, go to **www.nccer.org** or contact us at **1.888.622.3720**. Our website has information on the latest product releases and training, as well as online versions of our *Cornerstone* magazine and Pearson's product catalog.

Your feedback is welcome. You may email your comments to **curriculum@nccer.org,** send general comments and inquiries to **info@nccer.org**, or fill in the User Update form at the back of this module.

This information is general in nature and intended for training purposes only. Actual performance of activities described in this manual requires compliance with all applicable operating, service, maintenance, and safety procedures under the direction of qualified personnel. References in this manual to patented or proprietary devices do not constitute a recommendation of their use.

27104-13 Introduction to Construction Drawings, Specifications, and Layout

Objectives

When you have completed this module, you will be able to do the following:

1. List the types of drawings usually included in a set of plans and describe the information found on each type.
   a. Identify the different types of lines used on construction drawings.
   b. Identify selected architectural symbols commonly used to represent materials on plans.
   c. Identify selected electrical, mechanical, and plumbing symbols commonly used on plans.
   d. Identify selected abbreviations commonly used on plans.
   e. Describe the methods of dimensioning construction drawings.
   f. List the various types of construction drawings and describe each.
2. State the purpose of written specifications.
   a. Describe how specifications are organized.
   b. Explain the importance of building codes in construction.
3. Identify the methods of squaring a building.

Performance Tasks

Under the supervision of the instructor, you should be able to do the following:

1. Read and interpret foundation, floor, and other plan view drawings.
2. Read and interpret elevation view drawings.
3. Read and interpret section and detail drawings.
4. Read and interpret schedules.
5. Read and interpret written specifications.
6. Establish 90-degree angles using the 3-4-5 rule.

Trade Terms

Benchmark
Contour lines
Easement
Elevation view
Front setback
Monuments
Nominal size
Plan view
Property lines
Riser diagram
Topographical survey

Industry-Recognized Credentials

If you're training through an NCCER-accredited sponsor you may be eligible for credentials from NCCER's Registry. The ID number for this module is 27104-13. Note that this module may have been used in other NCCER curricula and may apply to other level completions. Contact NCCER's Registry at 888.622.3720 or go to **nccer.org** for more information.

Contents

Topics to be presented in this module include:

Figures

SECTION ONE

1.0.0 CONSTRUCTION DRAWINGS

Objective

List the types of drawings usually included in a set of plans and describe the information found on each type.

a. Identify the different types of lines used on construction drawings.
b. Identify selected architectural symbols commonly used to represent materials on plans.
c. Identify selected electrical, mechanical, and plumbing symbols commonly used on plans.
d. Identify selected abbreviations commonly used on plans.
e. Describe the methods of dimensioning construction drawings.
f. List the various types of construction drawings and describe each.

Performance Tasks 1 through 4

Read and interpret foundation, floor, and other plan view drawings.

Read and interpret elevation view drawings.

Read and interpret section and detail drawings.

Read and interpret schedules.

Trade Terms

Benchmark: A point established by the surveyor on or close to the building site. It is used as a reference for determining elevations during the construction of a building.

Contour lines: Imaginary lines on a site plan that connect points of the same elevation. Contour lines never cross each other.

Easement: A legal right-of-way provision on another person's property (for example, the right of a neighbor to build a road or a public utility to install water and gas lines on the property). A property owner cannot build on an area where an easement has been identified.

Elevation view: A drawing providing a view from the front or side of a structure.

Front setback: The distance from the property line to the front of the building.

Monuments: Physical structures that mark the location of a survey point.

Nominal size: Approximate or rough size (commercial size) by which lumber, CMUs, etc., is commonly known and sold, normally slightly larger than the actual size (for example, 2 × 4s).

Plan view: A drawing that represents a view looking down on an object.

Property lines: The recorded legal boundaries of a piece of property.

Riser diagram: A schematic drawing that depicts the layout, components, and connections of a piping system.

Topographical survey: An accurate and detailed drawing of a place or region that depicts all the natural and man-made physical features, showing their relative positions and elevations.

A special language of lines, symbols, abbreviations, and notes is used on construction drawings to deliver the architect's message as clearly and concisely as possible. This section describes the different types of lines, dimensioning, symbols, and abbreviations used on drawings, before delving into the actual drawings themselves.

1.1.0 Lines

Many different types of lines are used to draw and describe a structure. Lines are drawn wide or narrow, dark or light, and broken or unbroken, with each type of line conveying a specific meaning. *Figure 1* shows the most common lines used on construction drawings. A description of each type of line follows:

- *Object line* – Heavy line used to show the main outline of a structure, including exterior walls, interior partitions, porches, patios, sidewalks, parking lots, and driveways.
- *Dimension and extension lines* – Lightweight lines used to provide the dimensions of an object. An extension line is extended from an object at both ends of the object to be measured. Extension lines should not touch the object lines to avoid confusion with object lines. A dimension line is drawn at right angles between the extension lines and a number placed above, below, or to the side of it to indicate the length of the object being measured. Sometimes a gap is provided in the dimension line and the dimension is shown in the gap.
- *Center line* – Used to designate the center of an area or object and to provide a reference point

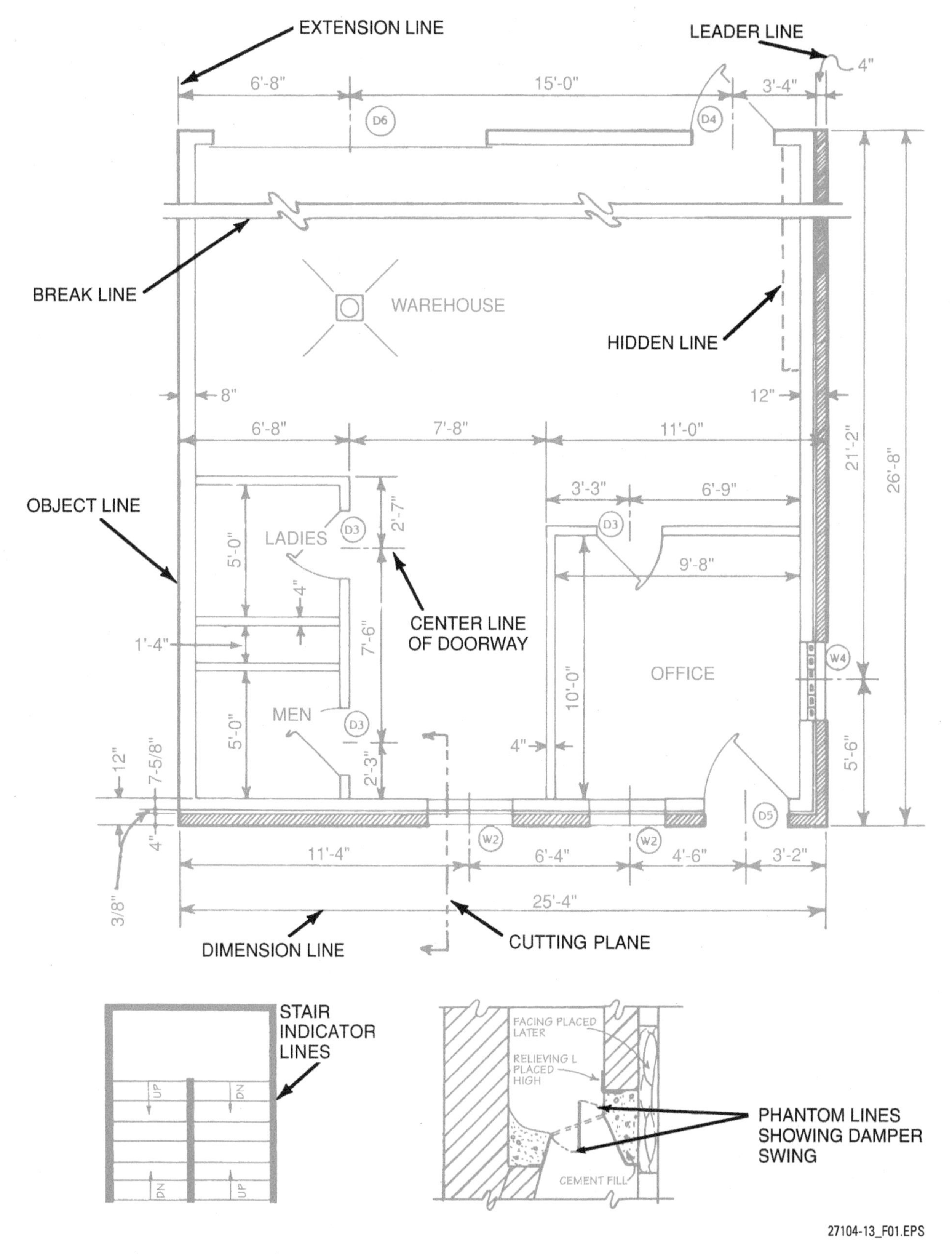

27104-13_F01.EPS

Figure 1 Lines used on construction drawings.

for dimensioning. Center lines are typically used to indicate the centers of objects such as columns, posts, footings, and door openings.

- *Cutting plane line* – Used to indicate an area that is being cut away and shown in a section view so that the interior features can be seen. The arrows at the ends of the cutting plane line indicate the direction in which the section is viewed. Letters identify the section view of that specific part of the structure. More elaborate methods of labeling cutting plane lines are used in larger sets of plans (*Figure* 2) where many sections are being used. The section view may be on the same page as the cutting plane line or on another page.
- *Break line* – Used to indicate that an object or area is not being shown in its entirety.
- *Leader line* – Used to connect a note or dimension to a related part of the drawing. Leader lines are usually curved or at an angle from the feature being distinguished to avoid confusion with dimensions and other lines.
- *Hidden line* – Used to indicate an outline that is invisible to an observer because it is covered by another surface or object that is closer to the observer.
- *Phantom line* – Used to indicate alternative positions of moving parts, such as a damper's swing or adjacent positions of related parts. It may also be used to represent repeated details.
- *Stair indicator line* – A short line with an arrowhead that shows the ascent or descent of a stairway on a floor plan.
- *Contour lines* – **Contour lines** are drawn on a site plan (covered in the *Site Plans* section) and **topographical survey** to show the changes in elevation and the contour of the land. The lines may be dashed or solid. Dashed lines are used to show the natural grade, while solid lines show the finish grade to be achieved during construction. Each line across the plot of land represents a different elevation relative to some point such as sea level or a local feature. Each line is drawn at a uniform change in elevation, called a contour interval, such as every 10'. The closer together the contour lines, the steeper the terrain will be.

1.2.0 Architectural Symbols

Symbols are used on construction drawings to indicate different kinds of materials, fixtures, and structural members. While some symbols are standard, the meaning of other symbols can vary from location to location. A set of drawings generally includes a sheet (usually Sheet 1) that identifies the specific symbols used on this set of drawings and their meanings.

> **NOTE**
> When using any set of drawings, always refer to the sheet containing the symbol legend first for reading and interpreting the drawings.

Some symbols are not intended to show part of an object, but are provided to guide you through the drawings themselves. Examples of this type of symbol include door and window designators that reference door and window schedules where additional information is provided. Other symbols are used to show the orientation of an object,

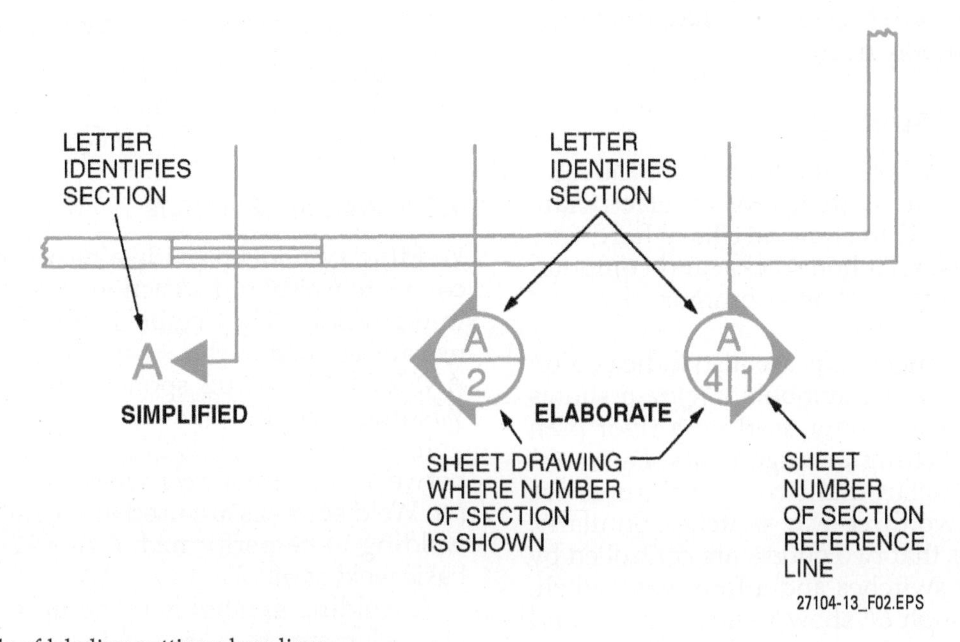

Figure 2 Methods of labeling cutting plane lines.

such as the north arrow, indicating the orientation of a structure on the building site. Two of the most common types of architectural symbols are material symbols and door and window symbols.

1.2.1 Material Symbols

Typically, materials are shown on drawings in two ways (*Figure 3*). One way that materials are shown is as a plan view, where the material is presented as you would see it when looking down on it. The second way materials are shown is as an elevation view or section view, where the symbol shows the material roughly as it would look when you are facing it.

1.2.2 Window and Door Symbols

Floor plans show a great deal of information in a small drawing area; therefore, symbols are used to convey needed information to a carpenter. Window and door symbols are usually shown with only a center line measurement to locate them from some reference point. However, more information about windows and doors is typically shown in separate window and door schedules contained in the drawing set. Window and door symbols are shown on drawings as either elevation or plan views (*Figure 4*).

1.3.0 Trade Symbols

A variety of symbols are shown on construction drawings that are useful for crafts other than carpentry. Each trade has its own symbols and abbreviations. For electrical, plumbing, and HVAC, there are symbols that show the types of equipment to be installed, such as switches, lavatories, and warm air supply ducts.

1.3.1 Electrical Symbols

Electrical symbols are used to indicate the location of outlets and switches for an electrician. They are also used to show auxiliary hardware such as buzzers, telephone jacks, and computer connections. *Figure 5* shows common electrical symbols.

Switching arrangements are also indicated on the floor plans using symbols. *Figure 6* shows some symbols commonly used on a floor plan to indicate switching arrangements. *Figure 6A* shows an electrician that two light fixtures are controlled by two three-way switches. Similarly, *Figure 6B* shows that two outlets are controlled by two three-way switches and a four-way switch. *Figures 6C* through *6F* show various outlets and fixtures controlled by a single-pole switch. (A single-pole switch is a device that opens and closes one side of the circuit only.)

1.3.2 Plumbing Symbols

Plumbing symbols indicate the hardware and fixtures required for the building. Floor plans usually indicate only where the different fixtures should be located and plumbed.

Figure 7 shows some common fixture symbols as they would appear on a floor plan. It also shows some common piping symbols that would be used on special plumbing diagrams.

1.3.3 HVAC Symbols

The location and types of HVAC equipment are shown on the floor plan(s), sections, and detail drawings using symbols. For HVAC systems, symbols are also used to show the direction of movement of the hot and cold air in the system. *Figure 8* shows some common HVAC symbols.

1.3.4 Site Plan and Topographical Survey Symbols

Site plans, also known as plot plans, show the position and sizes of all relevant structures on the site as well as the features of the terrain. It is difficult to show the amount of information required on these drawings if symbols are not used. A good site plan includes a legend of the symbols used, especially those that are nonstandard. *Figure 9* shows some of the symbols commonly used on site plans.

1.3.5 Structural Member Symbols

Structural steel shapes have a system of identification by the use of symbols. Some examples of these symbols are shown in *Figure 10*.

1.3.6 Welding Symbols

Welding symbols provide a means of conveying complete welding instructions from the designer to the welder. The symbols and their method of use are defined in the *American National Standard ANSI/AWS A2.4-2012*, sponsored by the American Welding Society.

In *ANSI/AWS A2.4-2012*, a distinction is made between the terms *weld symbol* and *welding symbol*. Weld symbols are used to indicate the type of welding to be performed. *Figure 11* shows some basic weld symbols.

A welding symbol is made up of as many as eight elements that are used together to provide

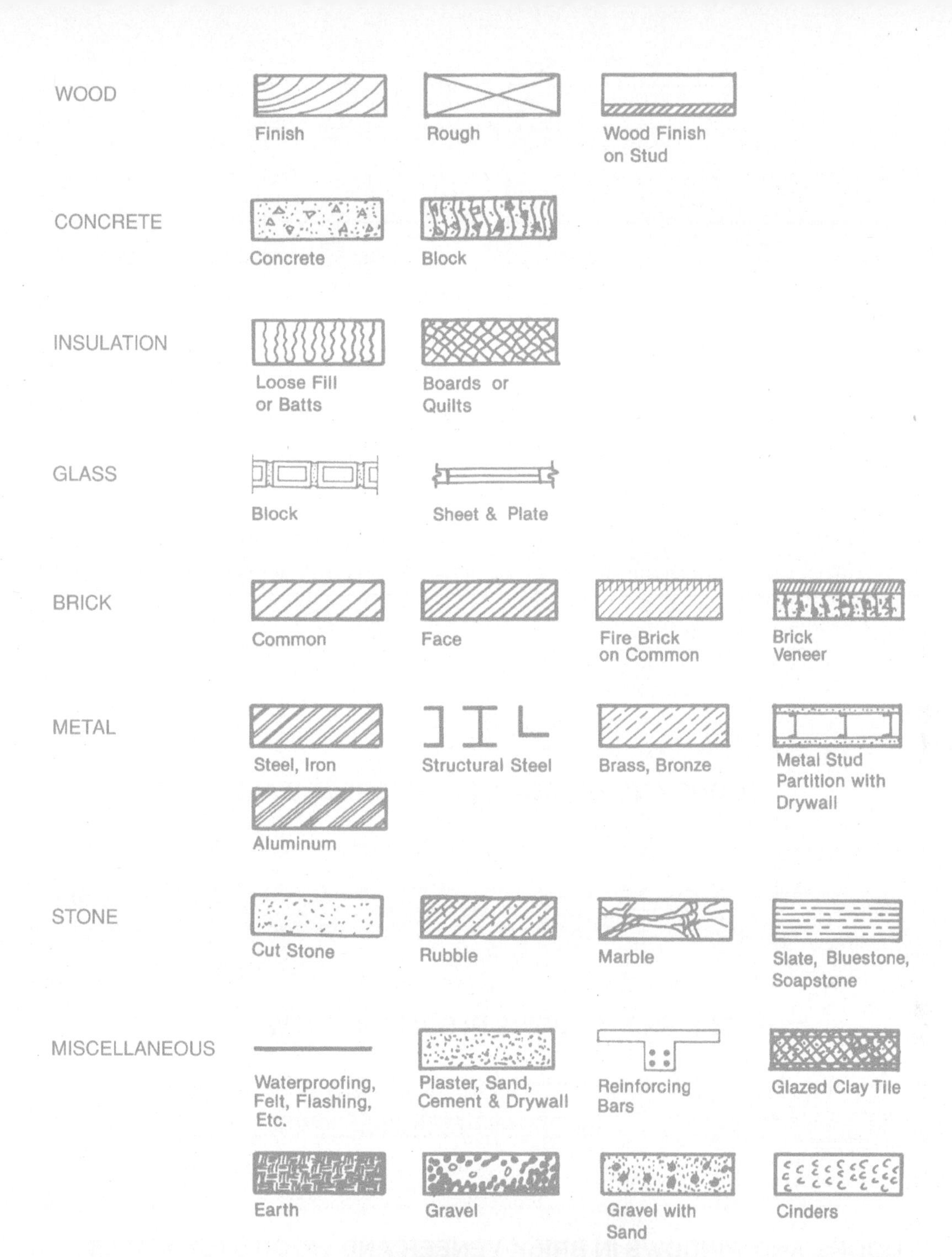

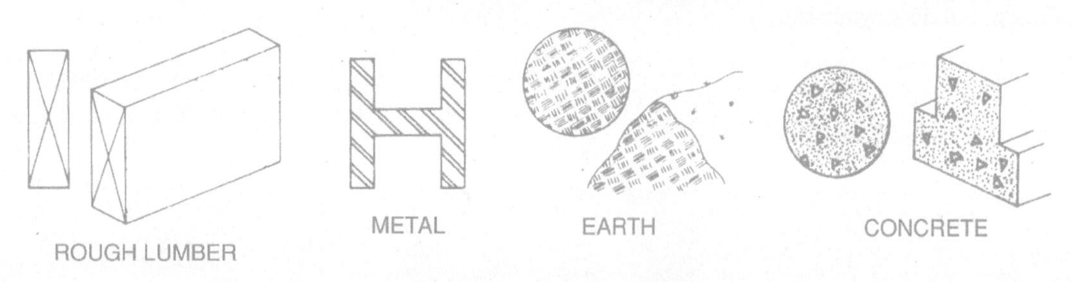

Figure 3 Typical material symbols.

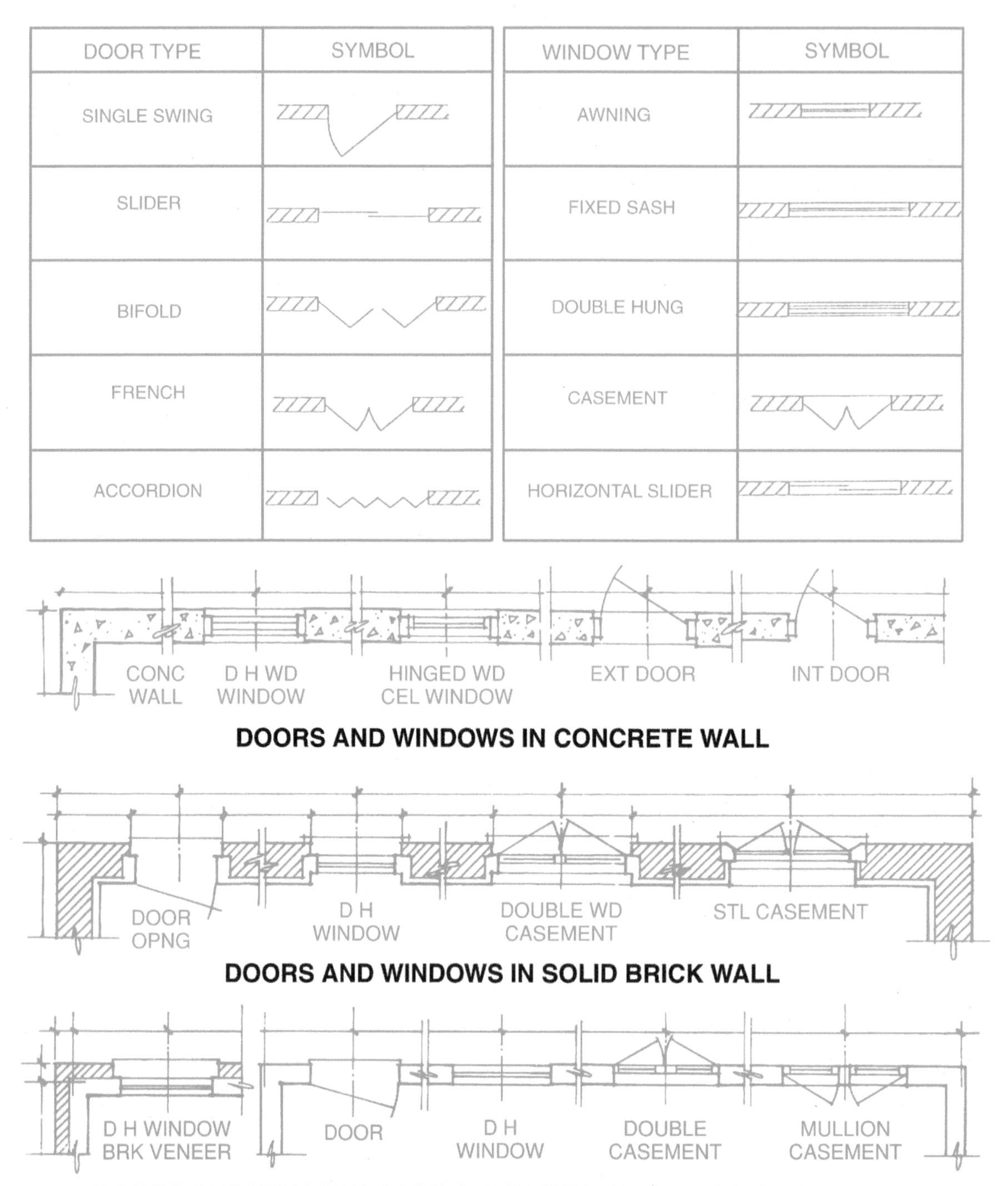

DOOR TYPE	SYMBOL	WINDOW TYPE	SYMBOL
SINGLE SWING		AWNING	
SLIDER		FIXED SASH	
BIFOLD		DOUBLE HUNG	
FRENCH		CASEMENT	
ACCORDION		HORIZONTAL SLIDER	

Figure 4 Window and door symbols.

GENERAL OUTLETS

Junction Box, Ceiling

Fan, Ceiling

Recessed Incandescent, Wall

Surface Incandescent, Ceiling

Surface or Pendant Single Fluorescent Fixture

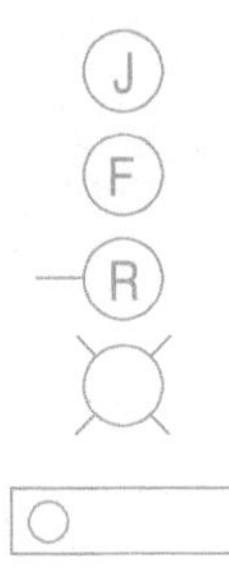

SWITCH OUTLETS

Single-Pole Switch	S
Double-Pole Switch	S_2
Three-Way Switch	S_3
Four-Way Switch	S_4
Key-Operated Switch	S_K
Switch w/Pilot	S_P
Low-Voltage Switch	S_L
Door Switch	S_D
Momentary Contact Switch	S_{MC}
Weatherproof Switch	S_{WP}
Fused Switch	S_F
Circuit Breaker Switch	S_{CB}

RECEPTACLE OUTLETS

Single Receptacle

Duplex Receptacle

Triplex Receptacle

Split-Wired Duplex Recep.

Single Special Purpose Recep.

Duplex Special Purpose Recep.

Range Receptacle

Switch & Single Receptacle

Grounded Receptacle

Duplex Weatherproof Receptacle

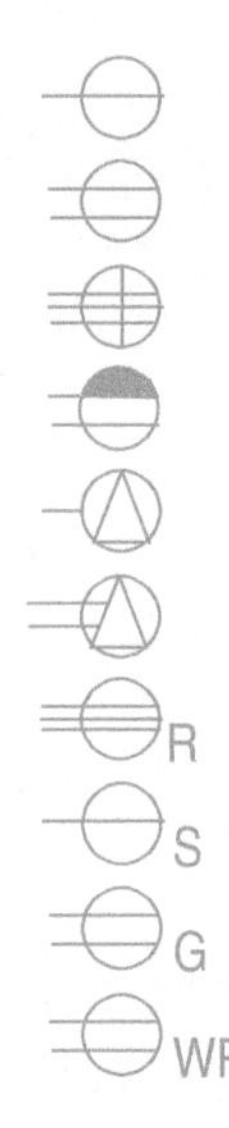

AUXILIARY SYSTEMS

Telephone Jack

Meter

Vacuum Outlet

Electric Door Opener

Chime

Pushbutton (Doorbell)

Bell and Buzzer Combination

Kitchen Ventilating Fan

Lighting Panel

Power Panel

Television Outlet

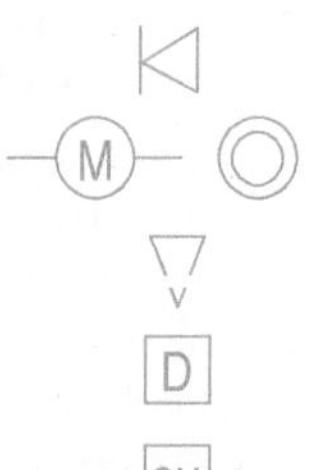

TV

27104-13_F05.EPS

Figure 5 Common electrical symbols.

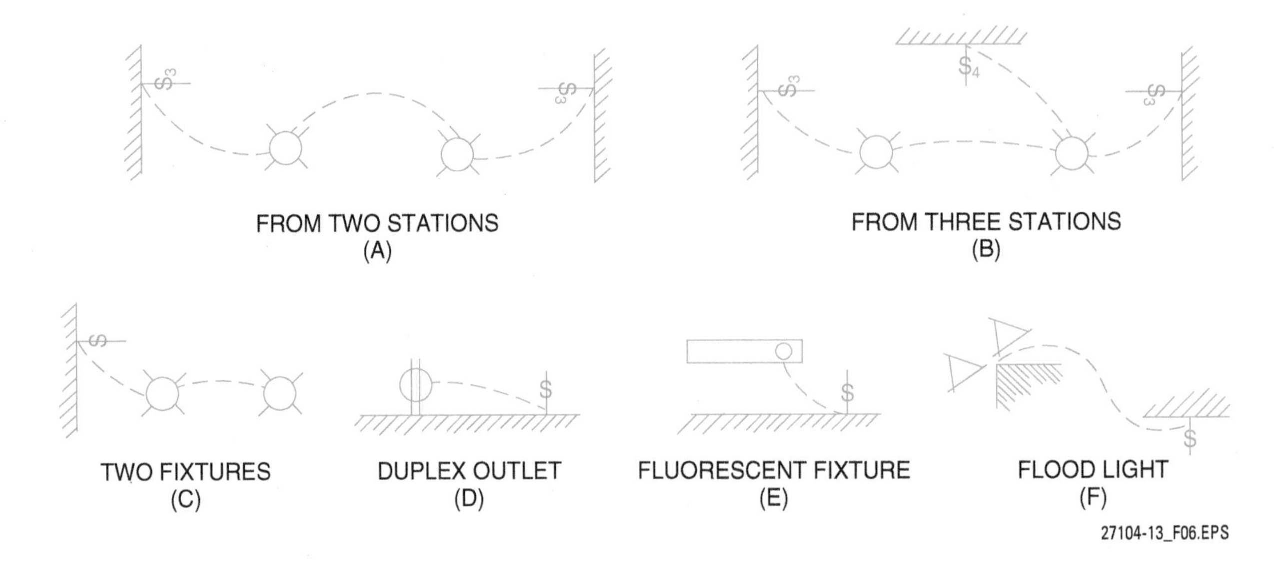

Figure 6 Electrical symbols showing control of light fixtures or an outlet.

exact welding instructions. *Figure 12* shows the standard elements of a welding symbol. As shown in *Figure 12*, the reference line is the key part of the welding symbol. All other elements are positioned with respect to this line. An arrow, which points to the location of the weld, is positioned on one end of the reference line. When necessary, a tail is positioned on the other end of the reference line. Other elements, such as the appropriate weld symbol used to show the type of weld, are placed on the reference line.

Information placed above the reference line indicates the weld is to be made on the other side of the joint from where the arrow points. Information placed below the reference line indicates the weld is to be made on the arrow side of the joint. Dimensions for the welds are drawn on the same side of the reference line as the weld symbol. Weld contour symbols are flush, concave, or convex. The contour symbol is located next to the weld symbol. Finish symbols usually accompany the contour symbol. The weld groove angle is shown on the same side of the reference line as the weld symbol. The size (depth) is placed to the left of the weld symbol, and the root opening is shown inside the weld symbol.

1.4.0 Abbreviations

Many written instructions are needed to complete a set of construction drawings. It is impossible to print out all such references, so a system of abbreviations is used. By using standard abbreviations, such as BRK for brick or CONC for concrete, the architect ensures that the drawings will be accurately interpreted. An extensive list of some commonly used abbreviations is contained in *Appendix A*. Note that some architects and engineers may use different abbreviations for the same terms. The title sheet in a drawing set commonly contains a list of abbreviations used in the set. For this reason, it is important to get a complete set of drawings and specifications including the title sheet(s).

Some practices for using abbreviations on drawings are as follows:

- Abbreviations are typically capitalized.
- Periods are used when abbreviations might look like a whole word.
- Abbreviations are the same whether they are singular or plural.
- Several terms have the same abbreviations and can only be identified from the context in which they are found.
- Many abbreviations are similar.

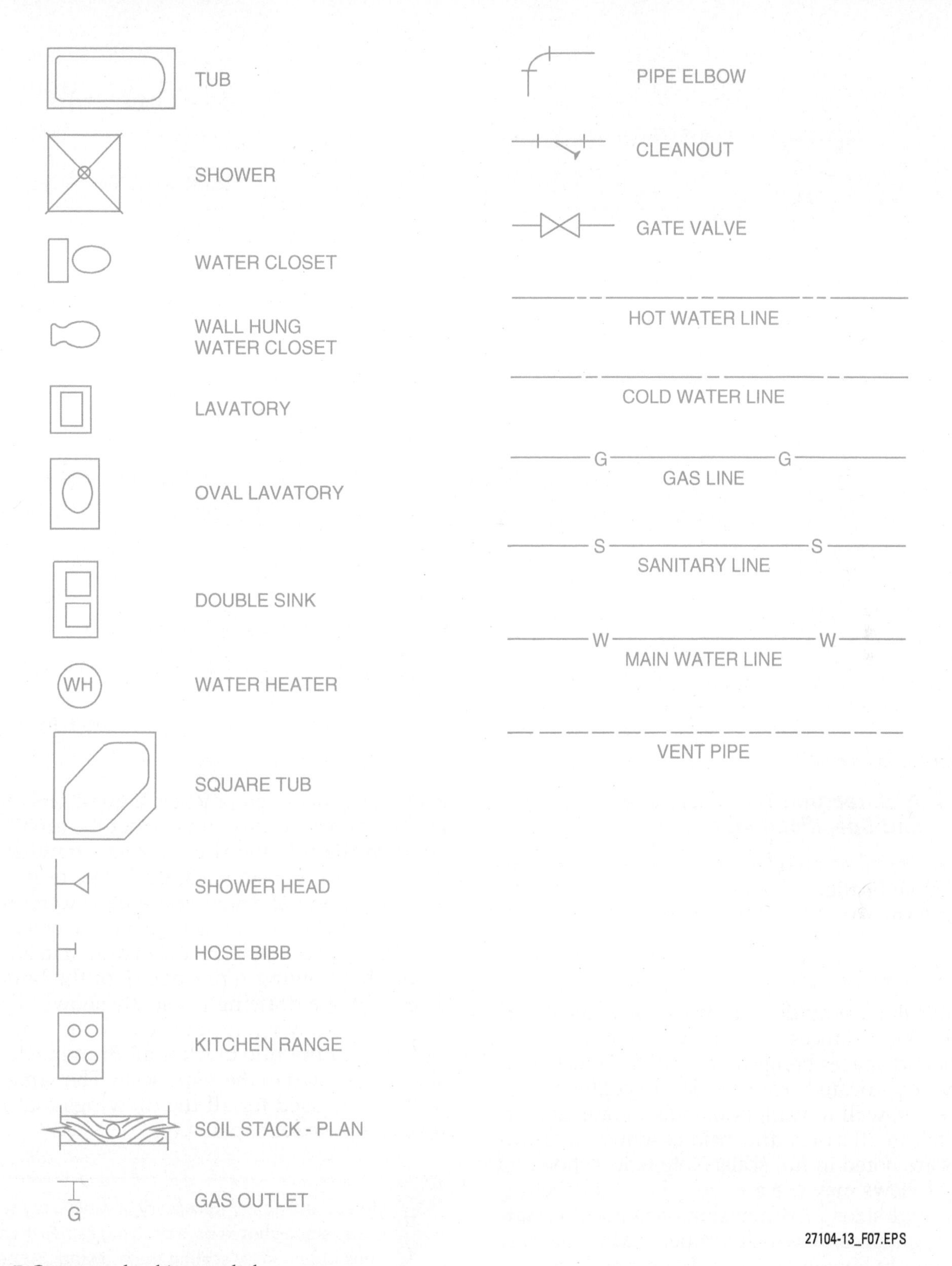

Figure 7 Common plumbing symbols.

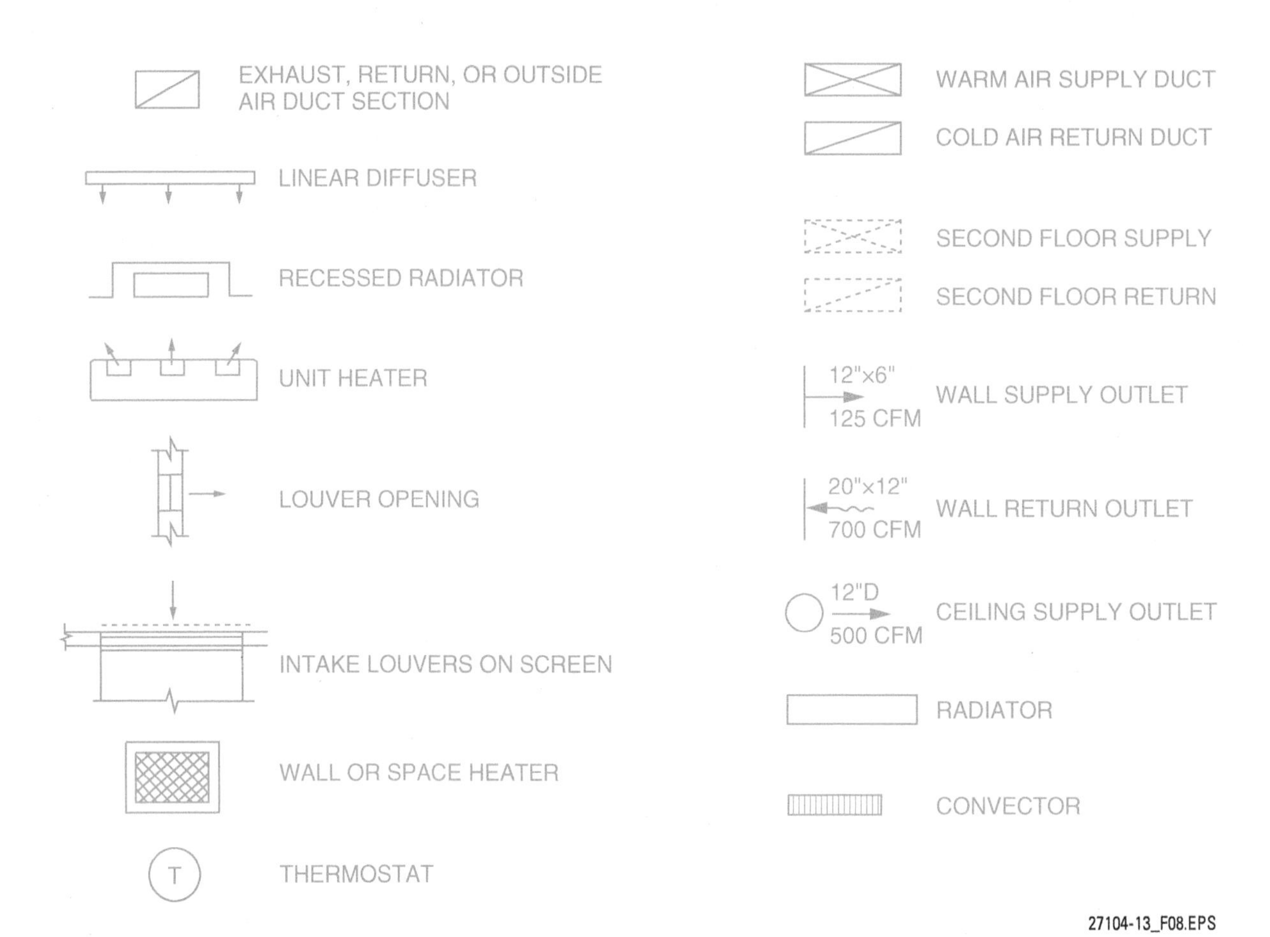

Figure 8 Common HVAC symbols.

1.4.1 Architectural Terms Used in Drawings and Specifications

There are many architectural terms used in plans and specifications. Some of the most commonly used terms are listed in *Appendix B.*

1.5.0 Dimensioning

Dimensions provided on drawings indicate actual sizes, distances, and measurements of the objects and spaces being represented. Dimensions may be indicated from outside to center, center to center, wall to wall, or outside to outside (*Figure 13*). In all cases, dimensions shown on drawings are noted in full scale. Note that section and detail views may use a **nominal size** in labeling. Nominal sizes or dimensions are approximate or rough sizes by which lumber, CMUs, etc., are commonly known and sold. For example, a 2 × 4 actually measures 1½" × 3½".

In some situations, it may be necessary to determine dimensions omitted from a drawing by measuring them with an architect's scale; however, this is not good practice. If questions regarding the dimensions occur, contact the architect to confirm the missing dimensions. Drawings are created using a specified scale. Inches or fractions of an inch on the drawing are used to represent feet and inches in the actual measurement of a building. For example, in a plan drawn to ¼" scale, ¼" on the drawing represents 1' of the building. The scale of a drawing is usually shown directly below the drawing. In some cases, the scale may be noted in the title block if all drawings on the sheet are drawn to the same scale. The same scale may not be used for all the drawings that make up a complete set of plans.

NOTE

Unless absolutely necessary, always use the dimensions shown on a drawing rather than ones obtained by scaling the drawing. Some print reproduction methods may slightly reduce or enlarge drawings, which can introduce errors if the print is scaled.

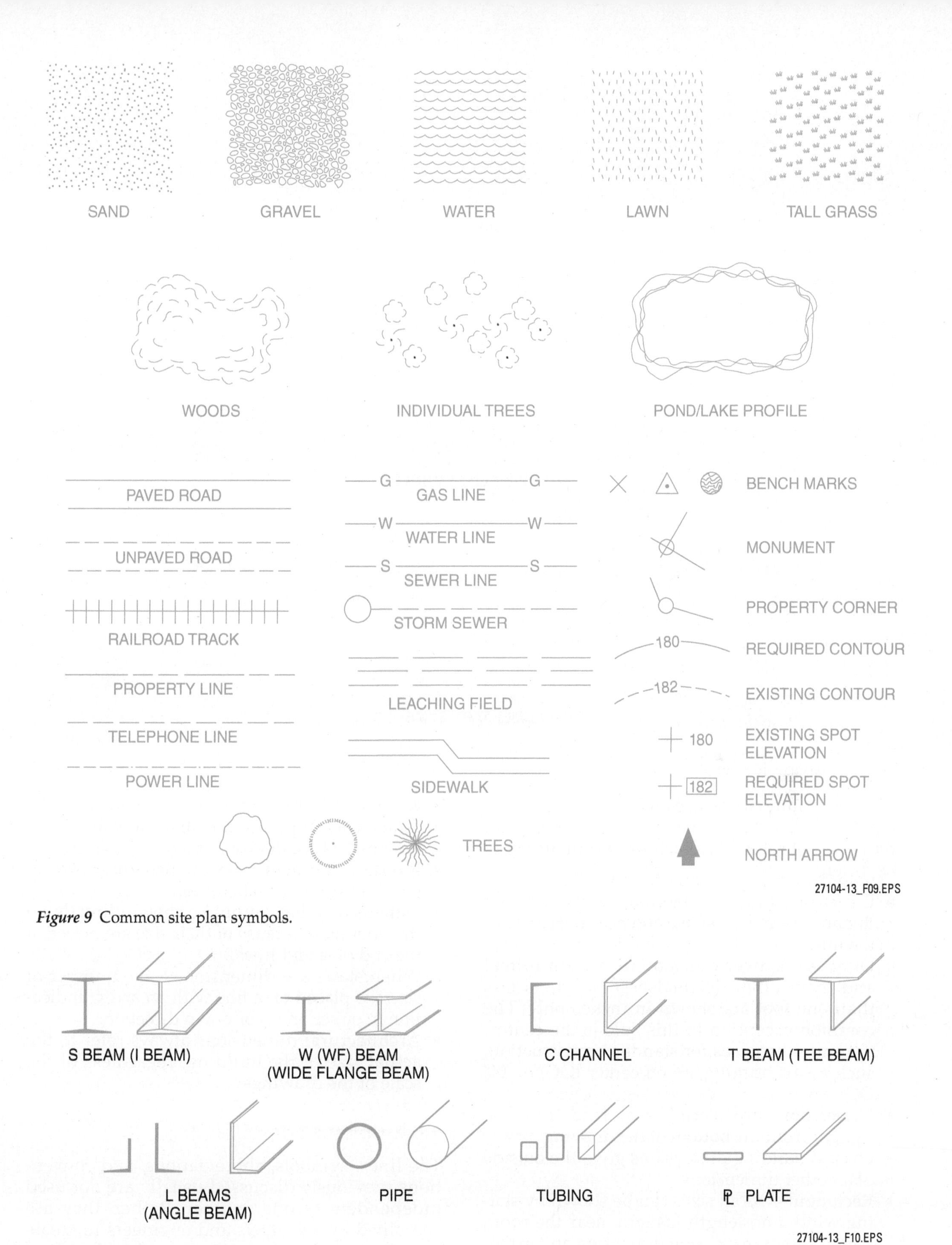

Figure 9 Common site plan symbols.

Figure 10 Structural member symbols.

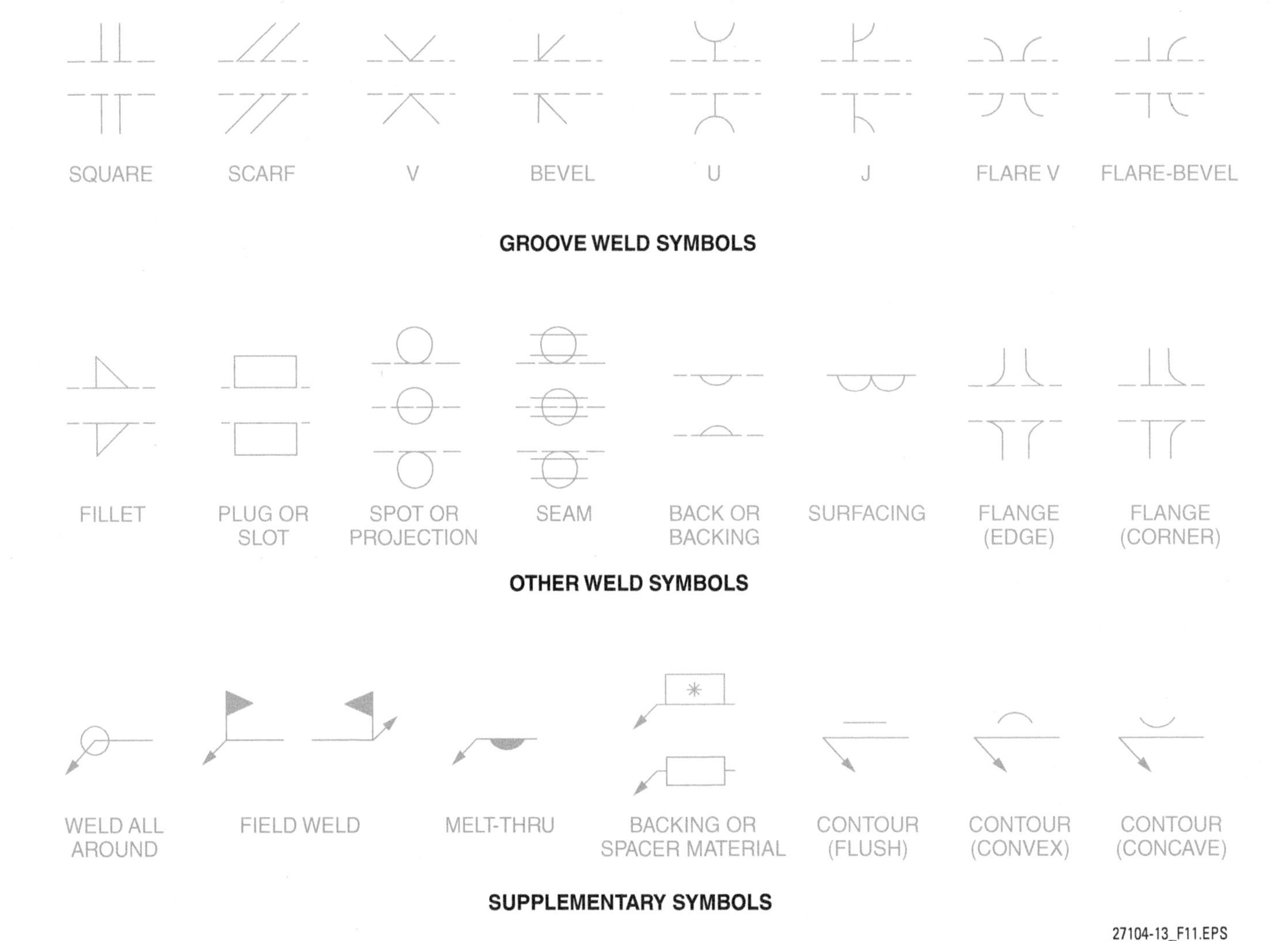

Figure 11 Basic weld symbols.

Some common practices used for dimensioning drawings are listed below. Keep in mind that these are not rules. The practices in your area may be different.

- Dimension lines are unbroken lines, with the dimensions placed above and near the center of the line.
- Dimensions over one foot are shown in feet and inches (not decimals). Dimensions less than one foot are shown in inches only. The common exception to this rule is the center-to-center distances for standard construction, such as for framing 16" on center (OC) or 24" OC.
- Dimensions are placed to be read from the right or from the bottom of the drawing.
- Overall building dimensions go to the outside of all other dimensions.
- Rectangular room sizes can be shown by stating width and length (usually near the room name), rather than using dimension and extension lines.
- Rooms are sometimes dimensioned from the center lines of partition walls, but wall-to-wall dimensions are more common.
- Window and door sizes are usually shown in window and door schedules.
- Dimensions that cannot be shown directly on the floor plan because of their size are placed at the end of leader lines.
- When stairs are dimensioned, the number of risers is placed on a line with an arrow indication in either an up or down direction.
- Architectural dimensions always refer to the actual size of the building, regardless of the scale of the drawings.

1.6.0 Construction Drawings

The lines, symbols, abbreviations, and conventions previously discussed usually are not used independent of one another. Rather, they are combined by architects and engineers to create construction drawings. Construction drawings

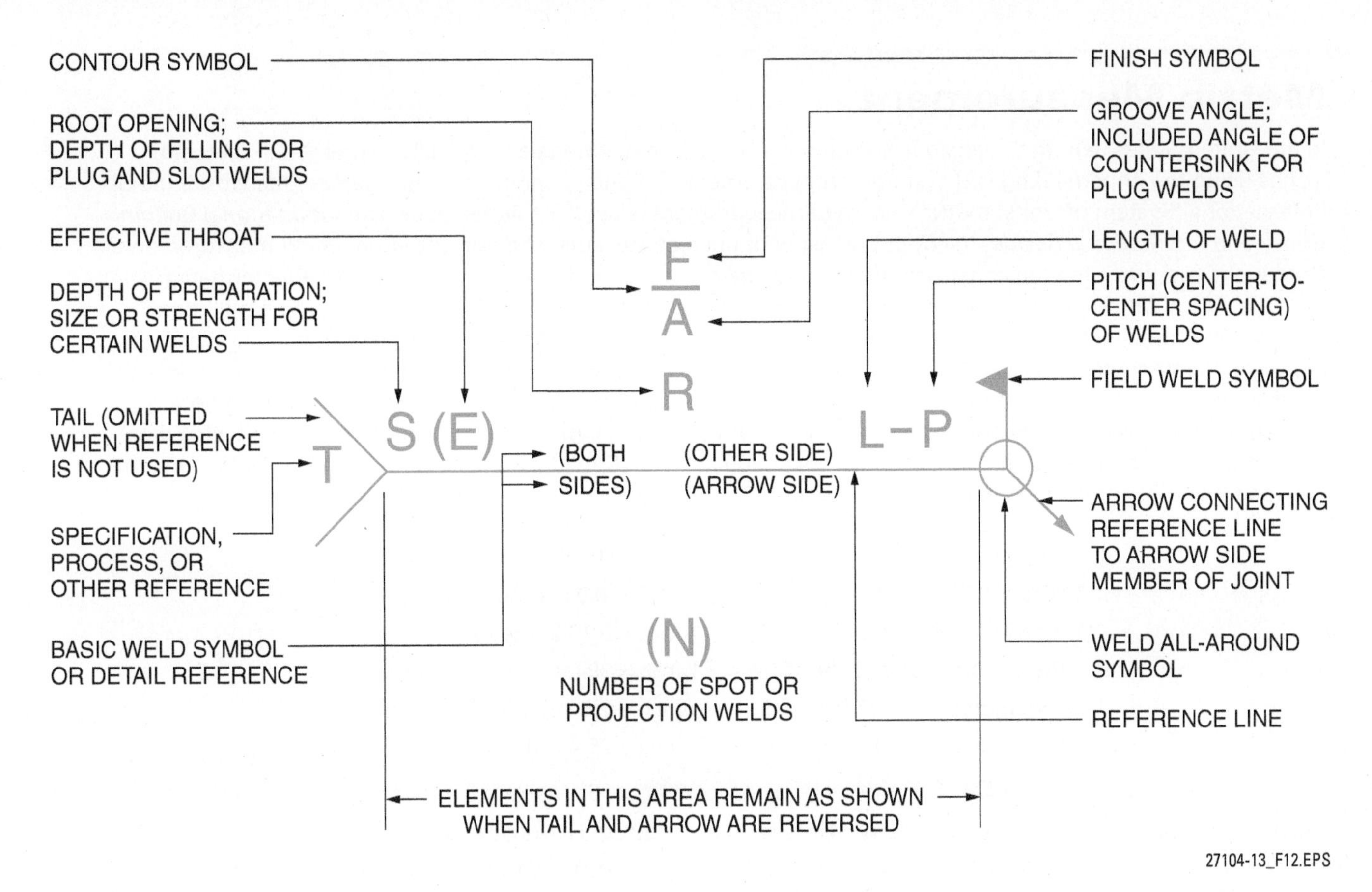

Figure 12 Elements of a welding symbol.

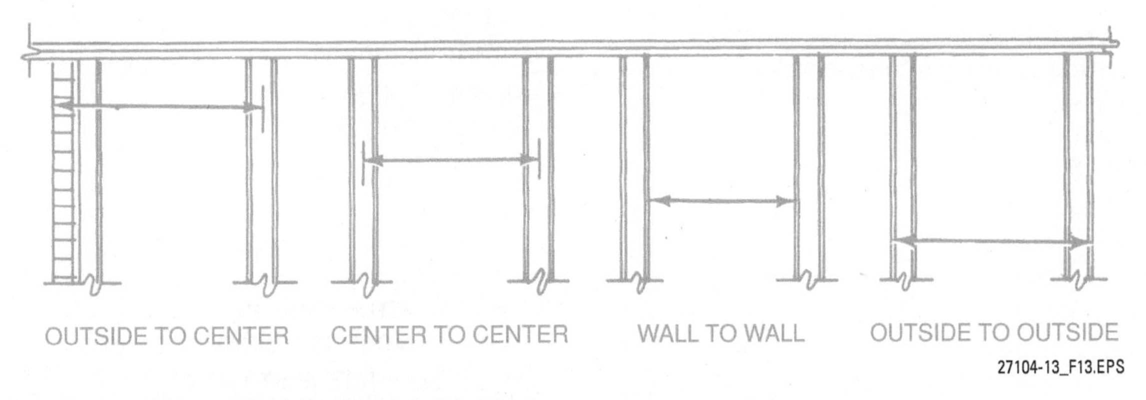

Figure 13 Common methods of indicating dimensions on drawings.

convey to carpenters and other skilled craftworkers the information needed to construct a specific building or structure. The construction drawings, together with the specifications, fully describe the building or structure about to be built. As a carpenter, it is extremely important that construction drawings and specifications are interpreted correctly. Failure to do so may result in costly rework and unhappy customers. Depending on the severity of the mistake, it can also expose you and your employer to legal liability.

The set of construction drawings, drawn to scale by an architect and/or engineer, show all of the information and dimensions necessary to build or remodel a structure. Copies of the original drawings are made for contractors and other vendors. These copies are commonly called prints. Most prints are produced on large engineering copiers (similar to photocopiers) from electronic files provided by the architect. Most prints show the building and other construction information using black lines on white paper.

Construction drawings consist of several different kinds of drawings assembled into a set (*Figure 14*). In order for complete information about a structure to be conveyed to the reader, various drawings in the set illustrate the structure using a variety of different views. A set of drawings also

Metric Measurement

In the United States, length dimensions shown on construction drawings are typically expressed in feet and inches using the imperial (English) system of measurement. In Canada and many other parts of the world, the SI (International System of Units) metric system of measurement is used for dimensions. The *International Building Code®* and *International Residential Code®* utilize both units of measure. The two common length measurements used in the metric system on construction drawings are the meter and millimeter, with the millimeter being 1⁄1000 of a meter. On drawings scaled between 1:1 and 1:100, a millimeter value is typically used. The millimeter abbreviation (mm) will not be shown, but there should be a note on the drawing indicating that all dimensions are given in millimeters unless otherwise noted. On drawings scaled between 1:200 and 1:2,000, a meter value is generally used. Again, the meter abbreviation (m) will not be shown and the drawing should have a note indicating that all dimensions are given in meters unless otherwise noted. Land distances (shown on site plans) are commonly expressed in meters or kilometers (1,000 meters). Conversion factors that can be used to convert between Imperial and metric linear units of measure are as follows:

1 kilometer = 0.62137 mile (mi)	1 mile = 1.609 kilometers (km)
1 meter = 39.37 inches (in)	1 yard = 0.9144 meter (m)
1 meter = 3.2808 feet (ft)	1 foot = 0.3048 meter (m)
1 meter = 1.0936 yards (yd)	1 foot = 304.8 millimeters (mm)
1 centimeter = 0.3937 inch (in)	1 inch = 2.54 centimeters (cm)
1 millimeter = 0.03947 inch (in)	1 inch = 25.4 millimeters (mm)

includes sheets that contain relevant written information, such as window or door schedules.

The types of written information and views normally contained in a drawing set include:

- Title sheets, title blocks, and revision blocks
- Architectural drawings consisting of:
  - Plan views
  - Elevation views
  - Sections
  - Details
  - Schedules
- Structural drawings
  - Foundation plans
- Mechanical plans
  - HVAC plans
  - Sprinkler plans
- Electrical plans
- Plumbing plans

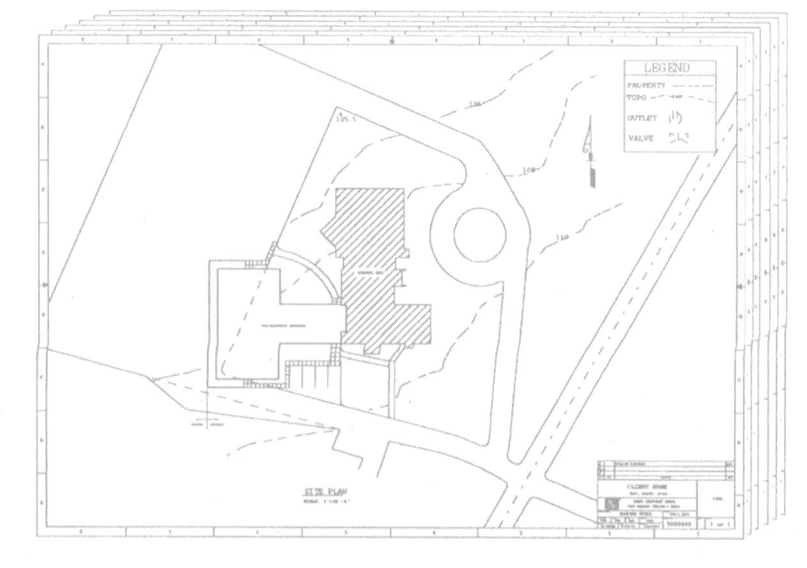

TITLE SHEET(S)
ARCHITECTURAL DRAWINGS
- SITE (PLOT) PLAN
- FOUNDATION PLAN
- FLOOR PLANS
- INTERIOR/EXTERIOR ELEVATIONS
- SECTIONS
- DETAILS
- SCHEDULES

STRUCTURAL DRAWINGS
MECHANICAL PLANS
ELECTRICAL PLANS
PLUMBING PLANS

27104-13_F14.EPS

Figure 14 Typical organization of a construction drawing set.

1.6.1 *Title Sheets, Title Blocks, and Revision Blocks*

A title sheet is typically placed at the beginning of a set of drawings or at the beginning of a major section of drawings. The title sheet provides an index to the other drawings, a list of abbreviations used on the drawings and their meanings, a list of symbols used on the drawings and their meanings, and various other project data such as the project location, size of the land parcel, and building size. The title sheet(s) should be carefully reviewed to understand the specific symbols and abbreviations used throughout the set of drawings. These symbols and abbreviations may vary from plan to plan.

A title block (*Figure 15*) is commonly placed on each sheet in a set of drawings. It is usually located in the lower-right corner of the sheet or along the right edge of the sheet.

A title block serves several general purposes in terms of communicating information. A title block typically contains the name of the firm that prepared the drawings, the owner's name, and the address and name of the project. It also provides locator information, such as the title of the sheet, the drawing or sheet number, the date the sheet was prepared, the scale, and the initials or names of the people who prepared and checked the drawing.

A revision block is typically shown on each sheet in a set of drawings, usually in the upper- or lower-right corner of the sheet near or within the title block. A revision block is used to record any changes (revisions) to the drawing. An entry in the revision block usually contains the revision

Think About It

Scaling Drawings

Measuring the length of an object or line on a drawing and converting the measurement to actual length using the given scale is called scaling a drawing. If a line on a floor plan measures 6" in actual length and the object being represented is drawn to a scale of ¼" = 1', what is the true length of the object?

Architect's Scale

Measurements are usually made on construction drawings using an architect's scale rather than a standard ruler. Architect's scales like those one shown on the right are divided into feet and inches and usually consist of several scales on one rule. Architect's scales are also available in other forms such as tapes or with wheels.

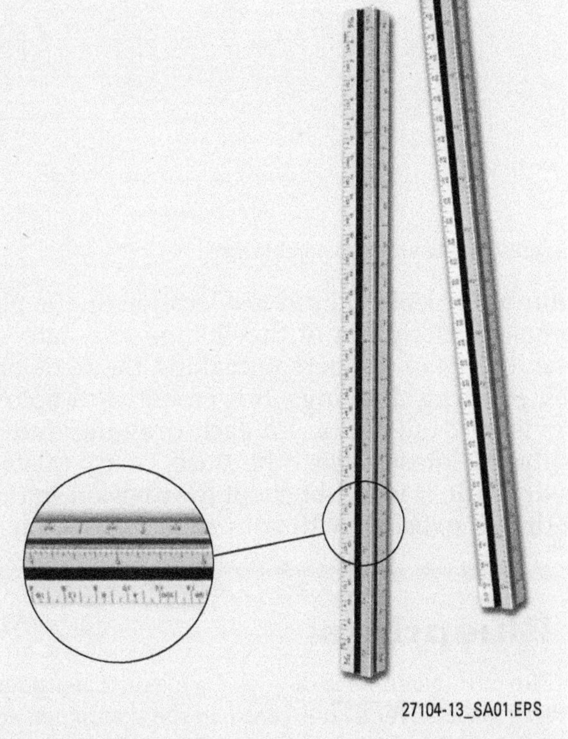

27104-13_SA01.EPS

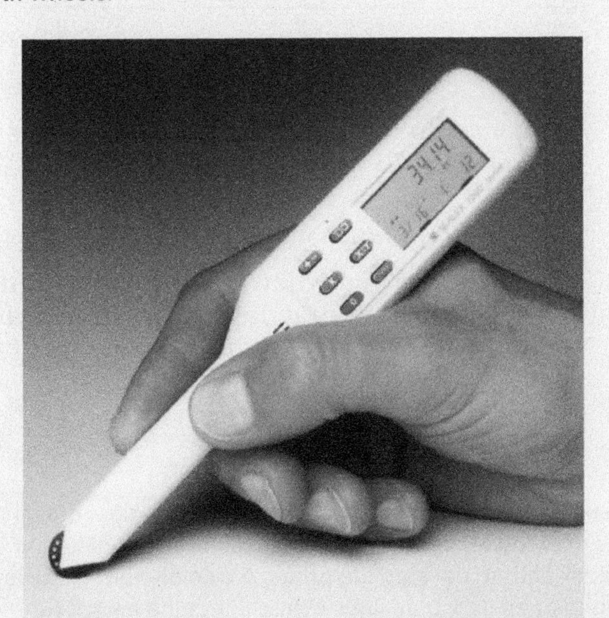

27104-13_SA02.EPS

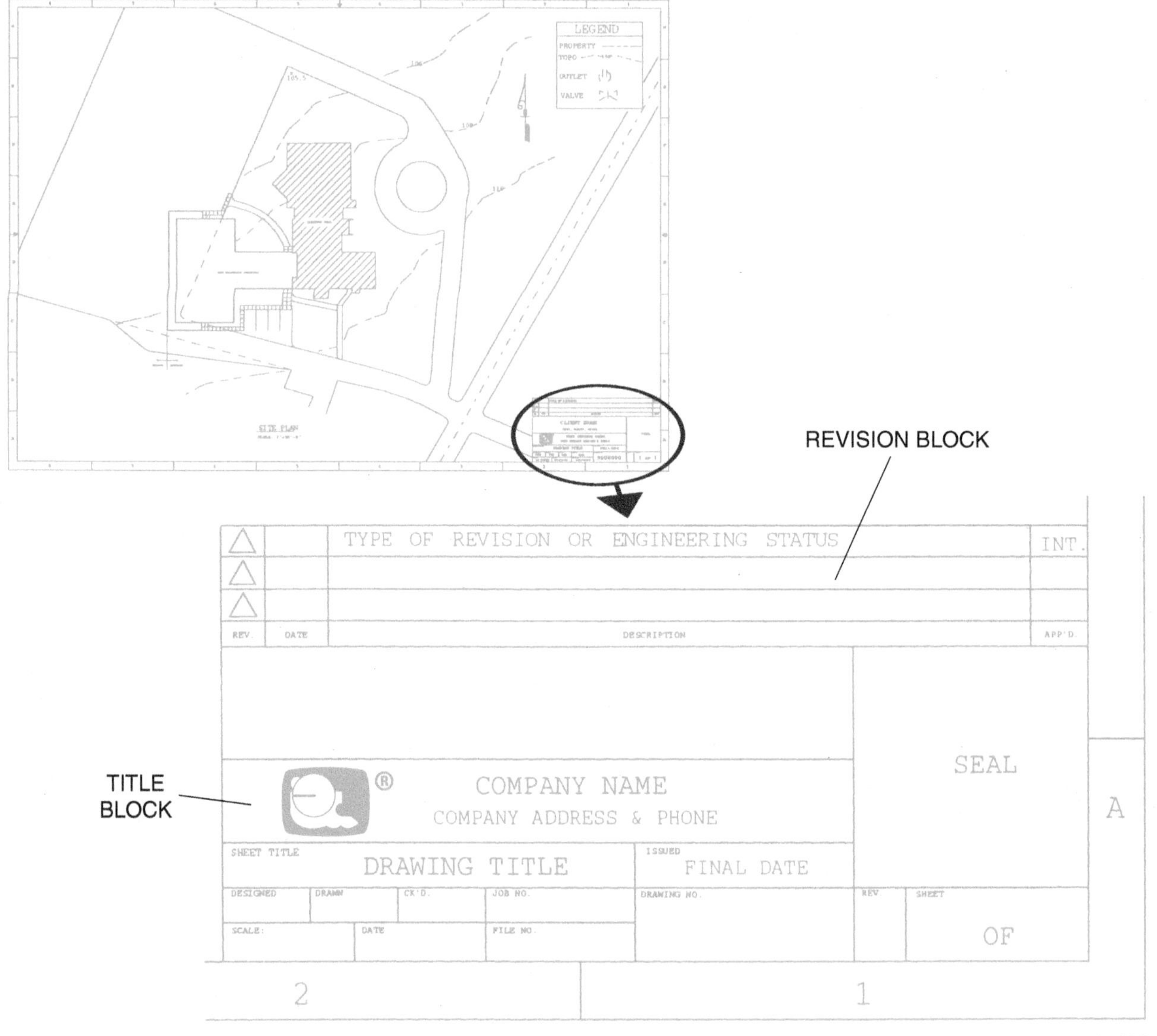

Figure 15 Title and revision blocks.

number or letter (keyed to a location on the plan), a brief description of the change, the date, and the initials of the person making the revision(s). When using drawings, it is essential to note the revision designation on each drawing and use only the latest issue; otherwise, costly mistakes will result. If in doubt about the revision status of a drawing, check with your supervisor to ensure the most recent version of the drawing is being used.

1.6.2 Site Plans

Several plan view drawings are included in a set of construction drawings. Plan view drawings show the structure from a vantage point above

Blueprints

The term *blueprint* is derived from a print reproduction method that was used in the past. The original drawing was placed over a light-sensitive sheet of paper, and the drawing and paper were exposed to ultraviolet light. The light-sensitive paper was then placed in an ammonia vapor bath to develop the prints. A true blueprint shows the details of the structure as white lines on a blue background or as dark blue lines over a light blue background. Even though the true blueprinting process is rarely (if ever) used anymore, the term *blueprint* is still widely used when referring to copies made from original drawings.

and looking down. The object is projected on to a horizontal plane. Typically, plan view drawings are made to show the overall construction site (site plan), the structure's foundation (foundation plan), and the structure's floor plans.

Man-made and topographical (natural) features and other relevant project information are shown on a site plan, including the information needed to properly locate the structure on the site. Man-made features include roads, sidewalks, utilities, and buildings. Topographical features include trees, streams, springs, and existing ground contours. Project information includes the building outline, general utility information, proposed sidewalks, parking areas, roads, landscape information, proposed contours, and any other information that conveys what is to be constructed or changed on the site. A prominently displayed north arrow is included for orientation purposes on site plans. Sometimes a site plan contains a large-scale map of the overall area that indicates where the project is located on the site. *Figures 16* and *17* show examples of basic site plans.

Typically, site plans include the following types of detailed information:

- Coordinates of control points or property corners
- Direction and length of **property lines** or control lines
- Description, or reference to a description, for all control and property **monuments**
- Location, dimensions, and elevation of the structure on site
- Finish and existing grade contours
- Location of utilities
- Location of existing elements such as trees and other structures
- Location and dimensions of roadways, driveways, and sidewalks
- Names of all roads shown on the plan
- Locations and dimensions of any **easements**

Like other drawings, site plans are drawn to scale. The scale used depends on the size of the project. A project covering a large area typically will have a small scale, such as 1" = 100', while a project on a small site might have a large scale, such as 1" = 10'.

The dimensions shown on site plans are typically expressed in feet and tenths of a foot (engineer's scale). However, some site plans state the dimensions in feet, inches, and fractions of an inch (architect's scale). Dimensions to the property lines are shown to establish code require-

Engineer's Scale

An engineer's scale is primarily used for land measurements, such as those included on site plans. An engineer's scale is incremented in multiples of 10 and measurements are expressed in decimals. The most common scales used on an engineer's scale are 10, 20, 30, 40, 50, and 60, which can all be combined on a triangular scale.

Engineer's scales and architect's scales are not interchangeable. Always check the scale noted on the print and select the proper scale to use when scaling the drawing.

Building Information Modeling (BIM)

Architectural plans have been used over the centuries to pictorially describe buildings and structures before they are actually built. In the past, draftspersons would draw these plans by hand. Today, most construction drawings are generated using a computer-aided design (CAD) system and architectural software. The CAD operator, working with architects and engineers, creates the drawings on the computer. The drawings can then be printed or plotted onto paper.

27104-13_SA03.EPS

Some CAD software is capable of building information modeling (BIM). BIM allows designers to prepare digital models of an entire structure and all its components. BIM allows people to virtually walk or even fly through a building before it has been built, in order to see how it will look when it has been completed. This allows designers to identify problems and conflicts with various components in time to correct them before construction has begun or is completed. BIM also allows designers to estimate the costs of constructing the building and even the cost of facilities maintenance over time.

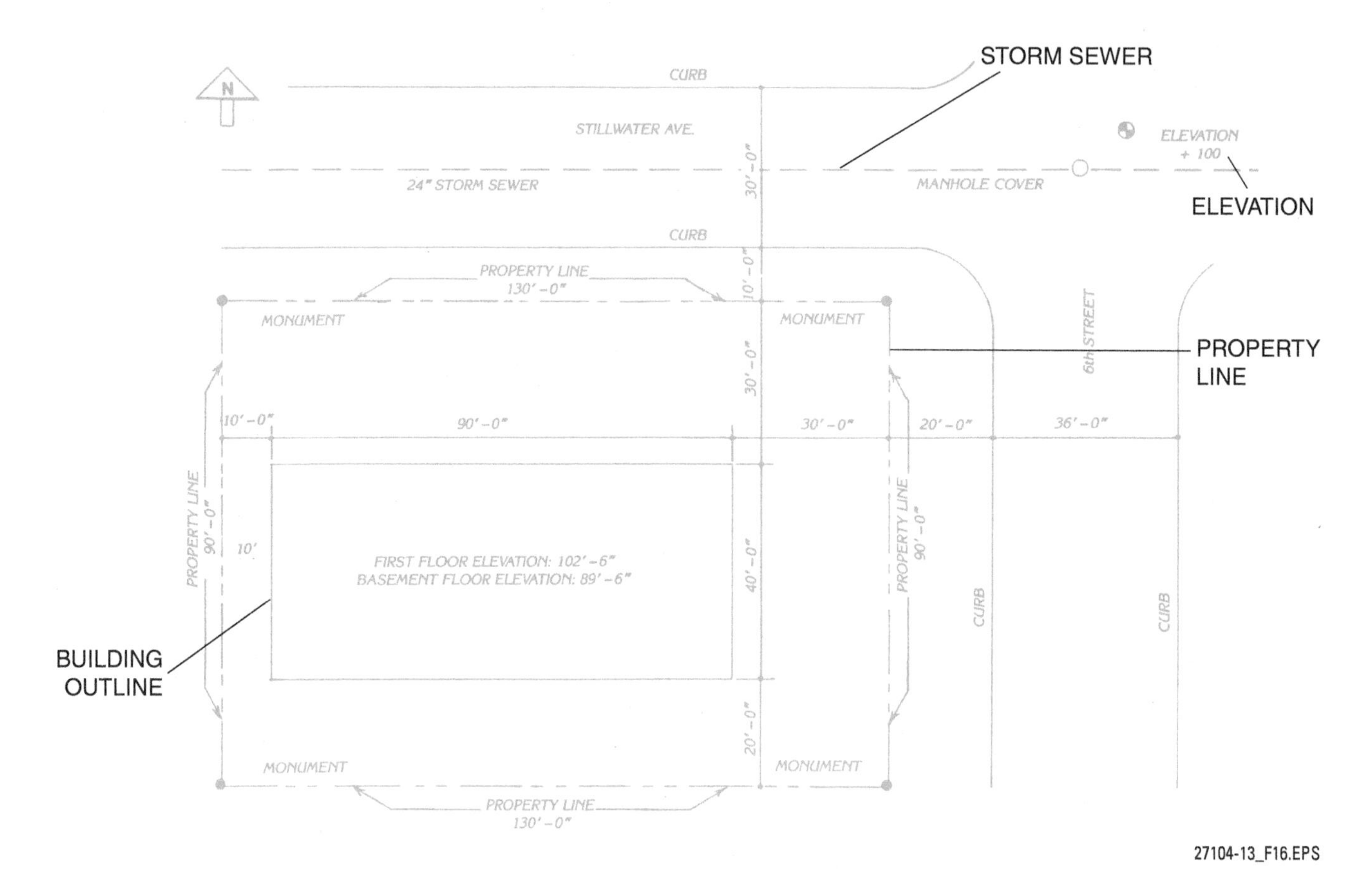

Figure 16 Typical site plan.

ments. Frequently, building codes require that nothing be built on certain portions of the land. For example, a local building code may have a **front setback** requirement that dictates the minimum distance that must be maintained between the street and the front of a structure.

Typically, side yards have an easement to allow for access to rear yards and to reduce the possibility of fire spreading to adjacent buildings. A property owner cannot build on an area where an easement has been identified. Examples of typical easements are the right of a neighbor to build a road; a public utility to install water, gas, or electric lines on the property; or an area set aside for groundwater drainage.

Site plans show finish grades (also called elevations) for the site based on data provided by a surveyor or engineer. It is necessary to know these elevations for grading the lot and for construction of the structure. Finish grades are typi-

Presentation Drawings

Presentation drawings are three-dimensional drawings that show a building from a desirable vantage point in order to display its most interesting features. Presentation drawings do not provide detailed information for construction purposes. They are used mainly by an architect or a contractor to sell the design of a house or building to a prospective customer. As a carpenter, you should be able to visualize and draw a rough three-dimensional sketch of a building after interpreting the construction drawings.

27104-13_SA04.EPS

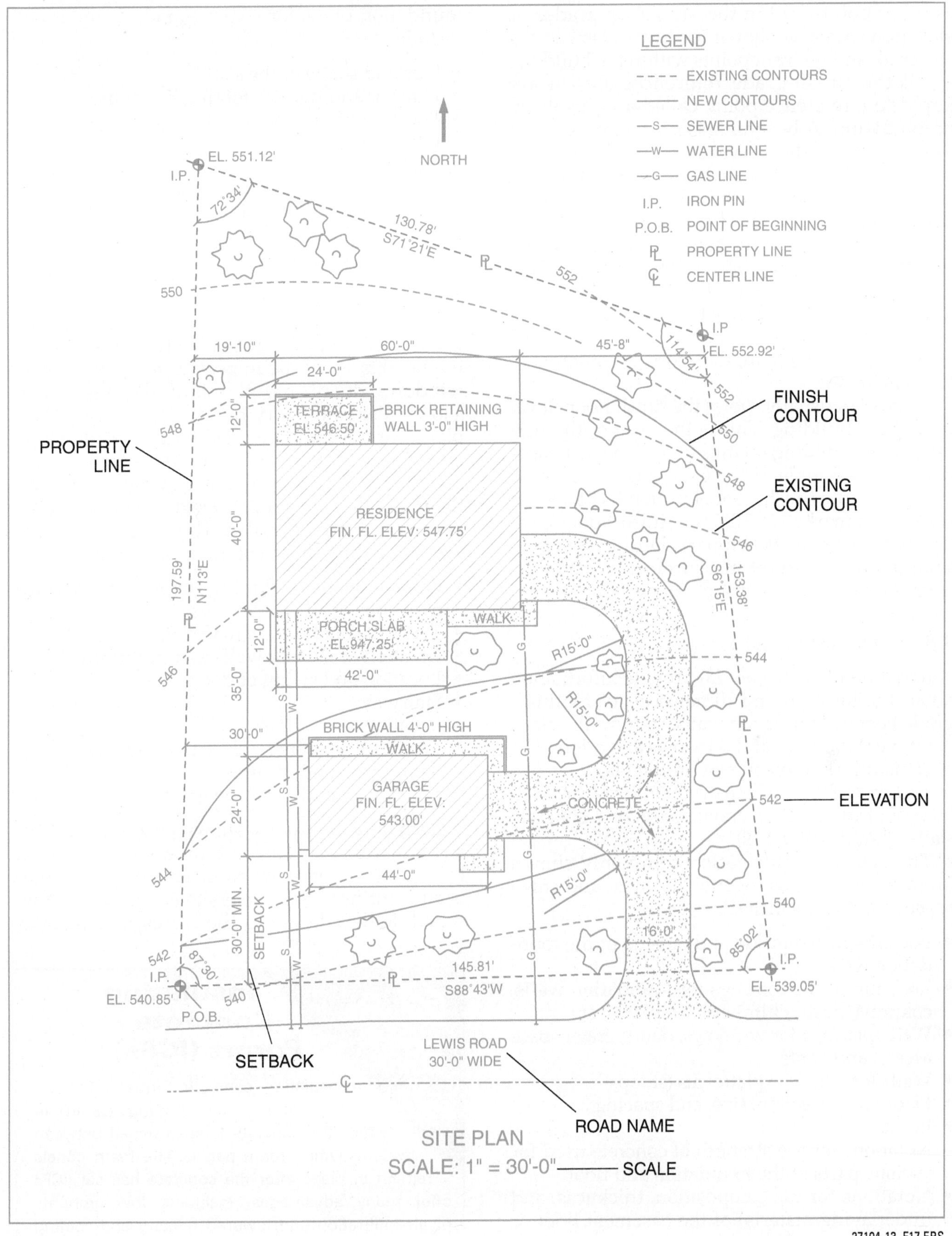

27104-13_F17.EPS

Figure 17 Typical site plan showing topographical features.

cally shown for all four corners of the lot as well as other points within the lot. Finish grades or elevations are also shown for the corners of the structure and relevant points within the building.

All the finish-grade references shown are keyed to a reference point, called a benchmark or job datum. A benchmark is a reference point established by the surveyor on or close to the property, usually at one corner of the lot. At the site, this point may be marked by a plugged pipe driven into the ground, a brass marker, or a wood stake. The location of the benchmark is shown on the site plan with a grade figure next to it. This grade figure may indicate the actual elevation relative to sea level, or it may be an arbitrary elevation of 100.00', 500.00', etc. All other grade points shown on the site plan, therefore, are relative to the benchmark.

A site plan usually shows the finish floor elevation of the building. This is the level of the first floor of the building relative to the job-site benchmark. For example, if the benchmark is labeled 100.00' and the finish floor elevation indicated on the plan is marked 105.00', the finish floor elevation is 5' above the benchmark. During construction, many important measurements are taken from the finish floor elevation point.

1.6.3 Foundation Plans

Foundation plans (*Figure 18*) provide information about the location and dimensions of footings, grade beams, foundation walls, stem walls, piers, equipment footings, and windows and doors. The specific information shown on the plan is determined by the type of construction involved: full-basement foundation, crawl space, or a concrete slab-on-grade (*Figure 19*).

The types of information commonly shown on foundation plans for full-basement and crawl space foundations include:

- Location of the inside and outside of the foundation walls
- Location of the footings for foundation walls, columns, posts, chimneys, fireplaces, etc.
- Wall openings for windows, doors, crawl space access, and vents
- Walls for entrance platforms (stoops)
- Floor joist direction, size, and spacing
- Location of stairways
- Notations for the strength of concrete used for various parts of the foundation and floor
- Notations for the composition, thickness, and underlaying material of the basement floor or crawl space surface
- Location of furnaces and other equipment

The types of information commonly shown on foundation plans for slab-on-grade foundations include:

- Size and shape of the slab
- Exterior and interior footing locations
- Location of fireplaces, floor drains, HVAC ductwork, etc.
- Notations for slab thickness
- Notations for wire mesh reinforcing, fill, and vapor barrier materials

1.6.4 Floor Plans

The floor plan is the main drawing of the entire set. For a floor plan view, an imaginary line is cut horizontally across the structure at varying heights so all the important features such as windows, doors, and plumbing fixtures can be shown. For multistory buildings, separate floor plans are typically provided for each floor. However, if several floors have the same layout (such as a hotel), one drawing may be used to show all the floors that are similar. *Figure 20* shows an example of a basic floor plan. The types of information commonly shown on floor plans include:

- Outside walls, including the location and dimensions of all exterior openings
- Types of construction materials
- Location of interior walls and partitions
- Location and swing of doors
- Stairways
- Location of windows
- Location of cabinets, electrical and mechanical equipment, and fixtures
- Location of cutting plane lines

Each door or window shown on a floor plan for a commercial building is typically accompanied by a number, letter, or both. This number/letter is an identifier, which refers to a door or window schedule that describes the corresponding size,

> GOING GREEN
>
> **Insulating Concrete Forms (ICFs)**
>
> Insulating concrete forms (ICFs) are a type of concrete forming system in which the concrete is sandwiched between two insulating foam panels. The foam panels remain in place after the concrete has set. ICFs offer many advantages, including low amounts of air infiltration and lowered heating and cooling loads, thus making the building very energy efficient.

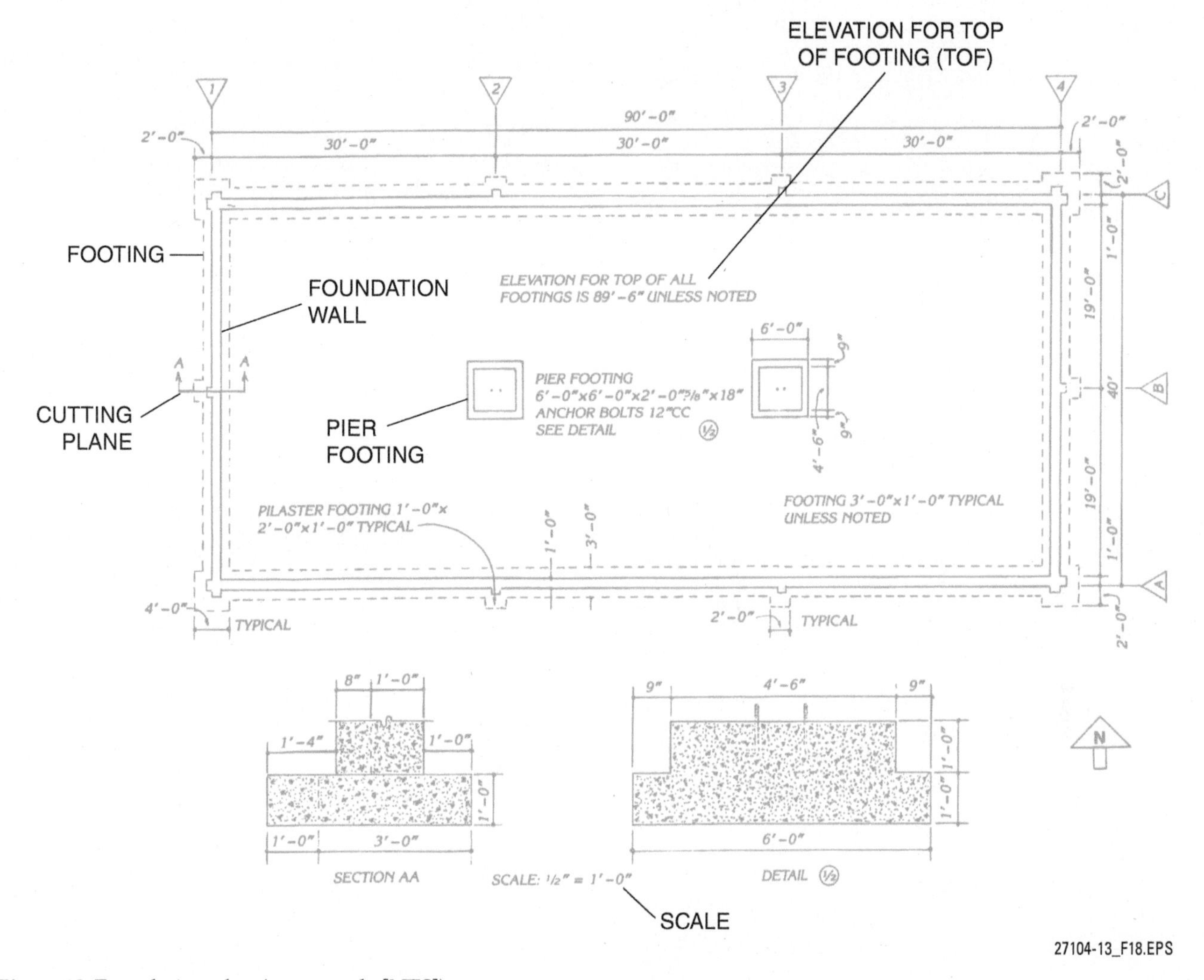

Figure 18 Foundation plan (not to scale [NTS]).

type of material, model number, etc., for the specific door or window. Door and window schedules are discussed in more detail later in this section. Residential floor plans often show door sizes directly on the floor plan.

1.6.5 Roof Plans

Roof plans provide information about the roof slope, roof drain placement, and other pertinent information regarding ornamental sheet metal work, gutters and downspouts, etc. Where applicable, the roof plan may also show information on the location of air conditioning units, exhaust fans, and other ventilation equipment. Not all sets of drawings include roof plans.

Some drawing sets may include a ceiling plan that shows the location of supply diffusers, exhaust grilles, access panels, and the location of structural components and other mechanicals. Some drawing sets have reflected ceiling plans, which show the details of the ceiling as though it were reflected by a mirror on the floor. Ceiling plans show features of the ceiling while keeping those features in proper relation to the floor plan. For example, if a vertical pipe runs from floor to ceiling in a room and is drawn in the upper-left corner of the floor plan, it is also shown in the upper-left corner of the reflected ceiling plan of that same room.

1.6.6 Elevation Views

Elevation views provide a view of the side of a structure. The structure's features are projected onto a vertical plane. Typically, elevation views are used to show the exterior features of a structure so the general size and shape of the structure can be determined. Elevation views clarify much of the information on the floor plan. For example, a floor plan shows where the doors and windows are located in the outside walls; an elevation view of the same wall shows actual representations of these doors and windows. *Figure 21* shows an example of basic elevation view drawings. The

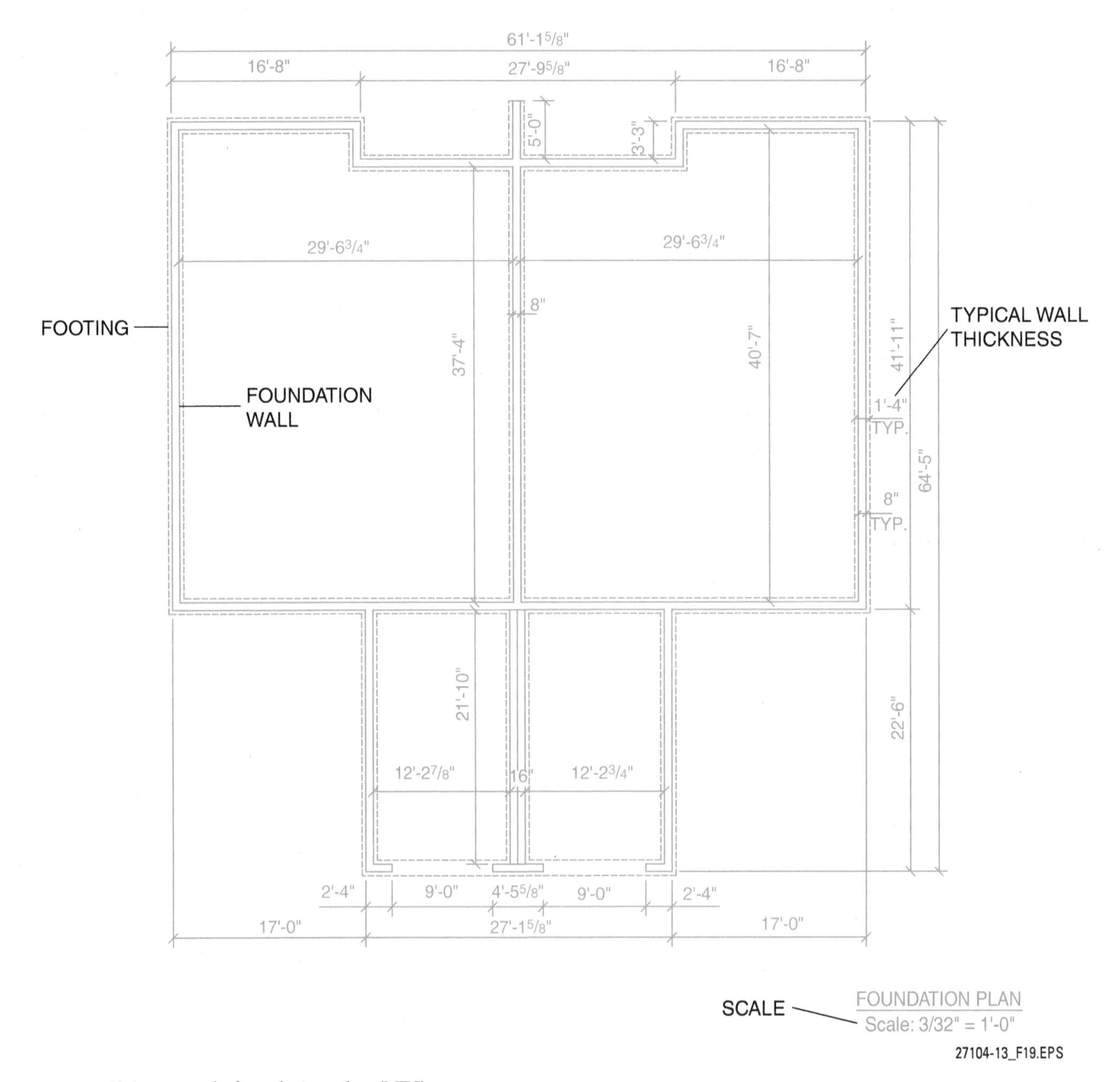

Figure 19 Slab-on-grade foundation plan (NTS).

types of information normally shown on elevation drawings include:

- Grade lines
- Floor height
- Window and door types
- Roof lines and slope, roofing material, vents, gravel stops, and projection of eaves
- Exterior-finish materials and trim

Unless one or more views are identical, four exterior elevation drawings are generally used to show the exterior of a building. More than four elevation drawings views may be required for complex buildings. Exterior elevation drawings are typically labeled in one of two ways: they may be labeled as front, right side, left side, and rear; or they may be designated by compass direction. For example, if the front of the building faces north, then this becomes the north elevation. The other elevations are then labeled accordingly (east, south, and west).

Drawing sets may also include interior elevation drawings for the walls in each partitioned area, especially for walls that have special features such as fireplaces.

1.6.7 Section Drawings

A section drawing or section view (*Figure 22*) shows the interior features of a wall or other structural feature. Section drawings are drawn as if a cut has been made through a feature at a

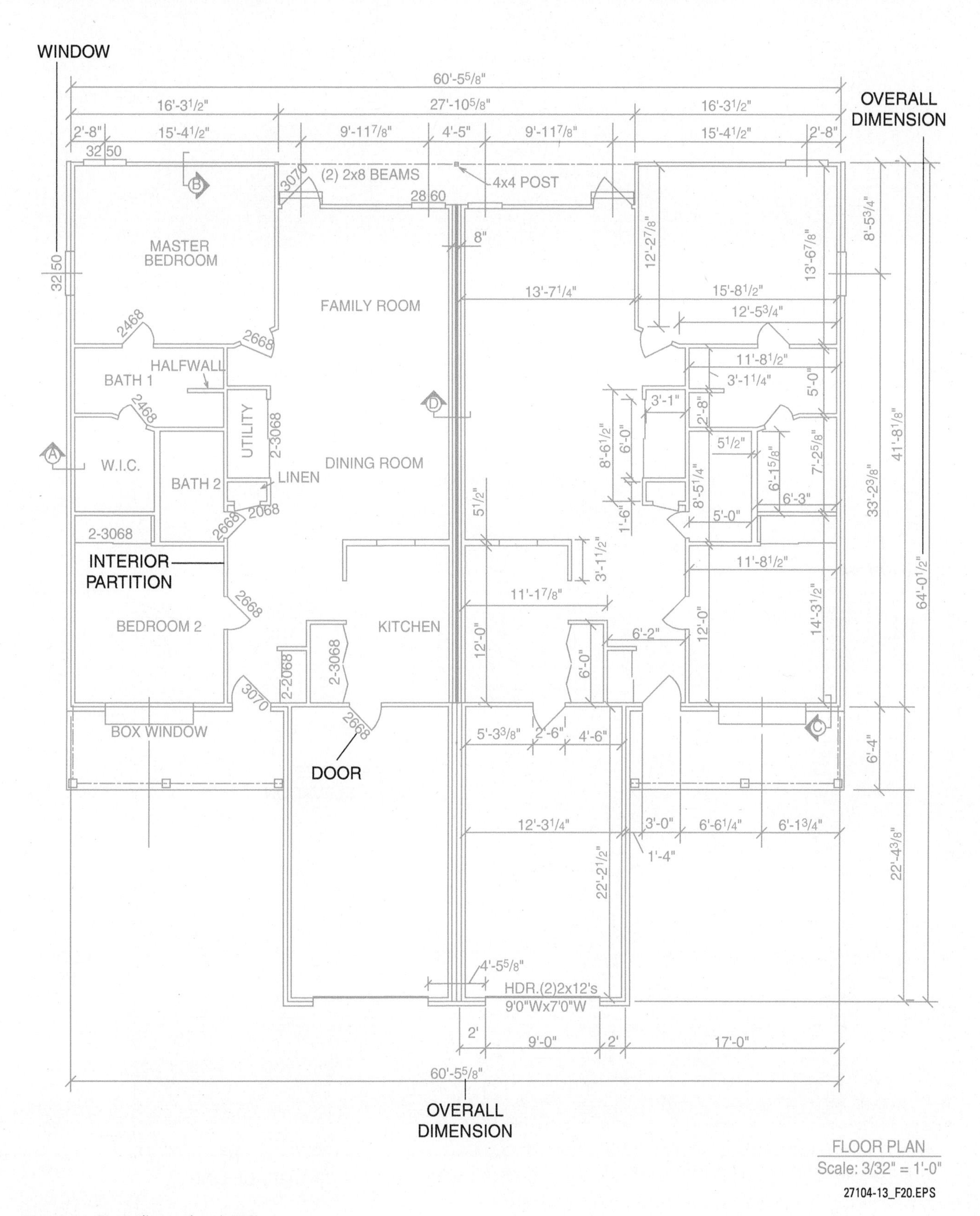

Figure 20 Basic floor plan (NTS).

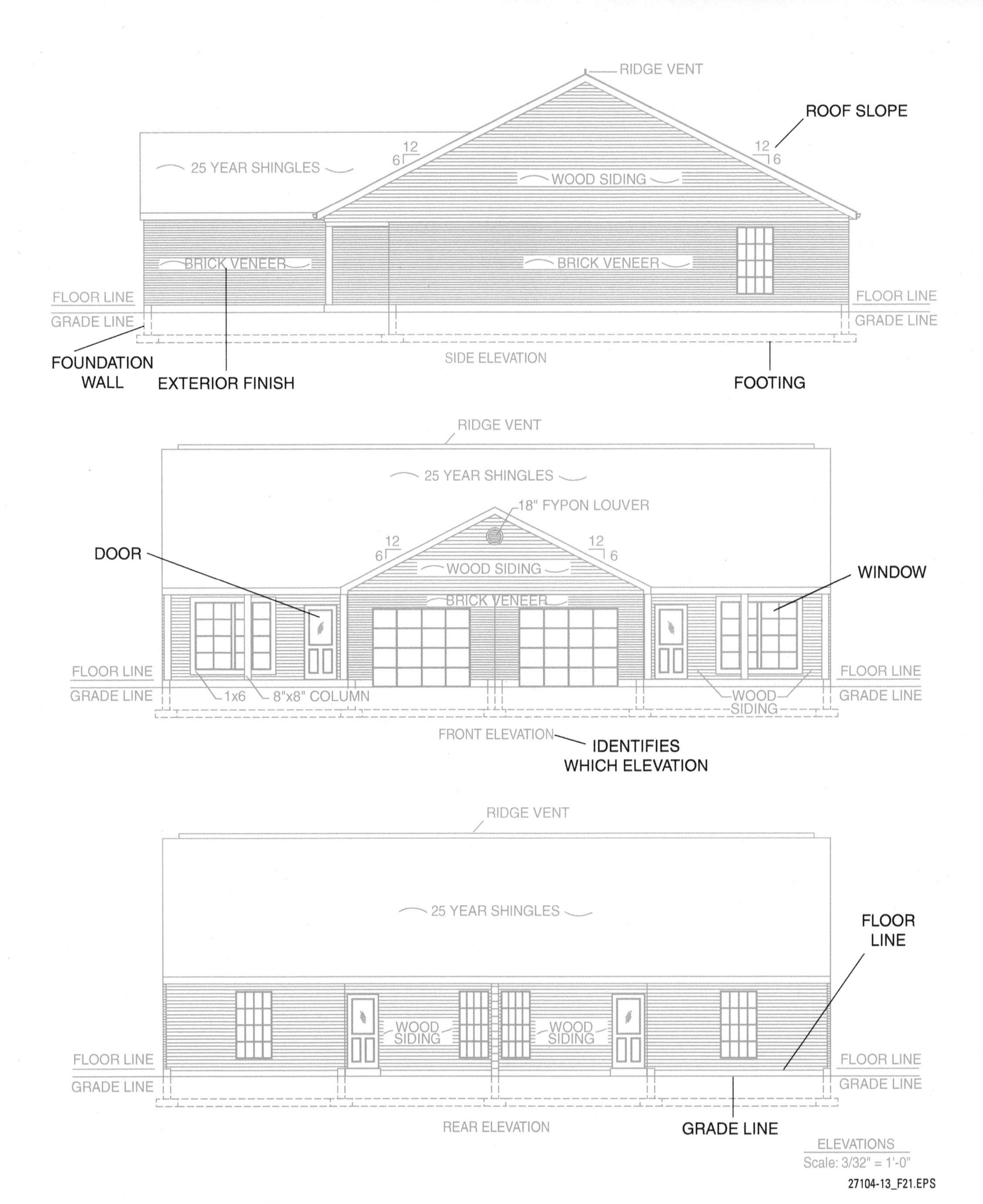

Figure 21 Elevation views (NTS).

GOING GREEN

Green Roofs

A green roof is a layer of vegetation grown on a rooftop. Green roofs provide shade and remove heat from the air through evapotranspiration, reducing temperatures of the roof surface and the surrounding air. On hot summer days, the surface temperature of a green roof can be cooler than the air temperature, whereas the surface of a conventional rooftop is considerably hotter.

Green roofs can be installed on a wide range of buildings, from industrial facilities to private residences. They can be as simple as a 2-inch covering of hardy groundcover or as complex as a fully accessible park complete with trees. Green roofs are becoming popular in the United States. Advantages of green roofs include:

- Reduced energy costs
- Reduced air pollution
- Better stormwater management

certain location. The location of the cut and the direction the section is to be viewed are shown on the related plan view.

Section drawings that show a view made by cutting through the length of a structure are referred to as longitudinal sections, while those showing the view of a cut through the width or narrow portion of the structure are referred to as traverse sections.

To show greater detail, section drawings are normally drawn to a larger scale than plan views. The types of information commonly shown by a section drawing include:

- Details of construction and information about stairs, walls, chimneys, or other parts of construction that may not show clearly on a plan view
- Floor levels in relation to grade
- Wall thickness at various locations
- Anchors and reinforcing steel

Floor Plans

As shown here, a floor plan is actually a horizontal section of the building with the cutting plane through the windows and doors.

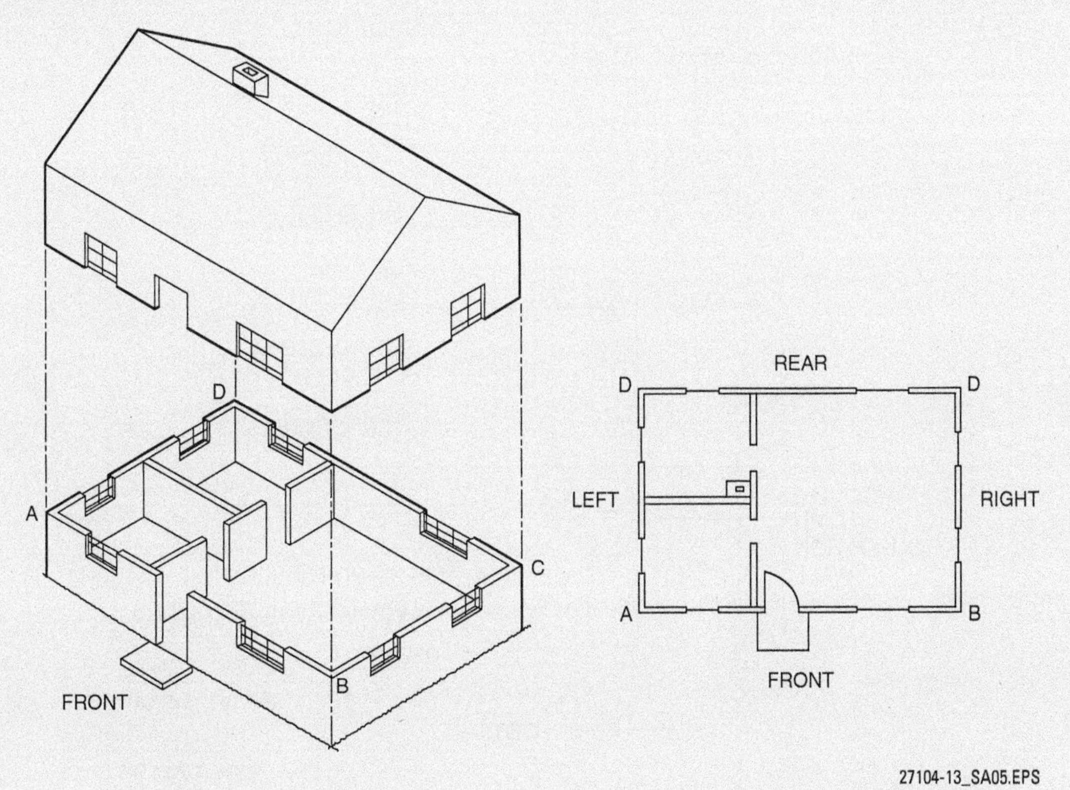

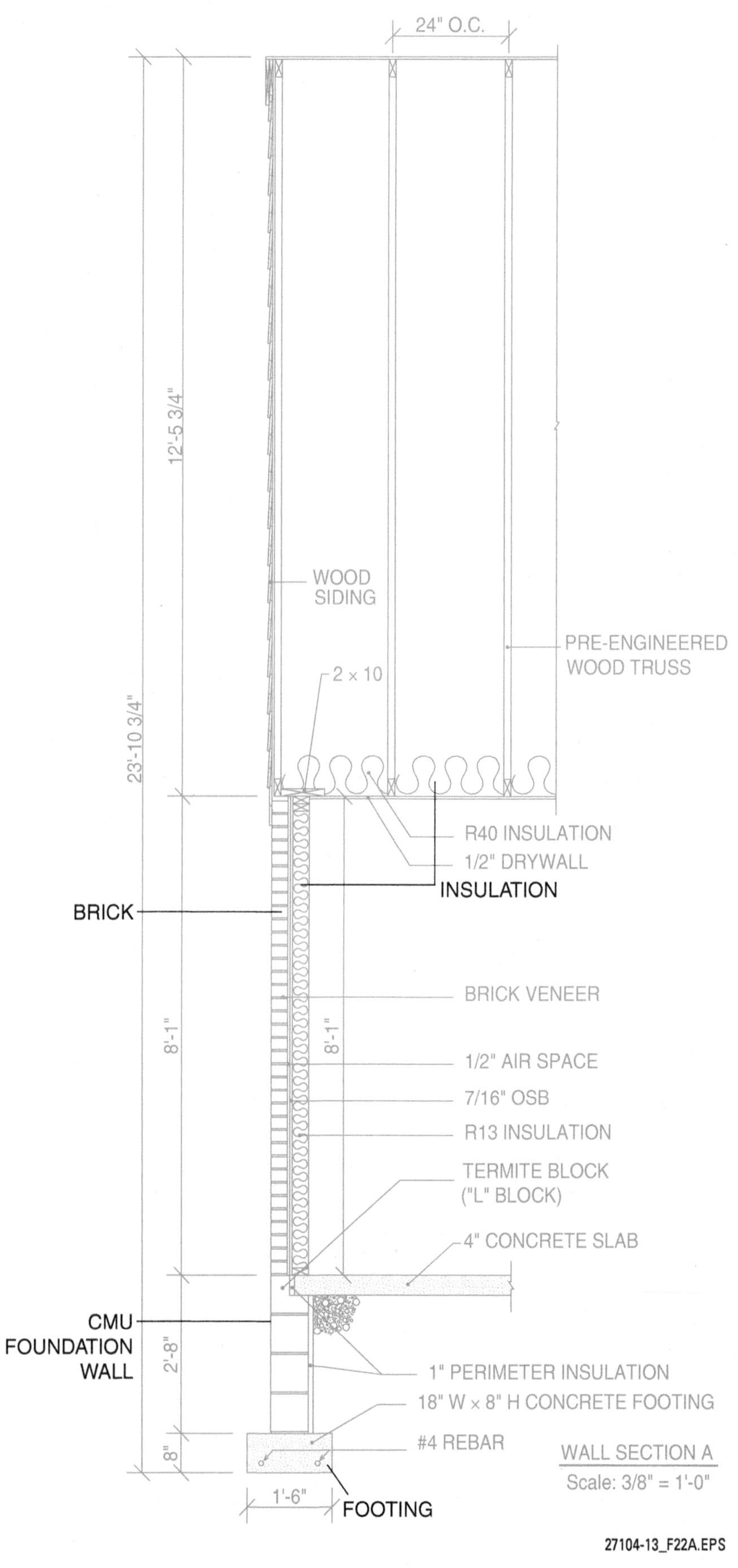

Figure 22 Section drawing (NTS, 1 of 3).

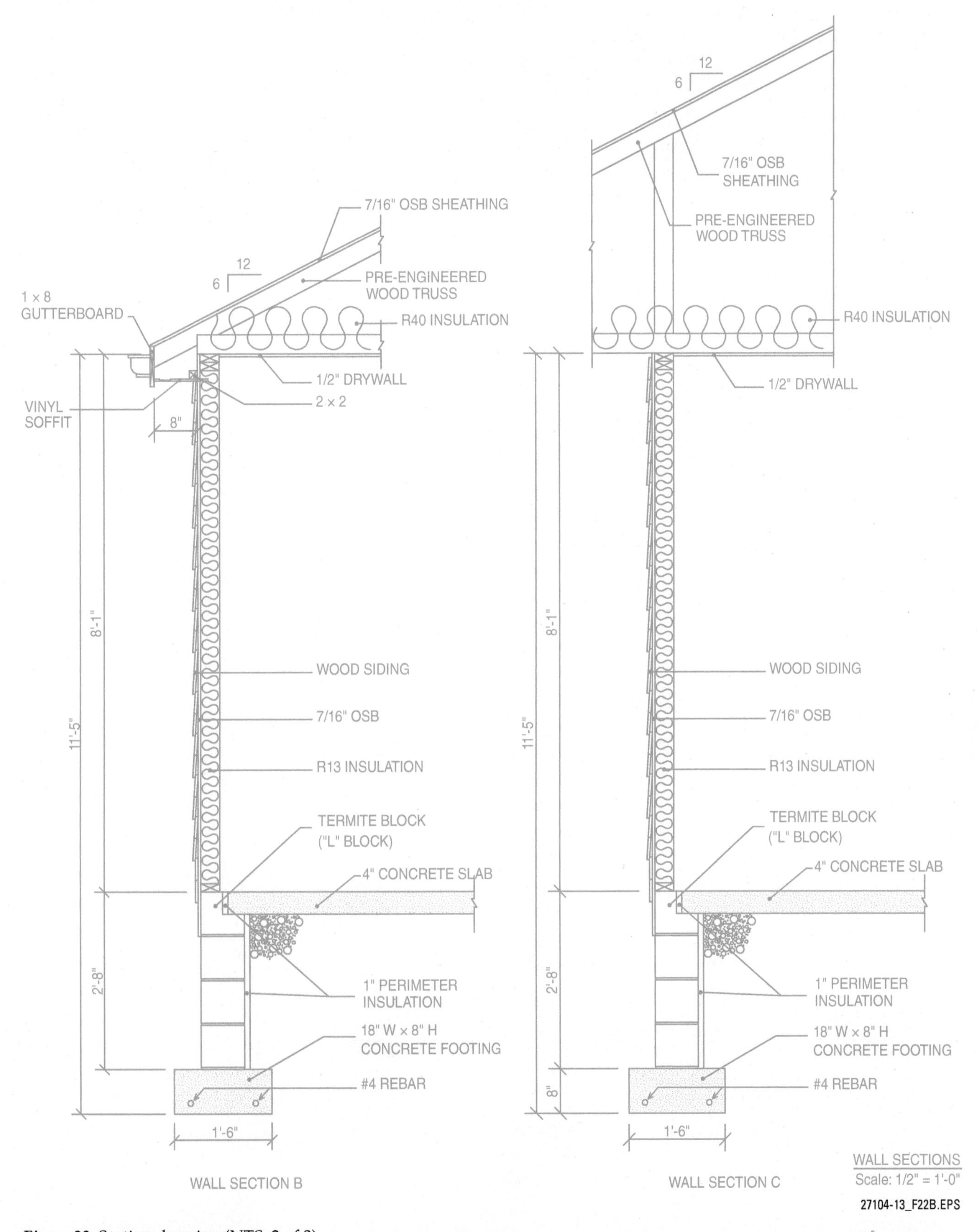

Figure 22 Section drawing (NTS, 2 of 3).

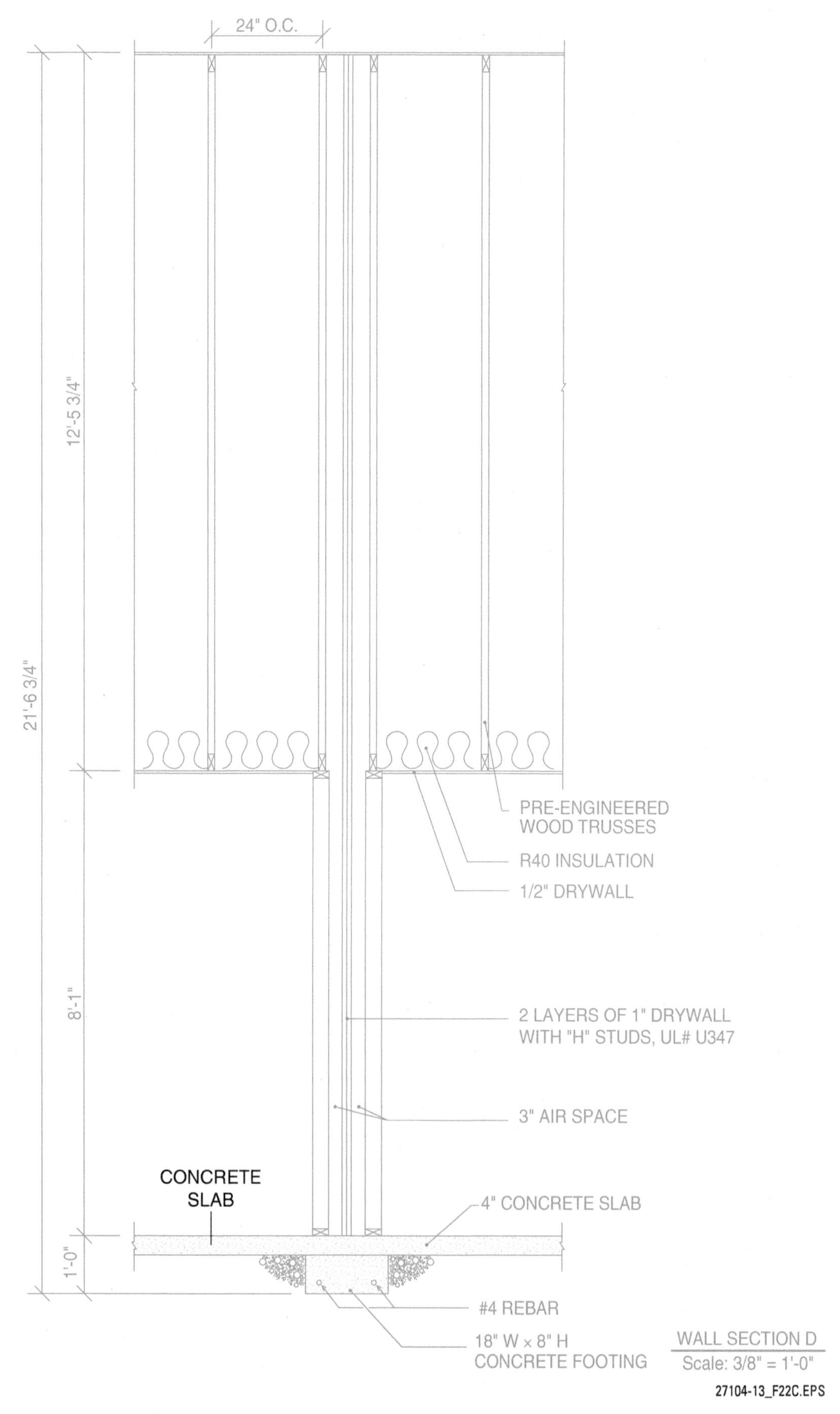

Figure 22 Section drawing (NTS, 3 of 3).

1.6.8 Detail Drawings

Detail drawings are enlargements of special features of a building or of equipment installed in a building. They are drawn to a larger scale in order to make the details clear. Detail drawings are often placed on the same sheet where the feature appears in the plan. *Figure 23* shows a series of detail drawings.

Typically, detail drawings are used for the following objects or situations:

- Footings and foundations, including anchor bolts, reinforcing, and control joints
- Beams, floor joists, bridging, and other support members
- Sills, floor framing, exterior walls, and vapor barriers
- Floor heights, thickness, expansion, and reinforcing
- Interior walls
- Windows, exterior and interior doors, and door frames
- Roofs, cornices, soffits, and ceilings
- Gravel stops, fascia, and flashing
- Fireplaces and chimneys
- Stairways and stair assemblies
- Millwork, trim, ornamental iron, and specialty items

1.6.9 Schedules

Schedules are tables that describe and specify the various types and sizes of construction materials used in a building. Door schedules, window schedules, and finish schedules are the most common schedules included on drawings (*Figure 24*). For commercial projects, additional schedules are provided for mechanical equipment and controls, plumbing fixtures, lighting fixtures, and any other equipment that needs to be listed separately.

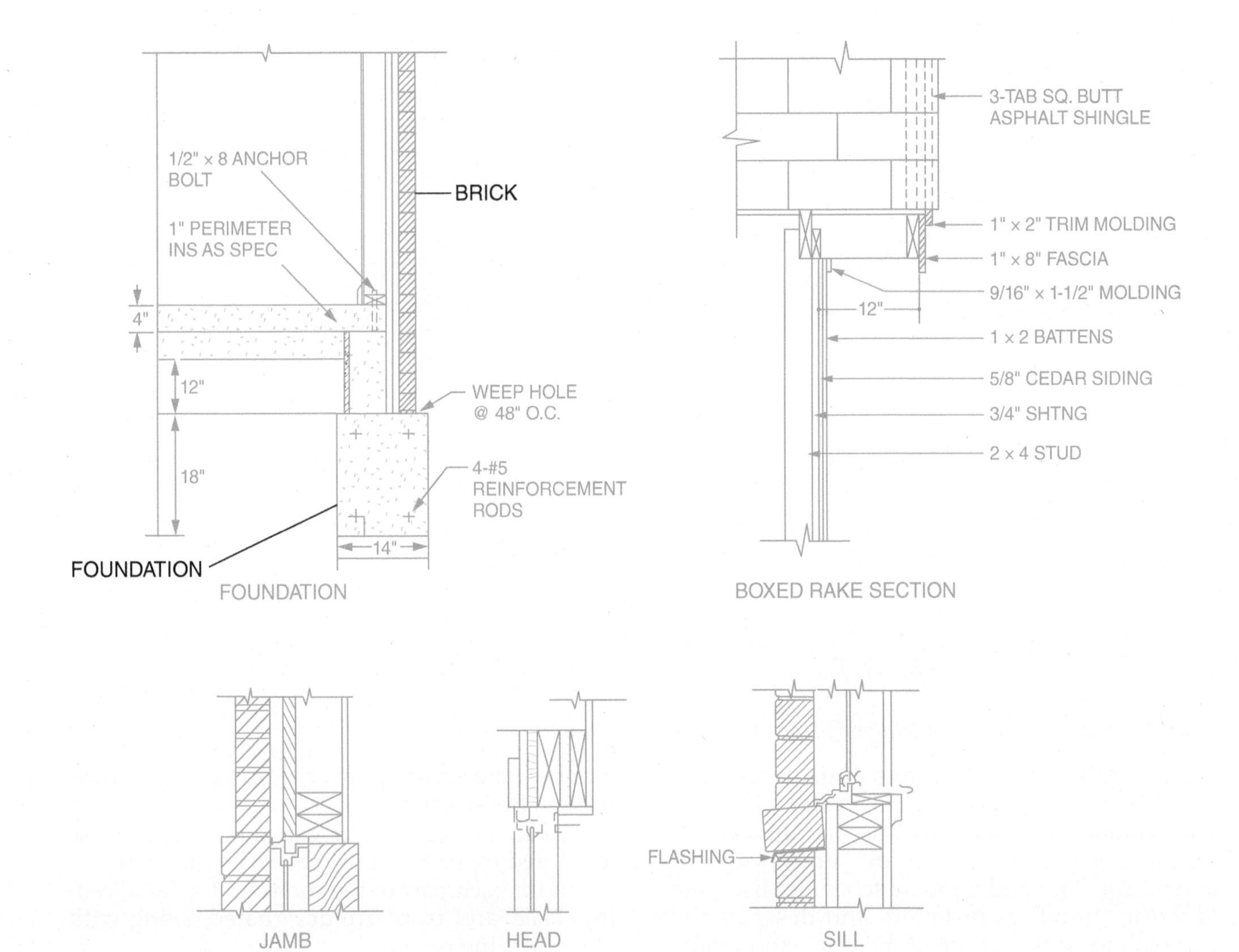

Figure 23 Detail drawings.

DOOR SCHEDULE

DOOR	WIDTH	HEIGHT	THICK-NESS	MAT'L	TYPE	STORM DOOR	QTY.	THRES-HOLD	REMARKS	MANUFACTURER
2068	2'-0"	6'-8"	1 3/8"	Wood-Ash	Hollow-core	NO	5	None	Oil Stain	LBJ Door Co.
2468	2'-4"	6'-8"	1 3/8"	Wood-Ash	Hollow-core	NO	1	None	Oil Stain	LBJ Door Co.
2668	2'-6"	6'-8"	1"	Wood-Ash	Cafe	NO	1 pr.	None	Oil Stain	LBJ Door Co.
2668	2'-6"	6'-8"	1 3/8"	Wood-Ash	Sliding Pocket	NO	1	None	Oil Stain	LBJ Door Co.
2668	2'-6"	6'-8"	1 3/8"	Wood-Ash	Hollow-core	1 Screen	6	Alum.	Screen door in garage	LBJ Door Co.
2868	2'-8"	6'-8"	1 3/4"	Metal Clad	Fireproof	YES	1	Alum.	Paint	LBJ Door Co.
3068	3'-0"	6'-8"	1 3/4"	Wood-Ash	Solid-core	NO	1	None	Oil Stain	LBJ Door Co.
3668	3'-6"	6'-8"	1 3/4"	Wood-Ash	Solid-core	YES	1	Alum.	Marine Varnish	LBJ Door Co.
6066	6'-0"	6'-6"	1/2"	Glass/Metal	Sliding	YES	1 pr.	Alum.	Sliding Screen	LBJ Door Co.
6068	6'-0"	6'-8"	1 1/4"	Wood-Ash	Bi-Fold	NO	2 sets	None	Oil Stain	LBJ Door Co.
1956	1'-9"	5'-6"	1/2"	Glass/Metal	Sliding Shower door	NO	2 sets	None	Frosted Glass	LBJ Door Co.

WINDOW SCHEDULE

SYMBOL	WIDTH	HEIGHT	MAT'L	TYPE	SCREEN & DOOR	QUANTITY	REMARKS	MANUFACTURER	CATALOG NUMBER
A	3'-8"	3'-0"	ALUM.	DOUBLE HUNG	YES	2	4 LIGHTS, 4 HIGH	LBJ Window Co.	141 PW
B	3'-8"	5'-0"	ALUM.	DOUBLE HUNG	YES	1	4 LIGHTS, 4 HIGH	LBJ Window Co.	145 PW
C	3'-0"	5'-0"	ALUM.	STATIONARY	STORM ONLY	2	SINGLE LIGHTS	H & J Glass Co.	59 PY
D	2'-0"	3'-0"	ALUM.	DOUBLE HUNG	YES	1	4 LIGHTS, 4 HIGH	LBJ Window Co.	142 PW
E	2'-0"	6'-0"	ALUM.	STATIONARY	STORM ONLY	2	20 LIGHTS	H & J Glass Co.	37 TS
F	3'-6"	5'-0"	ALUM.	DOUBLE HUNG	YES	1	16 LIGHTS, 4 HIGH	LBJ Window Co.	143 PW

HEADER SCHEDULE

HEADER SIZE	EXTERIOR		INTERIOR	
	26' + UNDER	26' TO 32'	26' + UNDER	26' TO 32'
(2) 2 x 4	3'-6"	3'-0"	USE (2) 2 x 6	
(2) 2 x 6	6'-6"	6'-0"	4'-0"	3'-0"
(2) 2 x 8	8'-6"	8'-0"	5'-6"	5'-0"
(2) 2 x 10	11'-0"	10'-0"	7'-0"	6'-6"
(2) 2 x 12	13'-6"	12'-0"	8'-6"	8'-0"

27104-13_F24.EPS

Figure 24 Examples of door, window, and header schedules.

Door, window, header, and finish schedules are of particular importance to a carpenter. Door and window types are identified on the various plan and elevation drawings by numbers and/or letters. The door and window schedules list these identifier numbers or letters and describe the corresponding size, type of material, and model number for each different type of door or window used in the structure.

In a finish schedule (*Figure 25*), each room is identified by name or number. The material and finish for each part of the room (walls, floor, ceiling, base, and trim) are designated, along with any clarifying remarks.

ROOM FINISH SCHEDULE																				
ROOMS	FLOOR				CEILING				WALL				BASE				TRIM			REMARKS
	CARPET	CERAMIC TILE	RUBBER TILE	CONCRETE	ACOUSTIC TILE	DRYWALL	PAINT	CERAMIC TILE	DRYWALL	PAINT	WALLPAPER	CERAMIC TILE	WOOD	RUBBER	CERAMIC TILE	STAIN	WOOD	STAIN	PAINT	
ENTRY		✓			✓				✓	✓	✓		✓			✓	✓	✓		See owner for all painting
HALL	✓				✓				✓	✓			✓			✓	✓	✓		
BEDROOM 1	✓				✓				✓	✓	✓		✓			✓	✓	✓		See owner for grade of carpet
BEDROOM 2	✓				✓				✓	✓			✓			✓	✓	✓		See owner for grade of carpet
BEDROOM 3	✓				✓				✓	✓			✓			✓	✓	✓		See owner for grade of carpet
BATH 1	✓	✓			✓			✓	✓	✓	✓	✓	✓			✓	✓	✓		Wallpaper 3 walls around vanity
BATH 2		✓			✓			✓	✓	✓	✓	✓			✓		✓	✓		Water-seal tile Wallpaper w/wall
UTIL + CLOSETS	✓		✓			✓	✓		✓	✓			✓			✓	✓	✓	✓	Use off-white flat latex
KITCHEN			✓		✓				✓	✓				✓			✓	✓		
DINING	✓				✓				✓	✓	✓		✓			✓	✓	✓		
LIVING	✓				✓				✓	✓			✓			✓	✓	✓		See owner for grade of carpet
GARAGE				✓		✓	✓					✓		✓			✓	✓		

27104-13_F25.EPS

Figure 25 Example of a finish schedule.

1.6.10 Structural Drawings

Structural drawings are created by a structural engineer and accompany the architect's plans. They are usually drawn for large structures such as office buildings or manufacturing facilities. Structural drawings show requirements for structural elements of the building, including columns, floor and roof systems, stairs, canopies, and loadbearing walls, and contain details such as:

- Heights of finished floors and walls
- Height and bearing of bar joists or steel joists
- Locations of bearing steel materials
- Height of steel beams, concrete plank, concrete Ts, and cast-in-place concrete
- Bearing plate locations
- Location, size, and spacing of anchor bolts
- Stairways

1.6.11 Mechanical, Electrical, and Plumbing Plans

For many construction jobs, information for the mechanical, electrical, and plumbing (MEP) craftworkers are included on the basic floor plans (*Figure 26*). However, for complex commercial projects, MEP information is generally shown on separate plan drawings.

Plumbing plans show the layout of water supply and sewage disposal systems and fixtures. Plumbing system plans generally have a separate plumbing **riser diagram** that shows the layout and identification of the piping and fixtures. A plumbing legend is usually located on the plan.

Mechanical plans show details for installation of the HVAC and other equipment. Such plans usually include a refrigerant piping schematic that shows the types and sizes of the piping and identification of the fittings. The plan may also in-

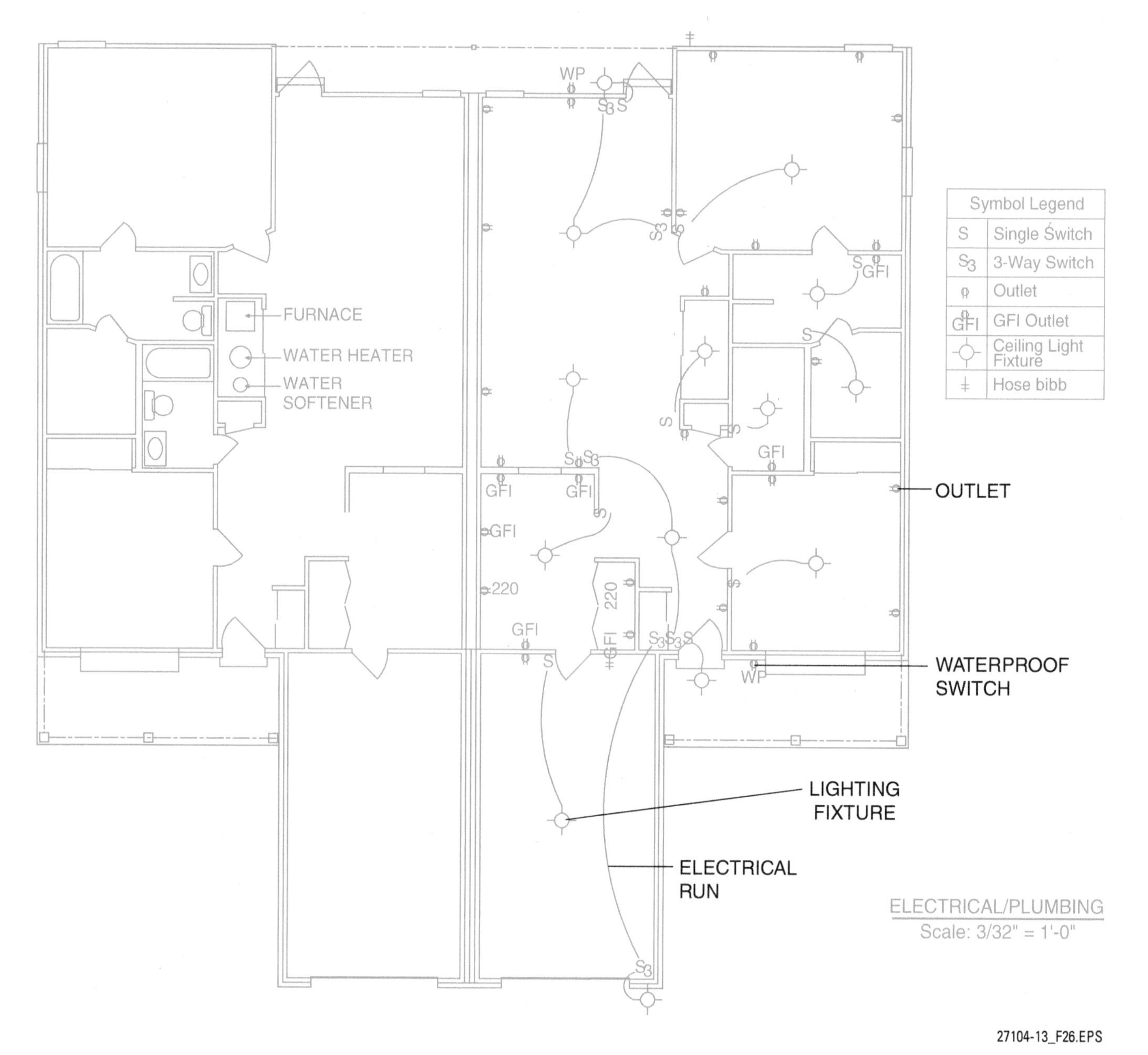

Figure 26 Example of an electrical/plumbing plan (NTS).

clude a legend defining the HVAC symbols used on the drawing.

Electrical plans show the details for installation of the electrical system and equipment. The locations of the meter, distribution panel, fixtures, switches, and special items are indicated, along with specifications for load capacities and wire sizes. The plan may also include a legend defining the symbols used on the drawing.

1.6.12 Shop Drawings

Shop drawings are specialized drawings that show how to fabricate and install components of a construction project. One type of shop drawing that may be created after an engineer designs the structure is a detail drawing showing the locations of all holes and openings, and provides notes specifying how the part is to be made. Assembly instructions may be included. This type of drawing is used principally for structural steel members.

Another type of shop drawing, commonly referred to as submittal drawings, pertains to special items or equipment purchased for installation in a building. Submittal drawings are usually prepared by equipment manufacturers. They show overall sizes, details of construction, methods of securing the equipment to the structure, and all pertinent data that the architect and contractor need to know for the final placement and installation of the equipment.

Shop drawings produced by a contractor or fabricator are usually submitted to the owner or architect for approval and revisions or corrections. The design drawing is often put on the same sheet as the shop drawing.

1.6.13 As-Built Drawings

As-built drawings are drawings that are formally incorporated into the drawing set to record changes. As-built drawings are created by the various trades by marking up a set of prints to show any differences between what was originally shown on a plan by the architect or engineer and what was actually built. Such changes result from the need to relocate equipment to avoid obstructions; to alter the location of a door, window, wall, or other feature, for some reason; or because the architect has changed a certain detail in the building design in response to customer preferences. On many jobs, any such changes to the design can only be made after a formal document called a change order has been generated and approved by the architect or other designated person. Depending on the complexity of the change, as-built drawings are typically outlined with a unique design or marked in red ink to make sure they stand out. Changes must be dated and initialed by the responsible party.

1.6.14 Soil Reports

Soil conditions are one of the factors that determine the type of foundation best suited for a structure. Building a structure on soil where the soil conditions can cause a large amount of uneven settlement to occur can result in cracks in the foundation and structural damage to the rest of the building. Therefore, in designing the foundation for a structure, an architect must consider the soil conditions of the building site. Typically, the architect consults a soil engineer who makes test bores of the soil on the building site and analyzes the samples. The results of the soil analysis are summarized in a soil report issued by the engineer. The soil report provides information about the loadbearing capacity of the soil and/or rock, and thereby provides direction to the architect. A soil report is often included as part of the drawing set.

1.6.15 Guidelines for Reading a Set of Drawings

The following general procedure is suggested as a method of reading a set of drawings for maximum understanding.

Step 1 Acquire a complete set of drawings and specifications—including the title sheet(s)—so you can better understand the abbreviations, symbols, etc., used throughout the drawings.

Step 2 Read the title block, taking note of the critical information such as the scale, date of last revision, drawing number, and architect or engineer. After you use a sheet from a set of drawings, be sure to refold the sheet with the title block facing up.

Step 3 Locate the north arrow. Always orient yourself to the structure. Knowing where north is enables you to more accurately describe the location of walls and other parts of the building.

Step 4 Be aware that the drawings in the set work together as a group. An architect or engineer draws multiple plans, elevations, and sections since more than one type of view is required to communicate the design concept. Learn how to use more than one drawing when necessary to locate needed information.

Step 5 Check the list of drawings in the set and note the sequence of the various plans. Some drawing sets have an index on the front cover. Notice that the prints in the set are of several categories:

- Architectural
- Structural
- Mechanical
- Electrical
- Plumbing

Using Mechanical Drawings

While carpenters usually work with architectural and structural drawings, useful notes and views are included on drawings in other sections, especially the mechanical section. For example, items not typically shown on architectural drawings that may be included on mechanical drawings include exposed HVAC ductwork, heating convectors, and fire sprinkler piping. Any one or all of these items may require the carpenter to build special framing, so make a habit of reviewing all drawings.

Step 6 Study the site plan to determine the location of the building to be constructed, the various utilities, roadways, and any easements.

Step 7 Check the floor plan to determine the orientation of the building. Observe the location and features of entries, corridors, offsets, and any special features.

Step 8 Check the foundation plan for the size and types of footings, reinforcing steel, and loadbearing substructures.

Step 9 Check the floor and wall construction and other details relating to exterior and interior walls.

Step 10 Study the features that extend for more than one floor, such as plumbing and vents, stairways, elevator shafts, and heating and cooling ductwork.

Step 11 Study the plumbing and/or mechanical plans for details of any framing or other carpentry work pertaining to the installation of the heating, cooling, and plumbing equipment. You must coordinate this type of work with members of the HVAC and plumbing trades.

Step 12 Study the electrical plans for details of any framing or other carpentry work pertaining to the installation of the electrical system, fixtures, or special equipment. You must coordinate this type of work with members of the electrical trade.

Step 13 Check the notes on various pages and compare the specifications against the construction details.

Step 14 Thumb through the sheets of drawings until you are familiar with all the plans and structural details.

Step 15 Recognize applicable symbols and their relative locations in the plans. Note any special construction details or variations that will affect the carpentry work.

1.0.0 Section Review

1. The types of lines used on a construction drawing to show parts that are obstructed by another surface or part that is closer to the viewer are _____.
   a. object lines
   b. leader lines
   c. center lines
   d. hidden lines

2. Additional information about doors and windows can be located _____.
   a. in a separate door and window schedule
   b. in the legend
   c. in the title block
   d. on the cover sheet of the drawings

3. The type of welding to be performed is indicated using a _____.
   a. welding symbol
   b. schematic
   c. weld symbol
   d. plumbing symbol

4. The same abbreviations are used by all architects and engineers.
   a. True
   b. False

5. If a line on a drawing actually measures 4" long and the drawing scale is ½" = 1'-0", the line represents an item that is _____.
   a. 4' long
   b. 8' long
   c. 12' long
   d. 16' long

6. The construction drawing that commonly includes the direction and length of property boundaries, location of utilities, and existing grade contours is the _____.
   a. foundation plan
   b. floor plan
   c. mechanical plan
   d. site plan

SECTION TWO

2.0.0 SPECIFICATIONS

Objective

State the purpose of written specifications.

a. Describe how specifications are organized.
b. Explain the importance of building codes in construction.

Performance Task 5

Read and interpret written specifications.

Specifications are written instructions provided by architectural and engineering firms to the general contractor and, consequently, to the subcontractors. Specifications are a contractual document used along with the construction drawings, and are just as important as the drawings themselves. Specifications furnish what the drawings cannot, in that they define the quality of work to be done and the materials to be used. Specifications serve several important purposes, including to:

- Clarify information that cannot be shown on the drawings
- Identify work standards, types of materials to be used, and the responsibility of various parties to the contract
- Provide information on details of construction
- Serve as a guide for contractors bidding on the construction project
- Serve as a standard of quality for materials and workmanship
- Serve as a guide for compliance with building codes and zoning ordinances
- Serve as the basis of agreement between the owner, architect, and contractors in settling any disputes

Specifications are legal documents. When there is a difference between construction drawings and specifications, the specifications usually take legal precedence over the working drawings. However, it is often the case that the plans are more specific to the job than the specifications. Therefore, notes on the plans may be considered by the architect/owner to be the true intent. Carefully watch for discrepancies between the plans and specifications and report them to your supervisor immediately.

2.1.0 Specifications Organization

Specifications consist of various elements, which may differ between construction jobs. For small projects, specifications may be simple; for large projects, complex. Basically, two types of information are contained in a set of specifications: special and general conditions, and technical aspects of construction.

2.1.1 Special and General Conditions

Special and general conditions cover the non-technical aspects of the contractual agreements. Special conditions cover topics such as safety and temporary construction. General conditions cover the following points of information:

- Contract terms
- Responsibilities for examining the construction site
- Types and limits of insurance
- Permits and payments of fees
- Use and installation of utilities
- Supervision of construction
- Other pertinent items

The general conditions section is the area of the construction contract where misunderstandings often occur. These conditions are usually much more explicit on large, complicated construction projects. Note that residential specifications often do not spell out general conditions and are basically material specifications only. An example of a typical residential material specification is shown in *Figure 27.*

2.1.2 Technical Aspects

The technical aspects section includes information on materials that are specified by standard numbers and by standard national testing orga-

Notes on Drawings

Sometimes the notes on drawings will contradict the specifications or be inconsistent with other requirements. Even though the specifications usually take precedence, the notes often are closer to the true intent. In any case, clarify any such discrepancy with a supervisor when it is noticed.

Jack DeMusse & Associates
East End Clinic Project

Issue Date: January 6, 2013

SECTION 061000 - ROUGH CARPENTRY

PART 1 - GENERAL

1.01 SECTION INCLUDES

A. Structural dimension lumber framing.

B. Exposed timber structural framing.

C. Roof-mounted curbs.

D. Roofing nailers.

E. Roofing cant strips.

F. Preservative treated wood materials.

G. Fire retardant treated wood materials.

H. Communications and electrical room mounting boards.

I. Concealed wood blocking, nailers, and supports.

J. Miscellaneous wood nailers, furring, and grounds.

1.02 RELATED REQUIREMENTS

A. Section 054000 - Cold Formed Metal Framing:

B. Section 061500 - Wood Decking.

C. Section 061800 - Structural Glued-Laminated Timber.

D. Section 076200 - Sheet Metal Flashing and Trim: Sill flashings.

1.03 SUBMITTALS

A. Product Data: Provide technical data on wood preservative materials, application instructions, and fasteners and adhesives.

B. Samples: For rough carpentry members that will be exposed to view, submit two samples, ____x____ inch (____x____ mm) in size illustrating wood grain, color, and general appearance.

C. Manufacturer's Certificate: Certify that wood products supplied for rough carpentry meet or exceed specified requirements.

1.04 DELIVERY, STORAGE, AND HANDLING

A. General: Cover wood products to protect against moisture. Support stacked products to prevent deformation and to allow air circulation.

B. Fire Retardant Treated Wood: Prevent exposure to precipitation during shipping, storage, or installation.

PART 2 - PRODUCTS

2.01 GENERAL REQUIREMENTS

A. Dimension Lumber: Comply with PS 20 and requirements of specified grading agencies.

ROUGH CARPENTRY 061000 - 1

27104-13_F27A.EPS

Figure 27 Typical specification (1 of 5).

nizations such as the American Society of Testing and Materials (ASTM). The technical data section of specifications can be of three types:

- *Outline specifications*—These specifications list the materials to be used in order of the basic parts of the job, such as foundation, floors, and walls.
- *Fill-in specifications*—This is a standard form filled in with pertinent information. It is typically used on smaller jobs.
- *Complete specifications*—For ease of use, most specifications written for large construction jobs are organized in the Construction Specifications Institute format called the *MasterFormat*™, which is discussed in the next section.

Jack DeMusse & Associates Issue Date: January 6, 2013
East End Clinic Project

1. If no species is specified, provide any species graded by the agency specified; if no grading agency is specified, provide lumber graded by any grading agency meeting the specified requirements.
2. Grading Agency: Any grading agency whose rules are approved by the Board of Review, American Lumber Standard Committee (www.alsc.org) and who provides grading service for the species and grade specified; provide lumber stamped with grade mark unless otherwise indicated.

B. Lumber fabricated from old growth timber is not permitted.

2.02 DIMENSION LUMBER FOR CONCEALED APPLICATIONS

A. See Structural Notes for additional information.

B. Grading Agency: West Coast Lumber Inspection Bureau (WCLIB). — DIMENSION LUMBER REQUIREMENTS

C. Sizes: Nominal sizes as indicated on drawings, S4S.

D. Moisture Content: S-dry or MC19.

E. Miscellaneous Framing, Blocking, Nailers, Grounds, and Furring:
1. Lumber: S4S, No. 2 or Standard Grade.
2. Boards: Standard or No. 3.

2.03 EXPOSED DIMENSION LUMBER

A. Sizes: Nominal sizes as indicated on drawings, S4S.

B. Moisture Content: S-dry or MC19.

2.04 CONSTRUCTION PANELS

A. Communications and Electrical Room Mounting Boards: PS 1 C-D Plugged or better plywood, or medium density fiberboard; 3/4 inch (19 mm) thick; flame spread index of 25 or less, smoke developed index of 450 or less, when tested in accordance with ASTM E 84. — PANEL REQUIREMENTS

B. Other Applications:
1. Plywood Concealed From View But Located Within Exterior Enclosure: PS 1, C-C Plugged or better, Exterior grade.
2. Plywood Exposed to View But Not Exposed to Weather: PS 1, A-D, or better.
3. Other Locations: PS 1, C-D Plugged or better.

2.05 ACCESSORIES

A. Fasteners and Anchors:
1. Metal and Finish: Stainless steel for exterior, high humidity or preservative-treated wood locations, unfinished steel elsewhere.

2.06 FACTORY WOOD TREATMENT

A. Treated Lumber and Plywood: Comply with requirements of AWPA U1 - Use Category System for wood treatments determined by use categories, expected service conditions, and specific applications. — TREATED LUMBER
1. Fire-Retardant Treated Wood: Mark each piece of wood with producer's stamp indicating compliance with specified requirements.
2. Preservative-Treated Wood: Provide lumber and plywood marked or stamped by an ALSC-accredited testing agency, certifying level and type of treatment in accordance with AWPA standards.

B. Fire Retardant Treatment:
1. Manufacturers:

ROUGH CARPENTRY 061000 - 2

27104-13_F27B.EPS

Figure 27 Typical specification (2 of 5).

Addenda and Change Orders

Addenda and change orders are contractual documents used to correct or make changes to original construction drawings and specifications. The difference between the two documents is a matter of timing. An addendum is written before the contract is awarded, while a change order is drawn up after the award of the contract.

Jack DeMusse & Associates   Issue Date: January 6, 2013
East End Clinic Project

a. Arch Wood Protection, Inc: www.wolmanizedwood.com.
b. Hoover Treated Wood Products, Inc: www.frtw.com.
c. Osmose, Inc: www.osmose.com.
d. Substitutions: See Section 016000 - Product Requirements.

2. Exterior Type: AWPA U1, Category UCFB, Commodity Specification H, chemically treated and pressure impregnated; capable of providing a maximum flame spread rating of 25 when tested in accordance with ASTM E 84, with no evidence of significant combustion when test is extended for an additional 20 minutes both before and after accelerated weathering test performed in accordance with ASTM D 2898.
   a. Kiln dry wood after treatment to a maximum moisture content of 19 percent for lumber and 15 percent for plywood.
   b. Treat exposed exterior rough carpentry items, including stairways, balconies, and covered walkways
   c. Do not use treated wood in direct contact with the ground.
3. Interior Type A: AWPA U1, Use Category UCFA, Commodity Specification H, low temperature (low hygroscopic) type, chemically treated and pressure impregnated; capable of providing a maximum flame spread rating of 25 when tested in accordance with ASTM E 84, with no evidence of significant combustion when test is extended for an additional 20 minutes.
   a. Kiln dry wood after treatment to a maximum moisture content of 19 percent for lumber and 15 percent for plywood at time of incorporation into the work.
   b. All interior rough carpentry items are to be fire retardant treated.
   c. Do not use treated wood in applications exposed to weather or where the wood may become wet.

C. Preservative Treatment:
1. Manufacturers:
   a. Arch Wood Protection, Inc; Product : www.wolmanizedwood.com.
   b. Chemical Specialties, Inc; Product : www.treatedwood.com.
   c. Osmose, Inc; Product : www.osmose.com.
   d. Conrad Forest Products .www.conradfp.com
   e. Substitutions: See Section 016000 - Product Requirements.
2. Preservative Pressure Treatment of Lumber Above Grade: AWPA U1, Use Category UC3B, Commodity Specification A using waterborne preservative to 0.25 lb/cu ft (4.0 kg/cu m) retention.
   a. Kiln dry lumber after treatment to maximum moisture content of 19 percent.
   b. Treat lumber exposed to weather.
   c. Treat lumber in contact with roofing, flashing, or waterproofing.
   d. Treat lumber in contact with masonry or concrete.
   e. Treat lumber less than 18 inches (450 mm) above grade.
3. Preservative Pressure Treatment of Plywood Above Grade: AWPA U1, Use Category UC2 and UC3B, Commodity Specification F using waterborne preservative to 0.25 lb/cu ft (4.0 kg/cu m) retention.
   a. Kiln dry plywood after treatment to maximum moisture content of 19 percent.
   b. Treat plywood in contact with roofing, flashing, or waterproofing.
   c. Treat plywood in contact with masonry or concrete.
   d. Treat plywood less than 18 inches (450 mm) above grade.

ROUGH CARPENTRY   061000 - 3

27104-13_F27C.EPS

Figure 27 Typical specification (3 of 5).

2.1.3 Specifications Format

For convenience in writing, speed in estimating, and ease of reference, the most suitable organization of the specifications is a series of sections dealing with the construction requirements, products, and activities, and that is easily understandable by the different trades. Those people who use the specifications must be able to find all information needed without spending too much time looking for it.

The most commonly used specification format in North America is *MasterFormat*™. This standard was developed jointly by the Construction Specifications Institute (CSI) and Construction Specifications Canada (CSC). Prior to 2004, the organization of construction specifications and supplier's catalogs was based on a standard with 16

Jack DeMusse & Associates
East End Clinic Project

Issue Date: January 6, 2013

PART 3 - EXECUTION

3.01 PREPARATION

A. Coordinate installation of rough carpentry members specified in other sections.

3.02 INSTALLATION - GENERAL

A. Select material sizes to minimize waste.

B. Reuse scrap to the greatest extent possible; clearly separate scrap for use on site as accessory components, including: shims, bracing, and blocking.

C. Where treated wood is used on interior, provide temporary ventilation during and immediately after installation sufficient to remove indoor air contaminants.

3.03 BLOCKING, NAILERS, AND SUPPORTS

A. Provide framing and blocking members as indicated or as required to support finishes, fixtures, specialty items, and trim.

B. In framed assemblies that have concealed spaces, provide solid wood fireblocking as required by applicable local code, to close concealed draft openings between floors and between top story and roof/attic space; other material acceptable to code authorities may be used in lieu of solid wood blocking.

C. In metal stud walls, provide continuous blocking around door and window openings for anchorage of frames, securely attached to stud framing.

D. In walls, provide blocking attached to studs as backing and support for wall-mounted items, unless item can be securely fastened to two or more studs or other method of support is explicitly indicated.

E. Where ceiling-mounting is indicated, provide blocking and supplementary supports above ceiling, unless other method of support is explicitly indicated.

F. Specifically, provide the following non-structural framing and blocking:
1. Cabinets and shelf supports.
2. Wall brackets.
3. Handrails.
4. Grab bars.
5. Towel and bath accessories.
6. Wall-mounted door stops.
7. Chalkboards and marker boards.
8. Wall paneling and trim.
9. Joints of rigid wall coverings that occur between studs.

3.04 ROOF-RELATED CARPENTRY

A. Coordinate installation of roofing carpentry with deck construction, framing of roof openings, and roofing assembly installation.

B. Provide wood curb at all roof openings except where prefabricated curbs are specified and where specifically indicated otherwise. Form corners by alternating lapping side members.

3.05 INSTALLATION OF CONSTRUCTION PANELS

A. Communications and Electrical Room Mounting Boards: Secure with screws to studs with edges over firm bearing; space fasteners at maximum 24 inches (610 mm) on center on all edges and into studs in field of board.

ROUGH CARPENTRY 061000 - 4

27104-13_F27D.EPS

Figure 27 Typical specification (4 of 5).

sections, known as divisions. The divisions and their subsections were individually identified by a five-digit numbering system. The first two digits represented the division number and the next three individual numbers represented successively lower levels of breakdown. For example, the number 13213 represented Division 13, subsection 2, sub-subsection 1 and sub-sub-subsection 3.

In 2004, the *MasterFormat*™ standard underwent a major change. What had been 16 divisions was expanded to four major groupings and 49 divisions with some divisions reserved for future expansion (*Figure 28*). The first 14 divisions of *MasterFormat*™ 2012 are essentially the same as the old format. Subjects under the old Division 15—Mechanical have been relocated to new Divisions 22 and 23. The basic subjects under old Division 16—Electrical have been relocated to new Divisions 26 and 27.

Jack DeMusse & Associates
East End Clinic Project
Issue Date: January 6, 2013

1. At fire-rated walls, install board over wall board indicated as part of the fire-rated assembly.
2. Where boards are indicated as full floor-to-ceiling height, install with long edge of board parallel to studs.
3. Install adjacent boards without gaps.
4. Size and Location: As indicated on drawings.
5. Paint all mounting boards. leave one copy of firetreatment stamp visible (unpainted) for building inspector

3.06 SITE APPLIED WOOD TREATMENT

A. Apply preservative treatment compatible with factory applied treatment at site-sawn cuts, complying with manufacturer's instructions.

B. Allow preservative to dry prior to erecting members.

3.07 TOLERANCES

A. Framing Members: 1/4 inch (6 mm) from true position, maximum. — CONSTRUCTION TOLERANCES

B. Variation from Plane (Other than Floors): 1/4 inch in 10 feet (2 mm/m) maximum, and 1/4 inch in 30 feet (7 mm in 10 m) maximum.

3.08 CLEANING

A. Waste Disposal: Comply with the requirements of Section 017419.
1. Comply with applicable regulations.
2. Do not burn scrap on project site.
3. Do not burn scraps that have been pressure treated.
4. Do not send materials treated with pentachlorophenol, CCA, or ACA to co-generation facilities or "waste-to-energy" facilities.

B. Do not leave any wood, shavings, sawdust, etc. on the ground or buried in fill.

C. Prevent sawdust and wood shavings from entering the storm drainage system.

END OF SECTION

ROUGH CARPENTRY 061000 - 5

27104-13_F27E.EPS

Figure 27 Typical specification (5 of 5).

In addition, the numbering system for the new *MasterFormat*™ organization was changed to six digits preceding the decimal point to allow for more subsections in each division. In the new numbering system, the first two digits represent the division number. The next two digits represent subsections of the division and the two remaining digits represent the third level sub-subsection numbers. The fourth level, if required, is a decimal and number added to the end of the last two digits. For example, the number 132013.04 represents Division 13, subsection 20, sub-subsection 13, and sub-sub-subsection 04.

2.2.0 Building Codes

Specifications are written to conform to building codes and the best practices of the construction trade. Model building codes provide minimum

DIVISIONS NUMBERS AND TITLES

PROCUREMENT AND CONTRACTING REQUIREMENTS GROUP	
Division 00	Procurement and Contracting Requirements
SPECIFICATION GROUP	
GENERAL REQUIREMENTS SUBGROUP	
Division 01	General Requirements
FACILITY CONSTRUCTION SUBGROUP	
Division 02	Existing Conditions
Division 03	Concrete
Division 04	Masonry
Division 05	Metals
Division 06	Wood, Plastics, and Composites
Division 07	Thermal and Moisture Protection
Division 08	Openings
Division 09	Finishes
Division 10	Specialties
Division 11	Equipment
Division 12	Furnishings
Division 13	Special Construction
Division 14	Conveying Equipment
Division 15	*Reserved*
Division 16	*Reserved*
Division 17	*Reserved*
Division 18	*Reserved*
Division 19	*Reserved*

27104-13_F28.EPS

Figure 28 Excerpt from the *MasterFormat*™ 2012 organization.

standards to guard the life, health, and safety of the public by regulating and controlling the design, construction, and quality of materials used in construction. Model building codes have also come to govern the use and occupancy, location of a type of building, and the ongoing maintenance of buildings and facilities. The model building codes are the legal instruments that enforce public safety in construction of human habitation and assembly structures. They are used not only in the construction industry but also by the insurance industry for compensation appraisals and claims adjustments, and by the legal industry for court litigation.

Until 2000, there were three model building codes. The three code-writing groups—Building Officials and Code Administrators (BOCA), Inter-

Specifications

Specifications supplement the related construction drawings in that they contain details not shown on the drawings. Specifications define and clarify the scope of the job. They describe the specific types and characteristics of the components that are to be used on the job and the installation methods for some of them. Many components are identified specifically by the manufacturer's model and part numbers. This type of information is used to purchase the various items of hardware needed to accomplish the installation in accordance with the contractual requirements.

national Conference of Building Officials (ICBO), and Southern Building Code Congress International (SBCCI)—combined their efforts under the umbrella of the International Code Council (ICC) with the purpose of writing one nationally accepted family of building and fire codes. The first editions of the *International Residential Code®* and *International Building Code®* were published in 2000, and continue to be updated on a three-year cycle.

- The *International Residential Code®* (*IRC*) addresses the design and construction of one- and two-family dwellings and townhouses.
- The *International Building Code®* (*IBC*) pertains to all other building construction not covered by the IRC.

Model building codes are written to be just that: models. Local jurisdictions, such as states, counties, or cities, model their building codes after the *IRC/IBC* and adopt them as their own. Once adopted by a local jurisdiction, the model building codes then become law. It is common for local jurisdictions to change and add new requirements to adopted model code requirements to provide more stringent requirements and/or meet local needs. The provisions of the building codes apply to the construction, alteration, movement, demolition, repair, structural maintenance, and use of any building or structure within the local jurisdiction. An important general rule about codes is that in almost every case the most stringent local code will apply.

In 2002, the National Fire Protection Association (NFPA) published its own building code, *NFPA 5000®*. There are now two nationally recognized model codes competing for adoption by the 50 states.

The format and chapter organization of the *IRC/IBC* and *NFPA 5000®* differ, but the content and subjects covered are generally the same. Both codes cover all types of occupancies from single-family residences to high-rise office buildings, as well as industrial facilities. They also cover structures, building materials, and building systems, including life safety systems.

A carpenter should be aware of the federal laws, building codes, and restrictions that affect the specific job being constructed. This should also include a basic understanding of the codes that pertain mainly to other trades, such as the NFPA gas and electrical codes.

Project Organization

Many people are involved in the design, planning, and execution of a project. The following description provides insight into the many people involved in project organization and the roles they play:

- *Building owner*—The owner can be either a public or private entity. Public projects are funded with public money from sources such as taxes, fees, and donations. Public owners can include federal, state, and local governments, school boards, park commissions, public universities, and transit authorities. Private projects are funded with private money from individuals, companies, and institutions.
- *Architect/engineer (A/E)*—The architect is the person or firm who does the design work for a project. They produce the drawings that govern every aspect of the project. In many cases, architectural firms also do engineering work because in commercial work a great deal of mechanical, structural, and electrical engineering, as well as other design work is required. Hence the term *A/E (architect/engineer)*. Some of the engineering activities include the following:

  Structural engineering — The structural engineer designs the frame of the building.
  Site engineering — The site engineer designs everything outside the footprint (perimeter) of the building.
  Soils engineer — The soils engineer tests the soil and may also perform concrete testing.
  Mechanical engineering — The mechanical engineer designs the HVAC and plumbing systems.
  Electrical engineering — The electrical engineer designs the lighting and electrical distribution systems.
- *General contractor (GC)*—In many cases, the GC manages the entire construction project, following the plans developed by the architect/engineer. The GC will schedule all the work on the project, buy the materials, and hire subcontractors to perform carpentry, plumbing, electrical, roofing, painting, and other work.
- *Construction manager (CM)*—The CM becomes involved earlier in the project than the GC. The CM provides construction expertise throughout the design phase of the project and acts as the owner's agent for the duration of the project. The GC often acts as the CM.

2.0.0 Section Review

1. The numbering system for *MasterFormat*™ 2012 contains _____.
   a. four digits
   b. six digits
   c. ten digits
   d. sixteen digits

2. The *International Building Code*® was first introduced in _____.
   a. 1986
   b. 1998
   c. 2000
   d. 2012

SECTION THREE

3.0.0 SQUARING A BUILDING

Objective

Identify the methods of squaring a building.

Performance Task

Establish 90-degree angles using the 3-4-5 rule.

Based on the information shown on the prints and included in the specifications, the building is located on the job site and the building corners are identified and staked. Most walls of buildings are perpendicular (90 degrees) to the adjacent wall. Thus, the walls must be square to one another when being laid out and constructed. The 3-4-5 rule has been used in construction for centuries to lay out or check a 90-degree angle without the use of a builder's level or transit-level.

The numbers 3-4-5 represent dimensions that describe the sides of a right triangle. When squaring smaller assemblies, the dimensions may be in inches. When squaring larger items, such as a building, the dimensions are in feet. The 3-4-5 rule is based on the Pythagorean theorem. It states that in any right triangle, the square of the longest side, called the hypotenuse (C), is equal to the sum of the squares of the two shorter sides (A and B). Stated mathematically:

$$C^2 = A^2 + B^2$$

Accordingly, for the 3-4-5 right triangle:

$$5^2 = 3^2 + 4^2$$

$$25 = 9 + 16$$

$$25 = 25$$

This theorem also applies if you multiply each number in the ratio (3, 4, and 5) by the same number. For example, if multiplied by the constant 3, it becomes a 9-12-15 triangle. For most construction layout and checking, right triangles that are multiples of the 3-4-5 triangle are used (such as 9-12-15, 12-16-20, 15-20-25, and 30-40-50). The specific multiple used is determined mainly by the relative distances involved in the job being laid out or checked. It is best to use the highest multiple that is practical because when smaller multiples are used, any error made in measurement will result in a much greater angular error.

Figure 29 shows an example of the 3-4-5 rule involving the multiple 48-64-80. In order to square or check a corner as shown in the example, first measure 48'-0" along one side from a corner, then 64' along the adjacent side from the same corner. The distance measured between the 48' and 64' points must be exactly 80' for the angle to be a perfect right angle. If the measurement is not exactly 80', the angle is not 90 degrees, which means that the direction of one of the lines or the corner point must be adjusted until a right angle exists.

It cannot be emphasized enough that exact measurements are necessary to obtain the desired results when using the 3-4-5 method of laying out or checking a 90-degree angle. Any error in the measurements of the distances will result in not establishing a right angle as desired, or if an existing 90-degree angle is being checked, inaccurate measurements may cause you to make an adjustment that might be unnecessary.

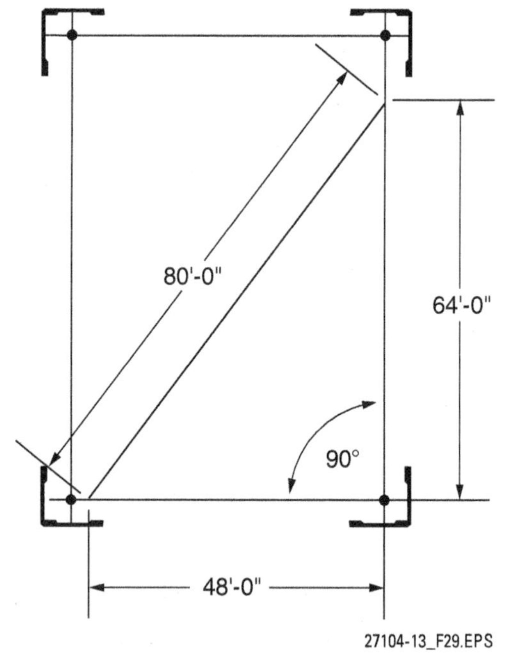

Figure 29 Example of checking building lines for square using the 48-64-80 multiple of a 3-4-5 triangle.

3.0.0 Section Review

1. When using the 3-4-5 method to create a 90-degree angle and the first two sides measure 21' and 28', the length of the hypotenuse is _____.
   a. 14'
   b. 24'
   c. 35'
   d. 42'

SUMMARY

Construction drawings show carpenters and other skilled craftworkers how to build a specific building or structure. Specifications are written instructions provided by architectural and engineering firms to the general contractor and, consequently, to the subcontractors. Specifications are just as important as the drawings in a set of prints; they furnish what the drawings cannot in that they define the quality of work to be done and the materials to be used. When a project has been awarded for a project, the building location must be laid out using the site plan as a guide. Since most buildings have 90-degree corners, the 3-4-5 method is used to lay out and square the building corners.

Review Questions

1. A drawing that shows the horizontal view of a building, the length and width of the building, and the floor layout of the rooms is called a(n) _____.
   a. elevation drawing
   b. foundation plan
   c. plot plan
   d. floor plan

Refer to *Review Question Figure 1* to answer questions 2–10.

2. The line marked A is called a(n) _____.
   a. object line
   b. hidden line
   c. cutting plane line
   d. break line

3. The material symbol marked B represents _____.
   a. common brick
   b. aluminum
   c. rubble
   d. concrete

4. The door or window symbol marked C represents a _____.
   a. double-hung window
   b. bifold door
   c. casement window
   d. sliding window

5. The electrical symbol marked D represents a _____.
   a. ceiling junction box
   b. split-wired duplex receptacle
   c. grounded receptacle
   d. waterproof receptacle

6. The plumbing symbol marked E represents a _____.
   a. hose bibb
   b. cleanout
   c. shower head
   d. wall-hung toilet

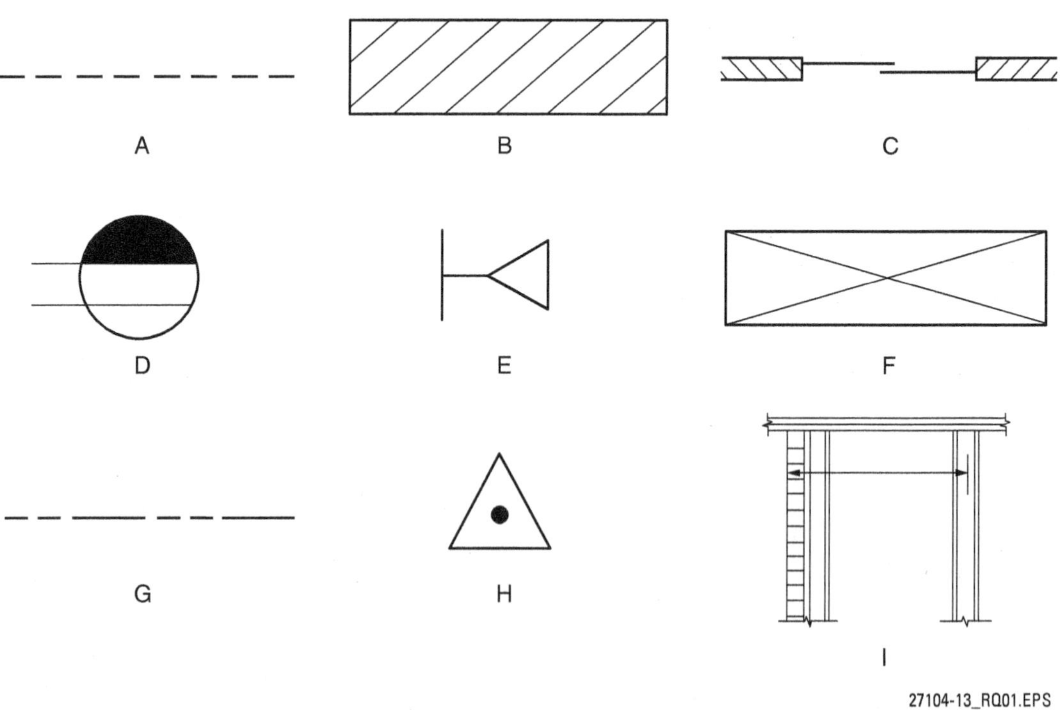

Figure 1

7. The HVAC symbol marked F represents a _____.

   a. supply duct
   b. return duct
   c. louver opening
   d. linear diffuser

8. When used on site (plot) plans, the line marked G represents a _____

   a. gas line
   b. power line
   c. contour line
   d. property line

9. When used on site (plot) plans, the symbol marked H represents a _____.

   a. property corner
   b. benchmark
   c. required spot elevation
   d. storm sewer

10. The dimension marked I represents a measurement made from _____.

    a. wall to wall
    b. center to center
    c. outside to center
    d. outside to outside

11. The distance required by building code from the street to the front of a structure is called a _____.

    a. street easement
    b. front setback
    c. porch line
    d. parkway

12. Roofs plans provide a side view of the structure.

    a. True
    b. False

13. Which three organizations worked together to develop the *International Building Code®*?

    a. ICBO, BOCA, NFPA
    b. SBCCI, NFPA, ICBO
    c. NFPA, BOCA, SBCCI
    d. BOCA, ICBO, SBCCI

14. The 3-4-5 rule, used when squaring a building, is based on the _____.

    a. isosceles triangle
    b. right triangle
    c. equilateral triangle
    d. acute triangle

15. Which of the following triangles does *not* follow the 3-4-5 rule?

    a. 10.5-14-17.5
    b. 21-28-35
    c. 18-24-36
    d. 15-20-25

Trade Terms Quiz

Fill in the blank with the correct term that you learned from your study of this module.

1. The different elevations on a site are shown as ________ on the site plan.
2. Official survey points may be marked using ________.
3. Local building codes may include a(n) ________ requirement in which the property owner cannot build.
4. A(n)________ may also influence the position of a structure on the depth of the site.
5. The legal boundaries of a site are known as ________.
6. All the finish grade references on the drawings are keyed to a(n) ________.
7. The ________ shows the structure from either the front or the side.
8. The ________ shows the structure from above.
9. Lumber is commonly sold by its ________ rather than its actual dimensions.
10. A detailed drawing of a region that depicts the natural and man-made features is a(n) ________.
11. A(n) ________ depicts the connections of a piping system.

Trade Terms

Benchmark
Contour lines
Easement
Elevation view
Front setback
Monuments
Nominal size
Plan view
Property lines
Riser diagram
Topographical survey

Mark Knudson

Superintendent/Safety Director
J.R. Abbott Construction Inc.
Seattle, Washington

How did you get started in the construction industry?
My great-grandfather was a carpenter. My grandfather was a carpenter. My dad was a carpenter and my oldest brother was a carpenter. I grew up in a family that built things. At a young age, I was exposed to building, and when I was in high school, I helped build many of the sets in the drama department. We built forts as kids and I helped my brother build his house next to the house I grew up in. I knew then that construction was the career I wanted to pursue after high school graduation. Two days after graduation (1975), I started working as a carpenter and have not taken any time off other than vacations.

Who or what inspired you to enter the industry?
I came from a family of carpenters and they were my inspiration.

What do you enjoy most about your career?
I enjoy the challenge and the fact that every project is different not only in regard to the structure of the building itself, but the structure of the organization. On every job, you have the opportunity to meet and work with new owners, architects, engineers, and craftworkers. There is nothing that even resembles being boring and monotonous in construction. The construction industry is ever changing and challenging, which makes every day interesting.

Why do you think training and education are important in construction?
Training and education are the foundation for success in any career path and equally so in our industry. Working on a construction site is difficult and demanding, and the more training and education that a person receives the better he or she will be prepared for the job at hand. Means and methods in construction are always evolving and therefore training and education in construction is, and should continue to be, a lifelong endeavor. Early training in an apprenticeship program is vital to the overall future of the construction industry—without it, we will not keep up with the demands of society in regard to construction.

Why do you think credentials are important in construction?
When I receive a doctor bill, it reads something like: John Smith DDS. MAGD, PS. For some reason credentials are important to the doctor—important enough to include it on his billing statement. Similarly, I believe it is important to have credentials for the training that we receive in construction. A doctor drills holes in teeth and fills them; we drill holes in concrete and fill them with epoxy and anchors. There is no difference—we both have been trained to do our jobs and we need credentials to support that training.

How has training/construction impacted your life and career?
Continued training has helped me advance in my career and brought me to the point of where I am today. Although formal classroom training has had an impact (and it does make a difference), on-the-job-training has really been the major factor in my career success. There is no substitute for on-the-job-training experience coupled with formal training along the way to move a person along in their career.

Would you recommend construction as a career to others? Why?
It really depends on the person. When I see a young person who has skills and a desire to build not only buildings, but also their own education and worth, I am quick to recommend the construction industry. Workers that come into the industry as a last resort because the job they trained for did not work out typically do not bring this desire and cannot live up to their employer's expectations, or their own expectations for that matter. The construction industry needs men and women who are passionate about construction and want to make it a lifelong career.

What does craftsmanship mean to you?
Craftsmanship means not being satisfied with "just okay." It means that the end-product is more important than getting the job done. It means leaving something behind that will last the test of time and that you and your family can be proud of.

Craftsmanship means not being too proud to ask for help if you need it to ensure the end-product is what the customer wants. It means doing the job right regardless of the cost. Ultimately, craftsmanship means that at the end of the day you go home with your head held high because you gave it your all.

Trade Terms Introduced in This Module

Benchmark: A point established by the surveyor on or close to the building site. It is used as a reference for determining elevations during the construction of a building.

Contour lines: Imaginary lines on a site plan that connect points of the same elevation. Contour lines never cross each other.

Easement: A legal right-of-way provision on another person's property (for example, the right of a neighbor to build a road or a public utility to install water and gas lines on the property). A property owner cannot build on an area where an easement has been identified.

Elevation view: A drawing providing a view from the front or side of a structure.

Front setback: The distance from the property line to the front of the building.

Monuments: Physical structures that mark the location of a survey point.

Nominal size: Approximate or rough size (commercial size) by which lumber, CMUs, etc., are commonly known and sold, normally slightly larger than the actual size (for example, 2 × 4s).

Plan view: A drawing that represents a view looking down on an object.

Property lines: The recorded legal boundaries of a piece of property.

Riser diagram: A schematic drawing that depicts the layout, components, and connections of a piping system.

Topographical survey: An accurate and detailed drawing of a place or region that depicts all the natural and man-made physical features, showing their relative positions and elevations.

Appendix A

Common Abbreviations

Common Abbreviations Used on Elevation Drawings

Aluminum AL
Asbestos ASB
Asphalt ASPH
Basement BSMT
Beveled BEV
Brick BRK
Building BLDG
Cast iron CI
Ceiling CLG
Cement CEM
Center CTR
Center line C or CL
Clear CLR
Column COL
Concrete CONC
Concrete block CONC B
Copper COP.
Corner COR
Detail DET
Diameter DIA
Dimension DIM.
Ditto DO.
Divided DIV
Door DR
Double-hung window DHW
Down DN or D
Downspout DS
Drawing DWG
Drip cap DC
Each EA
East E
Elevation EL
Entrance ENT
Excavate EXC
Exterior EXT
Finish FIN.
Flashing FL
Floor FL
Foot or feet ′ or FT
Foundation FND
Full size FS
Galvanized GALV
Galvanized iron GI
Gauge GA
Glass GL
Glass block GL BL
Grade GR
Grade line GL
Height HGT, H, or HT
High point H PT
Horizontal HOR
Hose bibb HB
Inch or inches ″ or IN.
Insulating (insulated) INS
Length LGTH, LG, or L
Length overall LOA
Level LEV
Light LT
Line L
Lining LN
Long LG
Louver LV
Low point LP
Masonry opening MO
Metal MET. or M
Molding MLDG
Mullion MULL.
North N
Not to scale NTS
Number NO. or #
Opening OPNG
Outlet OUT.
Outside diameter OD
Overhead OVHD
Panel PNL
Perpendicular PERP

Plate glass PL GL
Plate height PL HT
Radius R
Revision REV
Riser R
Roof RF
Roof drain RD
Roofing RFG
Rough RGH
Saddle SDL or S
Scale SC
Schedule SCH
Section SECT.
Sheathing SHTHG
Sheet SH
Shiplap SHLP
Siding SDG
South S
Specifications SPEC
Square SQ
Square inch SQ IN.
Stainless steel SST
Steel STL
Stone STN
Terra-cotta TC
Thick or thickness THK or T
Typical TYP
Vertical VERT
Waterproofing WP
West W
Width W or WTH
Window WDW
Wire glass W GL
Wood WD
Wrought iron WI

Common Abbreviations Used on Plan View Drawings

Access panel AP
Acoustic ACST
Acoustical tile AT.
Aggregate AGGR
Air conditioning AIR COND
Aluminum AL
Anchor bolt AB
Angle AN.
Apartment APT
Approximate APPROX
Architectural ARCH.
Area A.
Area drain AD
Asbestos ASB
Asbestos board AB
Asphalt ASPH
Asphalt tile AT.
Basement BSMT
Bathroom B
Bathtub BT
Beam BM
Bearing plate BRG PL
Bedroom BR
Blocking BLKG
Blueprint BP
Boiler BLR
Bookshelves BK SH
Brass BRS
Brick BRK
Bronze BRZ
Broom closet BC
Building BLDG
Building line BL
Cabinet CAB.
Caulking CLKG
Casing CSG
Cast iron CI
Cast stone CS
Catch basin CB
Cellar CEL
Cement CEM
Cement asbestos board CEM AB
Cement floor CEM FL
Cement mortar CEM MORT
Center CTR
Center to center C to C
Center line C or CL
Center matched CM
Ceramic CER
Channel CHAN
Cinder block CIN BL
Circuit breaker CIR BKR
Cleanout CO
Cleanout door COD.
Clear glass CL GL

Closet C, CL, or CLO
Cold air CA
Cold water CW
Collar beam COL B
Concrete CONC
Concrete masonry unit CMU
Concrete floor CONC FL
Conduit CND
Construction CONST
Contract CONT
Copper COP.
Counter CTR
Cubic feet CU FT
Cutout CO
Detail DET
Diagram DIAG
Dimension DIM.
Dining room DR
Dishwasher DW
Ditto DO.
Double acting DA
Double-strength glass DSG
Down DN
Downspout DS
Drain D or DR
Drawing DWG
Dressed and matched D & M
Dryer D
Electric panel EP
End to end E to E
Excavate EXC
Expansion joint EXP JT
Exterior EXT
Finish FIN.
Finished floor FIN. FL
Firebrick FBRK
Fireplace FP
Fireproof FPRF
Fixture FIX.
Flashing FL
Floor FL
Floor drain FD
Flooring FLG
Fluorescent FLUOR
Flush FL
Footing FTG
Foundation FND
Frame FR
Full size FS
Furring FUR.
Galvanized iron GI
Garage GAR.
Gas G
Glass GL
Glass block GL BL
Grille G
Gypsum GYP
Gypsum board GYP BD
Hardware HDW
Hollow metal door HMD
Hose bibb HB
Hot air HA
Hot water HW
I-beam I.
Inside diameter ID
Insulation INS
Interior INT
Iron I.
Jamb JB
Kitchen K
Landing LDG
Lath LTH
Laundry LAU
Laundry tray LT
Lavatory LAV
Leader L
Length L, LG, or LNG
Library LIB
Light LT
Limestone LS
Linen closet L CL
Lining LN
Linoleum LINO
Living room LR
Louver LV
Main MN
Marble MR
Masonry opening MO
Material MATL
Maximum MAX
Medicine cabinet MC
Minimum MIN
Miscellaneous MISC
Mixture MIX.

Modular MOD
Mortar MOR
Molding MLDG
Nosing NOS
Not to scale NTS
Obscure glass OBSC GL
On center OC
Opening OPNG
Outlet OUT.
Overall OA
Overhead OVHD
Pantry PAN.
Partition PTN
Plaster PL or PLAS
Plastered opening PO
Plate PL
Plate glass PL GL
Platform PLAT.
Plumbing PLBG
Porch P
Precast PRCST
Prefabricated PREFAB
Pull switch PS
Quarry tile floor QTF
Radiator RAD
Random RDM
Range R
Recessed REC
Refrigerator REF
Register REG
Reinforce or reinforcing REINF
Revision REV
Riser R
Roof RF
Roof drain RD
Room RM or R
Rough RGH
Rough opening RGH OPNG or RO
Rubber tile R TILE
Scale SC
Schedule SCH
Screen SCR
Scuttle S
Section SECT.
Select SEL
Service SERV
Sewer SEW.
Sheathing SHTHG
Sheet SH
Shelf and rod SH&RD
Shelving SHELV
Shower SH
Sill cock SC
Single-strength glass SSG
Sink SK or S
Soil pipe SP
Specifications SPEC
Square feet SQ FT
Stained STN
Stairs ST
Stairway STWY
Standard STD
Steel ST or STL
Steel sash SS
Storage STG
Switch SW or S
Telephone TEL
Terra-cotta TC
Terrazzo TER
Thermostat THERMO
Threshold TH
Toilet T
Tongue-and-groove T&G
Tread TR or T
Typical TYP
Unexcavated UNEXC
Unfinished UNF
Utility room URM
Vent V
Vent stock VS
Vinyl tile V TILE
Warm air WA
Washing machine WM
Water W
Water closet WC
Water heater WH
Waterproof WP
Weatherstripping WS
Weep hole WH
White pine WP
Wide flange WF
Wood WD
Wood frame WF
Yellow pine YP

Appendix B

Architectural Terms Commonly Found on Plans

Window Terms

Apron–A plain or molded piece of finish below the stool of a window that is installed to cover the rough edge of the wall finish.

Drip cap–A projection of masonry or wood on the outside top of a window to protect the window from rain.

Head jamb–Horizontal top post used in the framing of a window or doorway.

Light–pane of glass.

Lintel–Horizontal structural member supporting a wall over a window or other opening.

Meeting rail–The horizontal center rail of a sash in a double-hung window.

Mullion–A large, vertical division of a window opening.

Muntin–A strip of wood or metal that separates and supports the panes of glass in a window sash.

Sash–The part of a window in which panes of glass are set; it is generally movable, as in double-hung windows. The two side pieces are called stiles and the upper and lower pieces are called rails.

Side jambs – Vertical side posts used in the framing of a window or doorway.

Sill–Horizontal member at the bottom of a window or doorway.

Stool–A flat, narrow shelf forming the top member of the interior trim at the bottom of a window.

Stop bead–The strip on a window frame against which the sash slides.

Pitched Roof Terms

Flashing–Sheet metal, copper, lead, or tin that is used to cover open joints to make them waterproof.

Gable–The end of a ridged roof as distinguished from the front or rear side.

Louver–An opening for ventilation that is covered by sloping slats to exclude rain.

Ridge–The top edge of a roof where the two slopes meet.

Ridgeboard–A board that is placed on the edge at the ridge of a roof to support the upper ends of rafters.

Saddle–A tent-shaped portion of a roof between a chimney and the main part of the roof; built to support flashing and to direct water away from the chimney.

Valley–The intersection of two inclined sides of a roof.

Cornice Terms

Cornice–The part of a roof that projects beyond a wall.

Cornice return–The short portion of a cornice that is carried around the corner of a structure.

Crown molding–The molding at the top of the cornice and just under the roof; can also be used as interior trim.

Fascia–The outside flat member of a cornice.

Frieze–A trim member used just below the cornice.

Soffit–The underside of a cornice.

Stair Terms

Headroom–The distance between flights of steps or between the steps and the ceiling above.

Landing–The horizontal platform in a stairway.

Nosing–The overhanging edge of a stair tread.

Rise–The vertical distance from the top of a tread to the top of the next highest tread.

Riser–The vertical portion of a step.

Run–The horizontal distance that is covered by a flight of steps.

Stringer–The supporting member at the sides of a stairway.

Tread–The horizontal part of a step on which the foot is placed.

Structural Terms

Anchor bolt–A bolt with the threaded portion projecting from a structure; generally used to hold the frame of a building secure against wind load. Anchor bolts may also be referred to as hold-down bolts, foundation bolts, and sill bolts.

Batt–A type of insulation designed to be installed between framing members.

Battens–Narrow strips of wood or metal used to cover vertical joints between boards and panels.

Beam–One of the principal horizontal members of a building.

Bridging–The process of bracing floor joists by fixing lateral members between them.

Camber–The concave or convex curvature of a surface.

Expansion joint–The separation between adjoining parts to allow for small relative movements, such as those caused by temperature changes.

Footing–The foundation for a column or the enlargement at the bottom of a wall to distribute the weight of the superstructure over a greater area to prevent settling.

Furring–Strips of wood or metal applied to a wall or other surface to make it level, form an air space, or provide a fastening surface for a finish covering.

Girder–The main supporting beam (either timber or steel) that is used for supporting a superstructure.

Header–A wood beam that is set at a right angle to a joist to provide a seat or support.

Joist–Heavy horizontal member used to support the subfloor or ceiling finish material. Joists are laid edgewise to form a floor support; they rest on the wall or on girders.

Monolithic concrete–A continuous mass of concrete that is cast as a single unit.

Plate–A structural member with a depth that is substantially smaller than its length or width.

Rake–Trim members that run parallel to the roof slope and form the finish between the roof and the wall at the gable end.

Reinforced concrete–Concrete containing metal rods, wires, or other slender members. It is designed in such a manner that the concrete and metal act together to resist forces.

Sill plate–A horizontal member that is supported by a foundation wall or piers, and which in turn bears the upright members of a frame.

Slab–A cast-in-place concrete floor.

Stem wall–That portion of a foundation that rests on the footing.

Studs–The vertical, slender wood or metal members that are used to support the elements in walls and partitions.

Vapor barrier–A material that is used to retard the flow of vapor or moisture into the walls or floors and thus prevent condensation within them.

Veneer–The covering layer of material for a wall or facing materials applied to the external surface of steel, reinforced concrete, or frame walls.

Additional Resources

This module presents thorough resources for task training. The following reference material is suggested for further study.

Architectural Drawing and Light Construction. Upper Saddle River, NJ: Prentice Hall.

Code Check. Newtown, CT: Taunton Press.

Graphic Guide to Frame Construction. Newtown, CT: Taunton Press.

International Building Code® 2012. Falls Church, VA: International Code Council.

International Residential Code® 2012. Falls Church, VA: International Code Council.

MasterFormat™ 2012 Edition. Alexandria, VA: The Construction Specifications Institute (CSI) and Construction Specifications Canada (CSC).

Measuring, Marking, and Layout. Newtown, CT: Taunton Press.

Plan Reading & Material Takeoff. Kingston, MA: R.S. Means Company.

Reading Architectural Plans for Residential and Commercial Construction. Ernest R. Weidhaas. Upper Saddle River, NJ: Prentice Hall.

The Construction Specifications Institute. An organization that seeks to facilitate communication among all those involved in the building process, **www.csinet.org**

International Code Council. A membership organization dedicated to building safety and fire prevention through development of building codes, **www.iccsafe.org**

Figure Credits

SkillsUSA, Module Opener

Lake Mechanical Contractors, Inc., Figure SA03

Section Review Answer Key

Answer	Section Reference	Objective Reference
Section One		
1. d	1.1.0	1a
2. a	1.2.2	1b
3. c	1.3.6	1c
4. b	1.4.0	1d
5. b	1.5.0	1e
6. d	1.6.2	1f
Section Two		
1. b	2.1.3	2a
2. c	2.2.0	2b
Section Three		
1. c	3.0.0	3

NCCER CURRICULA — USER UPDATE

NCCER makes every effort to keep its textbooks up-to-date and free of technical errors. We appreciate your help in this process. If you find an error, a typographical mistake, or an inaccuracy in NCCER's curricula, please fill out this form (or a photocopy), or complete the online form at **www.nccer.org/olf**. Be sure to include the exact module ID number, page number, a detailed description, and your recommended correction. Your input will be brought to the attention of the Authoring Team. Thank you for your assistance.

Instructors – If you have an idea for improving this textbook, or have found that additional materials were necessary to teach this module effectively, please let us know so that we may present your suggestions to the Authoring Team.

NCCER Product Development and Revision
13614 Progress Blvd., Alachua, FL 32615
Email: curriculum@nccer.org
Online: www.nccer.org/olf

❑ Trainee Guide ❑ Lesson Plans ❑ Exam ❑ PowerPoints Other ____________

Craft / Level: Copyright Date:

Module ID Number / Title:

Section Number(s):

Description:

Recommended Correction:

Your Name:

Address:

Email: Phone:

27105-13

Floor Systems

Overview

The construction of a wood-frame floor begins with installation of the sill plate on the foundation. Once sill plates are installed, the carpenter lays out and marks the locations of the floor joists on the sill plates, and then installs the joists. Once all the joists are in place and header joists have been attached to the ends of the joists, a subfloor is placed over and attached to the joists. Bridging is installed between the joists to provide stability. These materials are stronger than comparable lengths of solid lumber and can therefore be placed across greater spans.

Module Five

Trainees with successful module completions may be eligible for credentialing through the NCCER Registry. To learn more, go to **www.nccer.org** or contact us at **1.888.622.3720**. Our website has information on the latest product releases and training, as well as online versions of our *Cornerstone* magazine and Pearson's product catalog.

Your feedback is welcome. You may email your comments to **curriculum@nccer.org,** send general comments and inquiries to **info@nccer.org**, or fill in the User Update form at the back of this module.

This information is general in nature and intended for training purposes only. Actual performance of activities described in this manual requires compliance with all applicable operating, service, maintenance, and safety procedures under the direction of qualified personnel. References in this manual to patented or proprietary devices do not constitute a recommendation of their use.

27105-13 FLOOR SYSTEMS

Objectives

When you have completed this module, you will be able to do the following:

1. Read and interpret specifications and drawings to determine floor system requirements.
   a. Explain the importance of specifications.
   b. List items commonly shown on architectural drawings.
   c. Describe information typically shown on structural drawings.
   d. Explain the importance of referencing mechanical, electrical, and plumbing plans.
   e. Describe the proper procedure for reading a set of prints.
2. Identify the different types of framing systems.
   a. Describe the general components of a platform-framed structure.
   b. List differences between platform framing and balloon framing.
   c. Describe the characteristics of post-and-beam framing.
3. Identify floor system components.
   a. Define *sill plate* and describe its role in floor framing.
   b. List and recognize different types of beams and girders and supports.
   c. List and recognize different types of floor joists.
   d. List and recognize different types of bridging.
   e. Explain the purposes of subfloor and underlayment.
4. Describe the construction methods for floor systems, and identify floor system materials.
   a. Describe how to check a foundation for squareness.
   b. Name the methods used to lay out and fasten sill plates to the foundation.
   c. Describe the proper procedure for installing a beam or girder.
   d. Describe how to lay out sill plates and girders for floor joists.
   e. Describe how to lay out and install floor joists for partitions and floor openings.
   f. Identify different types of bridging and describe how to properly install each type.
   g. Describe how to properly install subfloor.
   h. Explain how to install joists for projections or cantilevered floors.
5. Estimate the amount of material needed for a floor assembly.
   a. Describe how to estimate the amount of sill plate, sill sealer, and termite shield.
   b. Describe how to estimate the amount of beam or girder material.
   c. Describe how to estimate the amount of lumber needed for joists and joist headers.
   d. Describe how to estimate the amount of bridging required.
   e. Describe how to estimate the amount of subfloor material required.
6. Identify some common alternative floor systems.

Performance Tasks

Under the supervision of the instructor, you should be able to do the following:

1. Lay out and construct a floor assembly, including a rough opening and subfloor material.
2. Estimate the amount of material to frame a floor assembly from a set of plans.

Trade Terms

- Crown
- Dead load
- Dome pan
- End joist
- Fire stop
- Floor truss
- Foundation
- Header
- Header joist
- Joist hanger
- Lateral pressure
- Let-in
- Live load
- Long pan
- On center
- Pier
- Precast
- Rim joist
- Rough opening
- Scab
- Seasoned
- Shoring
- Single-layer floor system
- Soleplate
- Span
- Specifications
- Swing
- Tail joist
- Trimmer joist
- Two-layer floor system
- Underlayment

Industry-Recognized Credentials

If you're training through an NCCER-accredited sponsor you may be eligible for credentials from NCCER's Registry. The ID number for this module is 27105-13. Note that this module may have been used in other NCCER curricula and may apply to other level completions. Contact NCCER's Registry at 888.622.3720 or go to **nccer.org** for more information.

Contents

Topics to be presented in this module include:

Contents

Figures and Tables

SECTION ONE

1.0.0 READING SPECIFICATIONS AND PLANS

Objective

Read and interpret specifications and drawings to determine floor system requirements.

a. Explain the importance of specifications.
b. List items commonly shown on architectural drawings.
c. Describe information typically shown on structural drawings.
d. Explain the importance of referencing mechanical, electrical, and plumbing plans.
e. Describe the proper procedure for reading a set of prints.

Trade Terms

Pier: A structural column used to support other structural members, typically girders or beams.

Specifications: Written document included with a set of prints that clarifies information presented on the prints and provides additional information not easily presented on the prints.

Swing: The direction of door rotation.

Construction drawings, also known as plans or working drawings, and related written specifications contain the information and dimensions needed to build or remodel a structure. Proper reading and interpretation of the drawings is critically important in the construction of floor systems, walls, and roof systems. *Figure 1* shows the contents and sequence of a typical set of construction drawings. The drawings that apply when building a floor system are briefly reviewed in this section.

1.1.0 Specifications

Written specifications are equally as important as the drawings in a set of prints. Specifications furnish what the drawings cannot, in that they give detailed and accurate written descriptions of work to be done and the materials to be used. They include quality and quantity of materials, methods of construction, standards of construction, and manner of conducting the work. Specifications are a legally binding contract. The basic information found in a typical set of specifications includes:

- Contract
- Synopsis of the work
- General requirements
- Owner's name and address
- Architect's name
- Location of structure
- Completion date
- Guarantees
- Insurance requirements
- Methods of construction
- Types and quality of building materials
- Sizes

1.2.0 Architectural Drawings

Architectural drawings contain most of the detailed information needed by carpenters to frame a structure. The specific categories of architectural drawings commonly used include the following:

- Foundation plan
- Floor plan
- Section and detail drawings

1.2.1 Foundation Plans

The foundation plan is a view of the entire substructure below the first floor or frame of the building. It specifies the location and dimensions of footings, grade beams, foundation walls, stem walls, piers, equipment footings, and foundations. Generally, in a detail view, the foundation plan also shows the location of the anchor bolts or straps in foundation walls or concrete slabs.

1.2.2 Floor Plans

The floor plan is a cutaway plan view (top view) of a building, showing the length and width of the building and the room layout on that floor. Floor plans show the following information:

- Outside walls, including the location and dimensions of all exterior openings (windows/doors)
- Types of construction materials
- Location of interior walls and partitions
- Location and swing of doors
- Stairways
- Location of windows
- Location of cabinets, electrical and mechanical equipment, and fixtures
- Location of a cutting plane line

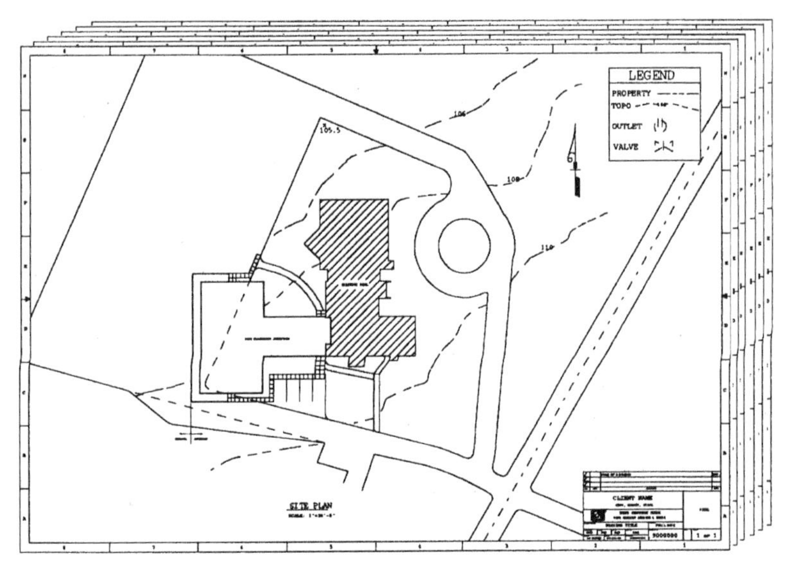

TITLE SHEET(S)

ARCHITECTURAL DRAWINGS

- SITE (PLOT) PLAN
- FOUNDATION PLAN
- FLOOR PLANS
- INTERIOR/EXTERIOR ELEVATIONS
- SECTIONS
- DETAILS
- SCHEDULES

STRUCTURAL DRAWINGS

MECHANICAL PLANS

ELECTRICAL PLANS

PLUMBING PLANS

27105-13_F01.EPS

Figure 1 Typical organization of a set of construction drawings.

1.2.3 Section and Detail Drawings

Section drawings are cutaway views through an object or wall to show its interior makeup. They are used to show the details of construction and provide information about walls, stairs, and other parts of construction that may not show clearly on a floor plan. A section view is limited to the specific portion of the building construction that the architect wishes to clarify. It may be drawn on the same sheet as an elevation or plan view or it may appear on a separate sheet. *Figure 2* shows a typical wall section.

Detail drawings are views that are typically drawn to a larger scale. They are often used to show aspects of a design that are too small to be shown in sufficient detail on a plan or elevation drawing. Like section drawings, detail drawings may be drawn on the same sheet as an elevation or plan drawing, or may appear on a separate sheet in the set of plans.

> **NOTE**
> The various drawings that are created and used for construction are the property of either the architect or the building owner. They cannot legally be used for any other purpose without express permission.

1.3.0 Structural Drawings

Structural drawings are created by a structural engineer and accompany the architect's plans. They are usually drawn for large structures such as office buildings or manufacturing facilities. Structural drawings show requirements for structural elements of the building including columns, floor and roof systems, stairs, canopies, and bearing walls. They provide such details as:

- Height of finished floors and walls
- Height and bearing of bar joists or steel joists
- Location of bearing steel materials
- Height of steel beams, concrete planks, concrete Ts and cast-in-place concrete
- Bearing-plate locations
- Location, size, and spacing of anchor bolts
- Stairways

1.4.0 Mechanical, Electrical, and Plumbing (MEP) Plans

Larger buildings commonly require mechanical, electrical, and plumbing (MEP) plans. Mechanical plans show temperature control and ventilation equipment including ducts, louvers, and registers. Electrical plans show all electrical equipment, lighting, and outlets. Plumbing plans show the size and location of water and gas systems if they are not included in the mechanical plans.

While carpenters primarily work with architectural and structural drawings, there are useful notes and views on drawings shown in other plans, especially the mechanical plans. For example, typical items not found on architectural drawings that may be found on mechanical plans are exposed heating, ventilating, and air conditioning (HVAC) ductwork; heating convectors; and fire-sprinkler piping. Any one or all of these

items may require special framing to be built, so make a habit of reviewing all drawings. Also, make sure to coordinate any such work with the appropriate other trades to ensure that the proper framing is done to accommodate ductwork, piping, wiring, etc.

1.5.0 Reading Prints

The following general procedure is suggested as a method of reading a set of prints:

Step 1 Read the title block; the title block provides general information about the drawing and contains critical information about the drawing such as the scale, date of last revision, drawing number, and architect or engineer of record.

Step 2 Locate the north arrow. Always orient yourself to the structure. Knowing where north is enables you to more accurately describe the locations of walls and other parts of the building.

Step 3 Be aware that different sheets in a set of prints work together as a team. The reason an architect or engineer draws plans, elevations, and sections is that it requires more than one type of view to communicate the whole project. Learn how to reference more than one drawing, when necessary, to find the needed information.

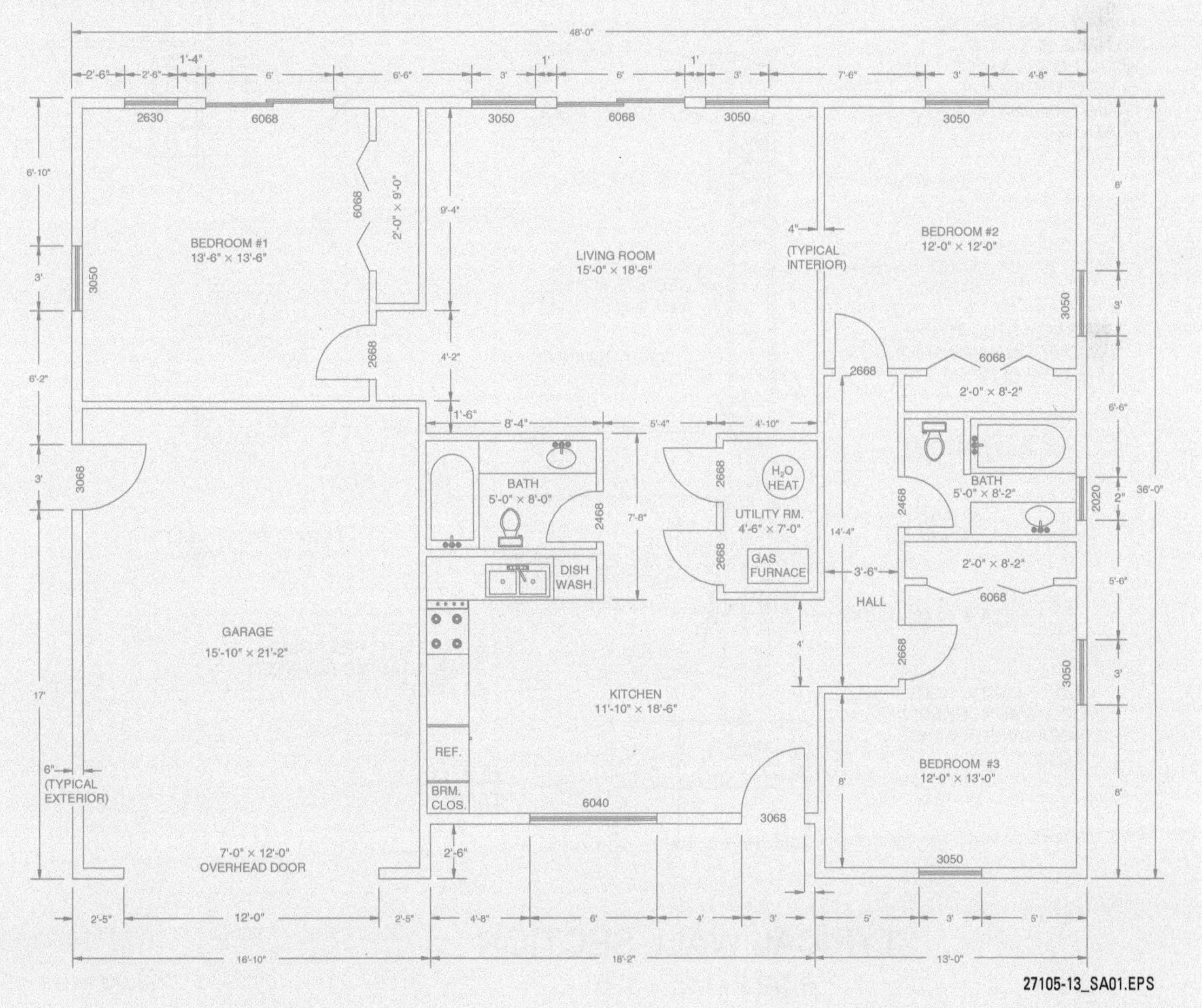

Floor Plan

A typical floor plan provides a top view that details the layout of the rooms for each floor in the building.

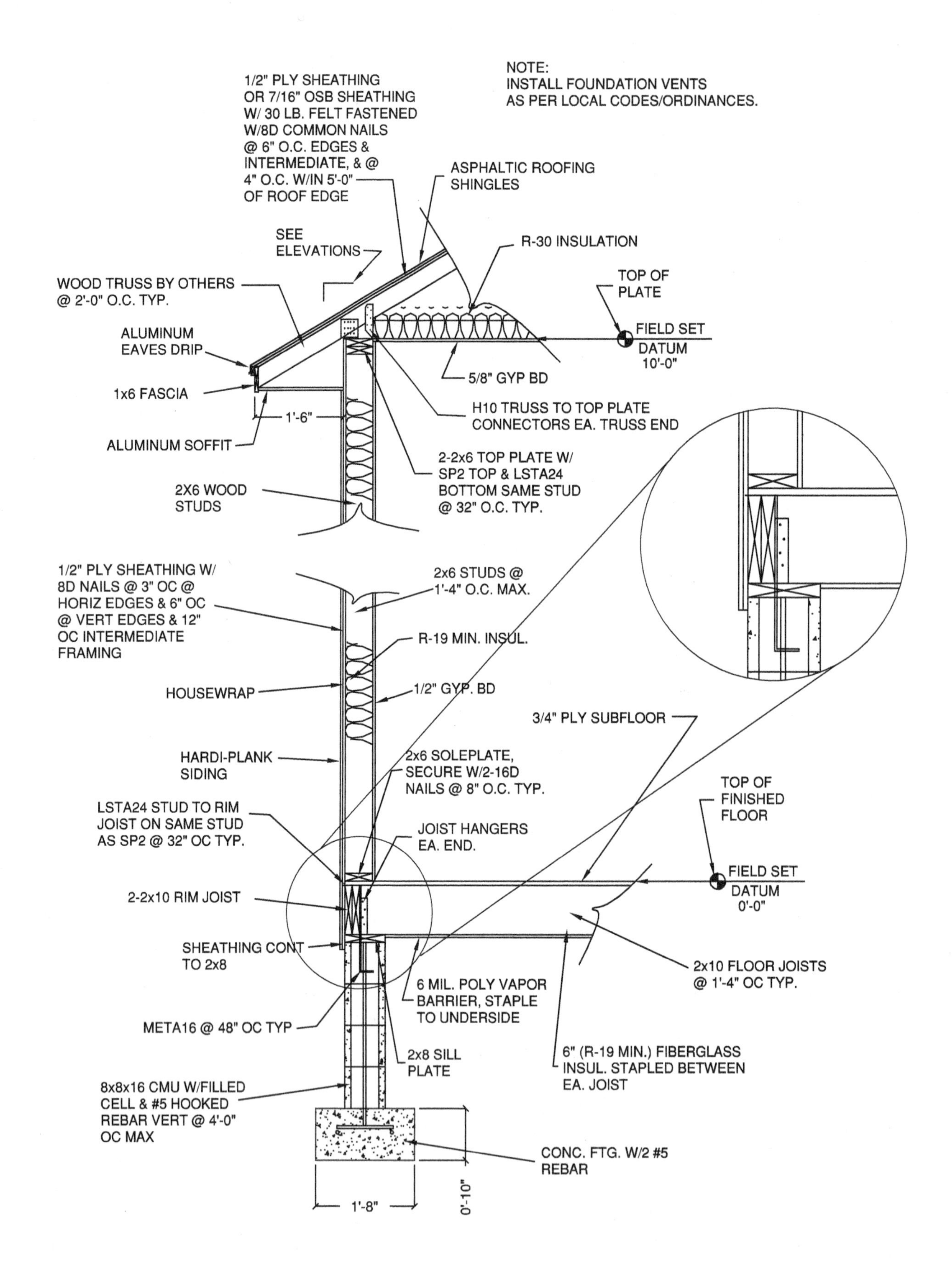

Figure 2 Typical wall section (varies by zone).

Step 4 Check the list of prints in the set; on a large set of prints, an index of the prints is typically included on Sheet 1. Notice that the prints are broken into several categories:

- Architectural
- Structural
- Mechanical
- Electrical
- Plumbing

Step 5 Study the site plan (plot plan) to observe the location of the building on the property. Notice that the geographic location of the building may be indicated on the site plan.

Step 6 Check the foundation and floor plans for the orientation of the building. Observe the location and features of entries, corridors, offsets, and any special features.

Dimensions vs. Scale

The scale of a drawing refers to the amount or percentage that a document has been reduced in relation to reality (full scale). Specific dimensions on documents should always take precedence over the scaled graphic representation shown on the prints. The drawing itself is meant to provide a general idea of the overall building layout. The dimensions on the drawing are what should be used when planning and building.

Step 7 Study the features that extend for more than one floor, such as plumbing and vents, stairways, elevator shafts, heating and cooling ductwork, and piping.

Step 8 Check the floor and wall construction and other details relating to exterior and interior walls.

Step 9 Check the foundation plan for size and types of footings, reinforcing steel, and loadbearing substructures.

Step 10 Study the mechanical plans for the details of the heating, cooling, and plumbing systems.

Step 11 Observe the electrical service entrance and distribution panels, and the installation of the lighting and power supplies for special equipment.

Step 12 Check the notes on the various sheets and compare the specifications against the construction details. Note any variations or discrepancies.

Step 13 Review the sheets of drawings until you are familiar with all the plans and structural details.

Step 14 Recognize applicable symbols and their relative locations in the plans. Note any special construction details or variations that affect your job.

1.0.0 Section Review

1. Specifications are a legally binding contract that provide written descriptions of the work to be performed and the materials to be used.

   a. True
   b. False

2. Which of the following is *not* shown on the foundation plans?

   a. Footings
   b. Room layout
   c. Foundation walls
   d. Grade beams

3. Which of the following is shown on the structural plans for a commercial building?

   a. Finish schedule
   b. Routing of electrical cables
   c. Underground plumbing
   d. Location, size, and spacing of anchor bolts

4. Ducts, louvers, and registers are typically shown on the _____.

   a. electrical plans
   b. structural plans
   c. mechanical plans
   d. plumbing plans

5. General information about a drawing in a set of prints is located in the _____.

   a. detail drawings
   b. section drawings
   c. title block
   d. architect's stamp

Section Two

2.0.0 Framing Systems

Objective

Identify the different types of framing systems.

a. Describe the general components of a platform-framed structure.
b. List differences between platform framing and balloon framing.
c. Describe the characteristics of post-and-beam framing.

Trade Terms

Fire stop: An approved material used to fill air passages in a frame to retard the spread of fire.

Let-in: Any type of notch in a stud, joist, etc., which holds another piece. The item that is supported by the notch is said to be let in.

On center: Distance between the center of one member (typically studs) to the center of the next member. Commonly abbreviated as OC.

Seasoned: Lumber with the appropriate amount of moisture removed to make it usable for construction.

Soleplate: The bottom horizontal member of a wall frame.

Span: The distance between structural supports such as walls, columns, piers, beams, or girders.

Various areas of the country have different methods of constructing wood-frame dwellings. This variation in methods can be attributed to economic conditions, availability of building materials, or climate in different parts of the country. This section covers the various types of framing systems and shows approved methods of construction today. Structures that are framed entirely of wood above the foundation can be categorized as follows:

- Platform frame
- Balloon frame
- Post-and-beam frame

2.1.0 Platform Frame

The platform frame, sometimes called the western frame, is used in most residential and light commercial construction. In platform frame construction, each floor of a structure is built as an individual unit (*Figure 3*). The subfloor is installed prior to the exterior walls being assembled and erected. Wall panels consisting of a soleplate, top plate, and studs are then constructed on top of the subfloor. Window and door openings are framed per the prints, and braces may be attached to or let in to the studs to keep the wall panels fairly square when raising them. Wall sheathing is commonly applied when the wall frames are lying on the subfloor.

The walls are built in sections while lying on the subfloor, and once a section of a wall is complete, it is then raised into place. The wall panel is nailed to the floor system at the bottom and the panel is plumbed. Another plate is added to the top plate when the wall panels are in position to tie adjacent walls together. This method of construction allows workers to work safely on the floor while constructing the wall panels.

Platform framing is subject to settling caused by the shrinkage of a large number of horizontal, loadbearing frame members. Settling can result in various problems such as cracked plaster, cracked gypsum-board joints, uneven ceilings and floors, poorly fitting doors and windows, and nail pops (nails that begin to protrude through the gypsum-board finish). Settling can be minimized by using seasoned lumber. The moisture content of framing lumber should be approximately 19 percent.

2.2.0 Balloon Frame

In hurricane-prone areas, balloon frame construction is used to counteract the wind loads placed upon the buildings. In balloon frame construction, the studs are continuous from the soleplate to the top (rafter) plate (*Figure 4*). The first-floor joists rest on a solid sill plate (usually a 2 × 6) and are covered with the subfloor. A soleplate is installed over the

Platform Framing

In platform framing, it is inevitable that the structure will begin to settle. While settling is unavoidable, the degree to which the structure settles can be minimized by not cutting any corners during the framing and subsequent construction process. Compacting the foundation soil and making sure the frame is properly braced during construction will also decrease the amount of settlement.

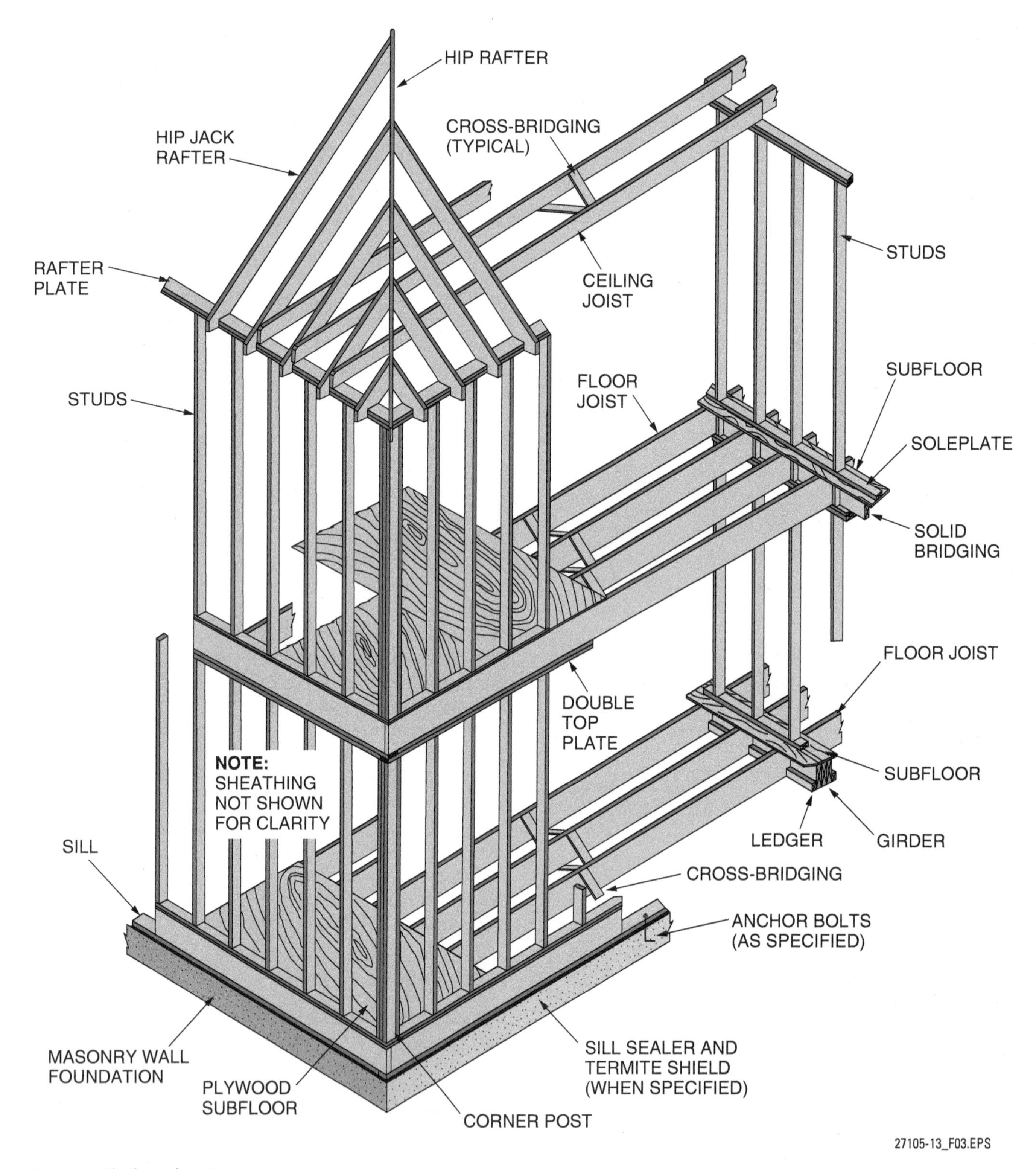

Figure 3 Platform framing.

subfloor and the full-length studs extend up to the rafter plate. The second-floor joists rest on a horizontal 1 × 4 or 1 × 6 board, called a ribbon, which is installed in notches (let-ins) in the edges of the studs. Diagonal braces (usually 1 × 4s) are let into the outside edges of the studs. Balloon frame construction is self-supporting; therefore, it does not need to rely on the sheathing for rigidity. Because balloon framing has studs that span multiple stories, alternative types of studs, such as engineered lumber or metal studs, may be required if solid wood studs of sufficient length are not available.

In balloon frame construction, fire stops must be installed in the walls in several locations. A fire stop is an approved material installed in the space between frame members to prevent the spread of

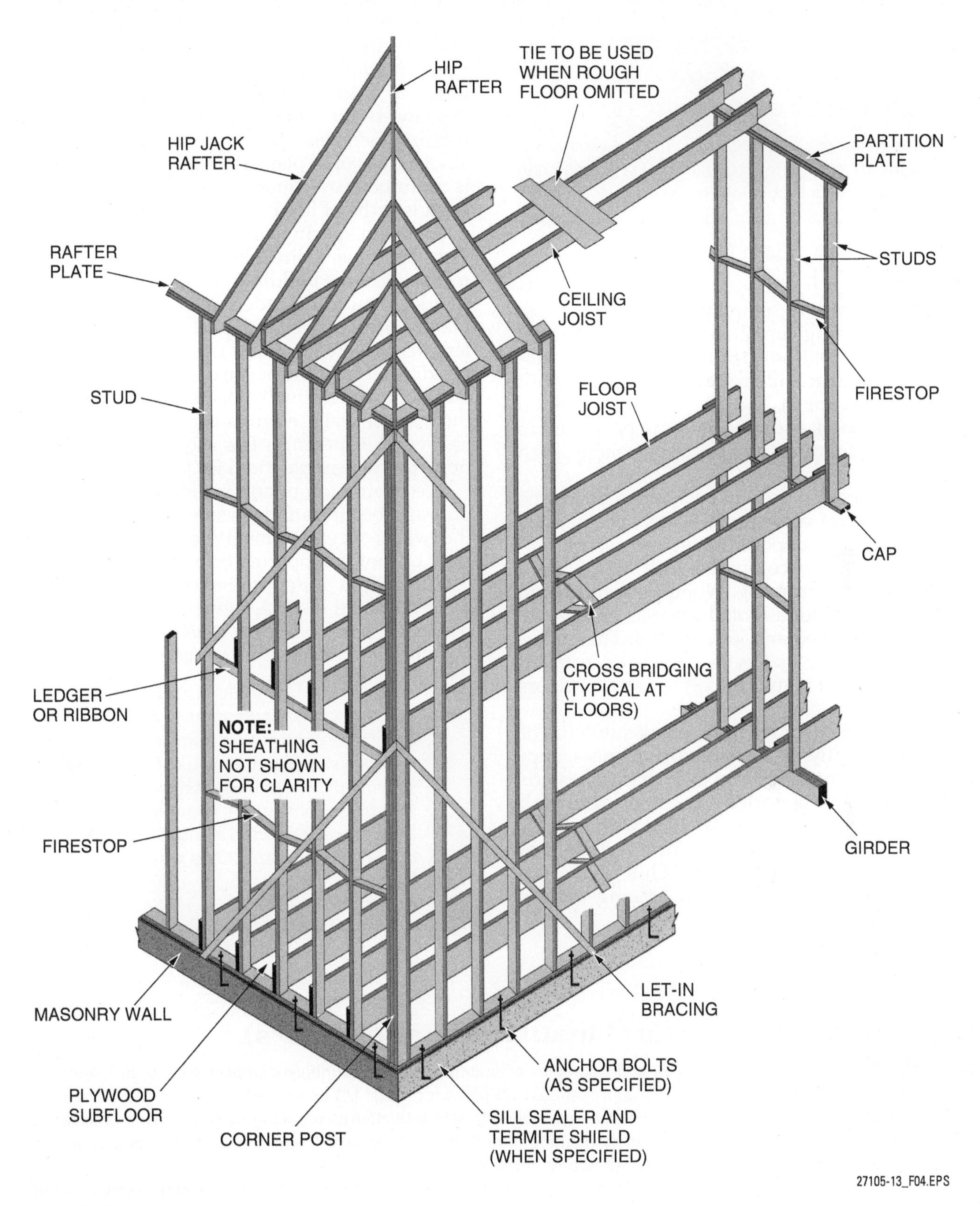

Figure 4 Balloon framing.

fire for a limited period of time. Fire stops must be the same width as the studs in the walls in which they are installed.

One advantage of balloon framing is that the shrinkage of the wood framing members is low, thus reducing the amount of settling. This is because wood shrinks across its width (but practically no shrinkage occurs lengthwise) providing for high vertical stability, and making the balloon frame adaptable for two-story structures.

Did You Know?

Balloon Framing

Balloon framing is frequently used in hurricane-prone areas for gable ends. In fact, this type of framing may be required by local building codes.

2.3.0 Post-and-Beam Frame

Post-and-beam framing, also known as plank-and-beam framing, employs much larger and sturdier framing members than conventional platform and balloon framing. The post-and-beam methods provide wide expanses of interior open space (*Figure 5*). The post-and-beam method of framing floors and roofs is commonly used in commercial buildings and places of worship, such as churches.

Since the beams used in post-and-beam frame construction are very sturdy, wider spacing may be used. Vertical supports are typically spaced 48" on center (OC), as compared with 16" OC used in conventional framing. When post-and-beam framing is used for a roof, the beams and planking can be finished and left exposed. The underside of the planking takes the place of an installed ceiling. Matched planks are often used for floors and roofs.

In post-and-beam framing, plank subfloors or roofs are usually of 2" nominal thickness, supported on beams spaced up to 8' apart. The ends of the beams are supported on posts or piers. Wall spaces between posts are provided with supplementary framing as required for attachment of exterior and interior finishes. This additional framing also provides lateral bracing for the building.

Compare post-and-beam framing to conventional platform framing that utilizes joists, rafters, and studs placed from 16" to 24" on center. Post-and-beam framing requires fewer but larger framing members spaced farther apart. The most efficient use of 2" planks occurs when the lumber is continuous over more an one span. When standard lengths of lumber such as 12', 14', or 16' are used, beam spacings of 6', 7', or 8' are indicated. This factor has a direct bearing on the overall dimensions of the building.

If local building codes allow end joints in the planks to fall between supports, planks of random lengths may be used and the beam spacing may be adjusted to fit the house dimensions. Windows and doors are normally located between posts in the exterior walls, eliminating the need for headers over the openings. The wide spacing between posts permits ample room for large glass areas.

A combination of conventional framing with post-and-beam framing is sometimes used where the two adjoin each other. Where a post-and-beam floor or roof is supported on a stud wall, a post is usually placed under the end of the beam to carry a conventional load. A conventional roof can be used with post-and-beam construction by installing a header between posts to carry the load from the rafters to the posts.

NOTE

Post-frame construction, commonly referred to as pole construction, consists of large trusses supported by large posts at the perimeter. Post-frame construction is commonly used for large open-air structures, such as agricultural buildings (pole barns). Girts provide posts with lateral stability and provide a nailing surface for siding.

GOING GREEN

Structural Insulated Panels (SIPs)

The construction and operation of buildings have a significant impact on the environment. Buildings account for approximately 39 percent of total US energy consumption and 38 percent of carbon dioxide emissions. Green buildings use less energy, reducing carbon dioxide emissions, and play an important role in combatting global climate change. Construction and operating buildings also uses a tremendous amount of natural resources.

Structural insulated panels (SIPs) consist of two outer structural panels and a thick inner core made of foam. SIPs are commonly used for roofs and walls in post-and-beam structures. An SIP system is one of the most airtight and well-insulated building systems available. An airtight SIP building uses less energy to heat and cool, allows for better control over indoor environmental conditions, and reduces construction waste.

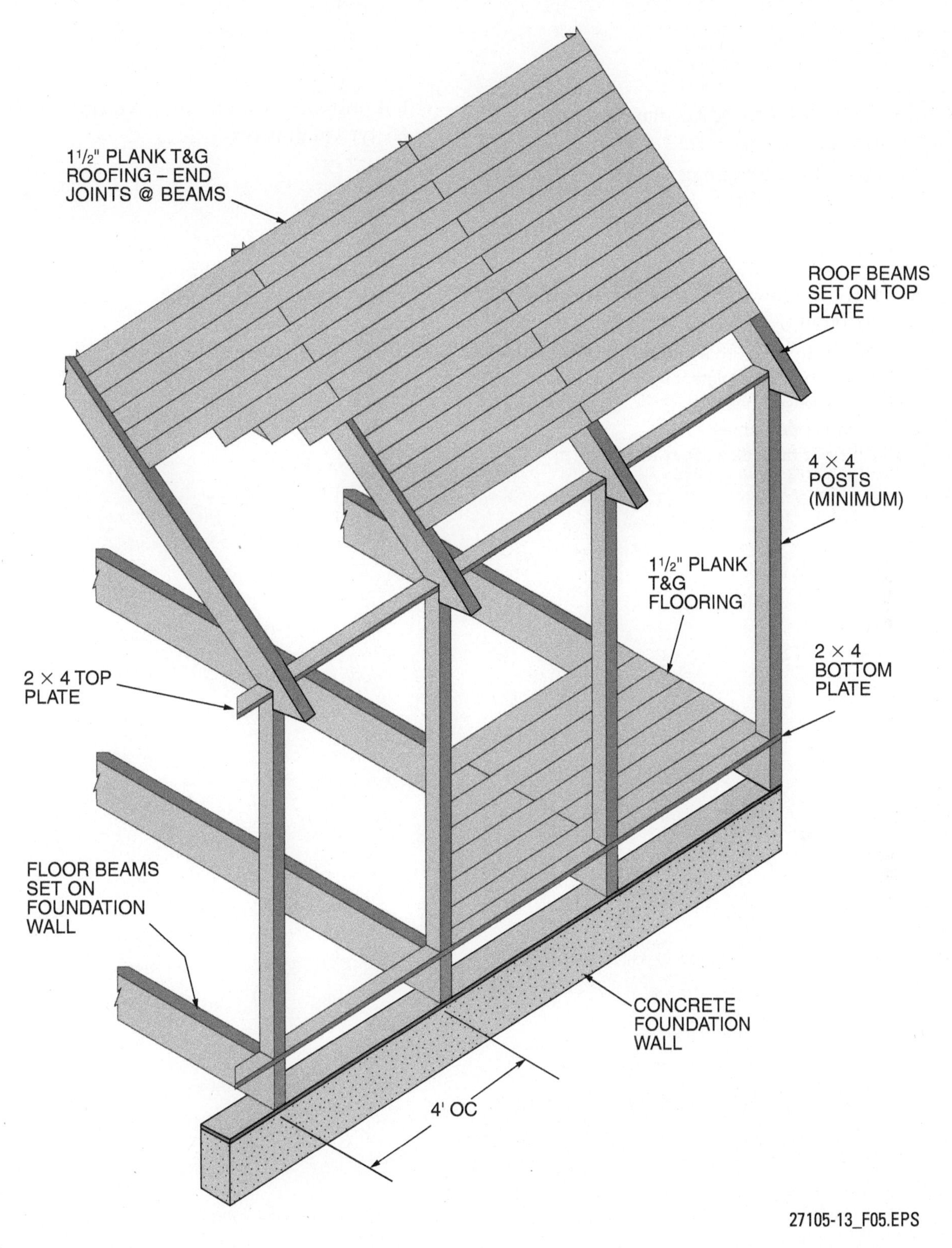

Figure 5 Post-and-beam framing.

2.0.0 Section Review

1. Most residential and light commercial construction uses _____.
   a. post-and-beam framing
   b. platform framing
   c. balloon framing
   d. post framing

2. Which of the following is a key characteristic of balloon framing?
   a. Full-length studs that extend from the soleplate to the rafter plate
   b. Large exposed beams
   c. Spacious open areas
   d. Each floor built as a separate platform

3. For post-and-beam framing, vertical supports are typically spaced _____.
   a. 12" OC
   b. 16" OC
   c. 19.2" OC
   d. 48" OC

SECTION THREE

3.0.0 FLOOR SYSTEMS

Objective

Identify floor system components.

a. Define *sill plate* and describe its role in floor framing.
b. List and recognize different types of beams and girders and supports.
c. List and recognize different types of floor joists.
d. List and recognize different types of bridging.
e. Explain the purposes of subfloor and underlayment.

Trade Terms

Crown: The highest point of the curved edge of a framing member.

Dead load: The weight of permanent, stationary construction and equipment included in a building.

Floor truss: An engineered assembly that is used to support floors.

Foundation: The supporting portion of a structure, including the footings.

Header joist: A framing member used in platform framing into which the common joists are fitted. Header joists are also used to support the free ends of joists when framing openings in a floor.

Joist hanger: A metal stirrup secured to the face of a structural member, such as a girder, to support and align the ends of joists flush with the member.

Live load: The total of all moving and variable loads that may be placed upon a structure.

Precast: Concrete structural elements that have been cast at a casting yard and raised into position using a crane.

Rough opening: Dimensions shown on the prints that indicate the distance from the inside edge of one trimmer joist to the inside edge of the opposing trimmer joist.

Single-layer floor system: Combined subfloor and underlayment; typically installed where direct application of carpet, tile, etc., to the floor is intended.

Tail joist: Short joists that run from an opening to a bearing.

Trimmer joist: A full-length joist that reinforces a rough opening in the floor.

Two-layer floor system: Floor system in which a layer of underlayment is installed over a layer of subfloor.

Underlayment: A material, such as particleboard or plywood, laid on top of the subfloor to provide a smoother surface for the finished floor.

Floor systems provide a base for the remainder of the structure. They transfer the weight of people, furniture, and materials, from the subfloor, to the floor framing, to the foundation wall, to the footing, then finally to the earth. Floor systems are constructed over basements or crawl spaces. Single-story structures built on slabs do not have wood- or metal-framed floor systems. However, multilevel structures may have both a slab and a floor system. *Figure 6* shows a typical platform-frame floor system and identifies the various parts.

When building a floor system, the prints and specifications should provide all the needed information about the floor system. Important information regarding the floor system includes:

- Type of wood or other materials used for sill plates, posts, girders, beams, joists, subfloors, etc.
- Size, location, and spacing of support posts or columns
- Direction of joists and girders
- Manner in which joists connect to girders
- Location of any loadbearing interior walls that run parallel to joists
- Location of any water closet (toilet) drains
- **Rough opening** sizes and locations of all floor openings for stairs, etc.
- Any cantilevering requirements
- Changes in floor levels
- Any special metal fasteners needed in seismic areas
- Types of blocking or bridging
- Clearances from the ground to girder(s) and joists for floors installed over crawl spaces

3.1.0 Sill Plates

Sill plates, also called sills, are the lowest members of a structure's frame. They rest horizontally on the **foundation** and support the floor joists.

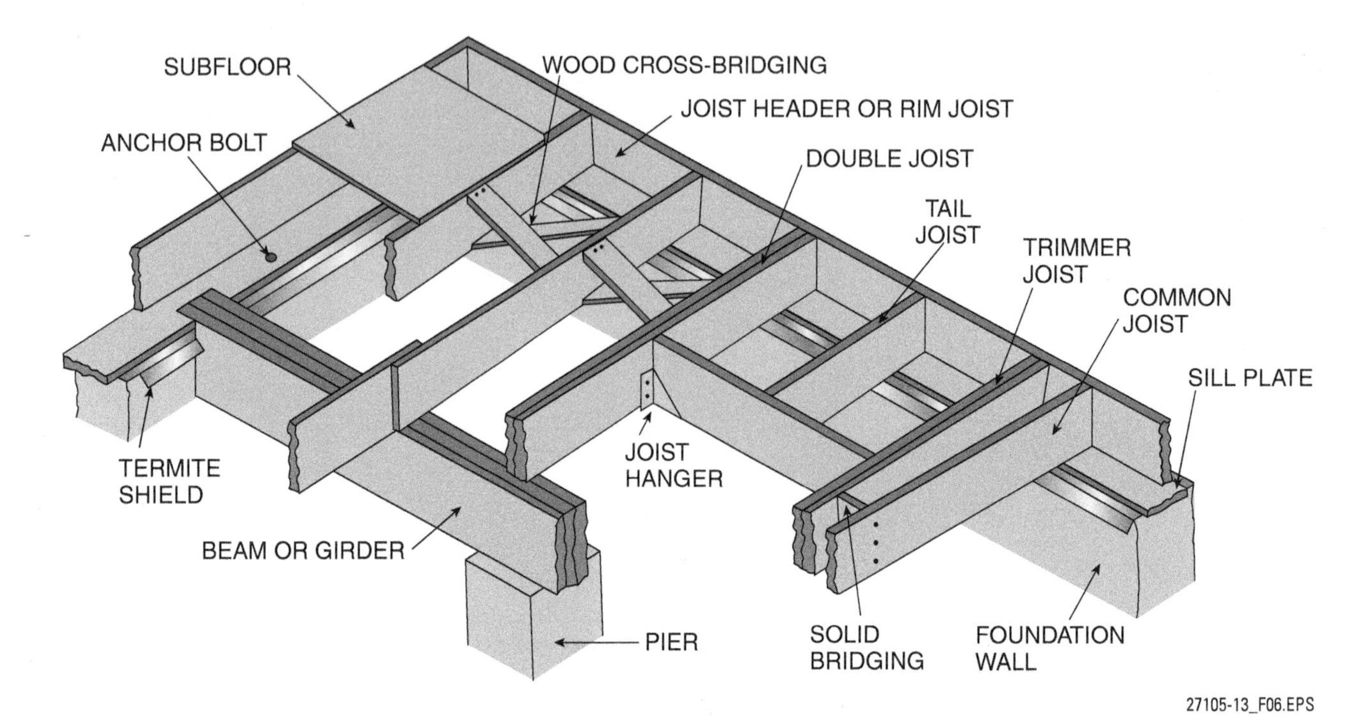

Figure 6 Typical platform-frame floor system.

The foundation is the supporting portion of a structure below the first-floor construction, and includes the footings. Sill plates serve as the attachment point to a concrete or CMU (concrete masonry unit) foundation for all of the other wood framing members. The sill plates provide a means of leveling the top of the foundation wall and also prevent the other wood framing lumber from making contact with the concrete or masonry, which can cause the lumber to rot.

Today, 2 × 8 lumber is typically used for the sill plates. Local codes typically require that treated lumber and/or foundation-grade redwood lumber be used for sill plates or other wood members whenever they come in direct contact with concrete or masonry.

Sill plates are attached to the foundation wall using either anchor bolts (*Figure 7*) or straps (*Figure 8*) embedded in the foundation. The exposed portion of the anchor strap is nailed to the sill plate. Some types of straps must be bent over the top of the sill plate, while others are nailed to the sides. The size, type, and spacing between the anchor bolts or straps must comply with local building codes. Location of the anchor bolts or straps and other related data are usually shown on the prints.

In many cases, such as in structures where the underfloor areas are used as part of the HVAC system or as a basement, a foam insulating material, called a sill sealer (*Figure 7*), is installed. Sill sealers account for irregularities between the

Foundation Checks

Always verify the foundation measurements and ensure that the foundation is square before beginning to frame the floor system. If the foundation is off by even a tiny amount, it can have a major impact on the framing.

Sill Plate Installation

Installing the sill plate is the first step in framing a building. Building codes require treated lumber for any components that are in direct contact with concrete, masonry, or the ground.

27105-13_SA02.EPS

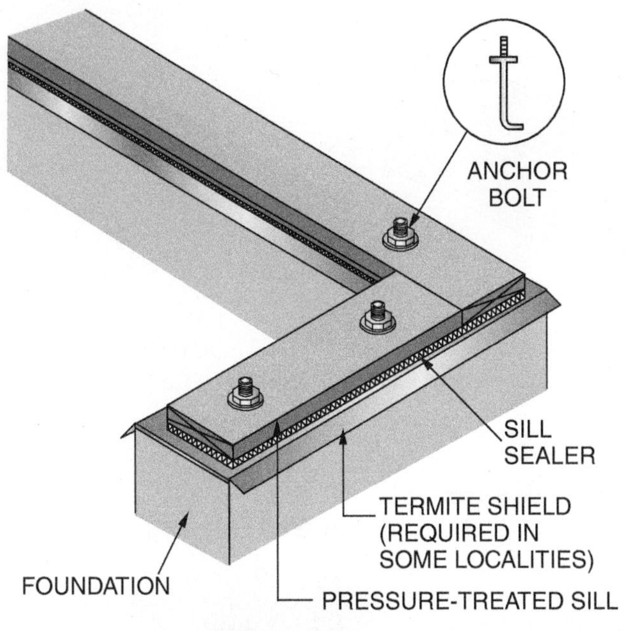

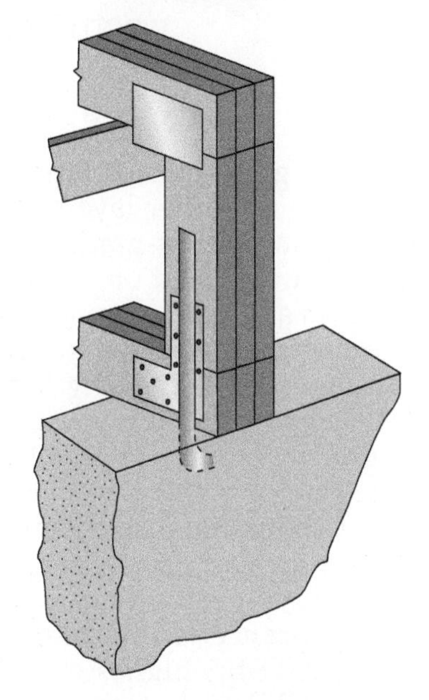

Figure 7 Typical sill plate installation.

foundation wall and the sill plate to seal against drafts, dirt, and insects. Sill-sealer material is available in standard widths for 2 × 4 and 2 × 6 walls. The sill sealer should be installed between the sill plate and the foundation wall, or between the sill plate and a termite shield (if used).

3.2.0 Beams/Girders and Supports

The distance between two outside walls (span) is frequently too great to be spanned by a single joist. When two or more joists are needed to cover the span, support for the inner joist ends must be provided by one or more beams, commonly called girders. Girders carry a very large portion of the weight of a building. They must be well designed, rigid, and properly supported at the foundation walls and on the supporting posts or columns. They must also be installed so they will properly support the joists. Girders are solid timbers, built-up lumber, engineered products (such as laminated veneer lumber [LVL], glued laminated lumber [glulam], or wood I-joists), or steel beams (*Figure 9*). Each type has advantages and disadvantages. Note that in some instances, precast reinforced concrete girders may also be used.

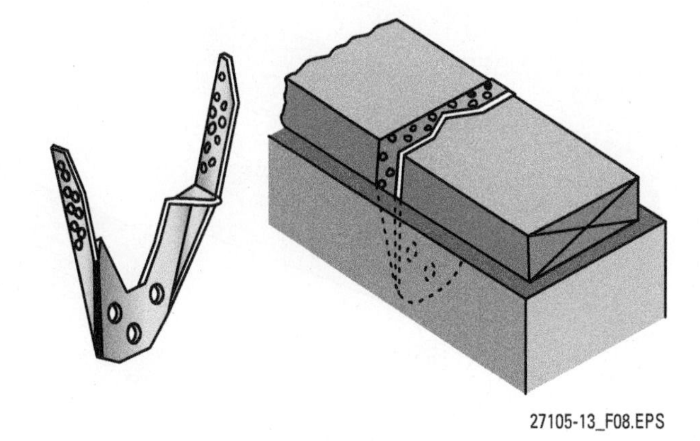

Figure 8 Typical sill plate anchor strap.

3.2.1 Solid Lumber Girders

Solid stock used for girders is available in various sizes, with 4 × 6, 4 × 8, and 6 × 6 being typical sizes. If straight, large timbers are available, their use can save time by not having to make built-up girders. However, solid pieces of large timber stock are often badly bowed and can create a rise in the floor unless the crown is pulled down. The crown is the high point of the curved edges of the framing members.

Termite Shields

In areas where there is a high risk of termite infestation, a sheet-metal termite shield should be installed below the sill plate. (In some areas of the country, this is a code requirement.) Termites live underground and come to the surface to feed on wood. They can enter through cracks in the masonry, tunnel through the hollow cells of CMUs, or build earthen tubes on the side of masonry to reach the wood. Termite shields reduce the possibility that the termites can make it to the structure.

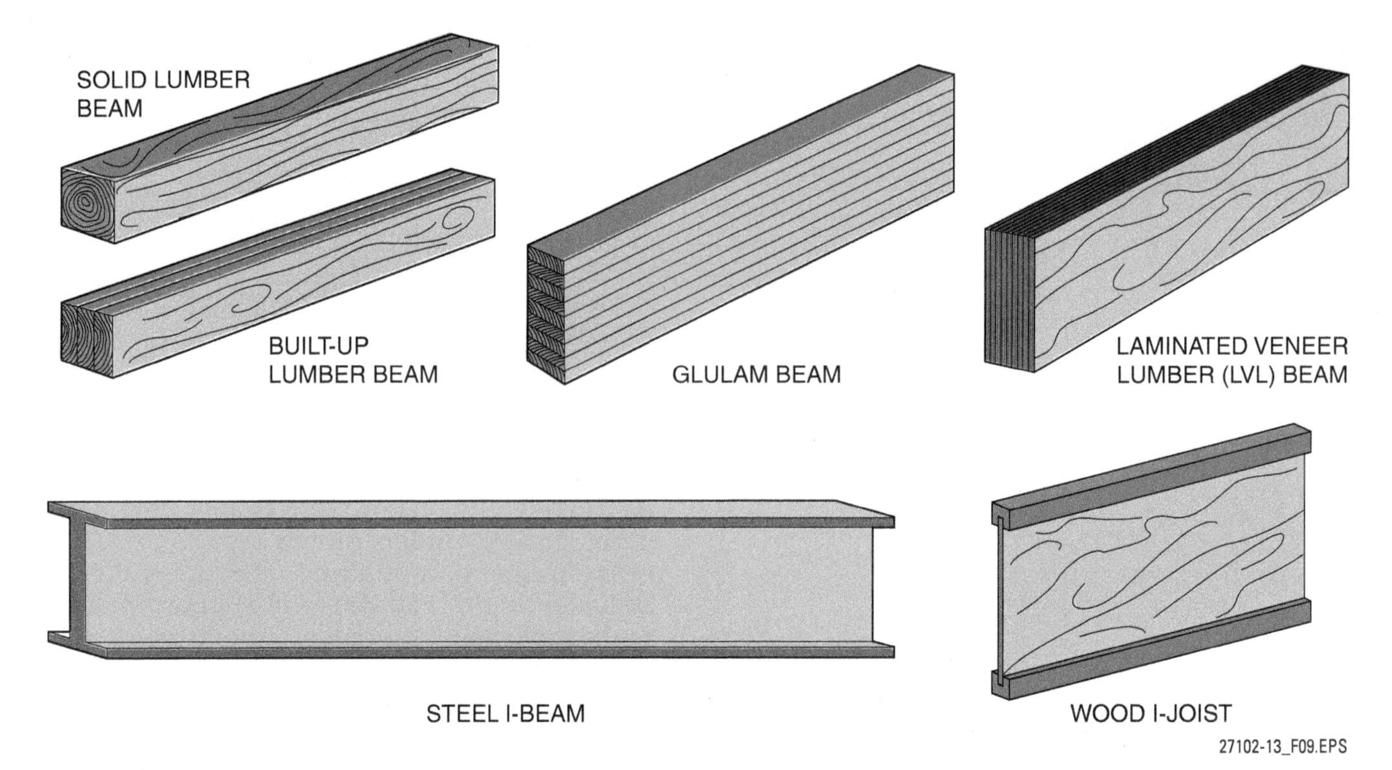

Figure 9 Types of girders.

3.2.2 Built-Up Lumber Girders

Built-up girders are usually constructed using nominal 2" stock (2 × 8s, 2 × 10s) nailed together so they act as one unit. Built-up girders have the advantage of not warping as easily as solid wood girders and are less likely to have decayed wood in the center. The disadvantage is that a built-up girder is not capable of carrying the same load as an equivalent-size solid wood girder. When constructing a built-up girder, the individual members must be fastened together according to code requirements. Butt joints of adjacent members should be staggered at least 4' in either direction. Construction of built-up girders is covered in more detail in the section, *Installing a Beam/Girder.*

3.2.3 Engineered Wood Girders

Laminated veneer lumber (LVL), glued laminated lumber (glulam), and wood I-joists are engineered wood products that are used for girders and other framing members. Their advantage is that they are stronger than the same-size solid lumber. For a given length, the greater strength of engineered wood products allows them to span a greater distance. Another advantage is that they are very straight, with no crowns or warps.

LVL girders are made from laminated wood veneers similar to plywood. The veneers are laid up in a staggered pattern with the veneers overlapping to increase the strength of the girder. Unlike plywood, the grain of each layer runs in the same direction as the other layers. The veneers are bonded with an exterior-grade adhesive, then pressed together and heated under pressure.

Glulam girders are made from lengths of solid, kiln-dried lumber that are glued together. As was discussed in the module *Building Materials, Fasteners, and Adhesives*, glulam is available in three appearance grades: industrial, architectural, and premium. For floor systems like those described in this module, industrial grade would normally be used because appearance is not a priority. Architectural grade is used where beams are exposed and appearance is important. Premium grade is used where the highest-quality appearance is needed. Glulam beams are available in various widths and depths and in lengths up to 40' long.

Larger sizes of wood I-joists may also be used to support the ends of joists. Wood I-joists consist of two-dimensional lumber flanges with a plywood or oriented strand board (OSB) web between them. Squash blocks, filler blocks, or web stiffeners may need to be installed at the point where the floor joists are supported to ensure the wood I-joist girders are not compacted or crushed when loads are placed on them.

3.2.4 Steel I-Beams

Steel I-beams can span the greatest distances and are often used when there are few or no piers or

interior supports. Steel I-beams can span greater distances with smaller beam sizes, thereby creating greater headroom in a basement or crawl space. For example, a 6"-high steel beam may support the same load as an 8"- or 10"-high solid wood beam, thereby providing 2" to 4" more headroom. Two types of steel beams are available: standard flange (S-beam) and wide flange (W-beam). The wide-flange beams are generally used in residential construction. Being metal, steel I-beams are more expensive than wood and are harder to work with, as they are heavier and more difficult to cut and drill. Steel I-beams are typically used only when the design or building code requires it.

Firecut Requirements

In the construction of concrete or masonry buildings, girders and/or joists should be firecut at a diagonal along the top of the member where it enters the wall. If the girder or joist burns through somewhere along its length, damage to the wall is prevented by allowing the member to fail and leave the concrete or masonry wall intact. Without firecutting, the burnt girders or joists would fail and the upper ends of them would rotate upwards, damaging the wall at the connection point and possibly pulling the wall inwards.

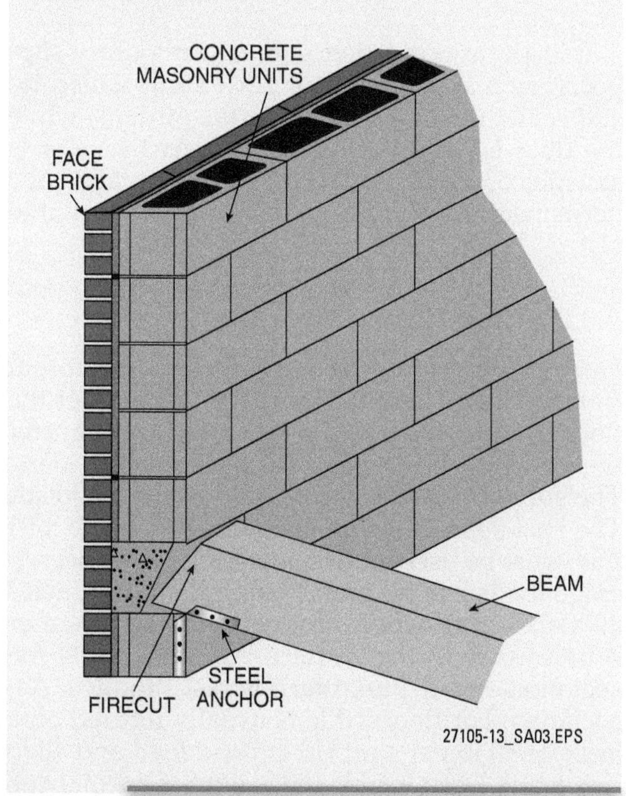

3.2.5 Beam/Girder Supports

Girders and beams must be properly supported at the foundation walls, and at the proper intervals in between, either by posts, columns, or piers (*Figure 10*). Solid or built-up 4 × 4 or 4 × 6 wood posts installed on pier blocks are commonly used to support floor girders, especially for floors built over a crawl space. However, all posts must be as wide as the girder they are supporting. Where a girder is jointed over a post, a 4 × 6 post is normally required. To secure wood posts to their footings, pieces of ½" reinforcing bar (rebar) or anchor bolts are often embedded in the support footings before the concrete sets. These rebar or bolts project into holes bored in the bottoms of the posts, and prevent the bottoms from sliding. The use of galvanized steel post anchors is another widely used method of fastening the bottoms of wood posts to their footings (*Figure 11*). The tops of the posts are normally fastened to the girder using galvanized steel post caps. In addition to securing the post to the girder, these caps also provide for an even bearing surface.

Four-inch round steel columns filled and reinforced with concrete (*Figure 10*), called lally columns, are commonly used as support columns in floors built over basements. Some types of lally columns must be cut to the required height, while others have a built-in jackscrew that allows the column to be adjusted to the proper height. Metal plates are installed at the top and bottom of the columns to distribute the load over a wider area. The plates normally have predrilled holes so that they may be fastened to the girder.

Support piers made of brick or CMUs (*Figure 10*) are more difficult to work with because their level cannot be adjusted. The height of the related footings must be accurate so that when using 4"-thick bricks or 8"-tall CMUs, their tops come out at the correct height to support the girder.

The spacing or interval required between the posts or columns is determined by local building codes based on the stress factor of the girder (i.e., how much weight is put on the girder beam). Note that the farther apart the support posts or columns are spaced, the larger the girder must be in order to carry the joists over the span between them. An example of a girder and supporting columns used in a 24' × 48' building is shown in *Figure 12*. In this example, column B supports one-half of the girder load existing between the building wall A and column C. Column C supports one-half of the girder load between columns B and D. Likewise, column D will share equally the girder loads with column C and the wall E.

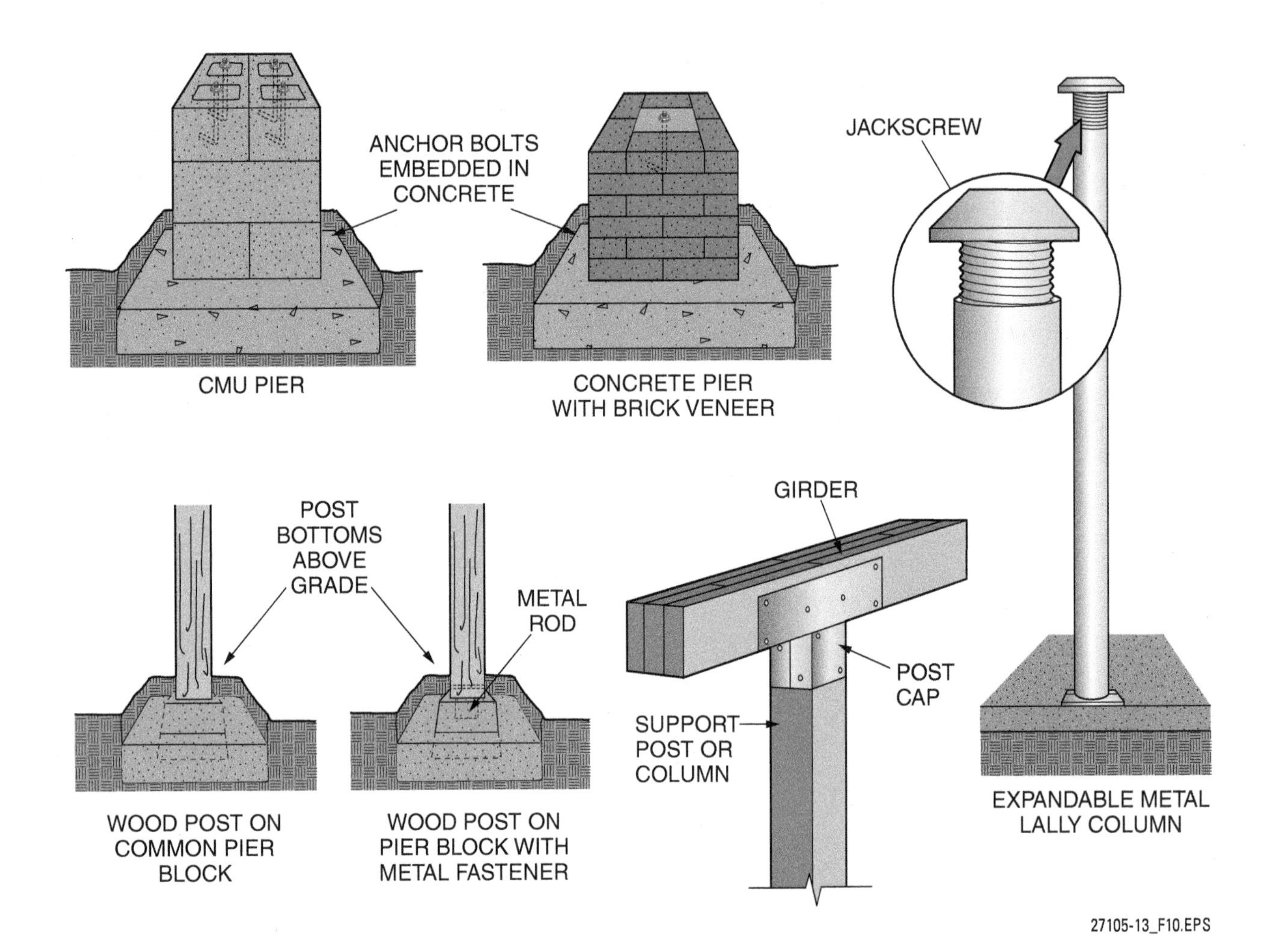

Figure 10 Typical methods of supporting girders.

As shown in *Figure 13,* support of girder(s) at the foundation walls can be done by constructing posts made from solid wood or piers made of CMUs. Another widely used method to support girders at outside walls is to construct girder (beam) pockets in the concrete or CMU foundation walls (*Figure 14*). Provide steel reinforcement as required by the job specifications.

The specifications for girder pockets vary with the size of the girder being used. A rule of thumb is that the pocket should be at least 1" wider than the beam and the beam must have at least 4" of bearing on the wall. Wood girders placed in a pocket should not be allowed to come in direct contact with the concrete or masonry foundation since the chemicals in the concrete or masonry can deteriorate the wood. The end of the wood girder should be placed on a steel plate that is at least ¼" thick. Some carpenters also use metal flashing to line the girder pocket to help protect the wood from the concrete. In some applications, any one of several types of galvanized steel girder hangers can be used to secure the girder to the foundation. *Figure 14* shows one common type.

It is a common practice to use temporary supports, such as jacks and/or 2 × 4 studs nailed together with braces, to support the girder(s) while the floor is being constructed. After the floor is assembled, but before the subfloor is installed, the permanent support posts or columns are put into place.

3.3.0 Floor Joists

Floor joists are a series of parallel, horizontal framing members that make up the body of the floor frame (*Figure 15*). Joists rest on and transfer the building load to the sill plates and girders. The subfloor is fastened to the tops of the joists. The joist span determines the length of the joist that must be used. Safe spans for joists under average loads can be found using the latest tables available from wood product manufacturers or sources such as the Western Wood Products Association and the Southern Forest Products Association. For floors, this is usually figured on a basis of 50 lb per sq ft (10 lb dead load and 40 lb live load). Dead load is the weight of permanent, stationary construction and equipment included

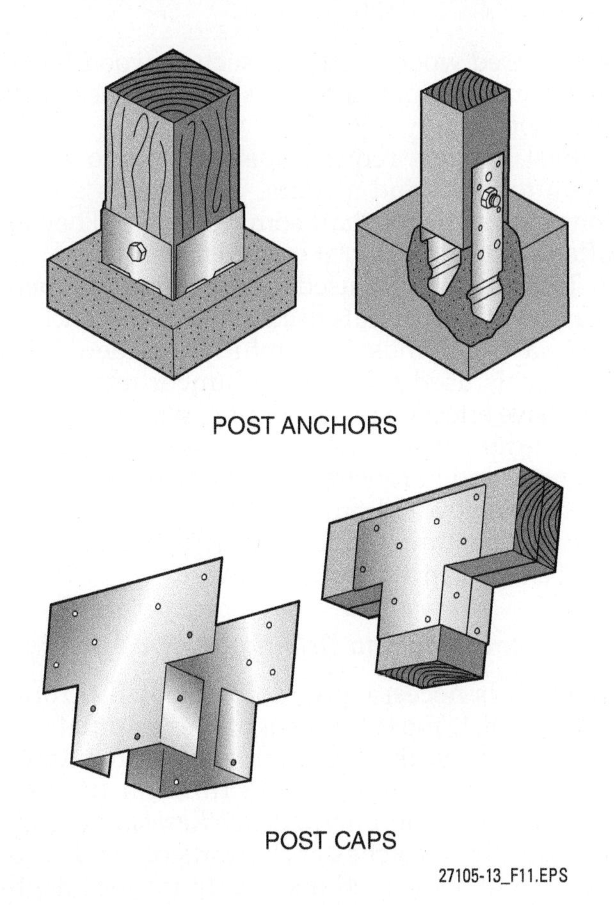

Figure 11 Typical post anchors and caps.

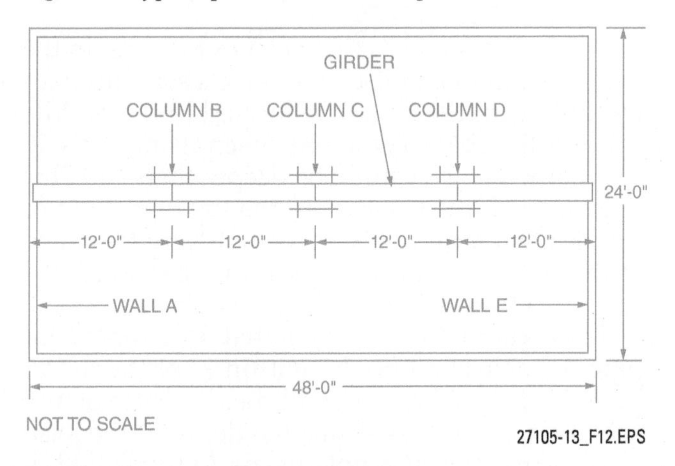

Figure 12 Example of column spacing.

stationary construction and equipment included in a building. Live load is the total of all moving and variable loads that may be placed upon a building. (See the *Appendix* for examples of joist span tables.)

Joists must not only be strong enough to carry the load that rests on them; they must also be stiff enough to prevent undue bending (deflection) or vibration. Too much deflection in joists is undesirable because it can make a floor noticeably springy. Building codes typically specify that the deflection downward at the center of a joist must not exceed $\frac{1}{360}$th of the span with normal live load.

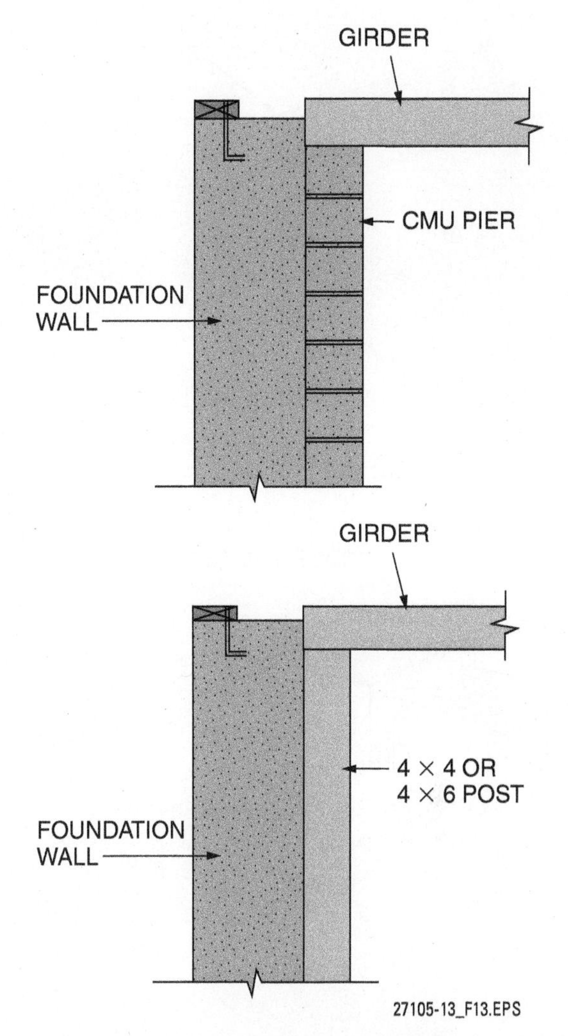

Figure 13 Post or pier support of a girder at the foundation wall.

For example, for a joist with a 15' span, this would equal a maximum of ½" of downward deflection (15' span × 12 = 180" ÷ 360 = 0.5").

Joists are typically installed 16" on center (OC) and are always placed crown up. However, in some applications, joists can be set as close as 12" OC or as far apart as 24" OC. All these distances (12", 16", and 24") are used because they accommodate 4' × 8' subfloor panels and provide a nailing surface where two panels meet. Joists can be supported by overlapping over the top of the girder, supported by a ledger, or supported by a joist hanger. *Figure 16* shows four methods for joist framing at the girder. Check your local code for applicability. Note that if joists are lapped over the girder, the minimum amount of lap is 4". *Figure 17* shows some examples of joist hangers that can be used to fasten joists to girders as well as other support framing members. Joist hangers are commonly used where the bottom of the girder must be flush with the bottoms of the joists. At the sill plate end of the joist, the joist should rest

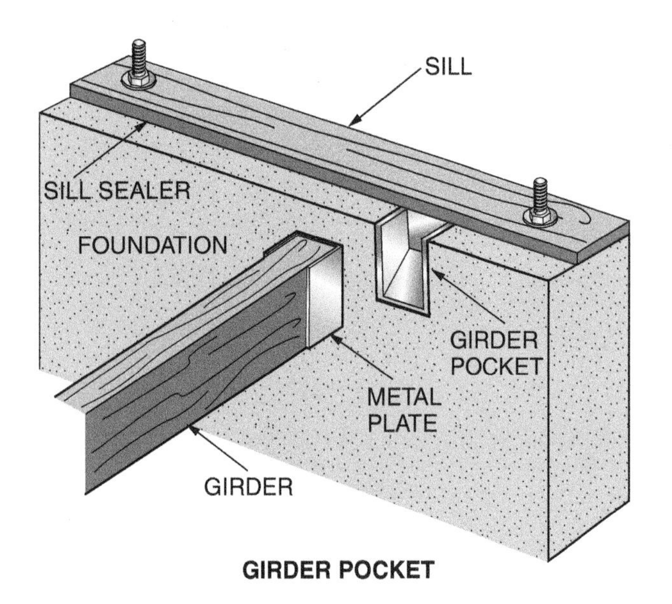

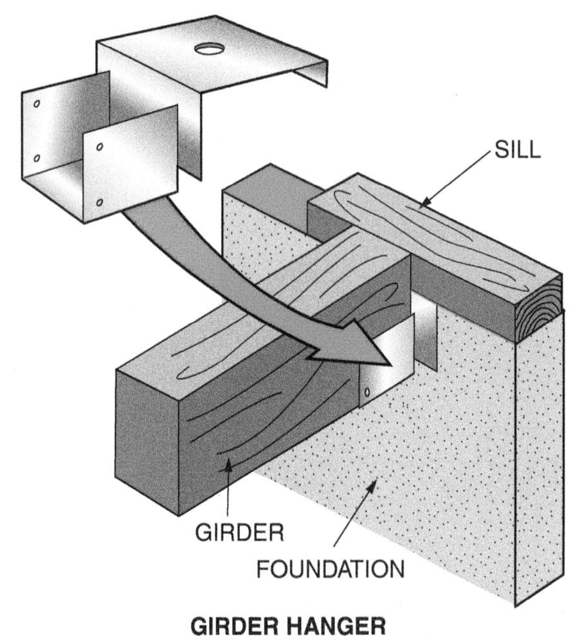

Figure 14 Girder pocket and girder hangers.

on at least 1½" of wood. In platform construction, the ends of all the joists are fastened to a header joist, also called a band joist.

Joists must be doubled where extra loads require additional support. When a partition runs parallel to the joists, a double joist is placed underneath. Joists must also be doubled around openings in the floor frame for stairways, chimneys, etc., to reinforce the rough opening in the floor. These additional joists used at such openings are called trimmer joists. They support the headers that carry short joists called tail joists. Double joists should be spread where necessary to accommodate plumbing.

In residential construction, floors traditionally have been built using solid wood joists. However, engineered wood products, such as wood I-joists and various types of trusses, are also becoming common.

Joist hangers require special nails to secure them to joists and girders. These nails are 1½" long and stronger than common nails. They are often referred to as joist or stub nails.

Treated lumber is used for many framing members, and the materials used for fasteners, anchors, and hardware must be carefully considered. The chemicals used for treating lumber may have a corrosive effect on metal coming into contact with it. Hot-dip galvanized or stainless steel fasteners, anchors, and hardware are recommended for use with treated wood products. The thicker the galvanized coating, the longer the expected life of the fastener, anchor, and hardware.

3.3.1 Notching and Drilling of Wood Joists

When it is necessary to notch or drill through a floor joist, most building codes stipulate the maximum depth of a notch and the maximum diameter of a hole passing through the joist. For example, the *International Residential Code®* specifies that notches at the ends of solid wood joists shall not exceed one-fourth the joist depth. Therefore, in a 2 × 10 floor joist, the notch must not exceed 2 ½" (see *Figure 18*).

The code also states that notches for pipes in the top or bottom of a joist shall not exceed one-sixth the depth, and shall not be located in the middle third of the span. Therefore, when using a 2 × 10 floor joist, a notch cannot be deeper than 1⅝". This notch can be made either in the top or bottom of the joist, but it cannot be made in the middle third of the span. This means that if the span is 12', the span area from 4' to 8' must not be notched.

The code further requires that holes bored for pipe or cable shall not be within 2" of the top or bottom of the joist, nor shall the diameter of any such hole exceed one-third the depth of the joist. This means that if a hole needs to be drilled, it may not exceed 3" in diameter if a 2 × 10 floor joist

Maintaining a Flush Top Surface

When securing a joist to a girder using a ledger, ensure that the top surface of the girder and joist are flush so the subfloor lies flat. To account for these small discrepancies, toenail the joists to the girders prior to installing the ledger. Once a smooth, flush surface is established, install the ledger under the joist.

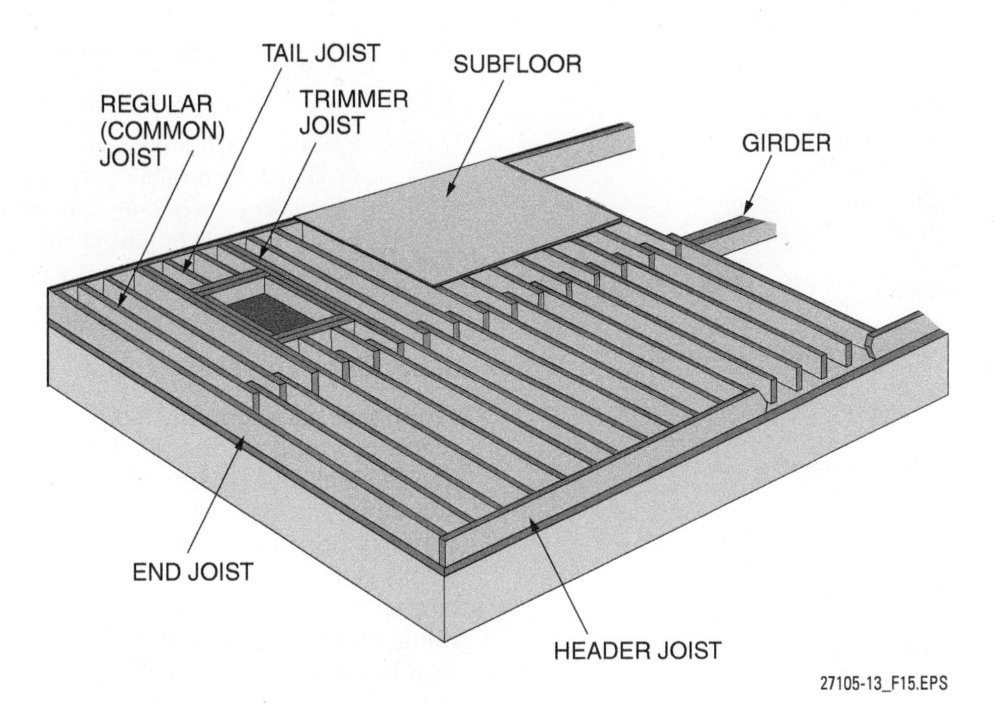

Figure 15 Floor joists.

is used. Always check the local codes for specific requirements in your jurisdiction.

3.3.2 Wood I-Joists

Wood I-joists are available in various depths and in lengths up to 80' (*Figure 19*). These joists are not prone to shrinking or warping like solid lumber. The design of wood I-joists provides them with a strength-to-weight ratio much greater than solid lumber. Because of their increased strength, wood I-joists can be used for greater spans than solid lumber joists. Wood I-joists can be cut to length, installed, and nailed like solid lumber. Special joist hangers and strapping, similar to those used with solid wood joists, are used to fasten wood I-joists to girders and other framing members. *Figure 20* shows a typical floor system constructed with wood I-beams.

For wood I-joists, cuts, notches, and holes are prohibited except where permitted by the manufacturer. Some I-joists are manufactured with perforated knockouts in the web, approximately 12" apart, that can accommodate electrical wiring. Never notch or drill through the beam flange or make cuts or holes in the web without checking the manufacturer's specification sheet. In addition, other engineered wood products such as LVL, parallel strand lumber (PSL), and glulam should not be cut or notched without first checking the specification sheets.

Wood I-Joists

In 1969, the first plywood I-joist was created. In 1977, the first I-joist was created using laminated veneer lumber (LVL). This new construction offered superior strength and stability. In 1990, oriented strand board (OSB) web material, constructed of interlocking fibers, began to be used in wood I-joists, as shown here. OSB is less expensive than plywood and is not as prone to warping or cracking. Engineered wood products were once only available through a handful of companies that pioneered the industry. Today, engineered lumber and lumber systems are offered by a wide variety of companies. In many areas of the country, the use of wood I-joists has surpassed the use of solid wood joists.

27105-13_SA04.EPS

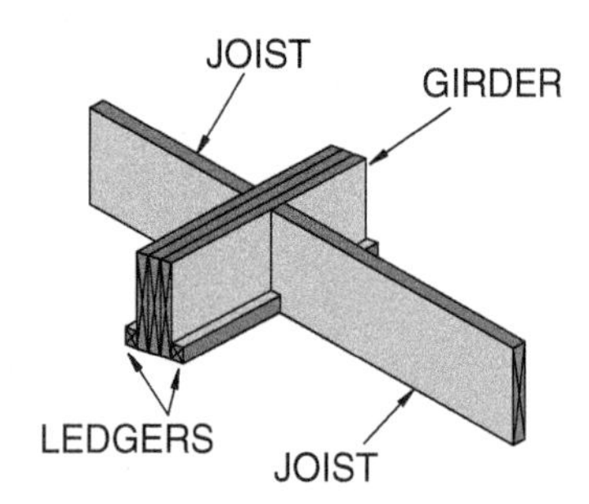

JOIST NOTCHED AROUND LEDGER

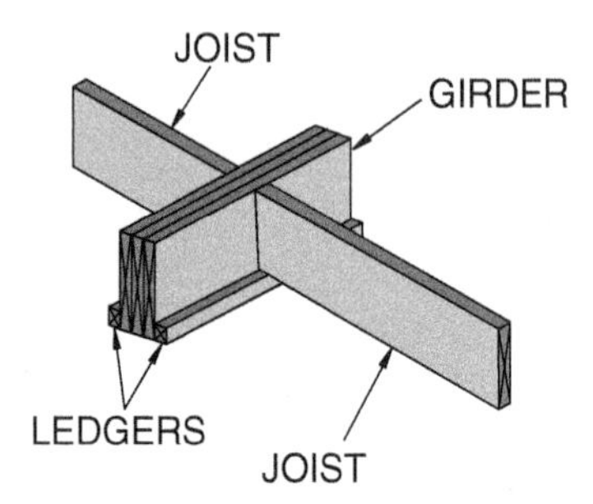

JOIST SITS ON LEDGER

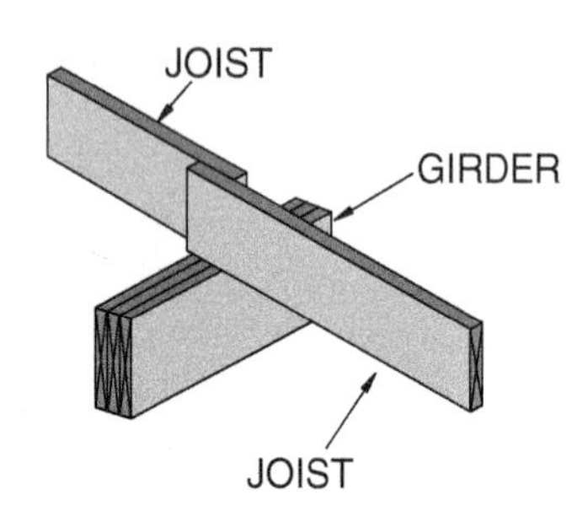

JOIST OVERLAP ON GIRDER

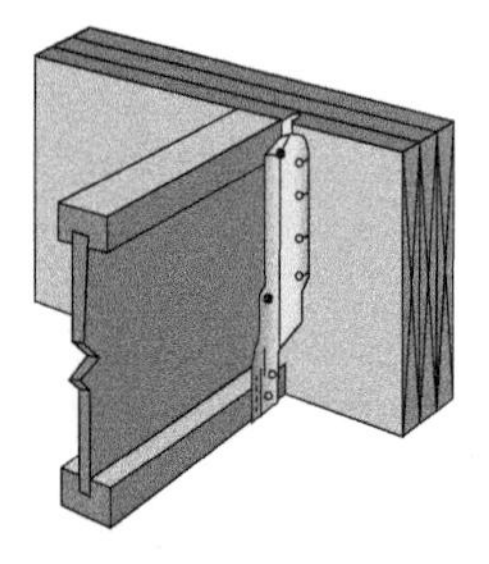
JOIST HANGER

27105-13_F16.EPS

Figure 16 Methods of joist framing at a girder.

Learn more about wood I-joists:

Wood I-Joist Manufacturers Association (WIJMA). An organization representing manufacturers of prefabricated wood I-joist and structural composite lumber.
www.i-joist.org

3.3.3 Floor Trusses

Floor trusses are manufactured joist assemblies made of wood and steel (*Figure 21*). Solid light-gauge steel and open-web steel trusses are also available, but are primarily used in commercial construction. Like the wood I-joists, trusses are stronger than comparable lengths of solid lumber, allowing them to be used over longer spans. Longer spans allow more freedom in building design because interior loadbearing walls and extra footings can often be eliminated. Trusses can generally be erected faster and easier with no need for trimming or cutting in the field. They also provide the additional advantage of permitting ducting, plumbing, and electrical wires to be run easily between the open webs.

Floor trusses consist of two primary components: chords and webs. The chords (horizontal members) are held rigidly in place using wood or steel webs that run vertically or diagonally between the chords. For parallel-chord wood trusses, connector plates tie the chords to the webs. Connector plates are toothed metal plates that fasten the webs and chords together at the intersecting points.

One type of truss commonly used in residential floor systems is the parallel-chord 4 × 2 truss. This name is derived from the chords consisting of 2 × 4 lumber with the lumber faces facing each other. Webs extend between the chords. Diagonal webs positioned at 45-degree angles to the chords mainly resist the shearing stresses in the truss. Vertical webs, which are placed at right angles to the chords, are used at critical load-transfer points where additional strength is required. While wood is used most frequently for webs, galva-

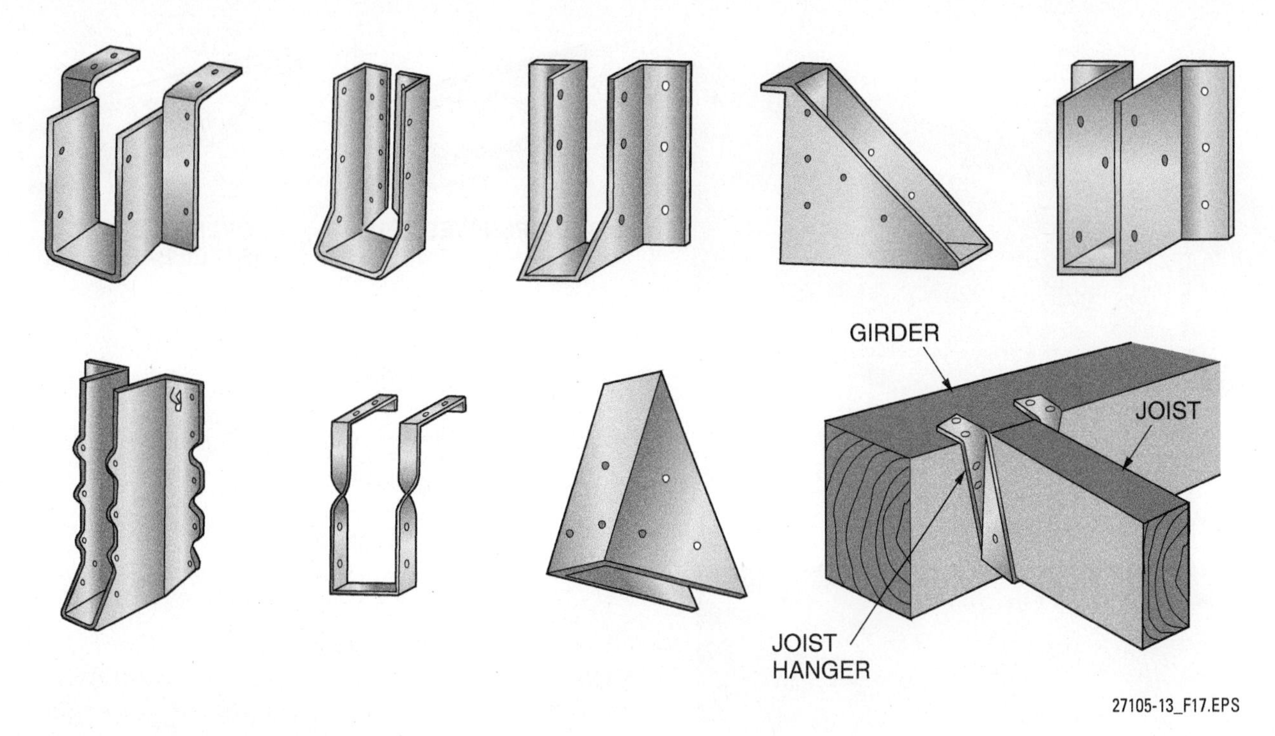

Figure 17 Typical types of joist hangers.

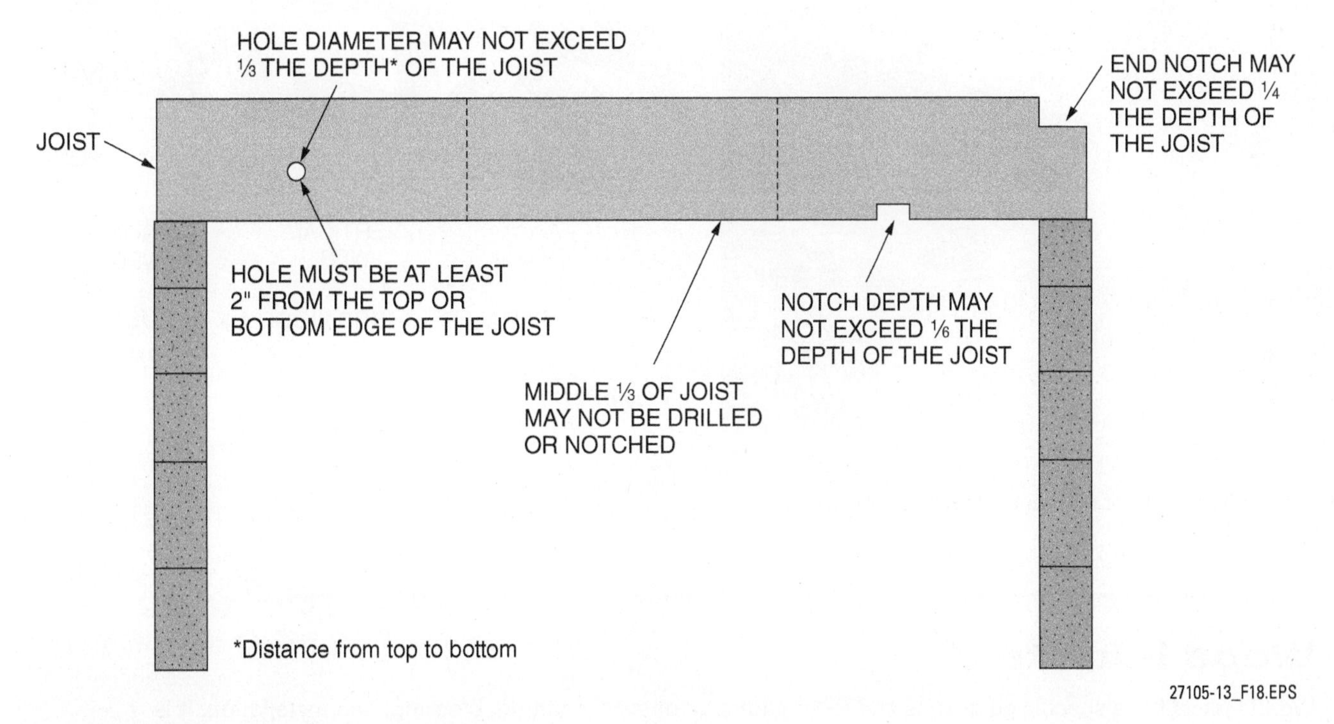

Figure 18 Notching and drilling of wood joists.

Figure 19 Wood I-joists.

Figure 20 Typical floor system constructed with wood I-joists.

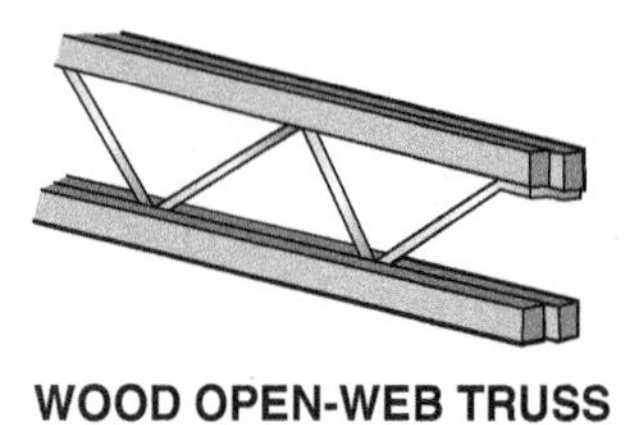

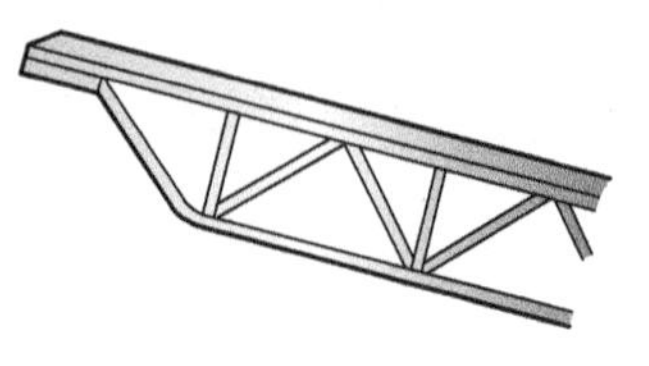

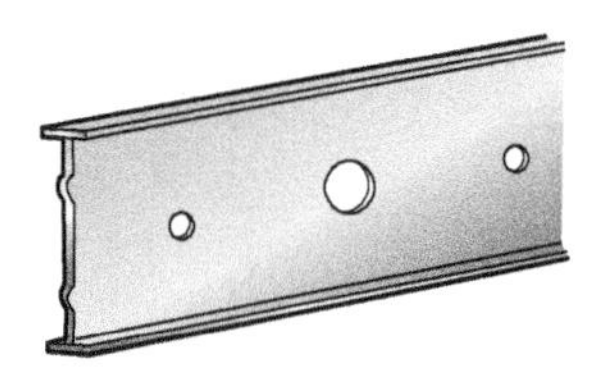

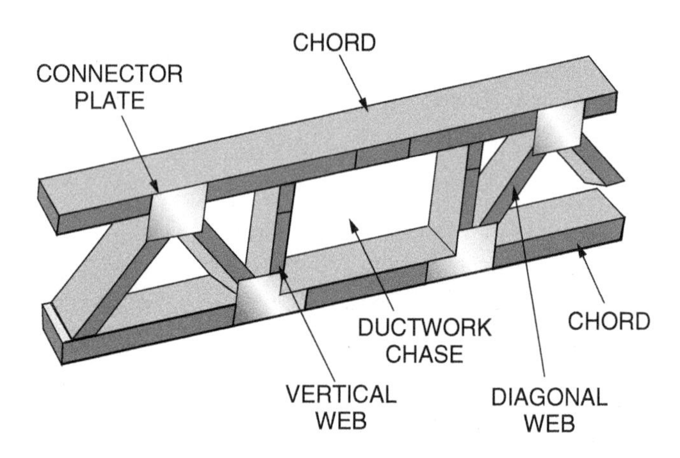

Figure 21 Typical floor trusses.

Wood I-Joists

Wood I-joists have specific guidelines and instructions to follow for cutting, blocking, and installation. It is important to always follow the manufacturer's instructions when installing these materials; otherwise, a very dangerous situation may be created by compromising the structural integrity of the joist. When wood I-joists are attached to a larger wood I-joist acting as a girder, backer blocks should be installed between the flanges of the girder to provide a flush nailing surface for the I-joist. Squash blocks may be installed along the flanges of joists when point loads bear directly on the wood I-joists. Backer blocks and squash blocks are dimensional lumber that is cut to length and installed between the flanges.

nized steel webs are also used. Trusses made with metal webs provide greater spans for any given truss depth than trusses with wood webs. The openings in the metal-web trusses are larger too, which provides more room for HVAC ductwork.

Note that there are several different kinds of parallel-chord trusses. What makes each one different is the arrangement of its webs. Typically, parallel-chord floor trusses with wood webs are available in depths ranging from 12" to 24" in 1" increments. The most common depths are 14" and 16". Some metal-web trusses are available with the same actual depth dimensions as 2 × 8, 2 × 10, and 2 × 12 solid wood joists, making them interchangeable with an ordinary joist-floor system. *Figure 22* shows a typical floor system constructed with trusses.

Learn more about trusses:

Building with Floor Trusses. Madison, WI: Wood Truss Council of America (11-minute DVD or video).

Storage, Handling, Installation & Bracing of Wood Trusses. Madison, WI: Wood Truss Council of America (69-minute DVD or video).

Wood Truss Council of America. An international trade association representing structural wood component manufacturers, **www.sbcindustry.com**

3.4.0 Bridging

Bridging is used to stiffen the floor frame to prevent unequal deflection of the joists and to enable an overloaded joist to receive some support from the joists on either side. Most building codes require that bridging be installed in rows between the floor joists at intervals of not more than 8'. For example, floor joists with spans of 8' to 16' need one row of bridging in the center of the span.

Three types of bridging are commonly used (*Figure 23*): wood cross-bridging, solid bridging, and metal cross-bridging. Cross-bridging is composed of pieces of wood or metal set diagonally between the joists to form an X. Wood cross-bridging is typically 1 × 4 lumber placed in double rows that cross each other in the joist space.

Metal cross-bridging is installed in a similar manner as wood cross-bridging. Metal cross-bridging is available in a variety of styles and different lengths for use with a particular joist size and spacing. It is usually made of 18-gauge steel and is ¾" wide. When using cross-bridging, nail the top of the bridging into place, but do not nail the bottom until the subfloor is installed. The cross-bridging should extend from the top edge of one joist to the lower edge of the adjoining joist.

Solid bridging, also called blocking, consists of solid pieces of lumber, usually the same size as the floor joists, installed between them. Solid bridging is installed in a staggered fashion to enable end-nailing of the bridging.

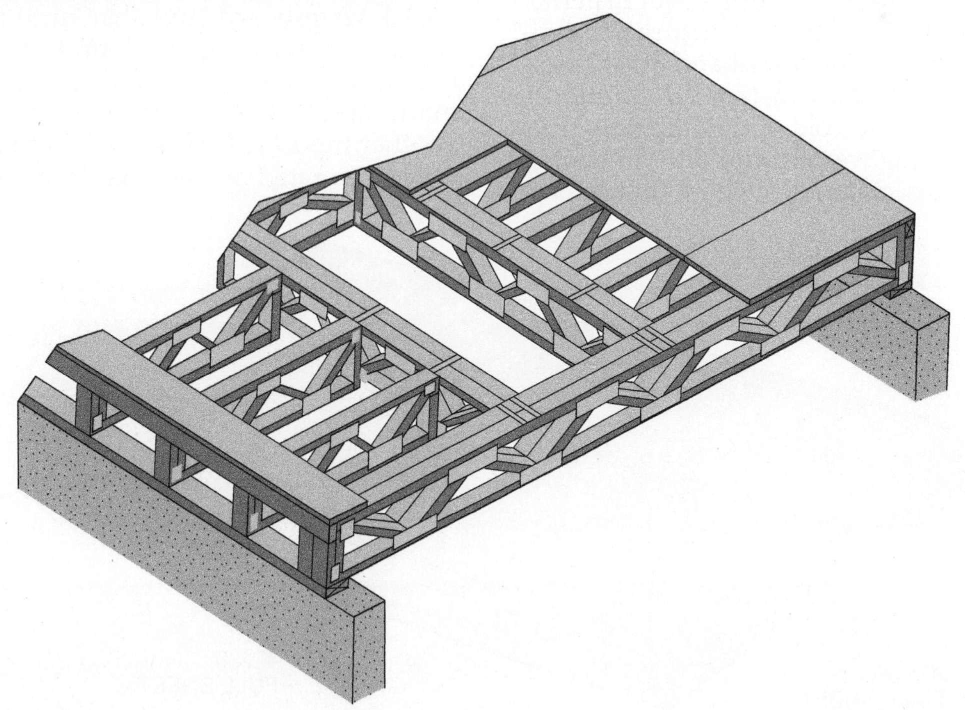

Figure 22 Typical floor system constructed with trusses.

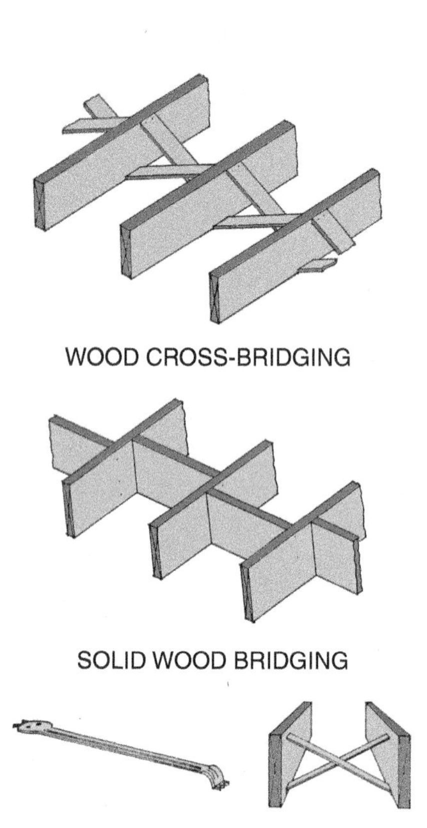

Figure 23 Types of bridging.

3.5.0 Subfloors

Subfloors consist of panels laid directly on and fastened to floor joists (*Figure 24*) to provide a base for underlayment and/or the finish floor material. A subfloor adds rigidity to the structure and provides a surface upon which wall and other framing can be laid out and constructed. Subfloors also act as a barrier to cold and dampness, thus keeping the building warmer and drier in winter. Underlayment is a material, such as OSB or particleboard, laid on top of the subfloor to provide a smoother surface for finished flooring. Underlayment is typically installed after the structure is built but before the finished floor is laid.

3.5.1 Plywood Subfloors

Butt-joint or tongue-and-groove plywood is widely used for residential subfloors. When the joists are placed 16" OC, ⅝" to ¾" thick, 4' × 8' panels are commonly used as the subfloor. The panels go down quickly and provide great rigidity to the floor frame. APA-rated sheathing plywood panels (*Table 1*) are generally used for subfloor in two-layer floor systems. APA-rated Sturd-I-Floor® tongue-and-groove plywood panels are commonly used in single-layer floor systems (combined subfloor-underlayment), where direct application of carpet, tile, etc., to the floor is intended.

Traditionally, plywood panels have been fastened to the floor joists using nails. Today, it is becoming more common to use a glued floor system in which the subfloor panels are both glued and nailed or screwed to the joists. A glued floor system stiffens the floors and helps to eliminate squeaks and nail popping. Procedures for installing plywood subfloors, including gluing, are described in the section, *Installing Subfloor.*

3.5.2 Manufactured-Panel Subfloors

Manufactured panels made of OSB, composite board, waferboard, and structural particleboard can also be used for subfloor. Detailed information on the construction and composition of these manufactured wood products is contained in another module. Panels made of these materials have been rated by the American Plywood Asso-

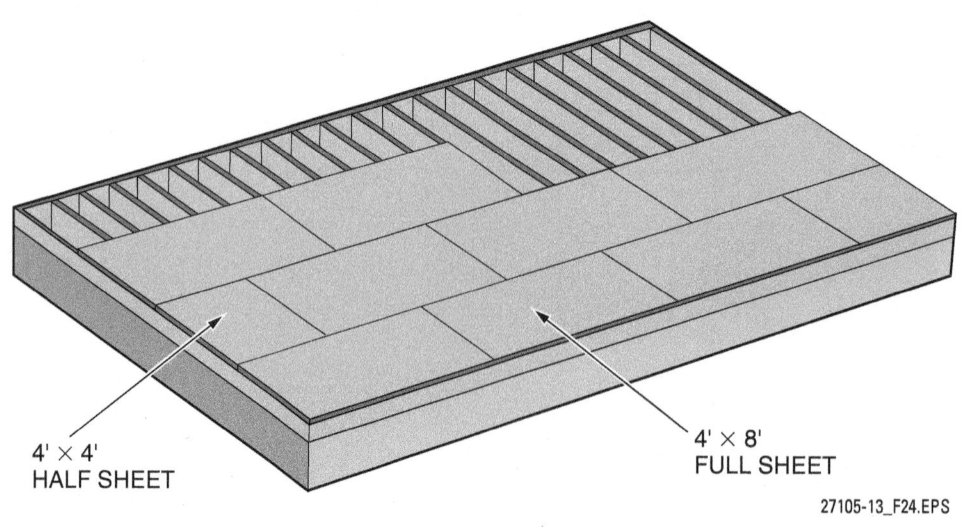

Figure 24 Subfloor installation.

Table 1 Guide to APA Performance-Rated Plywood Panels

Panel Grade	Thickness in Inches	Span Rating in Inches
Rated Sheathing	⅜	24/0
	7⁄16	24/16
	15⁄32	32/16
	19⁄32	40/20
	23⁄32	48/24
Rated Sturd-I-Floor®	19⁄32	20 OC
	23⁄32	24 OC
	⅞,1	32 OC
	1⅛	48 OC
Rated Siding	11⁄32	16 OC
	7⁄16	24 OC
	15⁄32	24 OC
T1-11	19⁄32	16 OC
	19⁄32	24 OC

ciation and meet all standards for subfloors. The method for installing manufactured panels is basically the same as that used for plywood.

3.5.3 Board Subfloors

There may be instances, especially in remodeling work, when 1 × 6 or 1 × 8 boards may be used as a subfloor. Boards can be laid perpendicular to the floor joists. However, it is more common to lay them diagonally across the floor frame at a 45-degree angle. This provides for more rigidity of the floor and also assists in bracing the floor joists. Also, if laid perpendicular to the joist in a subfloor where oak flooring is to be laid over it, the oak flooring (instead of the subfloor) would have to be laid diagonally to the floor joist. This is necessary to prevent the shrinkage of the subfloor from affecting the joints in the finished oak floor, which would cause the oak flooring to pull apart. Board subfloors are nailed at each joist. Typically, two nails are used in each 1 × 6 board and three nails for wider boards. Note that a subfloor made of boards is normally not as rigid as one made of plywood or other manufactured panels.

OSB Subfloors

Today, many builders prefer to use OSB for subfloors. It offers acceptable structural strength at a reduced price.

27105-13_SA05.EPS

Learn more about floor systems:

Builder Tips: Steps to Construct a Solid, Squeak-Free Floor System. Tacoma, WA: APA—The Engineered Wood Association.

Field Guide for Prevention and Repair of Floor Squeaks. Boise, ID: Trus Joist, a Weyerhauser business.

Quality Floor Construction. Tacoma, WA: APA—The Engineered Wood Association (15-minute video).

3.0.0 Section Review

1. The lowest members of a structure's frame are _____.

   a. joists
   b. beams
   c. sill plates
   d. soleplates

2. The distance between two outside walls is referred to as the _____.

   a. dead load
   b. live load
   c. span
   d. bearing capacity

3. Steel columns filled and reinforced with concrete are called _____.

   a. jackscrews
   b. lally columns
   c. joist hangers
   d. piers

4. The deflection at the center of a joist spanning 22 feet must not exceed ______.

   a. 0.25"
   b. 0.57"
   c. 0.73"
   d. 1.22"

5. In platform frame construction, the ends of all the joists are fastened to the _____.

   a. studs
   b. beam
   c. trimmer joists
   d. header joist

6. When installing cross-bridging, nail the bottom of the bridging into place first, but do not nail the top into place until the subfloor is installed.

   a. True
   b. False

7. Which of the following is *not* a purpose of the subfloor?

   a. To add rigidity to a structure
   b. To act as a barrier to cold and dampness
   c. To provide a surface upon which to build the wall frames
   d. To support the sill plate

SECTION FOUR

4.0.0 CONSTRUCTING FLOOR SYSTEMS

Objective

Describe the construction methods for floor systems, and identify floor system materials.

a. Describe how to check a foundation for squareness.
b. Name the methods used to lay out and fasten sill plates to the foundation.
c. Describe the proper procedure for installing a beam or girder.
d. Describe how to lay out sill plates and girders for floor joists.
e. Describe how to lay out and install floor joists for partitions and floor openings.
f. Identify different types of bridging and describe how to properly install each type.
g. Describe how to properly install subfloor.
h. Explain how to install joists for projections or cantilevered floors.

Performance Task 1

Lay out and construct a floor assembly, including a rough opening and subfloor material.

Trade Terms

End joist: The first and last joists in a floor system; they run parallel to the common joists.

Header: Used to frame floor openings; they run perpendicular to the direction of the joists and are typically doubled.

Lateral pressure: Sideways pressure against a structure, such as a foundation.

Rim joist: Consists of two header joists and the end joists.

Scab: A length of lumber applied over a joint to strengthen it.

After the foundation is completed and the concrete or mortar has properly set up, assembly of the floor system can begin. The first floor is usually framed before the foundation is backfilled to help the foundation withstand the lateral pressure placed on it by the soil. This section provides an overview of the procedures and methods used for laying out and constructing a basic platform floor assembly. When building any floor system, always coordinate your work with that of the other trades to ensure the framing is properly done to accommodate ductwork, piping, wiring, etc.

The construction of a platform floor assembly is typically performed in the following sequence:

Step 1 Check the foundation for squareness.

Step 2 Lay out and install the sill plates.

Step 3 Build and/or install the girders and supports.

Step 4 Lay out the floor joist locations on the sill plates and girders.

Step 5 Lay out the joist locations for partitions and floor openings.

Step 6 Cut and attach the joist headers to the sill plates.

Step 7 Install the joists.

Step 8 Frame the openings in the floor.

Step 9 Install the bridging.

Step 10 Install the subfloor.

4.1.0 Checking the Foundation for Squareness

Before installing the sill plates, ensure that the foundation wall meets the dimensions specified on the prints and that the foundation is square. However, keep in mind that parallel and plumb take precedence over square. The foundation is checked for squareness by making measurements of the foundation with a 100' measuring tape. First, the lengths of each of the foundation walls are measured and recorded (*Figure 25*). The measurements must be as exact as possible. Following this, the foundation is measured diagonally from one outside corner to the opposite outside corner. A second diagonal measurement is then made between the outsides of the remaining two corners. If the measured lengths of the opposite walls are equal and the diagonals are equal, the foundation is square. For buildings where the foundation is other than a simple rectangle, a good practice is to divide the area into two or more individual square or rectangular areas and measure each area as previously described.

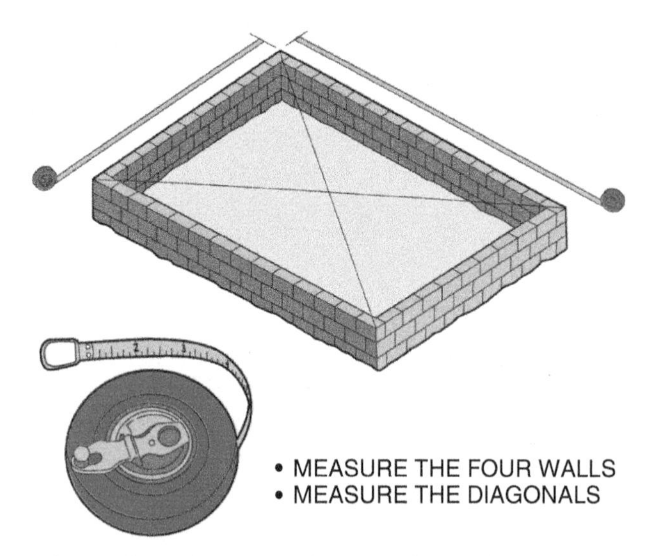

Figure 25 Checking the foundation for squareness.

4.2.0 Installing Sill Plates

For floors where the sill plate is installed flush with the outside of the foundation walls, sill plate installation begins by snapping chalklines on the top of the foundation walls in line with the inside edge of the sill plate (*Figure 26*). If the sill plate must be set in to accommodate the thickness of wall sheathing, brick veneer, etc., the chalklines may be snapped for the outside edge of the sill plates or the inside edge (if the foundation size allows). At each corner, the true location of the outside corner of the sill plate is used as a reference point to mark the corresponding inside corner of the sill plate on the foundation wall. To do this, the exact width of the sill plate stock being used must be determined. For example, if using 2 × 8 sill plates, 7¼" is the sill plate width measurement. After the exact inside corners are located and marked on the sill plate, chalklines are snapped between these points. This gives an outline on the top of the foundation wall of the exact inside edges of the sill plates. At this point, a good practice is to double-check the dimensions and squareness of these lines to make sure that they are accurate.

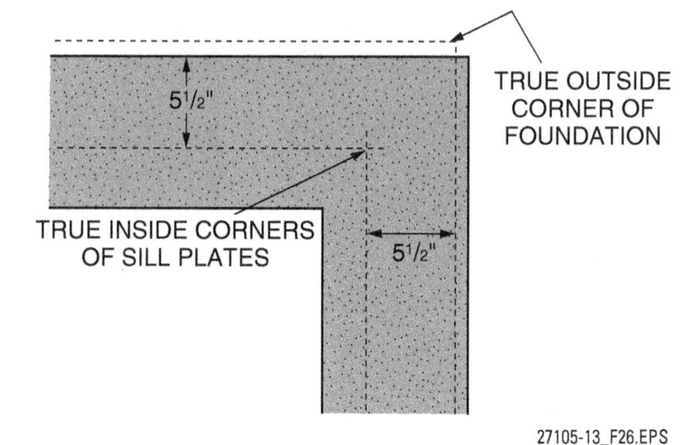

Figure 26 Inside edges of sill plates marked on the top of the foundation wall.

After the sill plate location is marked on the foundation, the stock can be measured and cut. Take into account that there must be an anchor bolt within 12" of the end of each plate. Also, sill plates cannot butt together over any opening in the foundation wall. When selecting the lumber, choose pieces that are as straight as possible for making the sill plates. Badly bowed pieces should not be used.

Holes must be drilled in the sill plates so that they can be installed over the anchor bolts embedded in the foundation wall. To lay out the location of these holes, hold the sill plate sections in place on top of the foundation wall against the anchor bolts (*Figure 27*). At each anchor bolt, use a square to scribe lines on the sill plate corresponding to both sides of the bolt. On the foundation, measure the distance between the center of each anchor bolt and the chalkline, then transfer this distance to the corresponding bolt location on the sill plate by measuring from the inside edge. After the hole layout is complete, the holes in the sill plate are drilled. They should be about ⅛" to ¼" larger than the diameter of the anchor bolt to allow for some adjustment of the sill plates, if necessary. Also, make sure all holes are drilled plumb through the sill plate.

Keep It Square

When it is time to install sill plates on the foundation, you may discover that the foundation wall is not exactly true and square. Don't use the foundation wall as a guide. Instead, ensure that the sill plates are square with each other by using a measuring tape to measure the four plates and the diagonals. If the opposite plates and the diagonals are equal, the sill plates are square with each other. This may mean that the outside edge of the sill plates may not align exactly with the outside edge of the foundation. If necessary, some sill plates may overlap or underlay the wall.

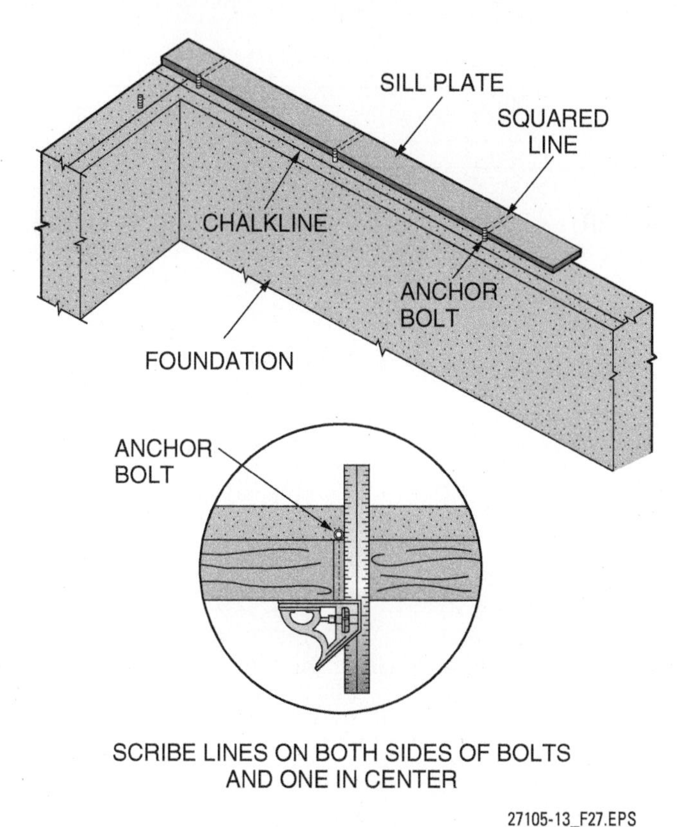

Figure 27 Square lines across the sill plate to locate the anchor bolt hole.

Before installing the sill plates, a termite shield (if used) and sill sealer are installed on the foundation. When these items are in place, the sill plate sections are positioned over the anchor bolts, making sure that the inside edges of the sill plate sections are aligned with the chalkline on top of the foundation wall and that the inside corners are aligned with their marks. The sill plates are then loosely fastened to the foundation with the anchor bolt nuts and washers, and the sill plate is checked to make sure it is level. An 8' carpenter's level can be used for this task. However, using a transit or builder's level and checking the level every 3' or 4' along the sill plate is more accurate. It cannot be emphasized enough how important it is that the sill plate be level. If the sill plate is not level, it will affect the level of the building's floors and walls. Low spots can be shimmed with plywood wedges or set into a small bed of grout or mortar. If the sill plate is too high, the high areas of the concrete foundation need to be ground or chipped away. After the sill plate has been leveled, the nuts for the anchor bolts can be fully tightened. Be careful not to overtighten the nuts, especially if the concrete is not thoroughly cured, because this can crack the wall.

4.3.0 Installing a Beam/Girder

Before installing a girder, use the job specifications to determine the details related to its installation. For example, assume a structure has a foundation that is 24' wide and 48' long (*Figure 28*). The foundation is cast concrete that is 12" thick with 6"-deep and 7"-wide girder (beam) pockets centered in the short walls. The girder is to be a built-up beam containing three thicknesses of 2 × 10s. Three 4" concrete-filled lally columns with ¼" steel plates at the top and bottom will be used to support the girder. The distance between these columns is 12'-0" OC.

To lay out the distances between each of the support columns, first use a measuring tape to measure from one end of the foundation to the other in the precise girder location. Using a plumb bob, hold the line at the 24'-0" mark. This locates the center of the middle support column. Then, measure 12'-0" to the left of center and 12'-0" to the right of center to locate the center of the other two support columns. The distance from the center of the two end columns into their girder pockets in the foundation walls is 11'-5½", allowing for ½" of space between the back of each girder pocket and the end of the girder (12'-0" – 6"-deep pocket – ½" space = 11'-5½"). Given these dimensions, the finished built-up girder for this example needs to measure 46'-11" long. When constructing this girder, remember that the joints should fall directly over the support columns and the girder crown must face upward.

> **NOTE**
> The *International Residential Code*® specifies the number and size of nails and the nailing pattern for framing members.

When framing floor systems, use the fasteners as indicated below:

- Joist to sill or girder, toe nail–Three 8d
- Rim joist or blocking to sill plate, toe nail–8d at 6" OC
- 1 × 6 subfloor or less to each joist, face nail–Two 8d or two 1¾" staples
- 2" subfloor to joist or girder, blind and face nail–Two 16d
- 2" planks (post-and-beam), at each bearing point–Two 16d
- Built-up girders and beams using 2" lumber, two nails at each end and at each splice–10d; nail each layer at 32" OC staggered at top and bottom
- Ledger strip for supporting joists, at each joist–Three 16d

46'-11"

11'-5½" | 12'-0" | 12'-0" | 11'-5½"
15'-5½" | 16'-0" | 15'-5½"
11'-5½" | 12'-0" | 12'-0" | 11'-5½"

BUILT-UP GIRDER

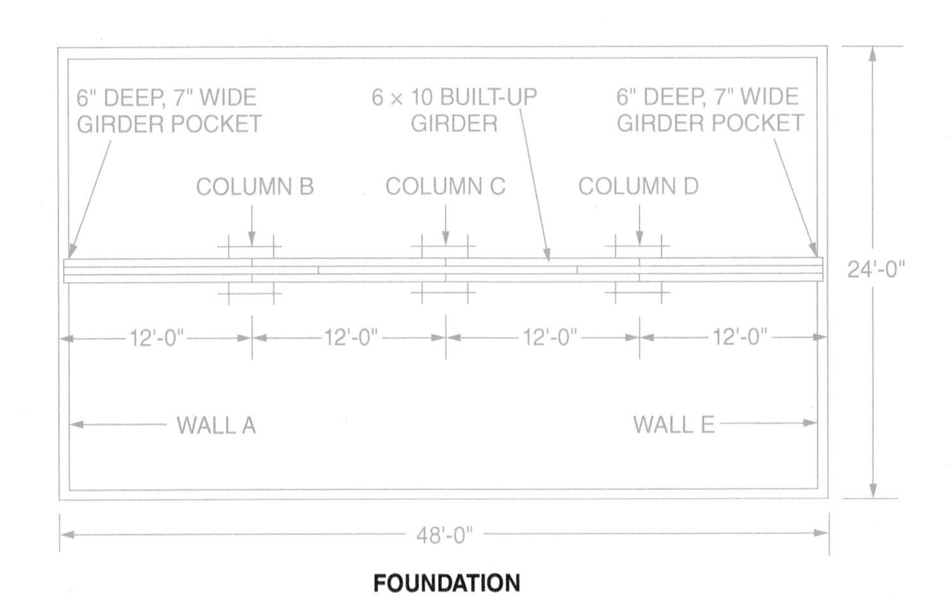

27105-13_F28.EPS

Figure 28 Example girder and support column data.

In addition, special nails are used to attach the joist hangers. Pneumatic, ring shank, and screw nails, or etched galvanized staples are used to apply the subfloor and underlayment. Construction adhesive may also be used when fastening subfloor and underlayment to the base materials.

As shown in *Figure 28*, the 6 × 10 built-up girder can be constructed using eight 12'-long 2 × 10s and three 16'-long 2 × 10s. Four of the 12'-long 2 × 10s are cut to 11'-5½" (so the splice falls directly over a column) and two of the 16'-long 2 × 10s are cut to 15'-5½". The 16' pieces are used in order to provide for an overlap of at least 4'-0" at the joints. To make the girder, nail the 2 × 10s together in the recommended pattern with 10d nails. Each layer should be nailed 32" OC at the top and bottom and staggered, and two nails should be placed at each end and at each splice. Note that the nailing schedule may vary in different jurisdictions, so make sure to consult local codes for the proper nailing schedule. Drive the nails at an angle for better holding power. Be sure to butt the joints together so they form a tight fit. Continue this process until the 46'-11" girder is finished. Once completed, the girder is put in place, supported by temporary posts or A-frames, and leveled. Note that the temporary supports are removed and replaced by the permanent lally columns after the floor system joists are installed.

GOING GREEN

Strawbale Construction

Strawbale construction uses both abundant and bio-based materials. The bales of straw used for this purpose are typically three to five times more densely packed than agricultural bales. Strawbale construction can be either structural or used as fill with a wood frame or other structure. Bales are stabilized using rebar or bamboo spikes. Bond beams are used to ensure a loadbearing surface along the tops of walls. Strawbale is becoming more widely used in certain parts of the United States. Arizona even has a special section of the building code for this construction method. Proper detailing to prevent moisture problems is critical for success. Deep overhangs and moisture barriers between bales and foundations are common. Strawbale construction has a high insulating value. It also has a two-hour or greater fire rating, depending on the exterior and interior finish.

4.4.0 Laying Out Sill Plates and Girders for Floor Joists

Joists should be laid out so that the edges of standard-size subfloor panels fall over the centers of the joists to allow the subfloor to be properly fastened along the edges. There are different ways to lay out a sill plate to accomplish this. One method for laying out floor joists 16" OC is described here. Begin by using a measuring tape to measure out from the end of the sill plates exactly 15¼" (*Figure 29*). At this point, use a square to square a line across the sill plate. To make sure of accurate spacing, drive a nail into the sill plate at the 15¼" line, then hook the measuring tape to the nail and stretch it the length of the sill plate. Make a mark on the sill plate at every point on the tape marked as a multiple of 16" (most tapes highlight these numbers). It is important that the spacing be laid out accurately; otherwise, the edges of the subfloor panels may not fall at the center of the joists.

After the sill plate is marked, use a square to square a line across the sill plate at each mark. Note that the lines marked on the sill plate mark the edge of the joists, not the center. Once the lines have been squared across the sill plate, one of two methods can be used to indicate on which side of the line the floor joist will rest. In one method, a line parallel to the initial line is made using a square. Some carpenters use a scrap piece of lumber equal in thickness to the floor joists to lay out the second line. If the layout has been started from the left side of the sill plate, as shown in *Figure 29*, the second line should be placed to the right of the line. If the layout has been started from the right side, the second line should be placed to the left of the initial line. The joists will then be placed between the two lines. This method is commonly used to eliminate confusion about the side of the line where the joist should be placed.

Another method used to indicate joist placement is by marking a narrow X next to each line on the sill plate to show the actual joist position. Note that the lines marked on the sill plate mark the edge of the joists, not the center. Be sure to mark the X on the proper side. If the layout has been started from the left side of the sill plate, the X should be placed to the right of the line. If the layout has been started from the right side, the X should be placed to the left of the line.

After the locations for all common 16" OC joists have been laid out, the locations for any double joists, trimmer joists, etc., should be marked on the sill plate and identified with a T (or other letter) instead of an X.

After the first sill plate has been laid out, the process is repeated on the girder and the opposite sill plate. If the joists are in line, Xs should

Measurement Tip

The reason 15¼", not 16", is used in *Figure 29* as the first measurement for joist layout is so the first subfloor panel will extend to the outside edge of the end joist (not the center of it). The edges of subfloor panels (except the first and last ones) need to fall on the center of the joists to provide a nailing surface. By reducing the first measurement by ¾" (half the thickness of the joist), the edge of the first subfloor panel will be shifted from the center of the first joist to the outside edge of it.

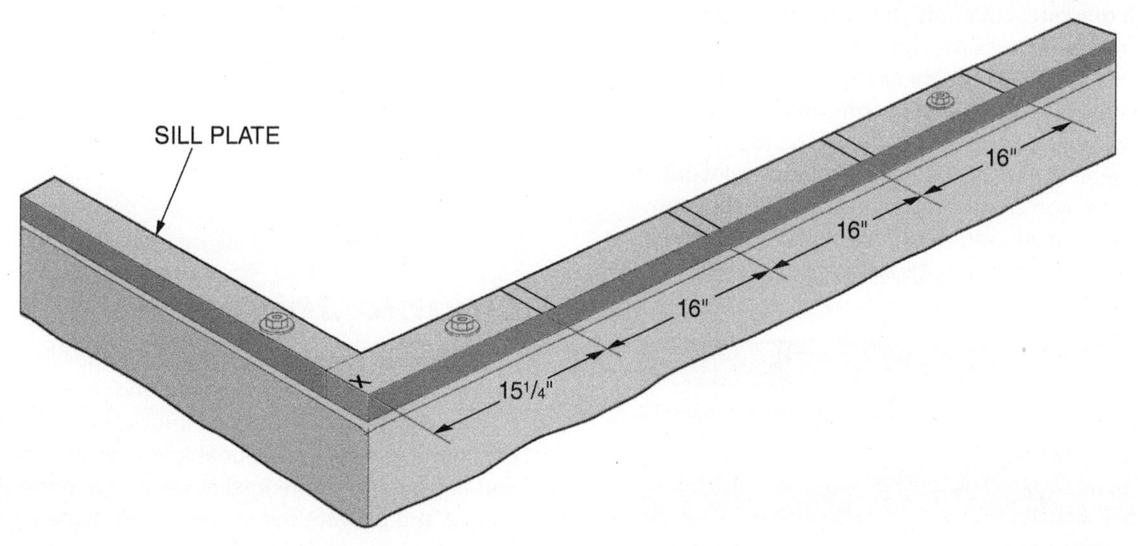

Figure 29 Marking the sill plate for joist locations.

be marked on the same side of the mark on both the girder and the sill plate on the opposite wall. If the joists are lapped at the girder, an X should be placed on both sides of the mark on the girder and on the opposite side of the mark on the sill plate on the opposite wall.

The location of the floor joists can be laid out directly on the sill plates, as previously described. However, in platform construction, some carpenters prefer to lay out the joist location on the header joists rather than the sill plate. If done on the header joists, the procedure is basically the same. Also, instead of making a series of individual measurements, some carpenters make a layout rod marked with the proper measurements, and use it to lay out the joist locations on the sill plates and girders.

4.5.0 Laying Out and Installing Floor Joists for the Partition and Floor Openings

After the locations of all the common floor joists are laid out, it is necessary to determine the locations of additional joists needed to accommodate loadbearing partitions, or floor openings, as shown on the drawings. Typically, these include:

- Double joists needed under loadbearing interior walls that are parallel with the joists (*Figure 30*). Depending on the structure, the double joists may need to be separated by 2 × 4 (or larger) blocks placed every 4' to allow for plumbing and electrical wires to pass into the wall from below. Loadbearing walls that run perpendicular to the joist system normally do not need extra joists added.
- Double joists needed for floor openings for stairs, chimneys, or other elements.

The sill plates and girder should be marked where the joists are doubled on each side of a large floor opening. Also, the sill plate and girder should be marked for the locations of the shorter tail joists at the ends of floor openings. They can be identified by marking their locations with a T instead of an X.

4.5.1 Cutting and Installing Rim Joists

After the joist locations are laid out on the sill plates and girder, the rim joist can be installed (*Figure 31*). The rim joist encloses the common joists and is constructed using the same size stock (typically 2 × 10s) as the joists. The rim joist assembly consists of the two header joists that run perpendicular to the joists and the first and last joists (end joists) in the floor system. Header joists are installed flush with the outside edges of the sill plate. This assembly is known as the box sill. Straight stock should be used so the header joists do not rise above the sill plates. Header joists need to sit flat on the sill plate so they do not push up the wall when a load is imposed. After the header joists are cut, they are toenailed to the sill plate and face-nailed where the end joists meet the header joist. Any header joist splices must be at the center of a joist or be joined by a scab. Note that some carpenters do not use header joists; they use blocks placed between the ends of the joists instead. Also, in areas subject to earthquakes, hurricanes, or tornadoes, codes may require that metal framing anchors be used to further attach the header and end joists to the sill plate.

Tape Measure Markings

Newer models of tape measures commonly include markings, in addition to the inch and feet measurements, to assist carpenters in common layout tasks. Many tape measures denote 16" increments in red to assist in common stud layout. Black diamonds are typically placed on tape measure at 19.2" increments to aid in truss layout. Trusses are commonly installed 19.2" OC, allowing five trusses to support each 8' roof sheathing panel. The red and black markings coincide at 8' (96") intervals.

27105-13_SA06.EPS

Double Joists

In addition to supporting parallel loadbearing interior walls, joists are doubled under extremely heavy objects. Whirlpools, bathtubs, and oversized refrigerators place additional loads on the floor and require a suitable joist system to support the weight and prevent the objects from damaging the structural integrity of the building.

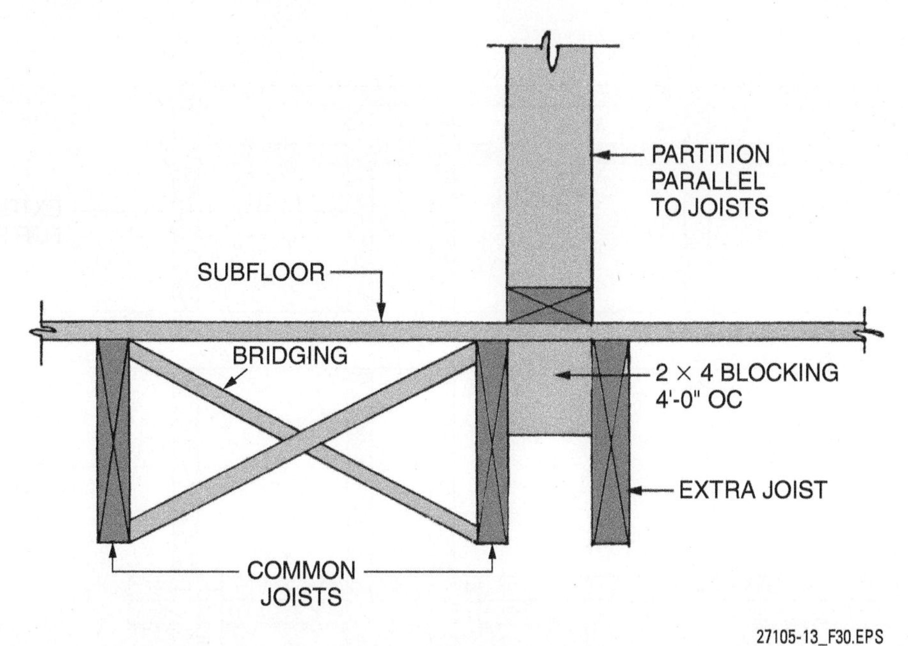

Figure 30 Double joists at a partition that runs parallel to the joists.

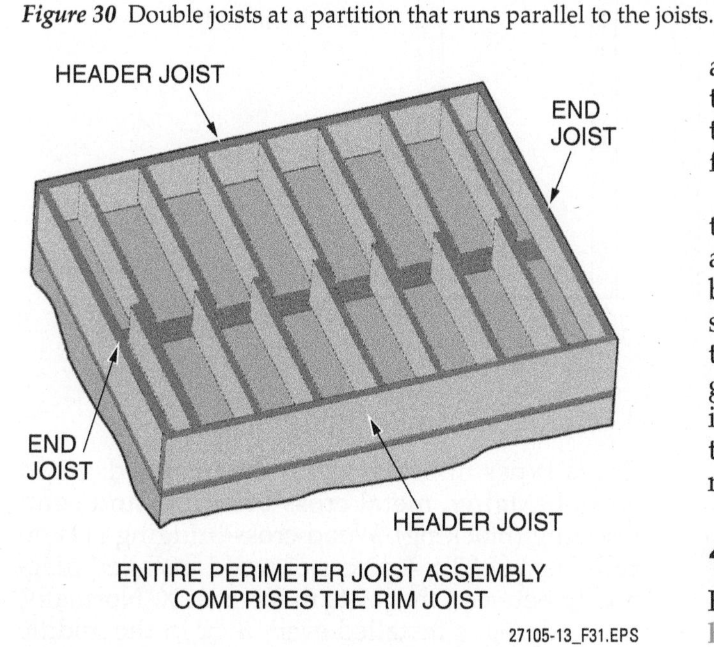

Figure 31 Rim joist installed on foundation.

4.5.2 Installing Floor Joists

With the rim joists in place, floor joists are placed at every spot marked on the sill plate. This includes the extra joists needed at the locations of partition walls and floor openings. When installing each joist, it is important to locate the crown and always point the crown up. With the joist in position at the header joist, hold the end tightly against the header joist and along the layout line so the sides are plumb, then end-nail it to the header joist and toenail it to the sill plate (*Figure 32*). Repeat this procedure until all the joists are attached to their associated header joist. To facilitate framing of openings in the floor, a good practice is to leave the full-length joists out where the floor openings occur.

After the joists are attached to the header joist, the joists are then fastened at the girder. If they are to be framed into the girder, the joists should be secured using a joist hanger or they can be supported with a ledger. In either case, toenail the joist to the girder so the top is flush with the girder. Then fasten the joist hangers or ledger into place. Where the joists overlap at the girder, the joists should be face-nailed together and toenailed to the girder.

4.5.3 Framing Openings in the Floor

Floor openings are framed by a combination of **headers** and trimmer joists (*Figure 33*). Headers run perpendicular to the direction of the common joists and are doubled. Full-length trimmer joists and short tail joists run parallel to the common joists. The drawings show the location of the opening in the floor frame and the size of the rough opening (RO). The RO dimensions are the dimensions from the inside edge of one trimmer joist to the inside edge of the opposing trimmer joist, and from the inside edge of one header to the inside edge of the opposing header. The method used to frame openings can vary depending on the particular situation. Trimmer joists and headers must be nailed together in a certain sequence so that there is never a need to nail through a double piece of stock.

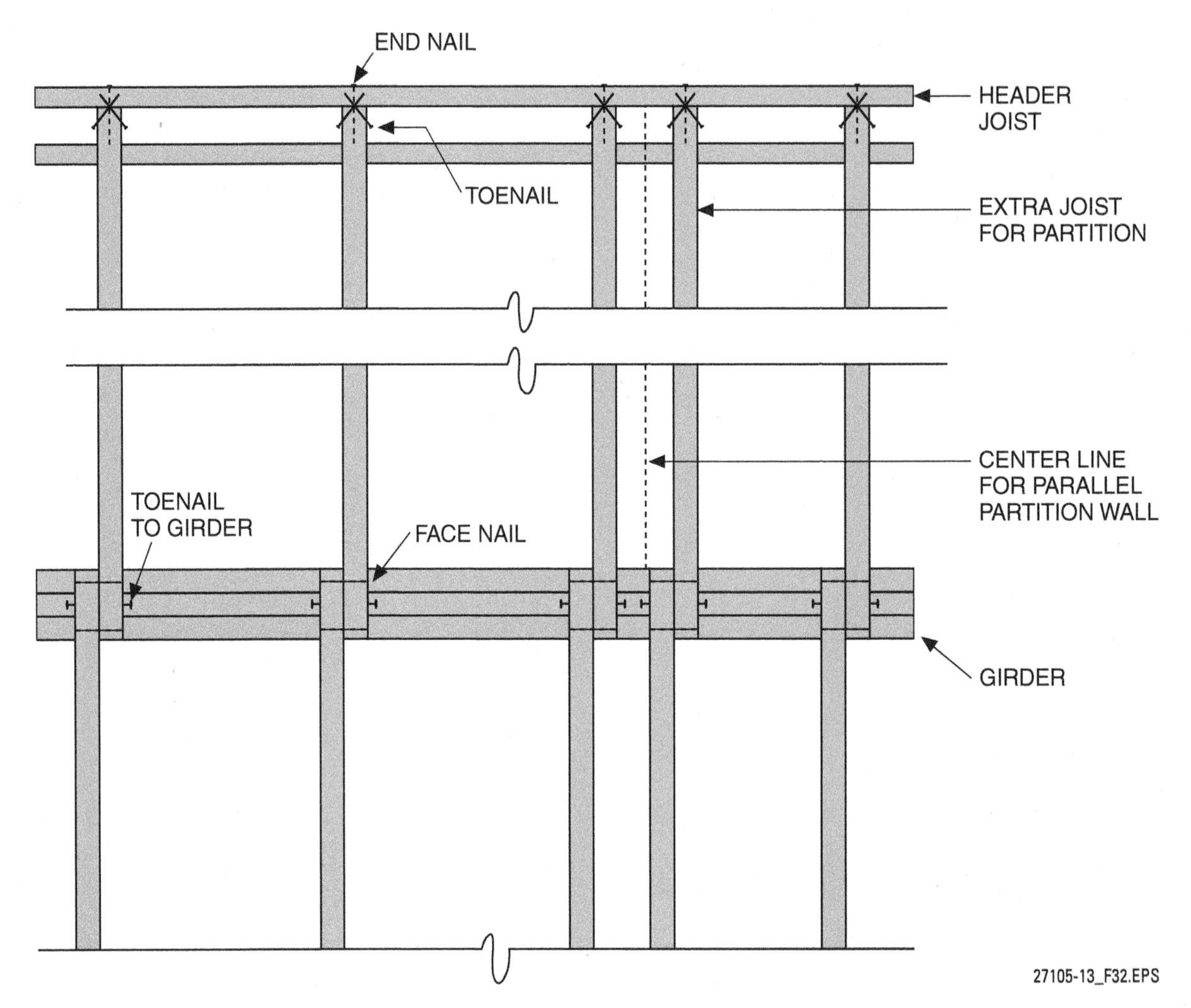

Figure 32 Installing joists.

A typical procedure for framing an opening like the one shown in *Figure 33* is given here:

Step 1 Install full-length trimmer joists A and C; then cut four header pieces with a length corresponding to the distance between the trimmer joists A and C.

Step 2 Nail two of these header pieces (headers 1 and 2) between trimmer joists A and C at the required distances.

Step 3 Following this, cut short tail joists X and Y and nail them to headers 1 and 2, as shown. Check the building code to determine if joist hangers are required.

Step 4 After headers 1 and 2 and tail joists X and Y are securely nailed, headers 3 and 4 can be installed and nailed to headers 1 and 2. Then, joists B and D can be placed next to and nailed to trimmer joists A and C, respectively.

4.6.0 Installing Bridging

Three types of bridging can be installed: wood cross-bridging, metal cross-bridging, and solid bridging (blocking). Wood cross-bridging is typically made of 1 × 4 pieces of wood installed diagonally between the joists to form an X. Normally, the bridging is installed every 8' or in the middle of the joist span. For example, joists with a 12' span would have a row of bridging installed at 6'.

A framing square can be used to lay out the bridging. First, determine the actual distance between the floor joists and the actual depth of the joist. For example, 2 × 10 joists at 16" OC measure 14½" between them. The actual depth of 2 × 10 joists is 9¼". To lay out the length and angles of the bridging for this situation, position the framing square on a piece of bridging stock, as shown in *Figure 34*. This will indicate the proper length and angle to cut the bridging. Make sure to use the same side of the framing square in both

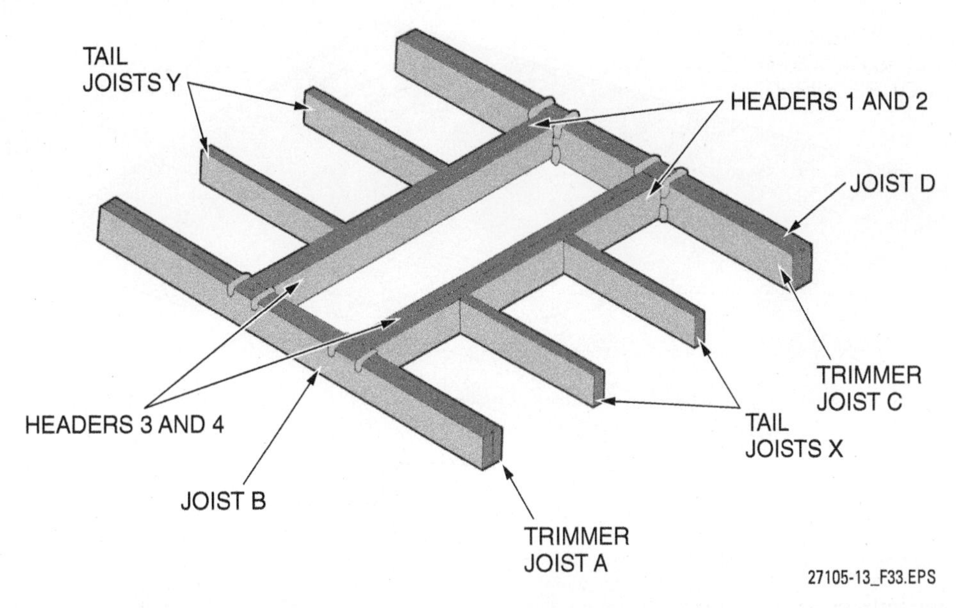

Figure 33 Floor-opening construction.

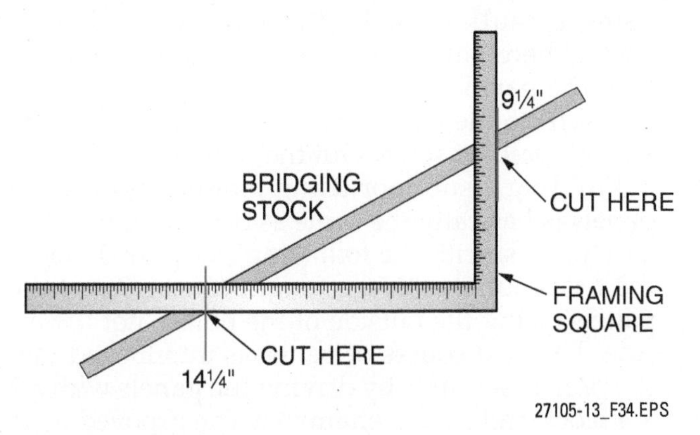

Figure 34 Framing square used to lay out wood cross-bridging.

places. Once the required length and angle of the bridging are determined, many carpenters set up a power miter saw or build a jig to ensure the bridges are cut to the proper length and at the correct angle.

Metal cross-bridging is available in a variety of styles and different lengths for use with a particular joist size and spacing. Metal cross-bridging should be installed according to the manufacturer's instructions. Layout of the joist system for the location of metal cross-bridging is done in the same manner as described for wood cross-bridging.

Solid bridging consists of solid pieces of joist stock installed between the joists. Layout for solid bridging is the same as for wood cross-bridging. The solid bridging is installed between pairs of joists, first on one side of the chalkline and then on the other side for the next pair of joists. This staggered method of installation enables end-nailing. Note that because of variations in lumber thickness and joist spacing, the length of the solid-bridging pieces may have to be adjusted.

4.7.0 Installing Subfloor

Subfloor installation begins by measuring 4' in from one side of the rim joist and snapping a chalkline across the tops of the floor joists from one end to the other. When installing 4' × 8' plywood, OSB, or similar floor panels, the long (8') dimension of the panels should run across (perpendicular) to the joists (*Figure 35*). Also, the panels must be placed so the joints are staggered in each successive course (row). (Never allow an intersection of four corners.)

Bridging Installation

To install bridging, a straight chalkline is snapped across the top of the joists at the center of their span before the subfloor is installed. Then, the top end of a piece of bridging is nailed flush with the top of the joist on one side of the line. (The bottom ends are not nailed until the subfloor is installed and its weight is applied.) Following this, nail another piece of bridging to the other joist in the same space. Ensure the bridging is flush with the top and on the opposite side of the line. Install bridging between the remaining pairs of joists until finished. When installing bridging between joists, make sure that the two pieces of bridging do not touch, as this can cause the floor to squeak.

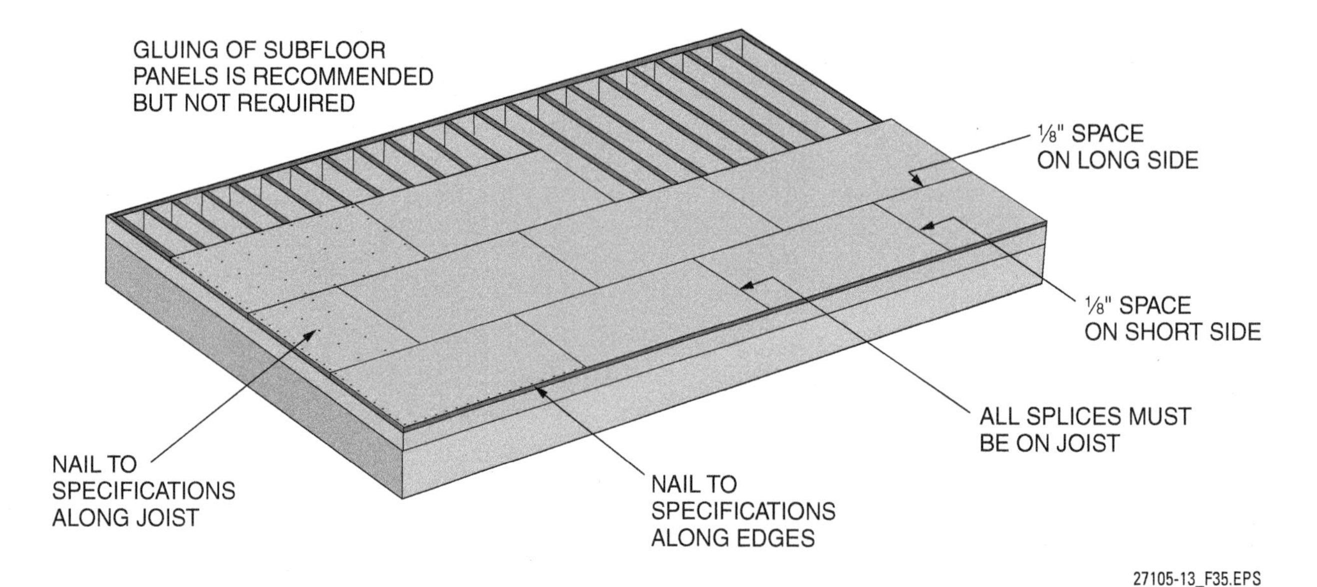

Figure 35 Installing butt-joint floor panels.

This is done by starting the first course with a full panel and continuing to lay full panels to the opposite end. Following this, start the next course using half a panel (4' × 4'), then continue to lay full panels to the opposite end. Repeat this procedure until the surface of the floor is covered. The ends of the panels that overhang the end of the building are then cut off flush with the rim joist. When butt-jointed plywood panels are used, at least ⅛" of space should be provided between each head joint and side joint for expansion. If installing a single-layer floor, blocking is also required under the joints of the butt-jointed panels. Specifications for nailing subfloor panels to floor members vary with local codes. To avoid fatigue, pneumatic nailers are commonly used to fasten subfloors to the joists. Some carpenters use power screw guns to fasten the subfloor to the joists.

Traditionally, plywood panels have been fastened to the floor joists using nails. Today, it is more common to use a glued floor system in which the subfloor panels are both glued and nailed (or screwed) to the joists, as previously discussed. Before each of the panels is placed, a ¼" bead of subfloor adhesive is applied to the joists using a caulk gun. Two beads are applied on joists where panel ends butt together. Following this, the panel should immediately be nailed (or screwed) to the joists before the adhesive sets. Be sure all nails or screws hit the floor joists.

Building a subfloor using tongue-and-groove panels is basically the same as described for butt-joint panels, with the following exceptions. Begin the first course with the tongue (*Figure 36*) of the panels facing the outside of the house, not the inside. The next course of panels is interlocked into the previous course by driving the panels with a 2 × 4 block and a sledgehammer. The grooved edge of the panels can take this abuse; the tongued edge cannot.

Subfloor Installation

When installing subfloor panels, stagger the layout of each course (row) for structural integrity. From the top view, the subfloor should have a traditional brick wall–style layout in which the vertical edges of the panels meet an adjoining row at the midpoint. There should never be an intersection of four corners.

4.8.0 Installing Joists for Cantilevered Floors

Porches, decks, and other projections from a building present some special floor-framing situations. Cantilevered floors overhang the foundation wall. When framing a cantilevered floor, it makes a difference whether the joists run parallel or perpendicular to the common joists in the floor system. If they are parallel, longer joists are simply run out past the foundation wall (*Figure 37*). If they run perpendicular to the common joists, the floor must be framed with cantilevered joists, meaning that the common joists need to be doubled up, then the cantilevered joists are tied into this double joist. As a rule of thumb only ⅓ of the total length of the joists can project past the foundation wall. Stated another way, the joist should extend inward a distance equal to at least twice the overhang. Check local building codes for exact requirements.

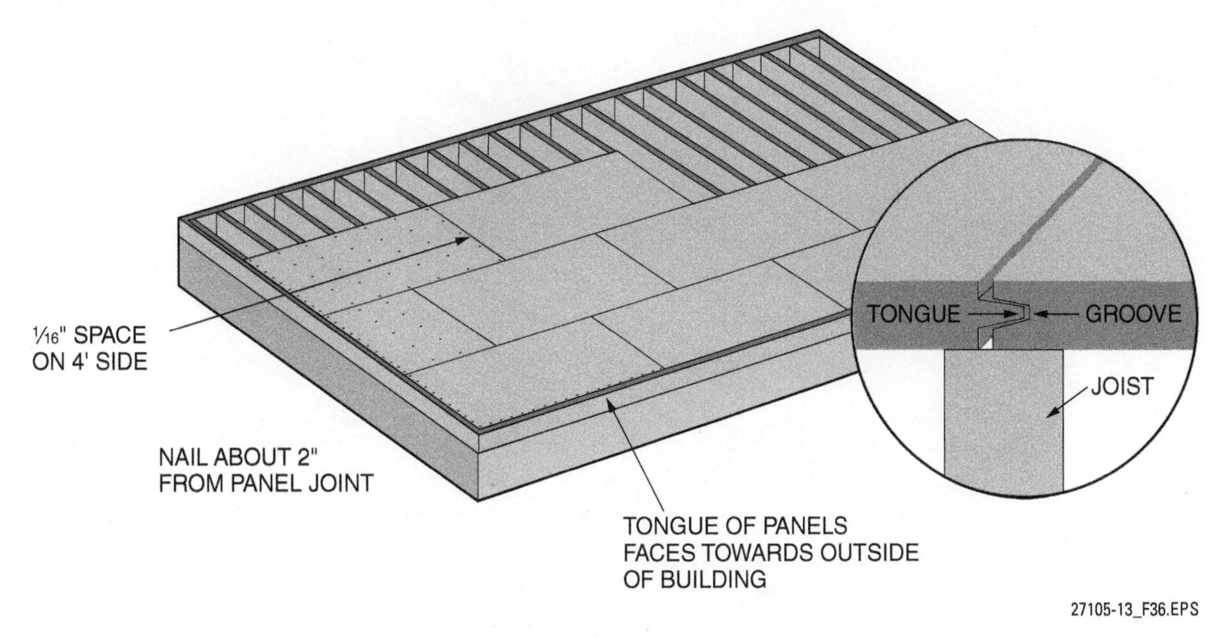

Figure 36 Installing tongue-and-groove subfloor panels.

Floor Joist Transitions

If floor joists are overlapped at a girder, the subfloor panel layout will differ from one side of the floor to the other by 1½". If the lap occurs at a natural break for the panels, the next course can be slid back 1½" so that it continues to butt on a joist. However, if the panels cover the joist lap, then 2 × 4 scabs must be nailed to the joists to provide a nailing surface for the panels at this transition.

Hangers and Ledgers

Be sure to check local construction codes when installing cantilevered joists. In some cases, the hangers are installed upside down (*Figure 37*) and ledgers are used to support the floor load.

Avoiding Panel Pinch

When nailing the individual tongue-and-groove panels, do not nail the last 6" of width along the groove. Nail this portion once the next course is properly driven into place or else the groove will be pinched, making seating of the tongue very difficult. If gluing the subfloor, apply a ⅛" bead of subfloor adhesive along the groove of the panel in addition to applying it to all the joist surfaces.

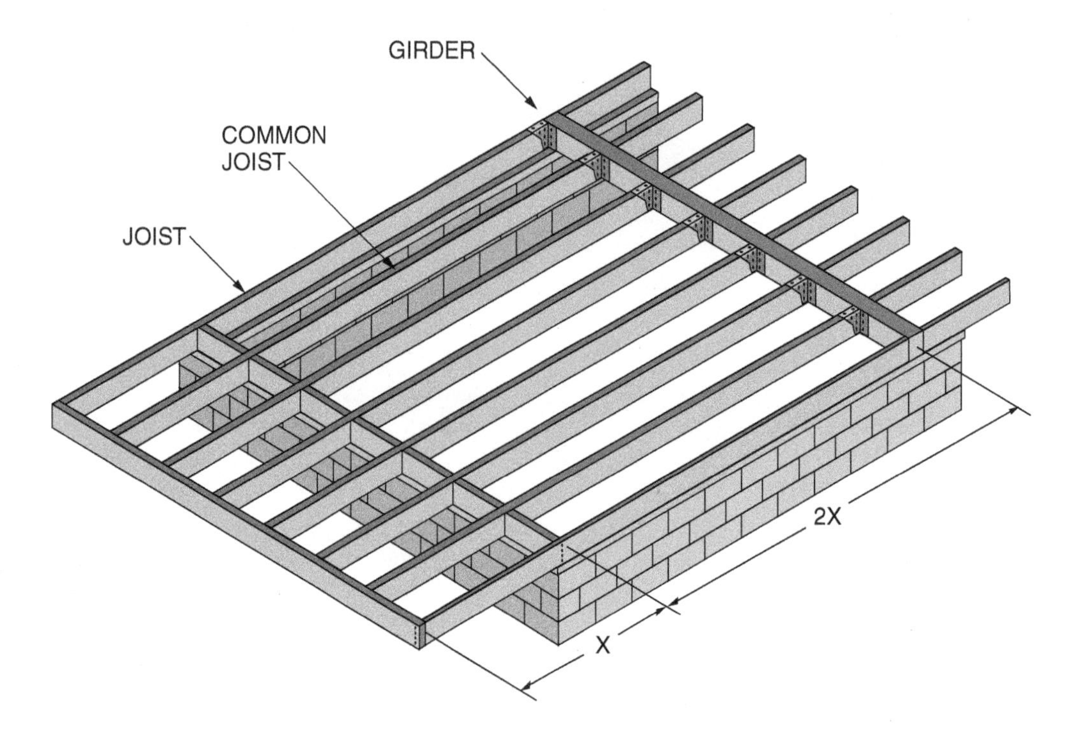

EXAMPLE OF CANTILEVERED JOISTS RUNNING IN SAME DIRECTION AS COMMON JOISTS

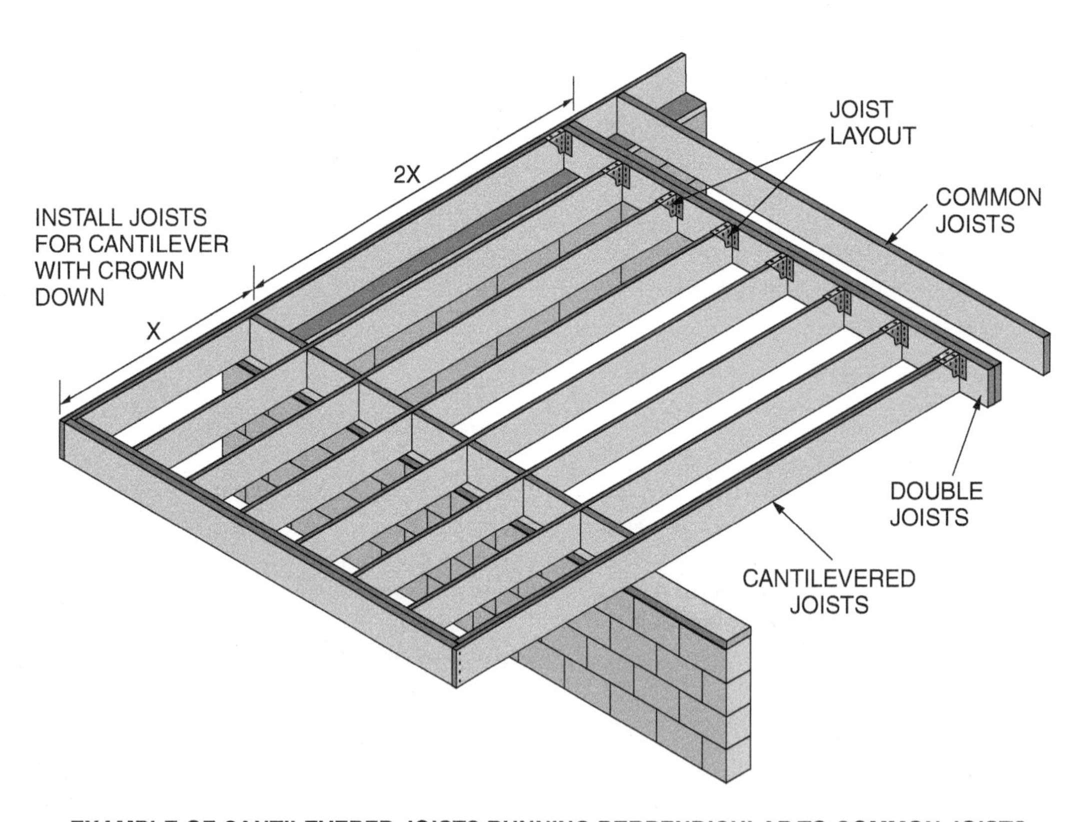

EXAMPLE OF CANTILEVERED JOISTS RUNNING PERPENDICULAR TO COMMON JOISTS

27105-13_F37.EPS

Figure 37 Cantilevered joists.

4.0.0 Section Review

1. When checking the foundation for squareness, if the measured lengths of the opposite walls are equal and the diagonal measurements are unequal, the foundation is square.

   a. True
   b. False

2. The distance between the end of a sill plate and the first anchor bolt should be no greater than _____.

   a. 12"
   b. 24"
   c. 36"
   d. 48"

3. The nail size to be used when constructing built-up girders is _____.

   a. 6d
   b. 8d
   c. 10d
   d. 16d

4. When laying out joist locations, mark an X on the sill plate to indicate _____.

   a. the center of the joist
   b. the actual joist position
   c. the actual stud location
   d. the center of the beam

5. When installing joists, ensure the crown points _____.

   a. to the left
   b. down
   c. up
   d. to the right

6. In a 15' joist span, bridging should be installed at _____.

   a. 4'
   b. 6'
   c. 7½'
   d. 8'

7. When installing a subfloor, the panels should be positioned so the joints are staggered.

   a. True
   b. False

8. If joists are to be cantilevered 4' beyond a foundation wall, the joists should be at least _____.

   a. 6' long
   b. 8' long
   c. 10' long
   d. 12' long

SECTION FIVE

5.0.0 ESTIMATING THE QUANTITY OF FLOOR MATERIALS

Objective

Estimate the amount of material needed for a floor assembly.

a. Describe how to estimate the amount of sill plate, sill sealer, and termite shield.
b. Describe how to estimate the amount of beam or girder material.
c. Describe how to estimate the amount of lumber needed for joists and joist headers.
d. Describe how to estimate the amount of bridging required.
e. Describe how to estimate the amount of subfloor material required.

Performance Task 2

Estimate the amount of material to frame a floor assembly from a set of plans.

For construction projects, a material takeoff is needed to properly determine the amount of materials required. Special structural elements such as floor openings, cantilevers, and partition supports that affect material requirements should be considered. The process begins by checking the specifications for the types and dimensions of materials to be used. It also requires that the drawings be checked or scaled to determine the dimensions of the various components needed. Once the types and dimensions of material are determined, a material takeoff can then be performed. The items included in a material takeoff for a floor system are:

- Sill sealer, termite shield, and sill plate
- Girders or beams
- Joists
- Joist headers
- Bridging
- Subfloor

For the purpose of an example, the sections below use the floor system shown in *Figure 38* to determine the quantity of floor framing materials required.

5.1.0 Estimating Sill Plate, Sill Sealer, and Termite Shield

To determine the amount of sill plate, sill sealer, and/or termite shield required, simply measure the perimeter of the foundation. For the example in *Figure 38*, the amount of material needed is 192 lineal feet [2 × (32' + 64') = 192'].

5.2.0 Beams/Girders

The quantity of girder material needed is determined by the type of girder and its length. For this example, a built-up beam made of three 2 × 12s is shown. The length of 2 × 12 material needed is 192 lineal feet (3 × 64 = 192'). If using a solid girder, the length of material needed would be 64 lineal feet.

5.3.0 Joists and Joist Headers

To determine the number of floor joists in a frame, divide the length of the building by the joist spacing and add one joist for the end and one joist for each partition that runs parallel to the joists. For the example, there are no partitions; therefore, the number of 2 × 8 joists is 49 [(64' × 12") ÷ 16" OC = 48 + 1 = 49]. Because there are two rows of joists (one on each side of the girder), the total number of joists needed is 98 (2 × 49 = 98). Each of these joists would be about 18' long. The amount of 2 × 8 material needed for the header joists is 128 lineal feet (2 × 64' = 128').

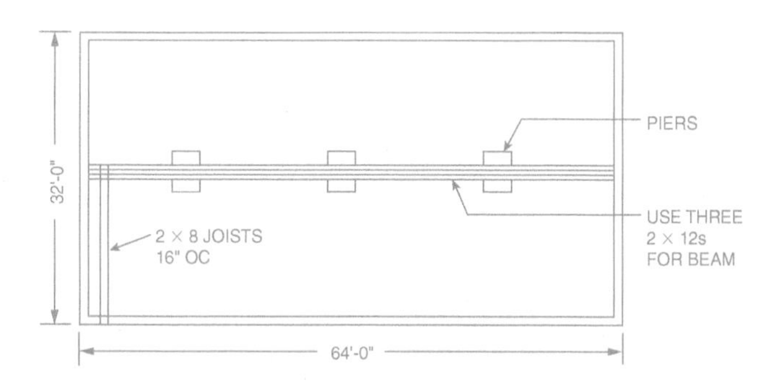

Figure 38 Determining floor system materials.

5.4.0 Bridging

Bridging and spacing requirements are dictated by the building code in effect in your jurisdiction. Some codes require one row of bridging in spans over 8' and less than 16' in length. Two rows of bridging are required in spans over 16'. In the example, each section of bridging has a 16' span so one row of bridging is required for each section.

To calculate the amount of wood cross-bridging needed, first multiply the number of rows of bridging needed by the length of each row of bridging to determine the total length (lineal footage) for the bridging rows; for the example, this is 128 lineal feet (2 rows × 64' = 128'). Next, multiply this value by the appropriate factor shown in *Table 2* to determine the total lineal feet of bridging needed. For the example, 2 × 8 joists are specified; therefore, the total amount of bridging needed is 256 lineal feet (2 × 128' = 256').

To determine the amount of solid bridging needed, multiply the number of rows of bridging needed by the length of each row of bridging. In the example, 128 lineal feet (2 × 64' = 128') of solid bridging are needed.

To calculate the amount of metal cross-bridging needed, first multiply the number of rows of bridging needed by the length of each row of bridging to determine the total length (lineal footage) for the bridging rows; for the example, this is 128 lineal feet (2 × 64' = 128'). Next, multiply this value by 0.75 (¾) to determine the number of 16" OC spaces between the joists. For the example, there are 96 spaces (128 × 0.75 = 96). Then, multiply the number of spaces by 2 to determine the total number of metal bridging pieces needed. For the example, we need 192 pieces (2 × 96 = 192).

Table 2 Wood Cross-Bridging Multiplication Factor

Joist Size	Spacing (Inches OC)	Lineal Feet of Material (per Foot of Bridging Row)
2 × 6, 2 × 8, 2 × 10	16	2
2 × 12	16	2.25
2 × 14	16	2.5

5.5.0 Subfloor

The number of subfloor panels required is determined by dividing the total floor area by the panel area. The area is calculated by multiplying the width by the length. To determine the number of 4' × 8' plywood or OSB panels required for the subfloor in the example, divide the total floor area by 32 (the area in square feet of one panel). For this example, the total floor area is 2,048 square feet (64' × 32' = 2,048 sq ft). Therefore, we need 64 panels (2,048 ÷ 32 = 64). For any fractional sheets, round up to the next whole sheet.

When using boards for a subfloor, first calculate the total floor area. To this amount, add a material waste factor. When using 1 × 6 lumber, a rule of thumb is to add ⅙ to the total area for waste; for 1 × 8 lumber, add ⅛ for waste.

5.0.0 Section Review

1. How many lineal feet of sill sealer are required for a building measuring 24' × 38'?

   a. 48 lineal feet
   b. 76 lineal feet
   c. 86 lineal feet
   d. 124 lineal feet

2. For the building described in Question 1, how many lineal feet of 2 × 10s are needed to construct a built-up beam consisting of three plies?

   a. 38 lineal feet
   b. 76 lineal feet
   c. 114 lineal feet
   d. 152 lineal feet

3. For a 20' × 40' building, how many 2 × 8 joists are required if they are spaced 16" OC and two rows of joists are required?

   a. 20
   b. 42
   c. 62
   d. 108

4. For the building described in Question 3, how many pieces of metal cross-bridging are required if one row of bridging is required for each span?

   a. 20
   b. 30
   c. 60
   d. 120

5. For the building described in Question 3, how many subfloor panels are required?

   a. 16
   b. 22
   c. 25
   d. 30

SECTION SIX

6.0.0 ALTERNATIVE FLOOR SYSTEMS

Objective

Identify some common alternative floor systems.

Trade Terms

Dome pan: Metal or fiberglass concrete form used for two-way joist systems.

Long pan: Metal or fiberglass concrete form used for one-way joist systems.

Shoring: Temporary bracing used to support above-grade concrete slabs while they set.

There are many types of alternative floor systems used in construction. Some commonly used systems are discussed here, but there are many other types of systems.

The structural framework for large buildings such as office buildings, apartment buildings, and hotels, is usually constructed with concrete and/or structural steel. Floors are commonly made of concrete that is cast into place on the job site using wood, metal, or fiberglass forms. Before the concrete is placed, provisions must be made for electrical, plumbing, and data-cabling pathways. Some floors, especially for upper levels of structures, may be precast and placed into position using a crane.

Once the formwork is in place, the concrete is cast into deck forms. **Shoring** is placed under the forms to support them until the concrete sets. *Figure 39* shows concrete cast over corrugated steel forms, which remain in place, providing channels though which cabling can be run. A section of metal, plastic, or fiber sleeve is often inserted vertically into the form before the concrete is poured to allow for electrical, communications, and other cabling to pass through the floor.

In some installations, underfloor duct systems are embedded in the concrete floor and are used to provide horizontal distribution of cables. Vertical access ports (handholes) are embedded in the concrete forms so the cable can be fished to various locations in the space. Trench ducts are metal troughs that are embedded in the concrete floor and used as feeder ducts for electrical power and telecommunication lines.

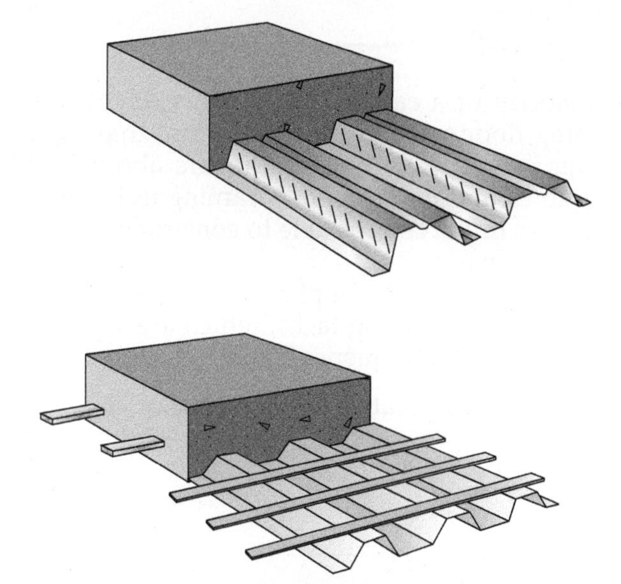

Figure 39 Corrugated steel forms.

Access floors consist of modular floor panels supported by pedestals. They may or may not have horizontal bracing in addition to the pedestals. This type of structure is used in computer rooms, intensive-care facilities, and other areas where a lot of cabling is required. In some applications, such as a factory, a trench may be formed in the concrete floor to accommodate cabling and other services.

In concrete joist systems, one method of forming concrete floor slabs along with concrete joists, beams, and girders is by using **long pans** or **dome pans**. Concrete joist systems allow thinner floor slabs with high bearing capacities to be created. The long pans or dome pans are fastened to the tops of supporting members and shoring and concrete is placed in the forms and leveled. The pans are then removed, leaving voids on the underside of the slab.

6.0.0 Section Review

1. Metal troughs that are embedded in the concrete floor and used as feeder ducts for electrical power and telecommunication lines are referred to as _____.
   a. trench ducts
   b. pedestals
   c. long pans
   d. dome pans

SUMMARY

A majority of a carpenter's time is devoted to framing floor systems. It is important that a carpenter not only be knowledgeable about both traditional and modern floor-framing techniques, but, more important, be able to construct modern floor systems.

The construction of a platform floor assembly involves the following tasks, which are typically performed in the sequence listed:

- Check the foundation for squareness.
- Lay out and install the sill plates.
- Build and install the girders and supports.
- Lay out the joist locations for partitions and floor openings.
- Cut and attach the joist headers to the sill plate.
- Install the joists.
- Frame the openings in the floor.
- Install the bridging.
- Install the subfloor.

Review Questions

1. Which of these items would *not* be found in the specifications?
   a. Methods of construction
   b. Detail drawings
   c. Types and qualities of building materials
   d. Location of the structure

2. On which construction drawing would you find the location and dimensions of grade beams?
   a. Floor plan
   b. Mechanical, electrical and plumbing plans
   c. Foundation plan
   d. Site plan

3. The drawings used in the construction of a building are legally the property of _____.
   a. the city or other local government body
   b. various subcontractors
   c. the general contractor
   d. the architect or building owner

4. An item that requires special framing, such as HVAC ductwork, is shown on _____.
   a. the architect's rendering
   b. the floor plan
   c. the mechanical, electrical, and plumbing plans
   d. the foundation plan

5. What would *not* be included in a typical sense of prints?
   a. architectural drawings
   b. structural drawings
   c. TSOL drawings
   d. MEP drawings

6. In balloon framing, the second floor joists sit on a _____.
   a. plate nailed to the top plate of the wall assembly below
   b. 2 × 4 ledger nailed to the studs
   c. sill attached to the top of the wall assembly below
   d. 1 × 4 or 1 × 6 ribbon let in to the wall studs

7. Shrinkage in wood framing members occurs mainly _____.
   a. lengthwise
   b. both lengthwise and across its width
   c. across its width
   d. in the middle

8. The platform frame used in modern residential and light commercial construction is also known as the _____.
   a. stick-built frame
   b. Chicago frame
   c. western frame
   d. stud-and-plate frame

9. The method of construction that experiences a relatively large amount of settling as a result of shrinkage is _____.
   a. platform framing
   b. brace framing
   c. balloon framing
   d. post-and-beam framing

10. The construction method that features widely spaced, heavy framing members is _____.
    a. brace framing
    b. balloon framing
    c. platform framing
    d. post-and-beam framing

11. In a set of construction drawings, the details about the floor used in a building most likely will be defined in the _____.
    a. architectural drawings
    b. structural drawings
    c. mechanical plans
    d. site plans

12. Large trusses, supported by posts at the building perimeter, are used in _____.
    a. post-frame construction
    b. clear-span construction
    c. post-and-truss construction
    d. tilt-up construction

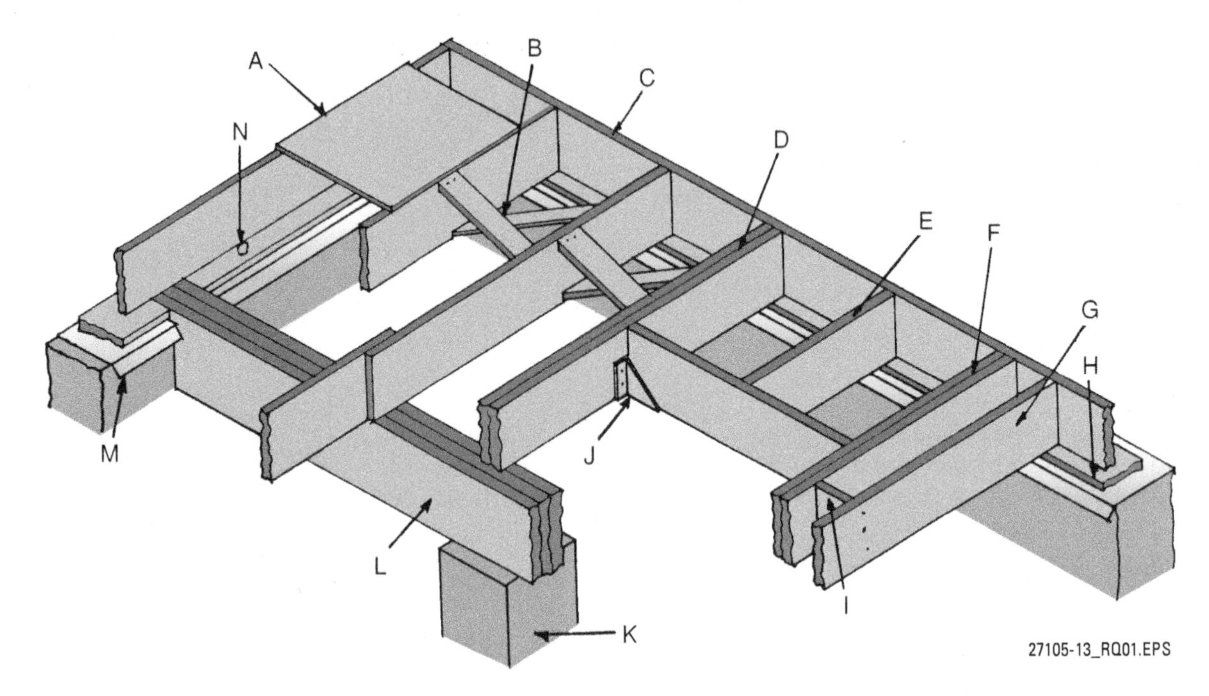

Figure 1

Use the *Review Question Figure 1* to answer questions 13–17.

13. Letter H is pointing to the _____.
    a. sill
    b. termite shield
    c. bearing plate
    d. sill sealer

14. Letter C is pointing to the _____.
    a. tail joist
    b. trimmer joist
    c. joist header
    d. common joist

15. Letter F is pointing to the _____.
    a. tail joist
    b. trimmer joist
    c. joist header
    d. common joist

16. Letter G is pointing to the _____.
    a. tail joist
    b. trimmer joist
    c. joist header
    d. common joist

17. Letter L is pointing to the _____.
    a. joist header
    b. beam or girder
    c. column
    d. triple joist

18. For a given size, the type of girder with the least loadbearing strength is the _____.
    a. built-up lumber girder
    b. solid lumber girder
    c. LVL girder
    d. glulam girder

19. For a given total load, if the span between its support columns is increased, the size of the girder _____
    a. can be decreased
    b. must be increased
    c. can remain the same
    d. must be decreased

20. Concrete joist systems allow thinner floor slabs with high bearing capacities to be created.
    a. True
    b. False

21. Which of the following panels creates the least rigid subflooring?
    a. Plywood
    b. Oriented strand board (OSB)
    c. Tongue-and-groove plywood
    d. 1 × 6 boards

22. In areas where termites are a problem, a metal shield should be installed _____.
    a. on top of the footing
    b. below the sill plate
    c. between the sill plate and rim joist
    d. under the bottom wall plate

23. Building codes specify that, with a live load, downward deflection at the center of a joist must not exceed _____.
    a. $\frac{1}{125}$th of the span
    b. $\frac{1}{240}$th of the span
    c. $\frac{1}{360}$th of the span
    d. $\frac{1}{450}$th of the span

24. Floor trusses consist of two primary components—chords and _____.
    a. webs
    b. flanges
    c. spacers
    d. stiffeners

25. A combined subfloor-underlayment system used where direct application of carpet or tile is intended is referred to as a(n) _____.
    a. single-layer floor system
    b. two-layer floor system
    c. raised floor system
    d. access floor system

26. The first task that should be performed when constructing a floor is to _____.
    a. lay out and install sill plates
    b. build and install girders and supports
    c. check the foundation for squareness
    d. lay out the sill plates and girders for joist locations

27. When laying out the sill for joist locations 16" OC, the first measurement on the sill should be at _____.
    a. 14$\frac{1}{4}$"
    b. 15$\frac{3}{8}$"
    c. 16"
    d. 15$\frac{1}{4}$"

28. Use special nails when fastening treated lumber, because treatment chemicals can cause ordinary fasteners to _____.
    a. become brittle
    b. corrode
    c. stain the wood
    d. shrink

29. The first floor should not be framed before the foundation is backfilled to allow the foundation to flex.
    a. True
    b. False

30. The number and size of nails and the nailing pattern for framing members is specified by the _____.
    a. construction contract
    b. *International Residential Code®*
    c. building inspector
    d. *American Residential Code*

31. Local codes regarding nailing schedules may vary.
    a. True
    b. False

32. Metal framing anchors may be used to attach the header and end joists to the sill plate in areas subject to _____.
    a. earthquakes
    b. hurricanes
    c. flash floods
    d. heavy snow

33. When installing solid bridging, why are pieces staggered from one joist to the next?
    a. To minimize squeaks
    b. To permit end-nailing
    c. To make most efficient use of material
    d. To provide greater support

34. How many lineal feet of sill material are needed for a building that measures 26' × 42'?
    a. 84
    b. 94
    c. 136
    d. 188

35. For the same structure, how many 4' × 8' panels of subflooring material are needed?
    a. 35
    b. 48
    c. 72
    d. 96

Trade Terms Quiz

Fill in the blank with the correct term that you learned from your study of this module.

1. The air-handling equipment placed in a commercial building would be considered part of the ________.
2. Studs and joists are commonly installed at 16" ________.
3. Before installing hardwood flooring, ________ would most likely be installed over the subfloor.
4. Masonry ________ may be used to support other structural members such as girders and beams.
5. A joist can be secured to the structure using a(n) ________.
6. The distance between structural supports of a floor system is known as the ________.
7. The ________ is created with the header joists and end joists.
8. ________ lumber should be used for construction lumber.
9. In balloon framing, the studs must be blocked with approved ________.
10. Lumber in a floor system must be installed with the ________ up.
11. In platform construction, the ends of the common joists are attached to a(n) ________ to form the band joist assembly.
12. A(n) ________ can be applied over a joint to strengthen it.
13. A metal or fiberglass ________ is used to create a one-way joist system.
14. Notches in studs or joists used to support another piece are known as a(n) ________.
15. In addition to the prints, the ________ should be referenced for additional information regarding the construction project.
16. ________ is exerted against a foundation by the ground around It.
17. Every building is supported by a(n) ________.
18. The bottom horizontal member of a wall frame is known as the ________.
19. A metal or fiberglass ________ is used to create a two-way joist system.
20. A rim joist assembly is created with the header joists and one ________ on each side.
21. When framing rough openings, a(n) ________ is installed at a 90-degree angle to the common joists and is commonly doubled.
22. People and furniture are considered to be a(n) ________ on a structure.
23. When carpet is the finished flooring material, a(n) ________ could be used.
24. Concrete structural members may be cast-in-place or ________ at a casting yard.
25. When framing a stairway opening in a floor system, the ________ dimension is typically provided on the prints.
26. Above-grade concrete slabs are supported by ________ until the concrete sets.
27. ________ run from an opening to a bearing.
28. Full-length ________ reinforce rough openings in a floor system.

29. _______ are created when underlayment is installed over a subfloor.

30. The door _______ determines how the door opens into a room.

Trade Terms

Crown
Dead load
Dome pan
End joist
Fire stop
Floor truss
Foundation
Header
Header joist
Joist hanger
Lateral pressure
Let-in
Live load
Long pan
On center
Pier
Precast
Rim joist
Rough opening
Scab
Seasoned
Shoring
Single-layer floor system
Soleplate
Span
Specifications
Swing
Tail joist
Trimmer joist
Two-layer floor system
Underlayment

John Yencho

Construction Technology Teacher
Trenton High School, Gilchrist County School District
Trenton, Florida

How did you get started in the construction industry and what inspired you to enter the industry?
I got started in the construction industry straight out of high school. I took wood shop while in high school and really enjoyed working with wood, which was my inspiration to start working in the industry. Things were a whole lot different in 1962, the year I graduated from high school. College, although important, did not have the influence on careers that it does today. There were plenty of jobs available and all you had to do to get a job was to show up at a job site ready to work.

That's exactly what I did; I showed up at a construction site with my tools in hand and asked for a job. They gave me a very important one—a gofor! If you don't know what a gofor is, it's the person on the crew that goes for coffee at break time, goes for materials when needed, and goes for anything anyone needs. Eventually, they actually allowed me to pick up a hammer and the rest is history.

What do you enjoy most about your career?
What I enjoy most about my career in construction is the variety of types of work and places to work in. I have built everything from birdhouses to skyscrapers in 42 of the 50 states, Puerto Rico, and Mexico over the last 50 years. After working as an apprentice, then moving on to journeyman, and finally a master carpenter, I was employed as construction superintendent for an architectural and engineering firm in Ashland, Ohio. One of the most interesting tasks I was assigned to do while employed by them was the development of a product for one of our subsidiaries. My employer went to Germany on vacation one summer, and when he came back, he tossed a set of plans on my desk and said "do something with this" and walked away. They were plans for something, but I had a hard time understanding them since all text was in German! As it turned out, the plans were for a wave machine developed by the German military prior to World War II. Hitler was not allowed to have certain ships prior to the war so they had built this extremely large machine to create scale waves to test scale models of their ships before they built them.

With task in mind I set out to create something we could sell. Eighteen months later (and a lot of frustration), my wife and I had developed a machine that was 3 feet long, 20 inches wide, and 30 inches high. Our first wave machine was built and Great Waves Incorporated was born. I spent the next eight years traveling the United States, Mexico, and Puerto Rico supervising the installation of our equipment in recreational swimming pools and water parks. If you've ever been in a wave pool, there's a good chance it was one of ours. For those of you who think I should be rich, guess again. The company held the patent and made all the money; I just earned a salary.

Why do you think training and education are important in construction?
As I said earlier, times were different back in the day. I only had a high school education and no real formal training when I got my first job in the construction industry. I don't believe I would have ever had the opportunity to become part of the industry today without the proper training and an education. Back then, all you needed was a dedicated work ethic and the willingness to work hard. Today, you still need a lot of hard work, dedication, and good work ethic, as well as training to get a good career, not just a job. And there is a difference between a career and a job.

Why do you think credentials are important in construction?
Having credentials is just as important as having a diploma. Without either of them, you are at a distinct disadvantage. Credentials tell prospective employers what you know, how hard you work, and how dedicated you are. That's what employers are looking for in employees. They don't have time in the competitive industry of construction to completely train every employee. They can't afford to take on employees who don't have experience, who have poor work histories, or who are accident risks through lack of knowledge. Credentials help employers determine

who the best candidates are for a position. In today's tough economic situations in a very competitive industry, companies need the best. Credentials identify you as one of the best.

How has training/construction impacted your life and career?
Eventually my training and experiences in the construction industry allowed me to be successful in my own career. There are always going to be better carpenters—I employed many carpenters who were better than me. What gave me the edge to be their boss (instead of vice versa)? My training, ability to communicate with others, and my willingness to learn, try new things, accept responsibility, and dedicate myself to getting the job done!

Today, as a high school teacher, I am able to teach others about the career I not only love, but that has made my life so much richer. Teaching students about construction is the icing on the chocolate cake of my life. The construction industry has given me much more than I ever imagined in 1962 when I walked up to that grizzly foreman as a nervous kid just looking for a job because I wanted money to buy a car. I have gone places I would never have been able to go; I have built things that will be around much longer than I will. It not only gave me the money I needed to buy that 1962 Chevy Nova II in burgundy with bucket seats I wanted so badly, but a whole lot more since.

Would you recommend construction as a career to others? Why?
I would and do recommend the construction industry to many others. Every May, I go to the middle schools of our school district to recruit students into my construction program at Trenton High School. If I didn't think a career in the construction industry was rewarding, exciting, and worthwhile, I couldn't recruit these young people. I put my name on it; that means something to me.

What does craftsmanship mean to you?
Being a craftsman in the construction industry brings me great pride and honor. Not everyone can do what I can do as a trained craftsman in the construction industry. It takes special skills, abilities, and training to do what we do. Being a craftsman is exciting and rewarding in many ways.

Trade Terms Introduced in This Module

Crown: The highest point of the curved edge of a framing member.

Dead load: The weight of permanent, stationary construction and equipment included in a building.

Dome pan: Metal or fiberglass concrete form used for two-way joist systems.

End joist: The first and last joists in a floor system; they run parallel to the common joists.

Fire stop: An approved material used to fill air passages in a frame to retard the spread of fire.

Floor truss: An engineered assembly that is used to support floors.

Foundation: The supporting portion of a structure, including the footings.

Header: Used to frame floor openings; they run perpendicular to the direction of the joists and are typically doubled.

Header joist: A framing member used in platform framing into which the common joists are fitted. Header joists are also used to support the free ends of joists when framing openings in a floor.

Joist hanger: A metal stirrup secured to the face of a structural member, such as a girder, to support and align the ends of joists flush with the member.

Lateral pressure: Sideways pressure against a structure, such as a foundation.

Let-in: Any type of notch in a stud or joist, which holds another piece. The item that is supported by the notch is said to be let in.

Live load: The total of all moving and variable loads that may be placed upon a structure.

Long pan: Metal or fiberglass concrete form used for one-way joist systems.

On center: Distance between the center of one member (typically studs) to the center of the next member. Commonly abbreviated as OC.

Pier: A structural column used to support other structural members, typically girders or beams.

Precast: Concrete structural elements that have been cast at a casting yard and raised into position using a crane.

Rim joist: Consists of two header joists and the end joists.

Rough opening: Dimensions shown on the drawings that indicate the distance from the inside edge of one trimmer joist to the inside edge of the opposing trimmer joist.

Scab: A length of lumber applied over a joint to strengthen it.

Seasoned: Lumber with the appropriate amount of moisture removed to make it usable for construction.

Shoring: Temporary bracing used to support above-grade concrete slabs while they set.

Single-layer floor system: Combined subfloor and underlayment; typically installed where direct application of carpet or tile, to the floor is intended

Soleplate: The bottom horizontal member of a wall frame.

Span: The distance between structural supports such as walls, columns, piers, beams, or girders.

Specifications: Written document included with a set of prints that clarifies information presented on the prints and provides additional information not easily presented on the prints.

Swing: The direction of door rotation.

Tail joist: Short joist that runs from an opening to a bearing.

Trimmer joist: A full-length joist that reinforces a rough opening in the floor.

Two-layer floor system: Floor system in which a layer of underlayment is installed over a layer of subfloor.

Underlayment: A material, such as particleboard or plywood, laid on top of the subfloor to provide a smoother surface for the finished floor.

Appendix

Guidelines for Sizing Girders and Joists

The sizes of the girders and joists are typically specified by the architect or structural engineer who designs the building. However, carpenters should be familiar with the general procedures used to size girders and joists.

Sizing Girders

Step 1 Determine the distance (span) between girder supports.

Step 2 Find the girder load width. The girder must be able to carry the weight of the floor on each side to the midpoint of the joist that rests upon it. Therefore, the girder load width is half the length of the joist span on each side of the girder multiplied by 2.

Step 3 Find the total floor load per square foot carried by the joists and bearing partitions. This is the sum of the loads per square foot with the exception of the roof load. Roof loads are not included because these are carried on the outside walls unless braces or partitions are placed under the rafters.

Step 4 Find the total load on the girder. This is the product of the girder span multiplied by the girder width multiplied by the total floor load.

Step 5 Select the proper size of girder according to local codes. *Table A-1* is typical. It indicates safe loads on standard-size girders for spans from 6' to 10'. Note that shortening the span is the most economical way to increase the load that a girder will carry.

Sizing Joists

After the girders are sized, then the joists can be sized.

Step 1 Determine the length of the joist span.

Step 2 Determine if there is a dead load on the ceiling.

Step 3 Select the proper size of joists according to local codes or by using the latest tables available from wood product manufacturers or sources such as the Western Wood Products Association and the Southern Forest Products Association. *Table A-2* indicates maximum safe spans for various sizes of wood joists under ordinary load conditions. For floors, this is usually figured on a basis of 50 pounds per square foot (10 pounds dead load and 40 pounds live load).

Table A-1 Typical Safe Girder Loads

Nominal Girder Size	Safe Load in Pounds for Spans Shown				
	6 ft	7 ft	8 ft	9 ft	10 ft
6 × 8 solid	8,306	7,118	6,220	5,539	4,583
6 × 8 built-up	7,359	6,306	5,511	4,908	4,062
6 × 10 solid	11,357	10,804	9,980	8,887	7,997
6 × 10 built-up	10,068	9,576	8,844	7,878	7,086
8 × 8 solid	11,326	9,706	8,482	7,553	6,250
8 × 8 built-up	9,812	8,408	7,348	6,554	5,416
8 × 10 solid	15,487	14,732	13,608	12,116	10,902
8 × 10 built-up	13,424	12,968	11,792	10,504	9,448

Table A-2 Safe Joist Spans

Nominal Joist Size	Spacing	30# Live Load	40# Live Load	50# Live Load	60# Live Load
2 × 6	12"	14'–10"	13'-2"	12'-0"	11'-1"
	16"	12'–11"	11'–6"	10'–5"	9'–8"
	24"	10'–8"	9'–6"	8'–7"	7'–10"
2 × 8	12"	19'–7"	17'–5"	15'–10"	14'–8"
	16"	17'–1"	15'–3"	13'–10"	12'–9"
	24"	14'–2"	13'–6"	11'–4"	10'–6"
2 × 10	12"	24'–6"	21'–10"	19'–11"	18'–5"
	16"	21'–6"	19'–2"	17'–5"	16'–1"
	24"	17'–10"	15'–10"	14'–4"	13'–3"
2 × 12	12"	29'–4"	26'–3"	24'–0"	22'–2"
	16"	25'–10"	23'–0"	21'–0"	19'–5"
	24"	21'–5"	19'–1"	17'–4"	16'–9"
3 × 8	12"	24'–3"	21'–8"	19'–10"	18'–4"
	16"	21'–4"	19'–1"	17'–4"	16'–0"
	24"	17'–9"	15'–9"	14'–4"	13'–3"
3 × 10	12"	30'–2"	27'–1"	34'–10"	23'–0"
	16"	26'–8"	23'–10"	21'–9"	20'–2"
	24"	22'-3"	19'–10"	18'–1"	16'–8"

Figure Credits

SkillsUSA, Module opener

John Yencho, Figure 2

Southern Forest Products Association, Figure SA03

Weyerhaeuser Wood Products, Figures SA05, 19, 20, and SA06

Section Review Answer Key

Answer	Section Reference	Objective Reference
Section One		
1. a	1.1.0	1a
2. b	1.2.1	1b
3. d	1.3.0	1c
4. c	1.4.0	1d
5. c	1.5.0	1e
Section Two		
1. b	2.1.0	2a
2. a	2.2.0	2b
3. d	2.3.0	2c
Section Three		
1. c	3.1.0	3a
2. c	3.2.0	3b
3. b	3.2.5	3b
4. c*	3.3.0	3c
5. d	3.3.0	3c
6. b	3.4.0	3d
7. d	3.5.0	3e
Section Four		
1. b	4.1.0	4a
2. a	4.2.0	4b
3. c	4.3.0	4c
4. b	4.4.0	4d
5. c	4.5.2	4e
6. c	4.6.0	4f
7. a	4.7.0	4g
8. d	4.8.0	4h
Section Five		
1. d*	5.1.0	5a
2. c*	5.2.0	5b
3. c*	5.3.0	5c
4. d*	5.4.0	5d
5. c.*	5.5.0	5e
Section Six		
1. a	6.0.0	6

* Please see the math calculations following this key for further explanation of these answers.

Section Three

4. $22' \times 12 = 264''$

   $264'' \div 360 = .73''$

Section Five

1. $(24 \times 2) + (38 \times 2) = 124$ lineal feet

2. $38 \times 3 = 114$ lineal feet

3. $(40 \times 12) \div 16 = 30 + 1 =$
   31 joists for one row
   $31 \times 2 = 62$ joists

4. 2 spans $\times$ $40' = 80' \times 0.75 =$
   60×2 pieces per space $= 120$ pieces

5. $(20' \times 40') = 800$ sq ft $\div$ 32 sq ft per panel
   $= 25$ panels

NCCER CURRICULA — USER UPDATE

NCCER makes every effort to keep its textbooks up-to-date and free of technical errors. We appreciate your help in this process. If you find an error, a typographical mistake, or an inaccuracy in NCCER's curricula, please fill out this form (or a photocopy), or complete the online form at **www.nccer.org/olf**. Be sure to include the exact module ID number, page number, a detailed description, and your recommended correction. Your input will be brought to the attention of the Authoring Team. Thank you for your assistance.

Instructors – If you have an idea for improving this textbook, or have found that additional materials were necessary to teach this module effectively, please let us know so that we may present your suggestions to the Authoring Team.

NCCER Product Development and Revision

13614 Progress Blvd., Alachua, FL 32615

Email: curriculum@nccer.org
Online: www.nccer.org/olf

❑ Trainee Guide ❑ Lesson Plans ❑ Exam ❑ PowerPoints Other ______________________

Craft / Level: Copyright Date:

Module ID Number / Title:

Section Number(s):

Description:

Recommended Correction:

Your Name:

Address:

Email: Phone:

27111-13

Wall Systems

Overview

The walls of most single-family dwellings are framed with 2 × 4 or 2 × 6 lumber. Exterior sheathing and siding, along with interior finishes such as gypsum board, are then attached to the framing. There are two critical steps in the framing process: accurate measuring and layout, and accurate leveling and plumbing of the walls. The use of steel studs for wall framing is common in commercial construction and is becoming increasingly popular in residential construction. While the layout process is essentially the same as that for lumber framing, different tools and fastening methods are used.

Module Six

Trainees with successful module completions may be eligible for credentialing through the NCCER Registry. To learn more, go to **www.nccer.org** or contact us at **1.888.622.3720**. Our website has information on the latest product releases and training, as well as online versions of our *Cornerstone* magazine and Pearson's product catalog.

Your feedback is welcome. You may email your comments to **curriculum@nccer.org**, send general comments and inquiries to **info@nccer.org**, or fill in the User Update form at the back of this module.

This information is general in nature and intended for training purposes only. Actual performance of activities described in this manual requires compliance with all applicable operating, service, maintenance, and safety procedures under the direction of qualified personnel. References in this manual to patented or proprietary devices do not constitute a recommendation of their use.

27111-13
Wall Systems

Code Note

Codes vary among jurisdictions. Because of the variations in code, consult the applicable code whenever regulations are in question. Referring to an incorrect set of codes can cause as much trouble as failing to reference codes altogether. Obtain, review, and familiarize yourself with your local adopted code.

Objectives

When you have completed this module, you will be able to do the following:

1. Identify the components of a wall system.
   a. Identify methods used to construct corner posts.
   b. Describe how to frame partition intersections.
   c. Explain the purpose of headers and describe how they are constructed.
   d. Describe how metal-framed walls are constructed.
2. Describe the procedure for laying out a wood frame wall, including plates, corner posts, door and window openings, partition Ts, bracing, and fire stops.
   a. Describe how to properly lay out a wood frame wall.
   b. Explain how to lay out wall openings.
3. Describe the correct procedure to assemble, erect, and brace exterior walls for a frame building.
   a. List the steps involved in assembling a wall.
   b. Identify where fire stops are to be installed and explain how they are installed.
   c. List the four steps involved in erecting a wall.
4. Describe wall framing techniques used in masonry construction.
5. Describe the correct procedure to estimate the materials required to frame walls.
   a. Explain how to estimate the amount of lumber required for soleplates and top plates.
   b. Describe how to estimate the number of studs required.
   c. Explain how to calculate the amount of material needed for a header.
   d. Describe how to estimate the amount of diagonal bracing required.
6. Identify alternative wall systems.
   a. Describe how concrete walls are constructed.
   b. Explain the difference between standard interior wall systems and alternative interior wall systems.

Performance Tasks

Under the supervision of the instructor, you should be able to do the following:

1. Lay out a wood frame wall, including plates, corner posts, door and window openings, partition Ts, bracing, and fire stops.
2. Assemble and erect a wood frame wall, including plates, corner posts, door and window openings, partition Ts, bracing, and fire stops.
3. Correctly install sheathing on a wall.
4. Estimate the materials required to frame walls.

27111-13
Wall Systems

Trade Terms

Blocking
Buck
Cripple stud
Double top plate
Drying-in
Furring strip
Header
Jamb
Monolithic
Rough sill
Sound transmission class (STC)
Tilt-up concrete construction
Top plate
Trimmer stud

Industry-Recognized Credentials

If you're training through an NCCER-accredited sponsor you may be eligible for credentials from NCCER's Registry. The ID number for this module is 27111-13. Note that this module may have been used in other NCCER curricula and may apply to other level completions. Contact NCCER's Registry at 888.622.3720 or go to **nccer.org** for more information.

Contents

Topics to be presented in this module include:

Figures and Tables

Section One

1.0.0 Wall Components

Objective

Identify the components of a wall system.

a. Identify methods used to construct corner posts.
b. Describe how to frame partition intersections.
c. Explain the purpose of headers and describe how they are constructed.
d. Describe how metal-framed walls are constructed.

Trade Terms

Blocking: A wood block used as a filler piece and a support between framing members.

Cripple stud: In wall framing, a short framing stud that fills the space between a header and a top plate or between the rough sill and the soleplate.

Double top plate: A plate made of two members to provide better stiffening of a wall. It is also used for connecting splices, corners, and partitions that are at right angles (perpendicular) to the wall.

Header: A horizontal structural member that supports the load over a window, door, or other wall openings.

Rough sill: The lower framing member attached to the top of the lower cripple studs to form the base of a rough opening for a window.

Top plate: The upper horizontal framing member of a wall used to carry the roof trusses or rafters.

Trimmer stud: A vertical framing member that forms the sides of rough openings for doors and windows. It provides stiffening for the frame and supports the weight of the header.

Figure 1 identifies the structural members of a wood frame wall. Each of the members shown on the illustration is then described. Carefully review the illustration and the following descriptions to familiarize yourself with the components before proceeding through this module.

- *Blocking* (*spacer*) – A wood block used as a filler piece and support between framing members.
- *Cripple stud* – In wall framing, a cripple stud is a short framing stud that fills the space between a header and a top plate or between the rough sill and the soleplate.
- *Double top plate* – A double top plate is a plate made of two members to provide better stiffening of a wall. It is also used for connecting splices, corners, and partitions that are at right angles (perpendicular) to the wall.
- *Header* – A horizontal structural member that supports the load over an opening such as a door or window.
- *King stud* – The full-length stud next to the trimmer stud in a wall opening.
- *Partition* – A wall that divides space within a building. A bearing partition or wall is one that supports the floors and roof directly above in addition to its own weight.
- *Rough opening* – An opening in the framing formed by framing members, usually for a window or door.
- *Rough sill* – The lower framing member attached to the top of the lower cripple studs to form the base of a rough opening for a window.
- *Soleplate* – The lowest horizontal member of a wall or partition to which the studs are nailed. It rests on the subfloor.
- *Stud* – The main vertical framing member in a wall or partition.
- *Top plate* – The upper horizontal framing member of a wall used to carry the roof trusses or rafters.
- *Trimmer stud* – A trimmer stud is a vertical framing member that forms the sides of rough openings for doors and windows. It stiffens the frame and supports the weight of the header.

1.1.0 Corner Assemblies

When framing a wall, solid corner assemblies that can support the weight of the structure are required. In addition to contributing to the strength of the structure, corner assemblies must provide a good nailing surface for sheathing and interior finish materials. Carpenters generally select the straightest, least defective studs when framing corner assemblies.

Figure 2 shows one common method of constructing corner assemblies in platform frame construction. In one corner assembly, there are two common studs with blocking between them. This arrangement provides a nailing surface for the first stud in the adjoining wall. Note the use of a double top plate to provide greater strength and stability to the wall frame.

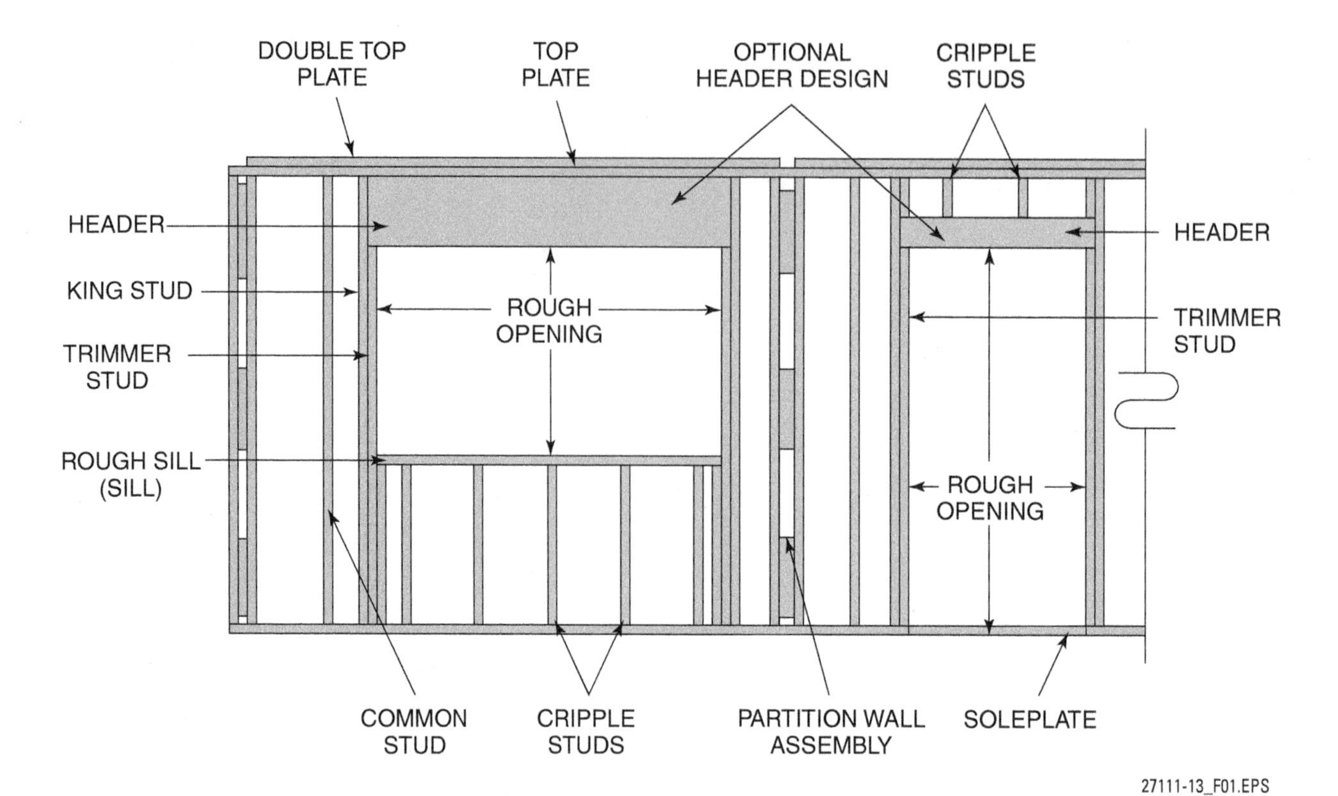

Figure 1 Wall and partition framing members.

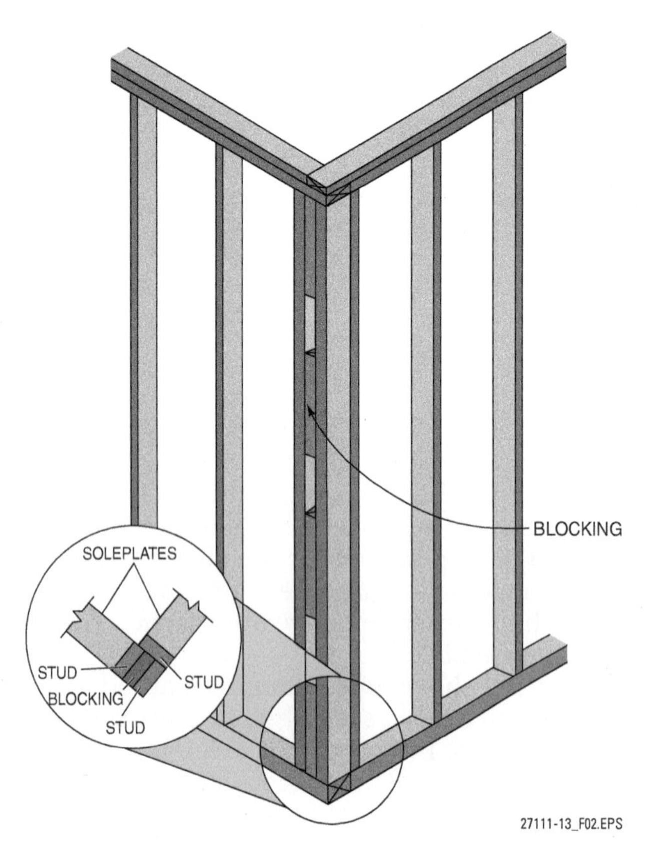

Figure 2 Corner construction typical of platform framing.

Figure 3 shows an alternative method to construct a corner. It has several advantages:

- It doesn't require blocking, which saves time and materials.
- It results in fewer voids in the insulation.
- It promotes better coordination among trades. For example, an electrician running wiring through the corner shown in *Figure 2* would need to bore holes through two or three studs and, possibly, a piece of blocking. However, an electrician wiring through the corner shown in *Figure 3* would need to bore through only two studs.

1.2.0 Partition Intersections

Interior partitions must be securely fastened to exterior walls. For that to occur there must be a solid nailing surface where the partition intersects the exterior wall frame. *Figure 4* shows a common way to construct a nailing surface for partition intersections, or Ts. The nailing surface can be a full stud nailed perpendicular between two other full studs, or it can be short pieces of 2 × 4 lumber, known as blocking, nailed between the two other full studs.

Figure 5 shows two other ways to prepare a nailing surface for partitions. Compare *Figure 4* and *Figure 5*. Notice that in *Figure 4* the spacing

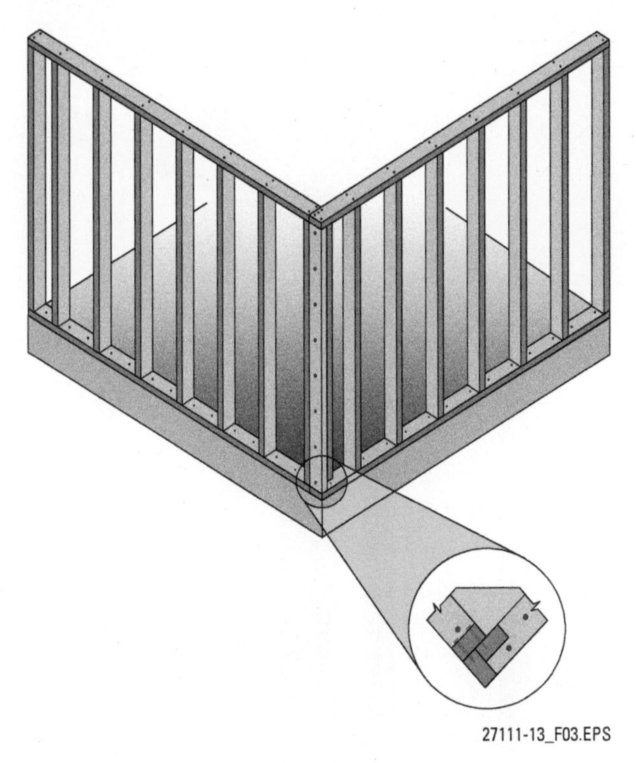

Figure 3 Alternative method of corner construction.

between studs differs, but in *Figure 5* the spacing between studs remains the same.

To lay out a partition location, measure from the end of the wall to the center line of the partition. Then, mark the locations for the partition studs on either side of the center line (*Figure 6*).

Although plans typically indicate on-center dimensions of studs, they sometimes show finish-to-finish dimensions. On-center dimensions are typically used in steel framing, which will be discussed in the section, *Framing with Steel Studs*.

1.3.0 Headers

When a wall frame is interrupted by an opening such as a window or door, a header is required to distribute the weight of the structure around the opening. Headers rest on the trimmer studs, which transfer the weight to the soleplate or subfloor, and then to the foundation. If an adequate header is not provided, the wall frame may have a tendency to sag and pinch the door or window, thus affecting its operation.

The width of a header should be equal to the rough opening plus the width of the trimmer studs. For example, if the rough opening for a 3'-wide window is 38", the width of the header would be 41" (1½" trimmer stud plus 38" rough opening plus the other 1½" trimmer stud, or 41" total).

1.3.1 Built-Up Headers

Headers are usually made of built-up lumber (although solid wood beams are sometimes used as headers). Built-up headers are generally constructed with two pieces of nominal 2" lumber,

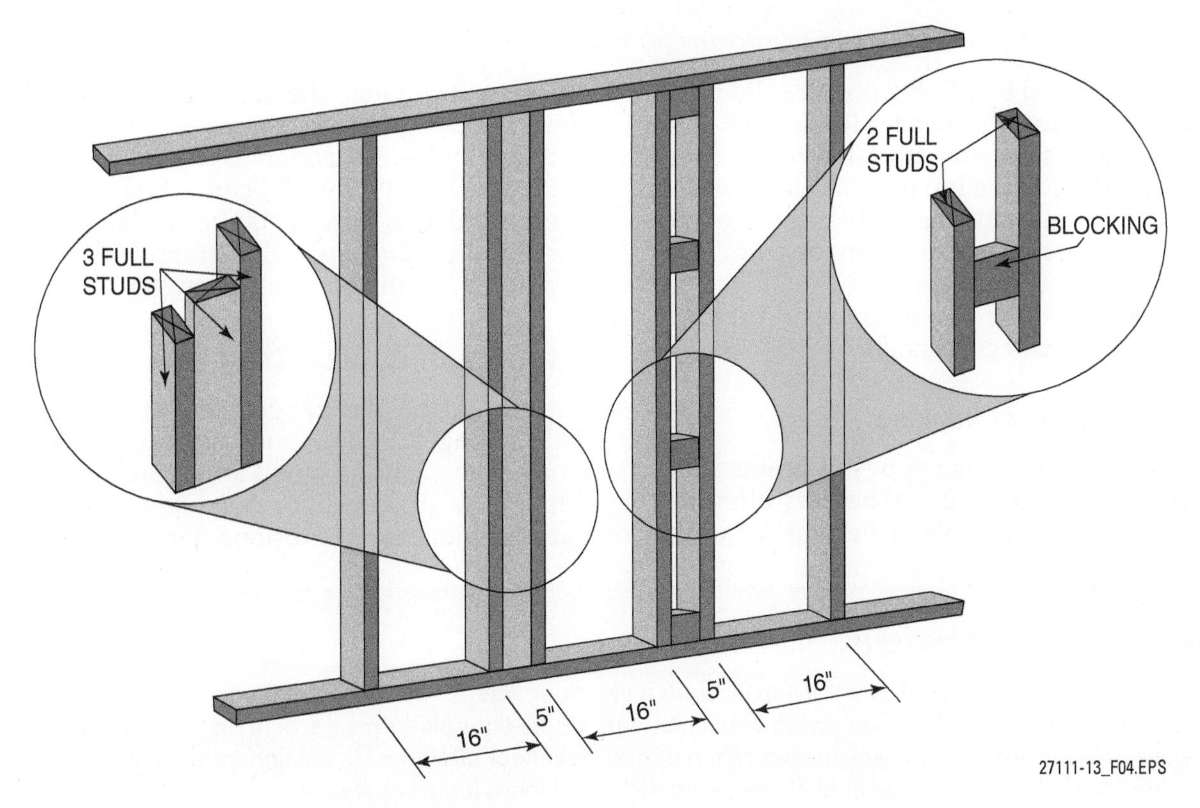

Figure 4 Constructing nailing surfaces for partitions.

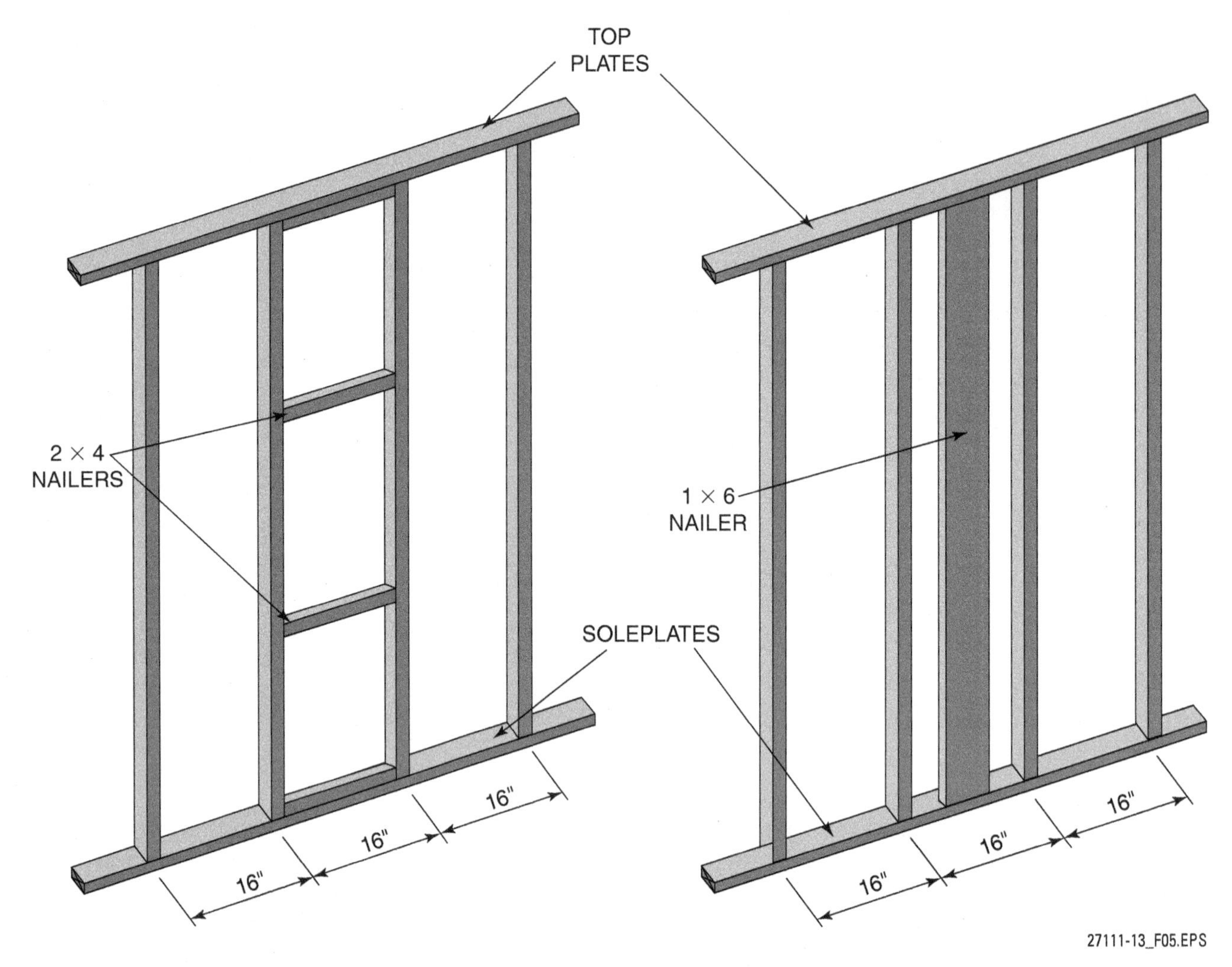

Figure 5 Additional ways to construct nailing surfaces for partitions.

which are separated by ½" plywood or solid wood spacers (*Figure 7*). A full header is used for large openings and fills the area from the rough opening to the bottom of the top plate. A small header with cripple studs is suitable for average-size windows and doors, and is usually made from 2 × 4 or 2 × 6 lumber with spacers.

Table 1 provides the maximum span typically used for various load conditions.

1.3.2 Other Types of Headers

Figure 8 shows some other types of headers that are used in wall framing. Carpenters often use truss headers when the load is very heavy or the span is extra wide. The architect's plans usually show the design of the trusses.

Other types of headers used for heavy loads are wood I-joists, steel I-beams, box beams, and engineered wood products such as laminated veneer lumber (LVL), parallel strand lumber (PSL), and glued laminated lumber (glulam).

1.4.0 Framing with Steel Studs

Depending on the gauge, steel studs are typically stronger, lighter, and easier to handle than wood studs. Unlike wood studs, steel studs will not split, warp, swell, or twist when exposed to varying moisture conditions. Furthermore, steel

Framing Methods

Platform framing is a method of construction in which a first-floor deck is built on top of the foundation walls. Then, the first-floor walls are erected on top of the platform. Upper-floor platforms are built on top of the first-floor walls, and upper-floor walls are erected on top of the upper-floor platforms. In balloon framing, the studs extend from the sill plate to the rafter plate, requiring the use of much longer studs.

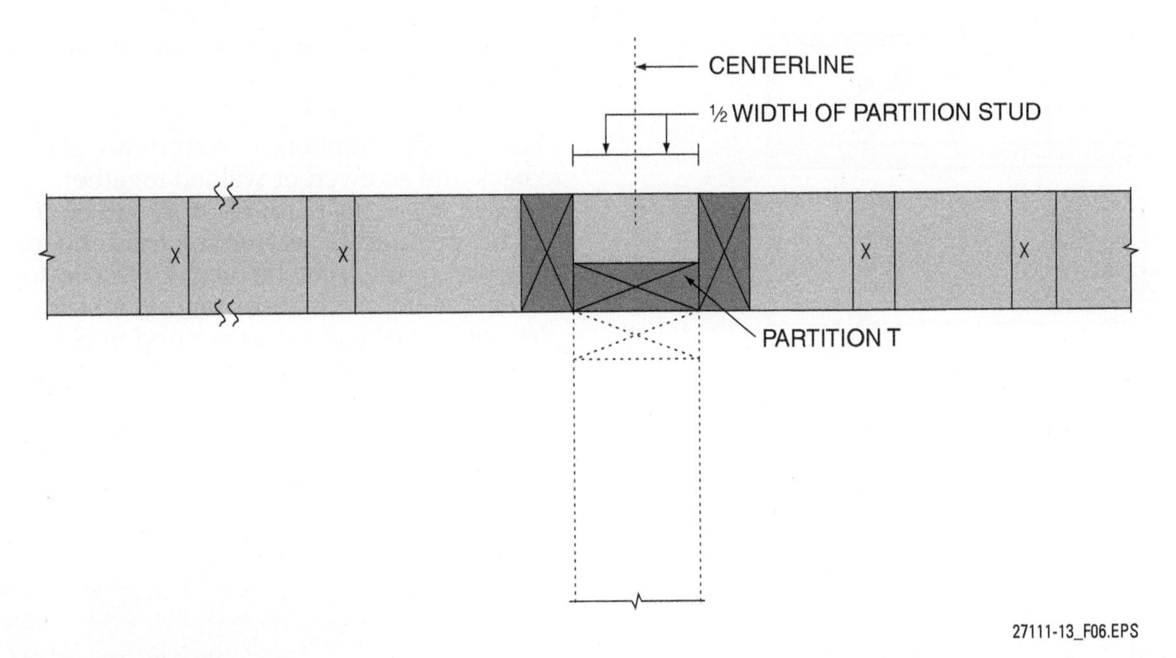

Figure 6 Partition T layout.

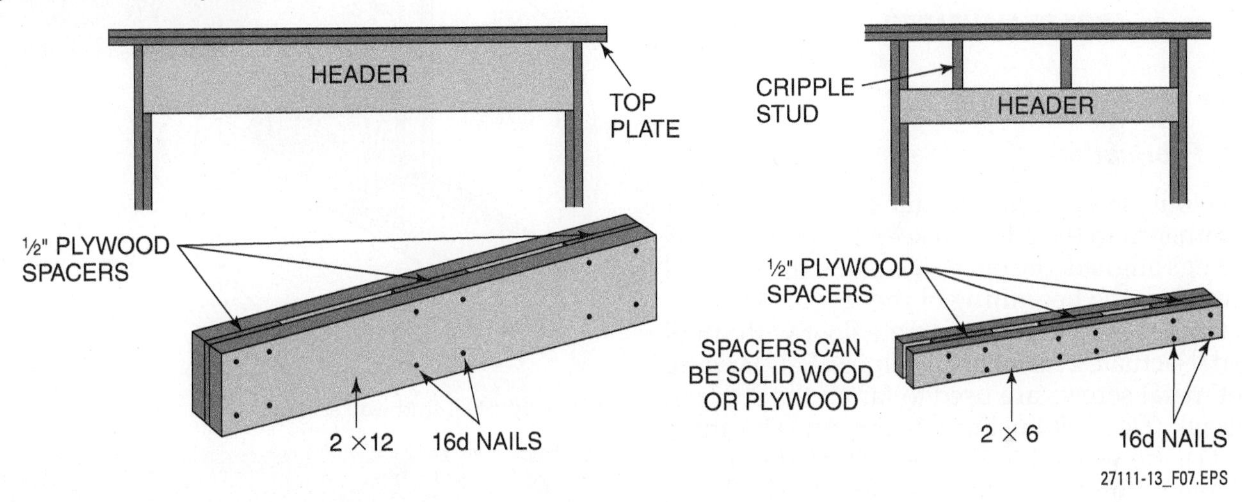

Figure 7 Two types of built-up headers.

Table 1 Maximum Span for Exterior Built-Up Headers

Built-Up Header Size	Single-Story Load	Two-Story Load	Three-Story Load
2 × 4	3'-6"	2'-6"	2'
2 × 6	6'	5'	4'
2 × 8	8'	7'	6'
2 × 10	10'	8'	7'
2 × 12	12'	9'	8'

studs will not burn as wood studs would. Steel studs are prepunched to permit quick installation of piping, wiring, and bracing members.

Steel studs are popular in residential, commercial, and industrial construction. Steel studs may be spaced 16" or 24" on center (OC). In wall framing, stud spacing is indicated in the drawings and specifications. Unlike wood (which has defects), steel studs are consistent in material composition.

There are three types of steel studs. The first is used for nonbearing walls that have facings to accept gypsum board. The second type will accept lath and plaster on both interior and exterior walls. The third type is a wide-flange steel stud, which is used for both loadbearing and nonbearing walls.

Similar to wood frame walls, steel frame walls are constructed using a number of framing members. Tracks are attached to floors and ceilings and studs are fastened to them. Tracks are also available for sills, fascia, and joint-end enclosures. Other accessories include channels, angles, and clips. For residential construction, steel trusses are also available.

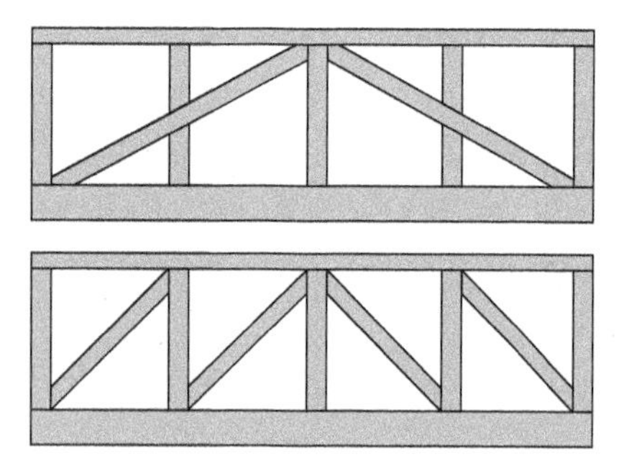

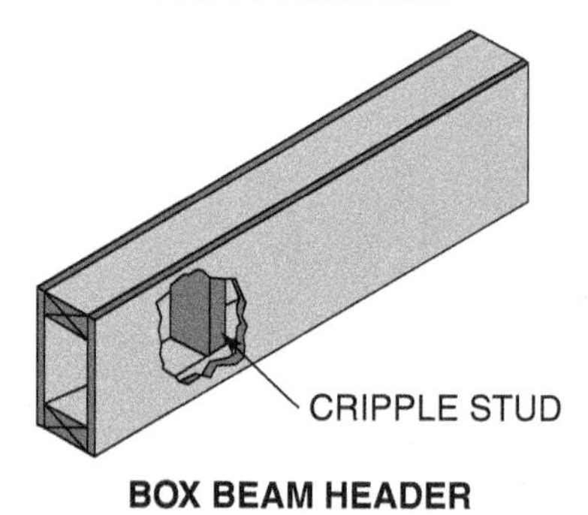

Figure 8 Other types of headers.

1.4.1 Fabrication

For layout, steel studs are laid out to the center line rather than the edge. To keep the prepunched openings aligned, the open side of the stud should always face the beginning of the layout. The bottom track is fastened to a concrete floor with small powder-actuated fasteners (*Figure 9*). Self-tapping sheet metal screws are used to fasten the studs to the track (*Figure 10*) using a screwgun. The studs may also be welded to the track instead of being attached with screws. Always check the building code in effect in your jurisdiction regarding approved fastening methods for steel framing members.

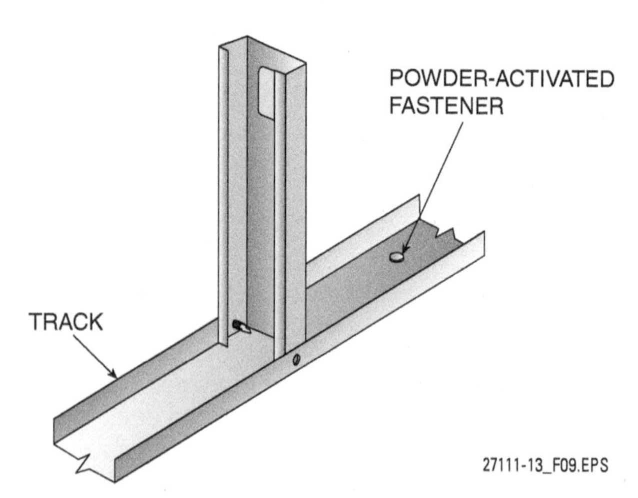

Figure 9 Metal track.

When constructing a rough opening using steel framing members, two studs are placed back to back and screwed or welded together. The stud that will act as the trimmer stud will be cut to the height specified to receive the header (*Figure 11*). A section of track can be used for the bottom part of the header, with short pieces of studs put in place over the header and secured in place. Blocking may be required to fasten millwork.

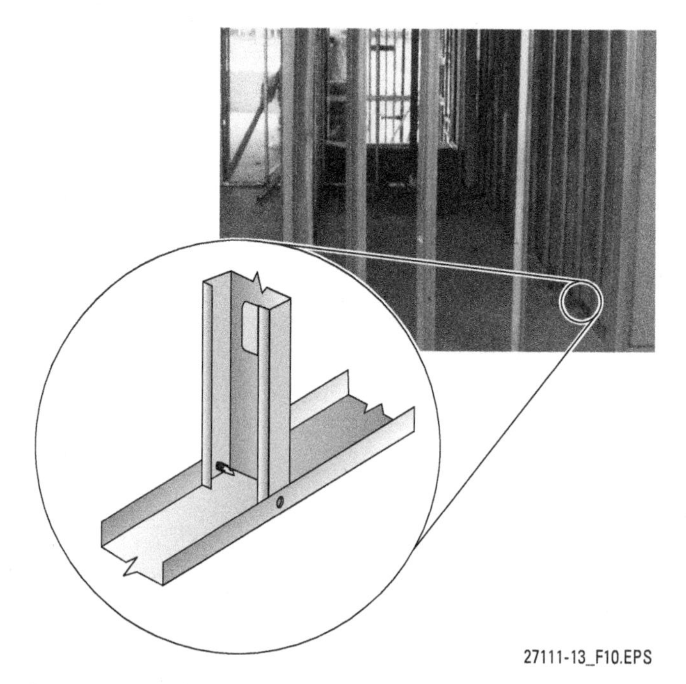

Figure 10 Steel stud wall section.

Figure 11 Window opening framed with steel studs.

Headers

A variety of materials can be used for headers. A header constructed of parallel strand lumber (PSL) is shown in the upper photo and a solid wood header is shown in the lower photo.

27111-13_SA01.EPS

1.0.0 Section Review

1. Which of the following is a purpose of double top plates?
   a. To provide support for headers.
   b. To support the floor joists.
   c. To provide greater stability to the wall frame.
   d. To provide a bearing surface for the sole-plate.

2. Partition intersections are commonly referred to as _____.
   a. Ts
   b. Is
   c. Ps
   d. Ls

3. If the rough opening for a window is 46" and 2 × 6s are used for framing members, the header width would be _____.
   a. 46"
   b. 49"
   c. 50"
   d. 58"

4. The bottom track of a steel frame wall is fastened to a concrete floor using _____.
   a. wood screws
   b. sheet metal screws
   c. powder-actuated fasteners
   d. lag bolts

Section Two

2.0.0 Wood Frame Wall Layout

Objective

Describe the procedure for laying out a wood frame wall, including plates, corner posts, door and window openings, partition Ts, bracing, and fire stops.

a. Describe how to properly lay out a wood frame wall.
b. Explain how to lay out wall openings.

Performance Task 1

Lay out a wood frame wall, including plates, corner posts, door and window openings, partition Ts, bracing, and fire stops.

Trade Terms

Jamb: The top (head jamb) and side members of a door or window frame that come into contact with the door or window.

This section covers the basic procedures for laying out wood frame walls with correctly sized window and door openings and partition Ts. In the section *Wall Framing in Masonry Construction*, you will be introduced to methods for framing window and door openings in masonry walls.

2.1.0 Wall Framing

Precise layout of framing members is extremely important. Finish material such as sheathing, gypsum board, paneling, etc., is sold in 4 × 8 panels. If the studs, rafters, and joists are not straight and evenly spaced for their entire length, it will be difficult to properly fasten the panel material to them. A tiny error on one end becomes a large error as you progress toward the other end. Spacings of 16" and 24" on center are used because they divide evenly into 48".

Walls are generally framed with 2 × 4 or 2 × 6 studs spaced 16" on center. In a one-story building, 2 × 4 spacing can be 24" on center. If 24" spacing is used in a two-story building, the lower floor must be framed with 2 × 6 lumber. Stud spacing is indicated on the drawings. The following provides an overview of the procedure for laying out a wall.

Step 1 Mark the locations of the soleplates by measuring in the width of the soleplate (e.g., 3½" for 2 × 4 studs) from the outside edge of the subfloor on each corner. Snap a chalkline to mark the soleplate location, and then repeat this for each wall.

Step 2 The top plate and soleplate are laid out together. Start by placing the soleplate next to the chalkline (*Figure 12*). Place the top plate against the soleplate so that the location of framing members can be transferred from the soleplate to the top plate. Note that the soleplate and top plate being laid out should be different lengths to prevent the butt joints from falling directly over one another. Some carpenters may tack the soleplate and top plate to the subfloor to prevent the plates from moving during layout.

Step 3 Lay out the common stud positions. To begin, measure and square a line 15¼" from one end of the soleplate. Subtracting this ¾" ensures that edges of the wall sheathing will fall at the center of the studs because the first sheet goes to the edge rather than to the center of the corner stud. Drive a nail at the 15¼" point and use a measuring tape or long tape measure to lay out and mark the stud locations every 16" (*Figure 13*). Align your framing square at each mark. Scribe a line along each side of the framing square tongue across both the soleplate and top plate. These lines will show the outside edges of each stud, centered on 16" intervals.

2.2.0 Laying Out Wall Openings

The floor plans for a building (*Figure 14*) show the locations of windows and doors in the walls. Notice that each window and door on the floor plan is identified by size. In this case, the widths are shown in feet and inches, but the complete dimension information is coded. Look at the window in bedroom #1 coded 2630. This means the window is 2'-6" wide (the width is always given first) and 3'-0" high. Similar codes indicate the widths and heights for the doors.

The door and window schedules (*Figure 15*), provided in the architectural drawings, list the window and door dimensions, along with types, manufacturers, and other information.

Placement of windows is important, and is normally dealt with on the architectural drawings. A good rule of thumb is to avoid placing horizontal

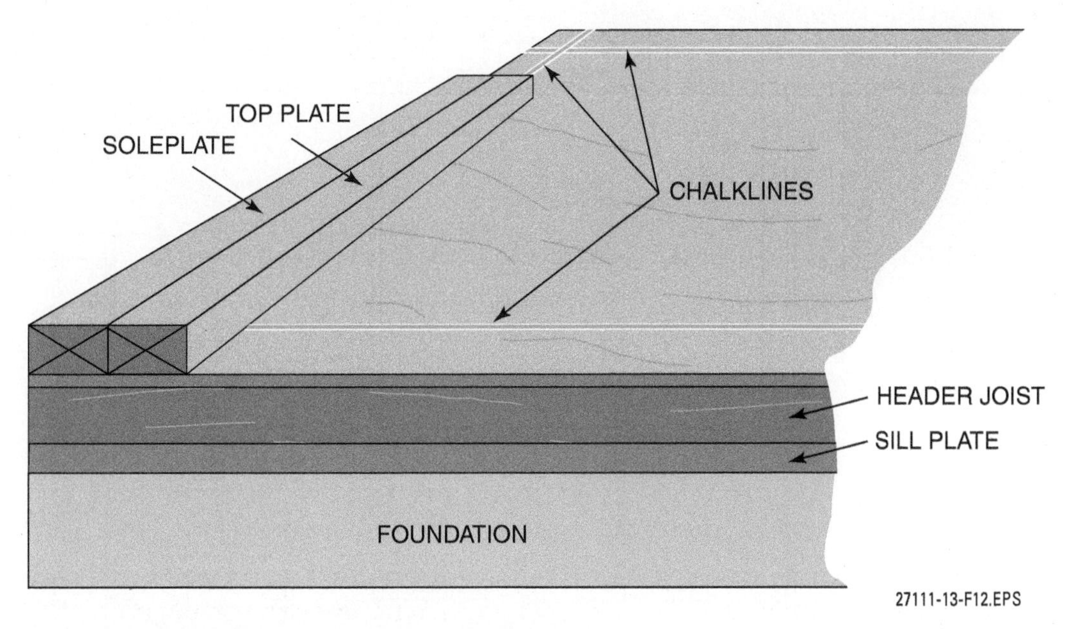

Figure 12 Soleplate and top plate positioned along the chalkline.

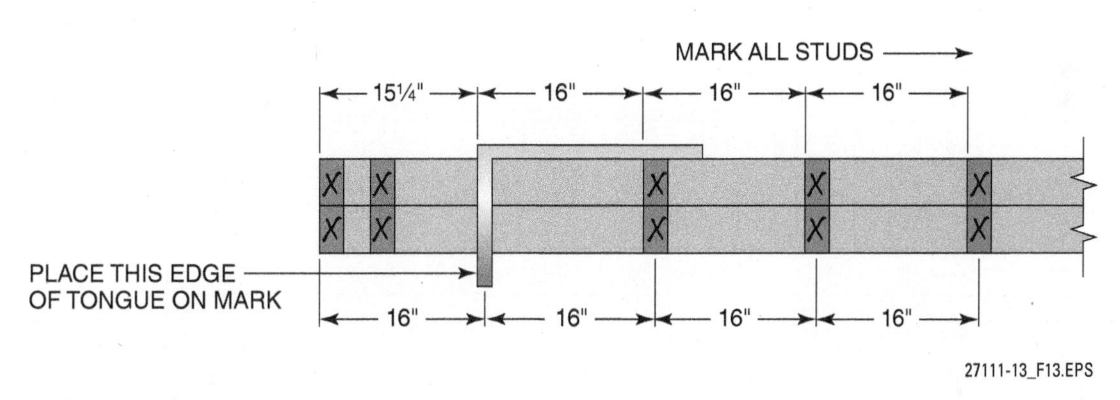

Figure 13 Marking stud locations.

framing members at eye level. Consider whether people will generally be sitting or standing when they look out the window. Unless they are architectural specialty windows, the tops of all windows should be at the same height. The bottom (rough sill) height will vary depending on the use. For example, the bottom of a window over the kitchen sink should be higher than that of a living room or dining room window. The standard height for a residential window is 6'-8" from the floor to the bottom of the window top (head) **jamb**.

The window and door schedules will provide the rough-opening dimensions for windows and doors. Another good source of rough-in information is the manufacturer's catalog. It will provide rough- and finish-opening dimensions, as well as the unobstructed glass dimensions.

When roughing-in a window, the rough-opening width equals the width of the window plus the thickness of the jamb material (this is usually 1½", i.e., ¾" on each side) plus the shim clearance (½" on each side). Therefore, the rough opening for a 3' window would be 38½" (36" + 1½" + 1" = 38½"). The rough-opening height is calculated in the same manner. Be sure to check the manufacturer's instructions for the dimensions of the windows you are using.

Chalkline Protection

If the chalkline will be exposed to the weather, spray it with a clear protective coating to keep it from being washed away.

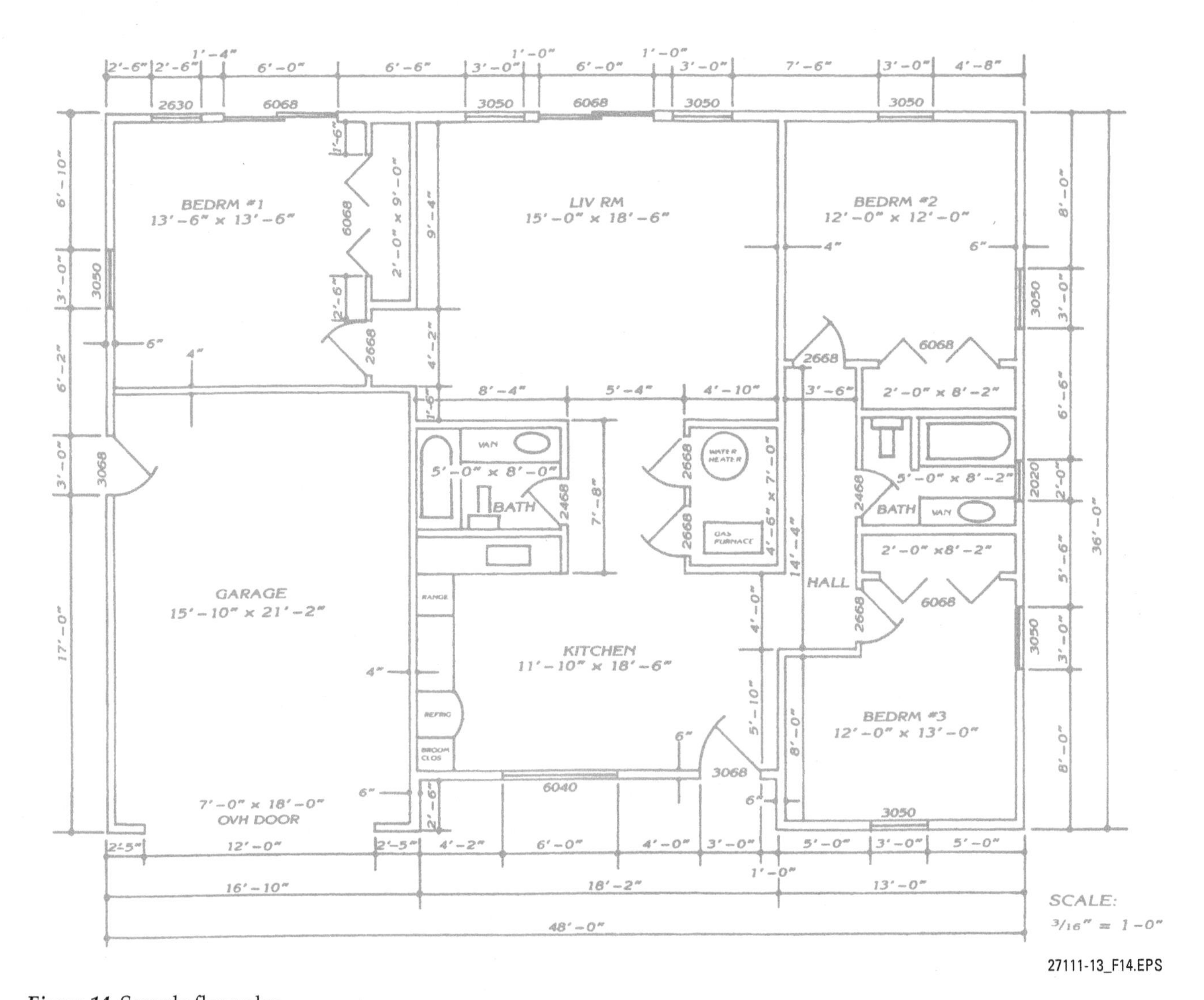

Figure 14 Sample floor plan.

To lay out a wall opening, proceed as follows:

Step 1 Measure from the corner to the start of the opening, and then add one-half the width of the window or door to determine the center line (*Figure 16*). Mark the locations of the full studs and trimmer studs by measuring in each direction from the center line. Mark the cripple studs 16" on center, starting with one trimmer stud.

Each window opening requires a king stud on each side, plus a header, cripple studs, and a sill (*Figure 17*). If the window is more than 4' wide, local codes may require a double sill. Door openings also require trimmer studs, king studs, and a header. Cripple studs will be needed unless the door is double-wide. In that case, a full header may be required.

Step 2 Mark the location of each common stud and king stud (*X*), trimmer stud (*T*), and cripple stud (*C*), as shown in *Figure 18*. (This is a suggested marking method. The important thing is to mark the locations with codes that you and other members of your crew will recognize.)

DOOR					FRAME					REMARKS
NO.	SIZE	MAT'L	TYPE	H.W.#	MAT'L	TYPE	HEAD	JAMB	SILL	
104A	PR. 3'-0" x 7'-8 5/8" x 1-3/4"	W.D.	IV	11	W.D.	—	8/16	4/16	3/16	
104B	PR. 3'-0" x 8'-0" x 1-3/4"	GLASS	VI	10	—	—	3/7	15/15, 16/15,	4/7	W/ FULL GLASS SIDELITE
105A	3'-0" x 7'-2" x 1-3/4"	W.D.	I	1	H.M.	II	5/15	6/7, 7/7,	—	
105B			I	1		I	SIM. TO 3/15,	SIM. TO 4/15,	—	
106			I	1		I	1/15	2/15	—	
107			I	1		I	1/15	2/15	—	
108	PR. 3'-0" x 7'-2" x 1-3/4"		II	4		II	8/5	9/15	—	"C" LABEL
109	3'-0" x 7'-2" x 1-3/4"		I	7		I	3/15	4/15	—	
110			I	7		I	3/15	4/15	—	
112			I	13		I	SIM. TO 3/15,	SIM. TO 4/15,	—	
113A			I	13		III	5/15	6/7	—	"C" LABEL
113B			I	13		I	SIM. TO 1/15,	SIM. TO 2/15,	—	
114	2'-6" x 7'-2" x 1-3/4"		I	13		I	SIM. TO 1/15,	SIM. TO 2/15,	—	
115A	PR. 3'-0" x 7'-0" x 1-3/4"	ALUM.	V	14	ALUM.	—	28/15	33/15	38/15	
115B	PR. 3'-0" x 7'-2" x 1-3/4"	W.D.	III	5	H.W.	I	10/15	11/15, 12/15,	—	"C" LABEL NOTE 1
121D			I	17		I	[illegible]/15	[illegible]/10	—	

WINDOW SCHEDULE

SYMBOL	WIDTH	HEIGHT	MAT'L	TYPE	SCREEN & DOOR	QUANTITY	REMARKS	MANUFACTURER	CATALOG NUMBER
A	3'-8"	3'-0"	ALUM.	DOUBLE HUNG	YES	2	4 LIGHTS, 4 HIGH	LBJ WINDOW CO.	141 PW
B	3'-8"	5'-0"	ALUM.	DOUBLE HUNG	YES	1	4 LIGHTS, 4 HIGH	LBJ WINDOW CO.	145 PW
C	3'-0"	5'-0"	ALUM.	STATIONARY	STORM ONLY	2	SINGLE LIGHTS	H & J GLASS CO.	59 PY
D	2'-0"	3'-0"	ALUM.	DOUBLE HUNG	YES	1	4 LIGHTS, 4 HIGH	LBJ WINDOW CO.	142 PW
E	2'-0"	6'-0"	ALUM.	STATIONARY	STORM ONLY	2	20 LIGHTS	H & J GLASS CO.	37 TS
F	3'-6"	5'-0"	ALUM.	DOUBLE HUNG	YES	1	16 LIGHTS, 4 HIGH	LBJ WINDOW CO.	141 PW

HEADER SCHEDULE

HEADER SIZE	EXTERIOR		INTERIOR	
	26' + UNDER	26' TO 32'	26' + UNDER	26' TO 32'
(2) 2 x 4	3'-6"	3'-0"	USE (2) 2 x 6	
(2) 2 x 6	6'-6"	6'-0"	4'-0"	3'-0"
(2) 2 x 8	8'-6"	8'-0"	5'-6"	5'-0"
(2) 2 x 10	11'-0"	10'-0"	7'-0"	6'-6"
(2) 2 x 12	13'-6"	12'-0"	8'-6"	8'-0"

27111-13_F15.EPS

Figure 15 Door, window, and header schedules.

Rough-Opening Dimensions

The residential plan shown in *Figure 14* shows dimensions to the sides of rough openings—that is, from the corner of the building to the near side of the first rough opening, then from the far side of the first rough opening to the near side of the second rough opening. However, it is more common for plans to show center-line dimensions, that is, from the corner of the building to the center of the first rough opening, then from the center of the first rough opening to the center of the second rough opening.

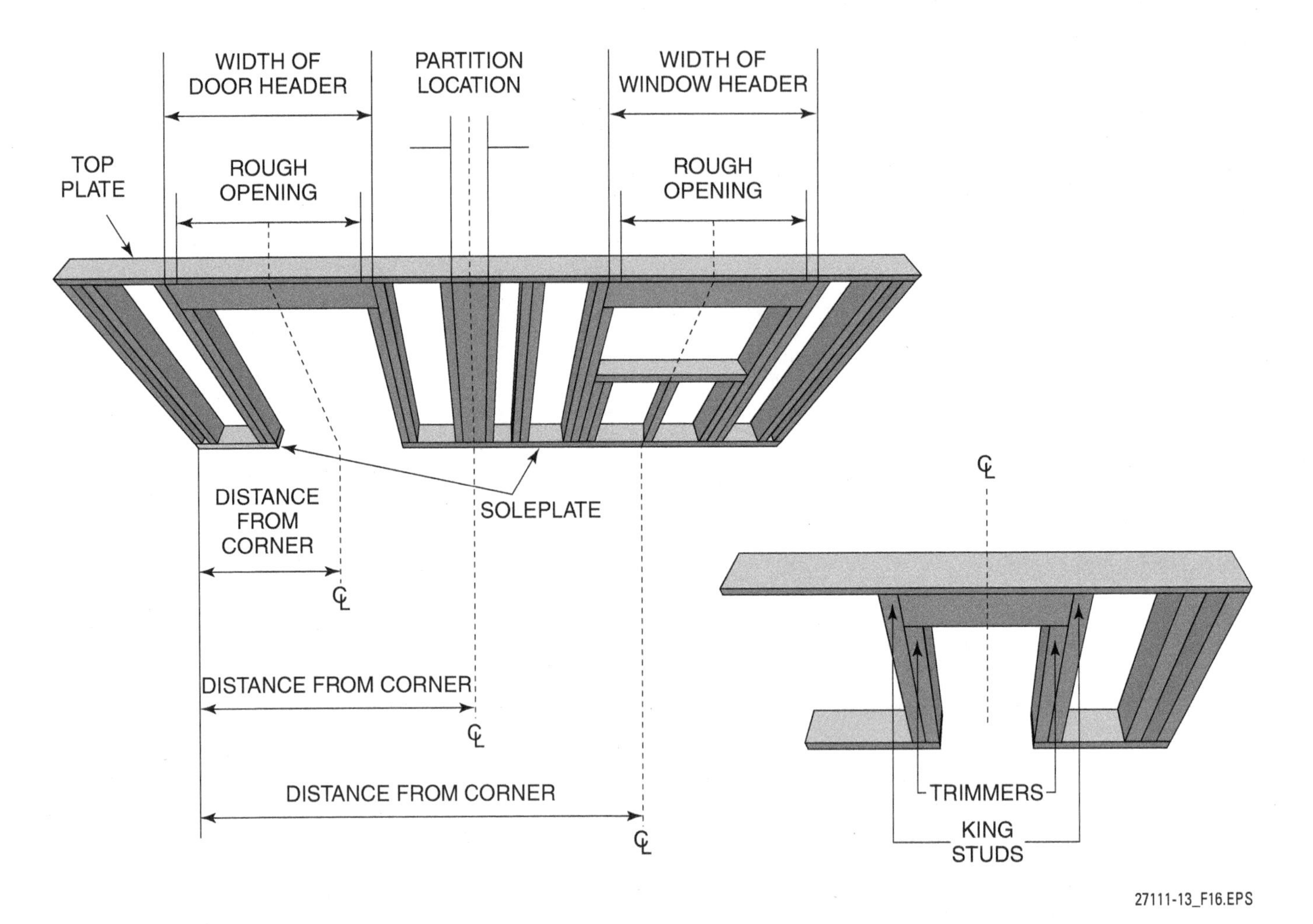

Figure 16 Laying out a wall opening.

GOING GREEN

Optimal Value Engineered (OVE) Framing

Optimal value engineered (OVE) framing for light wood frame construction is another way to use materials more efficiently. OVE framing is a good design solution because it allows more insulation to be incorporated in the walls of the building. This reduces the amount of heat that can escape through the walls and also increases the energy performance of the building. OVE framing also saves considerable materials during construction by using wood more efficiently. This means less lumber, fewer cuts, and less waste. According to the US Department of Energy, key principles of OVE framing include the following:

- Designing buildings on two-foot modules to make the best use of common panel sizes
- Spacing wall studs, roof joists, and rafters up to 24" on center
- Using in-line framing where floor, wall, and roof framing members are vertically in line with one another and loads are transferred directly downward
- Using two-stud corner framing and inexpensive drywall clips or scrap lumber for gypsum-board backing
- Eliminating headers in walls that are not loadbearing
- Using single lumber headers and top plates when appropriate

These techniques reduce the amount of wood used and create a better thermal envelope. They also make construction easier for crews that must install plumbing, electrical, and heating, ventilating, and air conditioning (HVAC) services in the walls. Fewer structural members mean less drilling during installation and less effort to install. All of these attributes make OVE framing a good strategy to green your building and save both first costs and life cycle costs.

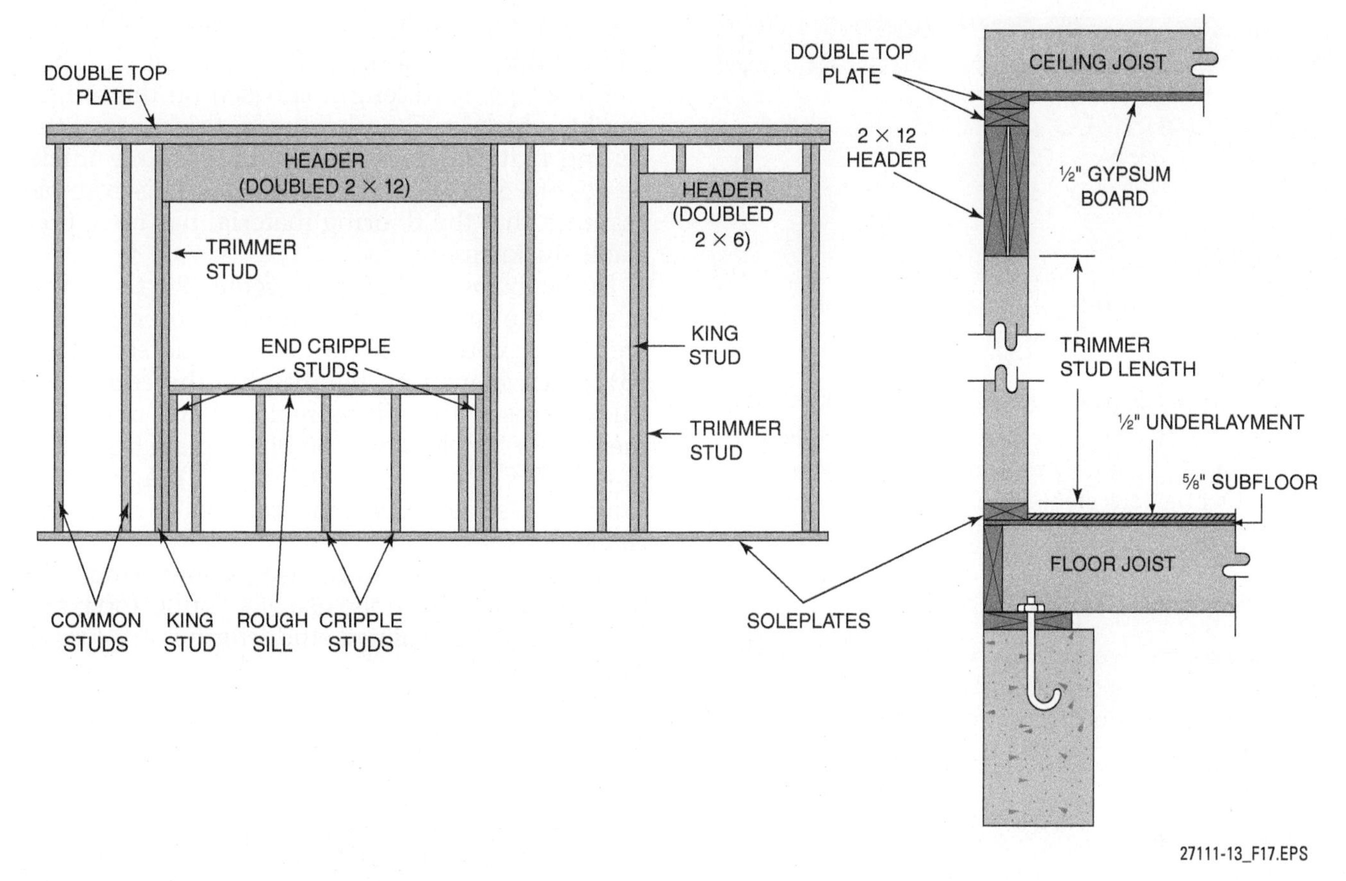

Figure 17 Window and door framing.

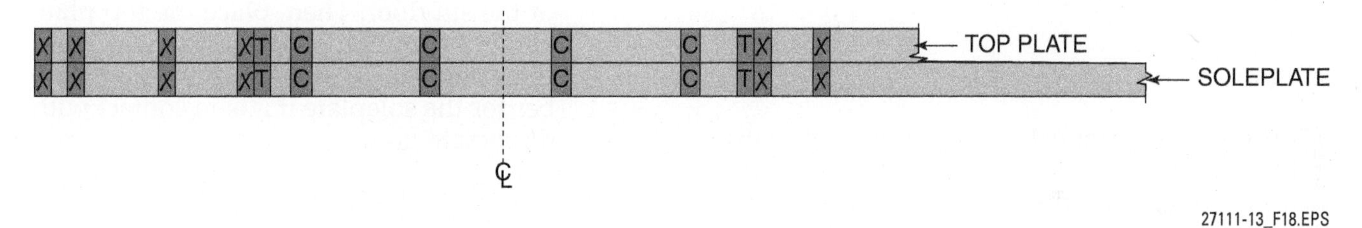

Figure 18 Example of soleplate and top plate marked for layout.

2.0.0 Section Review

1. When laying out stud positions, measure and square the first line _____.
   a. 12" from one end of the soleplate
   b. 15¼" from one end of the soleplate
   c. 16" from one end of the soleplate
   d. 19¼" from one end of the soleplate

2. The standard height for a residential window from the floor to the bottom of the window top jamb is _____.
   a. 6'-2"
   b. 6'-6"
   c. 6'-8"
   d. 7'-0"

SECTION THREE

3.0.0 ASSEMBLING AND ERECTING EXTERIOR WALLS

Objective

Describe the correct procedure to assemble, erect, and brace exterior walls for a frame building.

a. List the steps involved in assembling a wall.
b. Identify where firestops are to be installed and explain how they are installed.
c. List the four steps involved in erecting a wall.

Performance Tasks 2 and 3

Assemble and erect a wood frame wall, including plates, corner posts, door and window openings, partition Ts, bracing, and fire stops.

Correctly install sheathing on a wall.

Trade Terms

Drying-in: Applying sheathing, windows, and exterior doors to a framed building.

It is extremely important to precisely measure the first one of each type of stud that will be used (common, trimmer, and cripple) as a template for the others.

- *Common (king) studs–Figure 19* shows the methods for calculating the exact length of a common (king) stud for installation on a slab or wood floor.
- *Trimmer studs*–The length of a window or door trimmer stud is the distance between the top of the soleplate and the bottom of the header.
- *Cripple studs*–To determine the length of a cripple stud above a door or window, add the height of the trimmer studs and the thickness of the header, and then subtract that total from the length of a common stud. To determine the length of a cripple stud below a window, determine the rough-opening height from the floor, then subtract the combined thicknesses of the rough sill and soleplate.

To determine the stud length when erecting the wall frame directly on a concrete slab, simply subtract the thickness of the soleplate (1½") and double top plate (3") from the desired ceiling height and add the thickness of the ceiling material. In *Figure 19A*, the 93¼" stud length is based on the ceiling height, which is 97⅛", plus the ⅝" thickness of the ceiling material, less the combined plate thicknesses of 4½" (97⅛" + ⅝" – 4½" = 93¼"). This example assumes that the flooring material has no appreciable thickness.

In the example shown in *Figure 19B*, the thickness of the underlayment must also be considered. Therefore, the length of the stud should be 93¾"; that is, ceiling height plus the combined thicknesses of the ceiling material and underlayment (⅝" + ½"), less the combined plate thickness of 4½" (97⅛" + [⅝" + ½"] – 4½" = 93¾"). Again, this example assumes a flooring material of no appreciable thickness.

Figure 19C shows an interior, nonbearing wall that does not require the use of a double top plate. Therefore, the calculated stud length is 1½" longer (95¼").

3.1.0 Assembling Walls

The preferred procedure for assembling a wall is to lay out and assemble the wall on the subfloor with the interior side of the wall facing down.

Step 1 Start by laying the soleplate near the edge of the subfloor. Then, place the top plate about a common stud length away from the soleplate. Be sure to use treated lumber for the soleplate if it is in contact with a concrete floor.

Step 2 Assemble the corners and partition Ts using the straightest pieces of lumber to ensure that the corners are plumb. Also, save some of the straightest studs for placement in the wall where countertops or fixtures will hit the centers of studs (such as in kitchens, bathrooms, and laundry rooms).

Step 3 Lay a common stud at each X mark with the crown facing up. If a stud is bowed, replace it and use it to make cripple studs.

Step 4 Assemble the window and door headers and put them in place with the crowns facing up.

Step 5 Lay out and assemble the rough openings, making sure that each opening is the correct size and that it is square.

Step 6 Nail the framing members together. For 2 × 4 framing, drive two 16d nails through the soleplates and into the end of each stud. For 2 × 6 framing, use three nails. The use of a pneumatic nailer is recom-

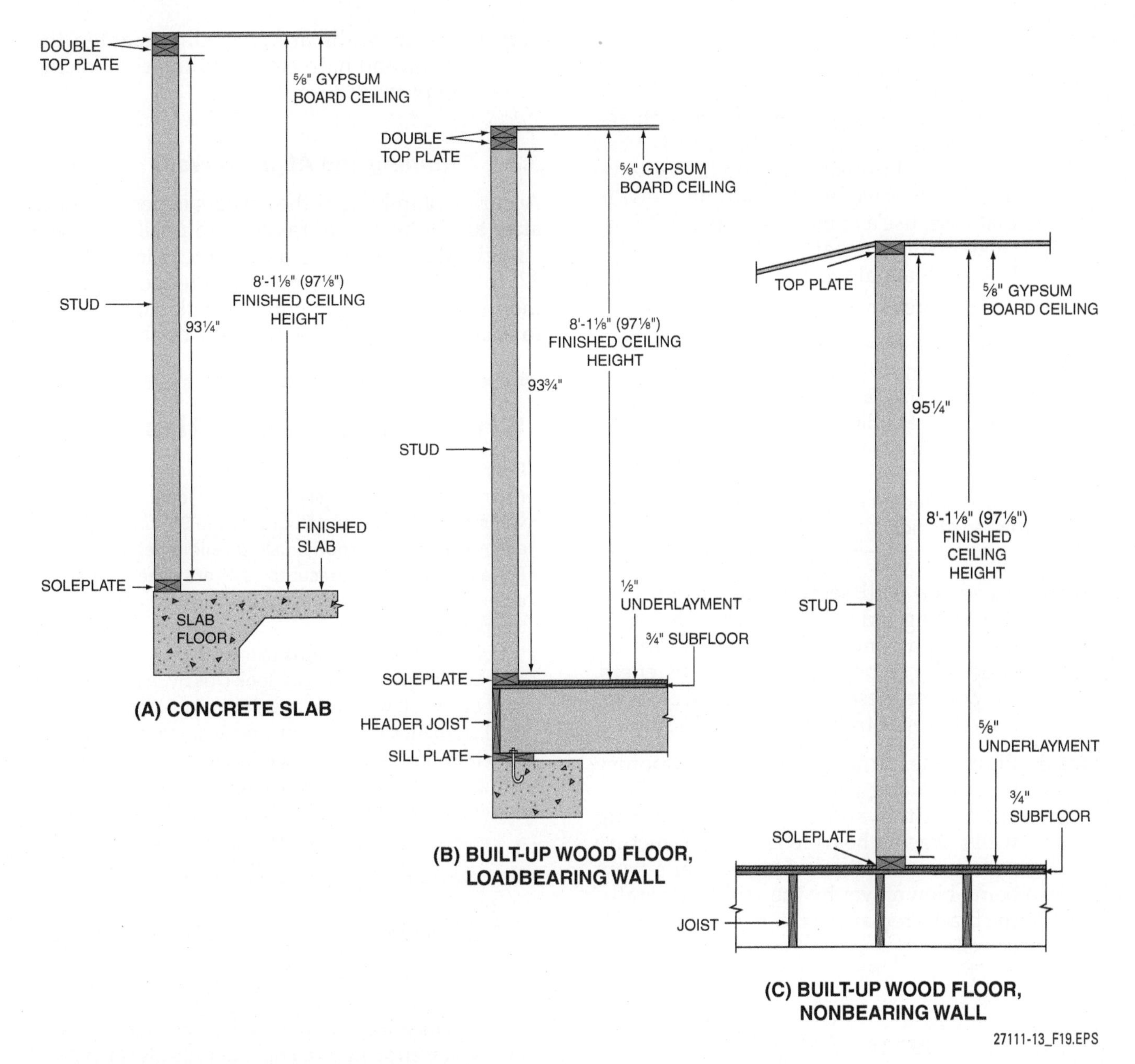

Figure 19 Calculating the length of a common stud.

mended for this purpose; however, do not use this tool if you have not received proper training.

3.2.0 Installing Fire Stops

In some areas, fire stops may be required by local building codes. Fire stops are short pieces of 2 × 4 blocking (or 2 × 6 pieces if the wall is framed with 2 × 6 lumber) that are nailed between studs (see *Figure 20*).

Without fire stops, the space between the studs will act like a flue in a chimney. Any holes drilled through the soleplate and top plate create a draft, and air will rush through the space. In a fire, air, smoke, gases, and flames can race through the chimney-like space.

The installation of fire stops has two purposes. First, it slows the flow of air, which feeds a fire through the cavity. Second, it can block flames (temporarily, at least) from traveling up through the cavity.

If the local building code requires fire stops, it may also require that holes through the soleplate and top plate (for plumbing or electrical runs) be plugged with a fire-stopping material to prevent airflow.

3.3.0 Erecting Walls

There are four primary steps in erecting a wall:

Step 1 If the sheathing was installed with the wall lying down, or if the wall is very long, it will probably be too heavy to be lifted into place by the framing crew. In that case, use a crane or the special lifting jacks made for that purpose (*Figure 21*). Install cleats along the edge of the subfloor to prevent the wall from sliding.

> **CAUTION**
>
> When erecting walls using a crane or special lifting jacks, always follow the manufacturer's instructions regarding the safe use of the equipment. In addition, ensure that the proper personal protective equipment (PPE) is worn, including appropriate protection for the head, eyes, and hands.

Step 2 Raise the wall section and fasten it in place using 16d nails on every other floor joist. On a concrete slab, use preset anchor bolts or powder-actuated fasteners. Do not use these tools if you have not received proper training and certification.

Step 3 Plumb the corners and apply temporary bracing along the wall's exterior. Then erect, plumb, and brace the remaining walls. Bracing helps to keep the structure square and will prevent the walls from being blown over by the wind. Generally, the braces remain in place until the roof is complete.

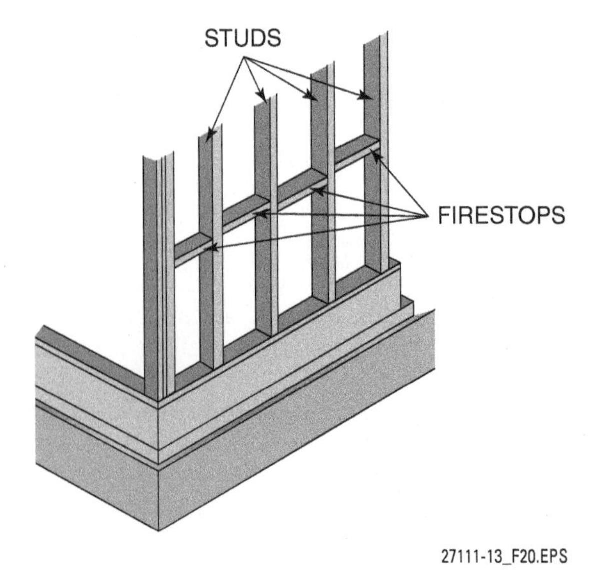

Figure 20 Fire stops.

Step 4 As the walls are erected, straighten the walls and nail temporary interior bracing in place.

3.3.1 Plumbing and Aligning Walls

Accurate plumbing of the corners is possible only after all the walls are raised into position. Use a straightedge along with a carpenter's level (*Figure 22*) to plumb the corners. Straightedges are commonly made of a 2 × 4 with ¾"-thick blocks nailed to each end. The blocks make it possible to accu-

> **Check Stud Lengths**
>
> Precut studs in various lengths are available from many lumberyards. Precut studs are ideal for walls on built-up wood floors for the 96" finished ceiling, a very common finished ceiling height. For example, studs precut to 92⅝" are used with exterior or loadbearing interior walls with double top plates.
>
> Sometimes lumberyards deliver the wrong size of precut studs. Unless you look closely, a precut stud doesn't look much different from a standard 8' (96") stud. Always check the precut studs to make sure they are the right length. Taking a few seconds to measure before you start might save hours of rebuilding later on.

> **Measure Twice, Cut Once**
>
> Before cutting all the studs, double-check your measurements, or have someone else double-check them. You do not want to find out there was a calculation error after you have cut 200 pieces of 2 × 6. When you are satisfied with the measurements, cut the required amounts for each type of stud and stack them neatly near the wall locations. Be sure to allow ample room to assemble the walls.

> **Cripple Studs**
>
> It's good to know how to calculate the proper lengths as the text describes; however, in practice, lumber dimensions and assembly tolerances can vary. Therefore, it's better to hold off until after the headers, trimmer studs, and rough sills are assembled, and then actually measure and cut the cripple studs to the required lengths.

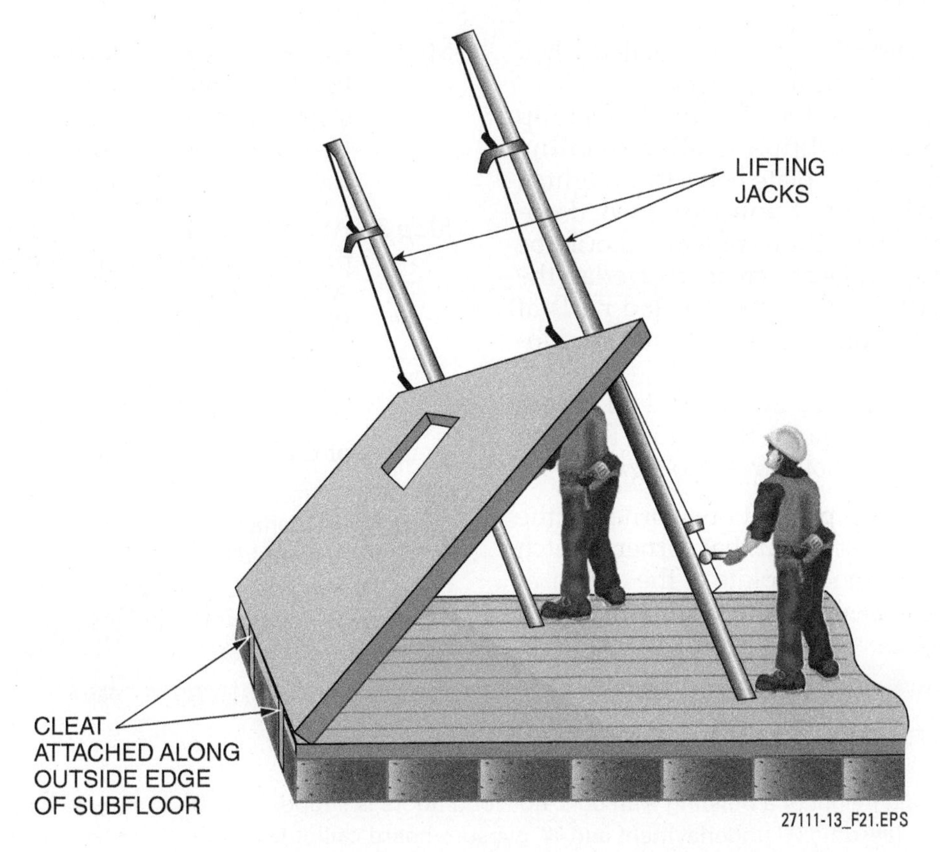

Figure 21 Wall-lifting jack.

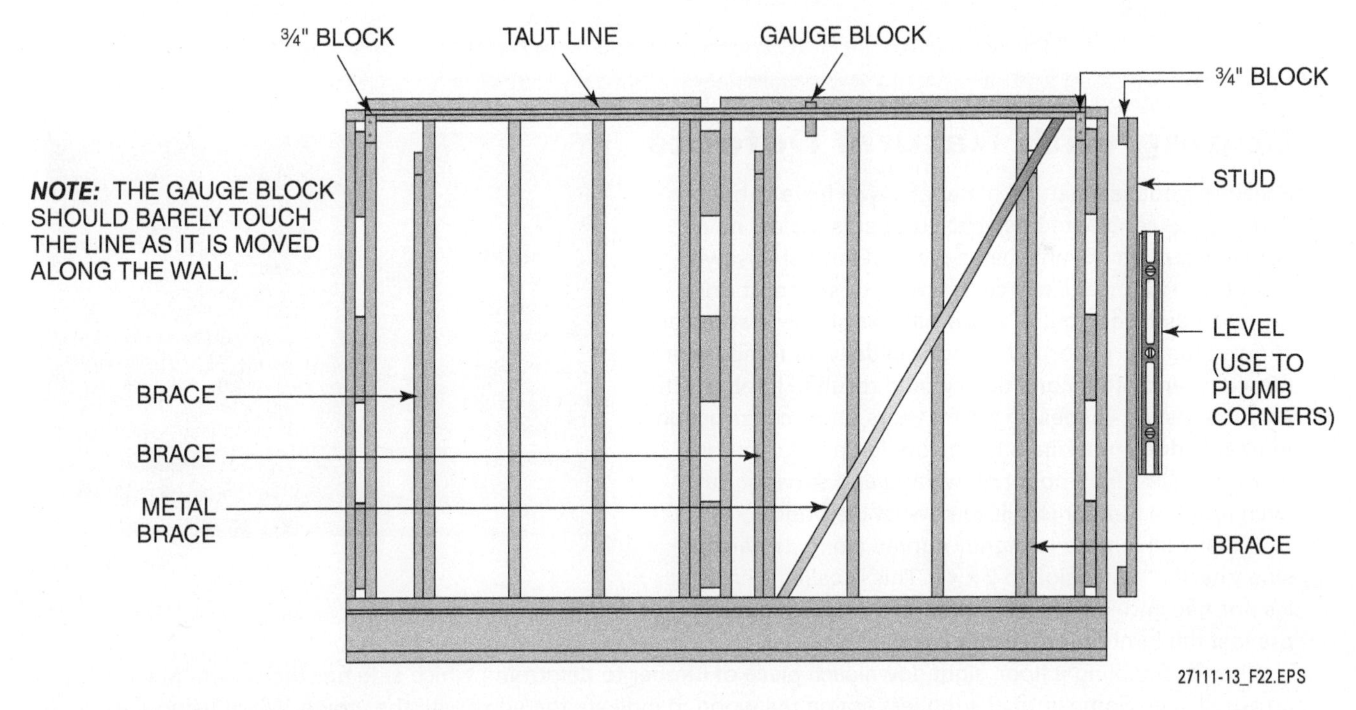

Figure 22 Plumbing and aligning a wall.

rately plumb the wall from the soleplate to the top plate. (If you just placed the level directly against the wall, any bow or crown in the end stud would indicate a false reading.) Long framer's levels or plumb bobs can also be used to plumb and align walls.

The plumbing of corners requires two carpenters working together. One carpenter releases the nails at the bottom end of the corner assembly brace so that the top of the wall can be moved in or out. At the same time, the second carpenter watches the level. The bottom end of the brace is renailed when the level shows that the wall is plumb.

Install the second plate of all the double top plates (*Figure 23*). In addition to strengthening bearing walls, the second plate helps to straighten a bowed or curved wall. If bows are turned opposite each other, intersections of walls should be double-plated after walls are erected. Overlap the corners and partition Ts. Drive two 16d nails at each end, then drive one 16d nail at each stud location.

After you have plumbed and braced all corner assemblies, align the tops of the walls. To align the walls, proceed as follows (refer to *Figure 22*):

Step 1 Start at the top plate at one corner of the wall. Fasten a string at that corner. Stretch the string to the top plate at the corner at the opposite end of the wall, and fasten the string.

Step 2 Cut three small blocks from 1 × 2 lumber. Place one block under each end of the string so that the line is clear of the wall. Use the third block as a gauge to check the wall at 6' or 8' intervals.

Step 3 At each checkpoint, nail one end of a temporary brace near the top of a wall stud. Also attach a short 2 × 4 block to the subfloor. Adjust the wall (by moving the top of the wall in or out) so the string is barely touching the gauge block. When the wall is in the right position, nail the lower end of the brace to the floor block.

NOTE

Do not remove the temporary braces until you have completed the framing—particularly the floor or roof diaphragm that sits on top of the walls—and sheathing for the entire building.

Determining Stud Length

1. If the finish ceiling height of a building with built-up wood floors is supposed to be 9'-6", how long are the studs in exterior walls (assume ½" underlayment and ⅝" gypsum board ceiling)?
2. If the top of the rough sill of a window is supposed to be 32" above the floor, how long are the cripple studs below the sill?

Coping with Natural Defects

Because wood is a natural material, it will have variations that may be considered defects. Lumber is almost never perfectly straight. Even when a piece of lumber is sawn straight, it will most likely curve, twist, or split as it dries.

In the past, wood cut at a sawmill was simply stacked and allowed to air dry. Normal changes in daily air temperature, humidity, and other conditions would result in lumber with very interesting shapes. In recent years, most construction lumber is dried in a kiln (a large, low-temperature oven), which reduces the amount of twisting and curving. But even modern lumber is still somewhat distorted.

27111-13_SA02.EPS

Virtually all lumber is slightly curved along its narrow side (the 1½" dimension in 2 × 4s). This is called a crown. It's normal, and, unless it is extreme, the crown doesn't prevent the lumber from being used.

When assembling a floor, sight down each piece of lumber to determine which side has the crown. Mark the crowned side. Some lumber suppliers stamp the wood to indicate the edge with the crown. When using the lumber in a floor, position all the lumber crown side up. Weight on the floor will cause the crown to flatten out.

In a wall, there's no force that will flatten the crown. However, for a more uniform nailing surface for the sheathing, position the lumber so all crowns face the same direction.

A curve along the wide side of lumber (the 3½" dimension in 2 × 4s) is called a bow. If a noticeably bowed stud is used in a wall, sheathing could not be nailed to it in a straight line. Either the nail would miss the bowed stud, or you would have to take extra time to lay out a curved line to nail through. Since that is not good use of your time, discard the bowed stud and use a straight one instead.

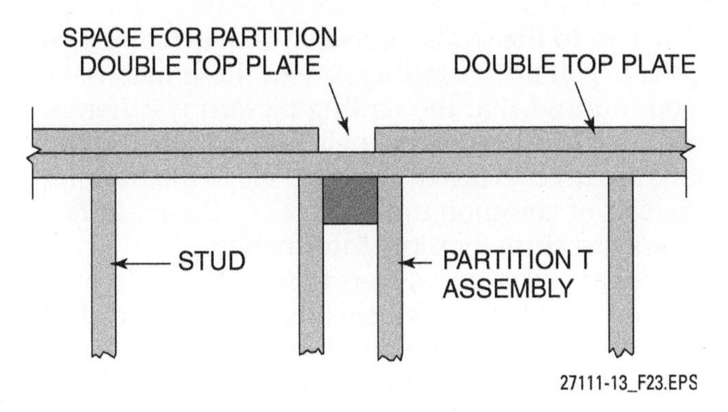

Figure 23 Double top plate layout.

3.3.2 Bracing Walls

Permanent bracing is important in the construction of exterior walls. Many local building codes require bracing when certain types of sheathing are used. In some areas where high winds are a factor, lateral bracing is a requirement even when ½" plywood is used as the sheathing.

Several methods of bracing can be used. One method is to let in a 1 × 4 or 1 × 6 at a 45-degree angle at each corner of the exterior walls. Another method is to cut 2 × 4 braces at a 45-degree angle for each corner. Still another type of bracing used (where permitted by the local code) is galvanized metal wall bracing (*Figure 24*).

Metal wall bracing is easier to use than let-in wood bracing. Instead of notching out the studs for a 1 × 4 or 2 × 4, use a circular saw to make a diagonal groove in the studs, top plate, and soleplate for the rib of the bracing and nail the bracing to the framing members.

In some areas of the country, plywood is used as corner bracing instead of wood or metal diagonal bracing. However, along with plywood came different types of sheathing that are by-products of the wood industry and do not have sufficient strength to withstand wind loads. When these materials are used, permanent bracing is also required. Building codes in some areas permit ½" plywood panels to be used on each corner of the structure in lieu of diagonal bracing, when the balance of the sheathing is fiberboard. In other areas of the country, the codes require the use of bracing except where ½" plywood is used throughout for sheathing. Still, in other areas, the use of bracing is required regardless of the type of sheathing used. Always refer to the building code in effect in your jurisdiction to determine bracing requirements.

Figure 24 Use of metal bracing.

Fire-Stopping Caulk

Fire-stopping caulk is commonly used to seal holes in walls or structural members. Two types of caulk are available: intumescent and endothermic. Intumescent caulk rapidly expands when exposed to heat and closes off voids left by burning or melting construction materials. Endothermic caulk releases water vapor when exposed to heat and slows the spread of fire in the building. Fire-stopping caulk must be properly applied according to the manufacturers' instructions.

Raising a Wall

When a wall is being raised, either by hand or with a lifting jack, the bottom of the wall can slide forward. If the wall slides off the floor platform, the wall or objects on the ground below it can be damaged and workers can be injured. Use wood blocks or cleats securely nailed to the outside of the header joists or end joists to prevent the wall from sliding off the subfloor.

Some carpenters use metal banding (used to bundle loads of wood from the mill) to achieve the same effect. One end of a short length of banding is nailed to the floor platform. The other end is bent up 90 degrees and nailed to the bottom of the soleplate. When the wall is raised, the flexible band straightens out horizontally, much like a hinge, but the wall cannot slide forward.

3.3.3 Sheathing Walls

Sheathing is used to close in the walls. APA-rated panels, such as plywood, oriented strand board (OSB), and other reconstituted wood products, are generally used for sheathing.

Some carpenters prefer to apply the sheathing to a squared wall while the wall frame is still lying on the subfloor. Although this helps to ensure that the wall is square, it has two drawbacks:

- It may make the wall too heavy for the framing crew to lift.
- If the floor is not perfectly straight and level, it will be a lot more difficult to square and plumb the walls once they are erected.

When plywood is used, the panels will range from 5⁄16" to 3⁄4" thick. A minimum thickness of 3⁄8" is recommended when siding is to be applied. Thicker (3⁄4") panels are recommended when the sheathing is used as the exterior finish surface. The panels may be placed with the grain running horizontally or vertically. If the panels are placed horizontally, local building codes may require that blocking be used between the studs along the horizontal edges.

Typical nailing schedules require 6d nails for panels that are 1⁄2" thick or less and 8d nails for thicker panels. Nails are spaced 6" apart at the panel edges and 12" apart at intermediate studs.

Other materials that may be used as sheathing include fiberboard (insulation board), exterior-rated gypsum board, and rigid foam sheathing. A major disadvantage of these materials is that siding cannot be nailed directly to them. It must either be nailed to the studs or special fasteners must be used. If you are installing any of these materials, keep in mind that the nailing pattern is different from that of APA-rated panels. In addition, roofing nails are used to fasten these panels to the framing (instead of common nails). Refer to the manufacturer's literature for more information.

In some situations, two types of panels may be used as exterior wall sheathing. Fiberboard or rigid foam sheathing is used for the general areas and APA-rated sheathing is installed vertically at the corners to eliminate the need for corner bracing. Always refer to the drawings and local building codes for details.

3.3.4 Panelized Walls

Instead of building walls on the job site, the walls can be prefabricated in a shop and transported to the job site where they are raised into position. The walls, or panels, are either set with a small crane or by hand. The wall sections or panels will vary in length from 4' to 16'.

When working with a pre-engineered structure, the **drying-in** time is much quicker than with a field-built structure. A 1,200-square-foot residence, for example, can be dried-in within two working days. The siding would be applied at the factory and the walls erected the first day. The soffit and fascia would be installed on the morning of the second day, and the roof dried-in and ready to shingle by the morning of the third day. The residence would be ready for rough-in plumbing, electrical, heating, and cooling by the third day.

3.0.0 Section Review

1. When nailing framing members together for a 2 × 6 wall, use _____.
   a. two 8d nails
   b. three 8d nails
   c. three 12d nails
   d. three 16d nails

2. Which of the following is *not* true about fire stops?
   a. They slow the flow of air through a wall cavity.
   b. They are sized to match the width of the framing lumber.
   c. They are installed directly below the top plate.
   d. They temporarily block the spread of flames.

3. What is the spacing for nails when fastening plywood panels to the wall frame?
   a. 12" apart at panel edges and at intermediate studs
   b. 12" apart at panel edges and 6" apart at intermediate studs
   c. 6" apart at panel edges and 12" apart at intermediate studs
   d. 6" apart at panel edges and 6" apart at intermediate studs

SECTION FOUR

4.0.0 WALL FRAMING IN MASONRY CONSTRUCTION

Objective

Describe wall framing techniques used in masonry construction.

Trade Terms

Furring strip: Narrow wood strip nailed to a wall or ceiling as a nailing base for finish material.

Masonry walls are commonly covered with interior wall finish materials. Carpenters have little control over the quality of the masonry structure. However, it is important that the walls are put up square and plumb. If a structure starts off level and plumb, little difficulty will be encountered with measuring, cutting, and fitting. If the walls or floor are not plumb and level, problems can be expected throughout the structure.

Masonry walls must be furred-in before installing the interior finish (*Figure 25*). As a general rule, masonry walls should be furred-in by installing 1 × 2 furring strips on 16" centers. (Some contractors will apply 1 × 2 furring strips at 24" OC; while this may save material, it does not provide the same quality as walls constructed with 1 × 2s on 16" centers.) Metal furring channels may also be used to fur-in masonry walls.

Wood furring strips are fastened to masonry walls using masonry nails measuring 1¼" to 1¾" long. Remember, all wood that comes in contact with concrete or masonry must be pressure treated.

Backing for partitions that butt into a masonry wall is accomplished using one of the following methods. In the first method, determine the location of the partition and then fasten a 1 × 2 to the masonry wall, centered on the partition (i.e., locate the center of the partition at the bottom corner of the wall, move back 2¾", and mark the wall by plumbing with a straightedge and level). Fasten the 1 × 2 to the wall with one edge on the mark. This will allow an even space on either side of the partition to receive the gypsum board, as shown in *Figure 25A*.

In the second method, 16"-long, 2 × 4 blocks are fastened to the bottom, center, and top of the masonry wall where the partition is located (*Figure 26*), and a furring strip is installed on each side of the block. The partition stud is then fastened to the blocks. This arrangement provides an adequate nailing surface for the gypsum board. (The proper nailing surface for gypsum board is ¾" to 1". If the blocks were not installed against the masonry wall, only ¼" of nailing space would be provided.)

Corners should receive special consideration when installing the furring strips and gypsum board. Ensure that the furring strips toward the corners are positioned to allow the gypsum board to slip by the furring strips. The furring strips should be installed approximately ⅝" from the corner in both directions (*Figure 27*).

A 1 × 2 is used at floor level for attachment of the baseboard. Either a narrow or wide baseboard can be used. Some carpenters will install a simple furring strip at floor level and depend on the vertical strips for baseboard nailing. Once the gypsum board has been installed, it is difficult to find the strips when nailing the baseboard.

A sequence of installation should be established for fastening the furring strips to the wall. Either start from the right and work to the left or work from left to right. This sequence will allow the trim carpenter to locate the furring strips. When applying the furring strips to the wall, remember that the strips must be placed 16" OC for the gypsum board to be nailed properly. Similar to laying out wall studs, lay out the first furring strip at 15¼" and then lay out the second strip 16" from the first mark.

Metal door frames and metal window frames or channels are used almost exclusively with masonry walls. However, in rare instances, wood framing is used. Each installation is unique and a complete examination of this subject is beyond the scope of this module. There are two general rules, however:

- Use treated wood when it is in contact with masonry.
- Use cut nails, expansion bolts, anchor bolts, or other fasteners to attach the wood to the masonry securely.

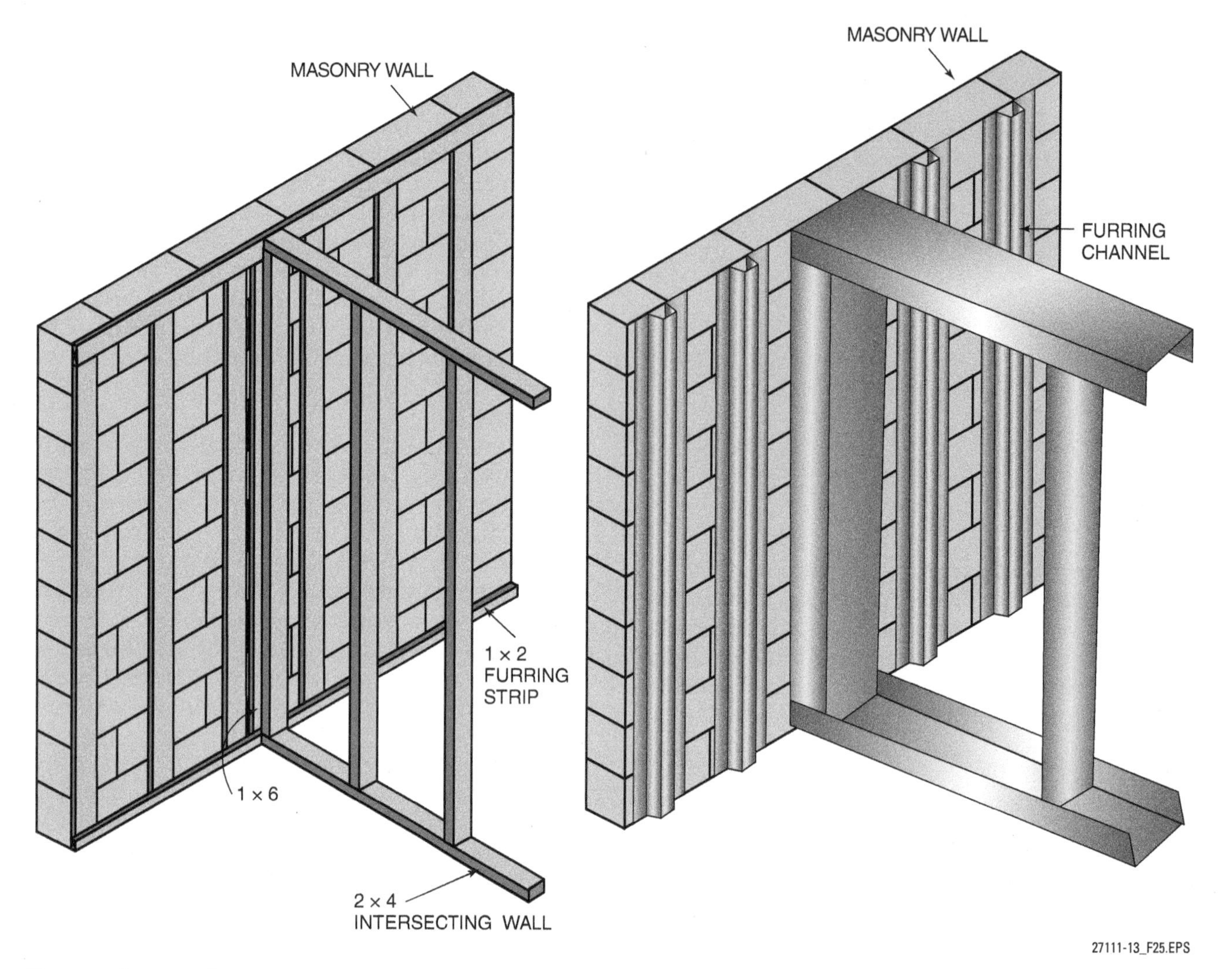

Figure 25 Partition T backing using a metal or wood furring channel.

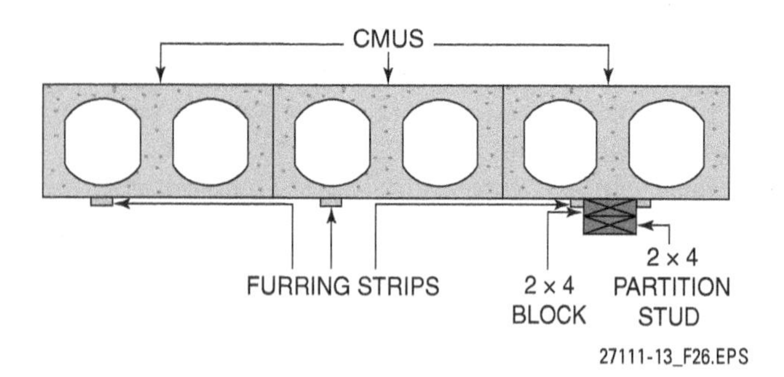

Figure 26 Partition backing using 2 × 4 blocks.

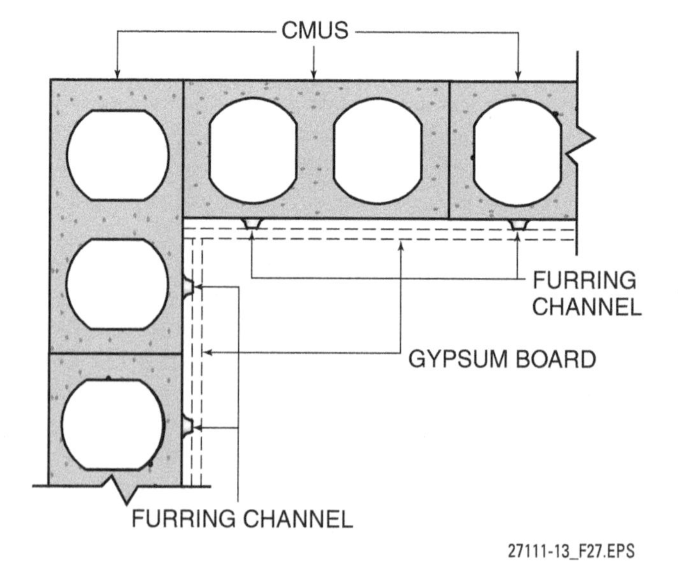

Figure 27 Placement of furring strips at corners of masonry walls.

Masonry Curtain Wall

This masonry curtain wall panel combines 2"-thick architectural precast concrete brick with a heavy-gauge stainless steel frame and insulated stainless steel anchor doors.

27111-13_SA03.EPS

4.0.0 Section Review

1. Wood furring strips are attached to a masonry wall using _____.
   a. lag bolts
   b. sheet metal screws
   c. masonry nails
   d. 16d finish nails

SECTION FIVE

5.0.0 ESTIMATE WALL FRAMING MATERIALS

Objective

Describe the correct procedure to estimate the materials required to frame walls.

a. Explain how to estimate the amount of lumber required for soleplates and top plates.
b. Describe how to estimate the number of studs required.
c. Explain how to calculate the amount of material needed for a header.
d. Describe how to estimate the amount of diagonal bracing required.

Performance Task 4

Estimate the materials required to frame walls.

In this section, the basic steps needed to estimate the amount of lumber to frame the walls of a building will be discussed. Here's some important information that you need for this example of an estimate:

- The structure measures 24' × 30' (*Figure 28*).
- The walls are 8' tall.
- The building is framed with 2 × 4s at 16" OC.
- The building has two 24"-wide windows on each narrow side, and two 48"-wide windows and one 36"-wide door on each wide side.
- The interior of the building is divided into three rooms: one is 12' × 30', two are each 11'-1½" × 12', separated by a 4'-10"-wide hallway.
- The long wall that separates the large room from the two smaller rooms and hallway is a loadbearing partition.
- Each of the smaller rooms has a 36" door leading to the hallway.
- A 36"-wide opening, without a door, leads from the large room to the hallway.

5.1.0 Estimating Soleplates and Top Plates

Assume that the plate stock is ordered in 12' lengths.

Step 1 Determine the length of the walls in feet. For loadbearing walls, multiply the length by 3 to account for the soleplates and double top plate (remember that a double top plate is made of two 2 × 4s stacked together).

Step 2 Divide the result by 12 to determine the number of 12' pieces of stock. Round up to the next full number, and allow for waste.

Example:

For the exterior walls, add the lengths of the two long walls and the two narrow walls:

30' + 30' + 24' + 24' = 108'

Also add the length of the interior partitions:

30' + 12' + 12' = 54'

Add those numbers together:

108' + 54' = 162'

Multiply by 3 to account for the soleplate and double top plate:

162' × 3' = 486'

Divide by 12 to determine the number of 12' pieces needed:

486' ÷ 12' = 40 ½ pieces

Since you can buy only full pieces, you will need to round up to the next whole number, which is 41. So, for all of the soleplates, top plates, and double top plates in this building, you need 41 pieces of 12' 2 × 4s.

NOTE

The example estimate here for soleplate and top plate material is a very simple one and works for this small, simple structure. Actually, only the center loadbearing partition (and the exterior walls, of course) will require double top plates. However, the extra amount of lumber from this simpler estimate is small, and you need to calculate more lumber for waste anyway. In a larger, more complex structure, calculate the different quantities of lumber for loadbearing and nonbearing partitions to get a more precise estimate.

5.2.0 Estimating Studs

Step 1 Determine the length in feet of all the walls.

Step 2 The general industry standard is to allow one stud for each foot of wall length, even when you are framing 16" on center. This

should cover any additional studs that are needed for openings, corners, partition Ts, and blocking.

When you were estimating for soleplate and top plate material, you found that there were 162' of exterior walls and interior partitions. You will need one stud for every foot of wall, or 162 studs.

5.3.0 Estimating Headers

Assume that 2 × 12 headers with plywood spacers are used. This might cost a little more than using narrower headers with cripple studs, but you won't need to lay out, cut, and assemble cripple studs above the headers.

Step 1 Use the architectural drawings (to be safe, check both the floor plans and the door and window schedules to make sure they agree) to identify the number and size of each window and door.

Step 2 The header rests on the trimmer studs in the rough opening. Therefore, 3" (1½" trimmer studs on each side) must be added to the rough-opening dimension to account for the trimmer studs.

Step 3 Double the length of each header.

Step 4 Combine the lengths obtained in Step 3 into convenient lengths for ordering.

Step 5 Order enough ½" plywood for spacers.

Example:

In *Figure 28*, there are four 4830 windows (each 48" wide):

48" + 3" = 51"; 51" × 4 = 204"

Two 2430 windows (each 24" wide):

24" + 3" = 27"; 27" × 2 = 54"

Four 3680 doors (each 36" wide):

36" + 3" = 39"; 39" × 4 = 156"

One 3'-0" opening with no door:

36" + 3" = 39"

Add those numbers together:

204" + 54" + 156" + 39" = 453"

Then double that total:

453" × 2 = 906"

Divide by 12 to get feet:

906" ÷ 12 = 75'-6"

Round up to 76'.

5.4.0 Estimating Diagonal Bracing

For the example, diagonal bracing will be installed at each end of all exterior walls. You can use diagonal metal bracing or 1 × 4 lumber. Braces extend from the top plate to the soleplate at a 45-degree angle. The walls for this building are 8' high, so 12'-long braces are required. (Review the *Core* module, *Introduction to Construction Math,* to determine how to find the length of the brace using the formula for the hypotenuse of a right triangle.)

Step 1 Determine the number of outside corners.

Step 2 Determine the length of each brace based on the wall height, then multiply by the number of corners.

Example:

This part is pretty easy. *Figure 28* shows that the building is a rectangle with four corners. Each corner has two sides, so there are eight places where diagonal braces are installed. Since 12' of bracing is required at each location, 8 × 12' = 96' of bracing altogether.

Remember that let-in or diagonal bracing is typically applied to the interior side of the exterior walls.

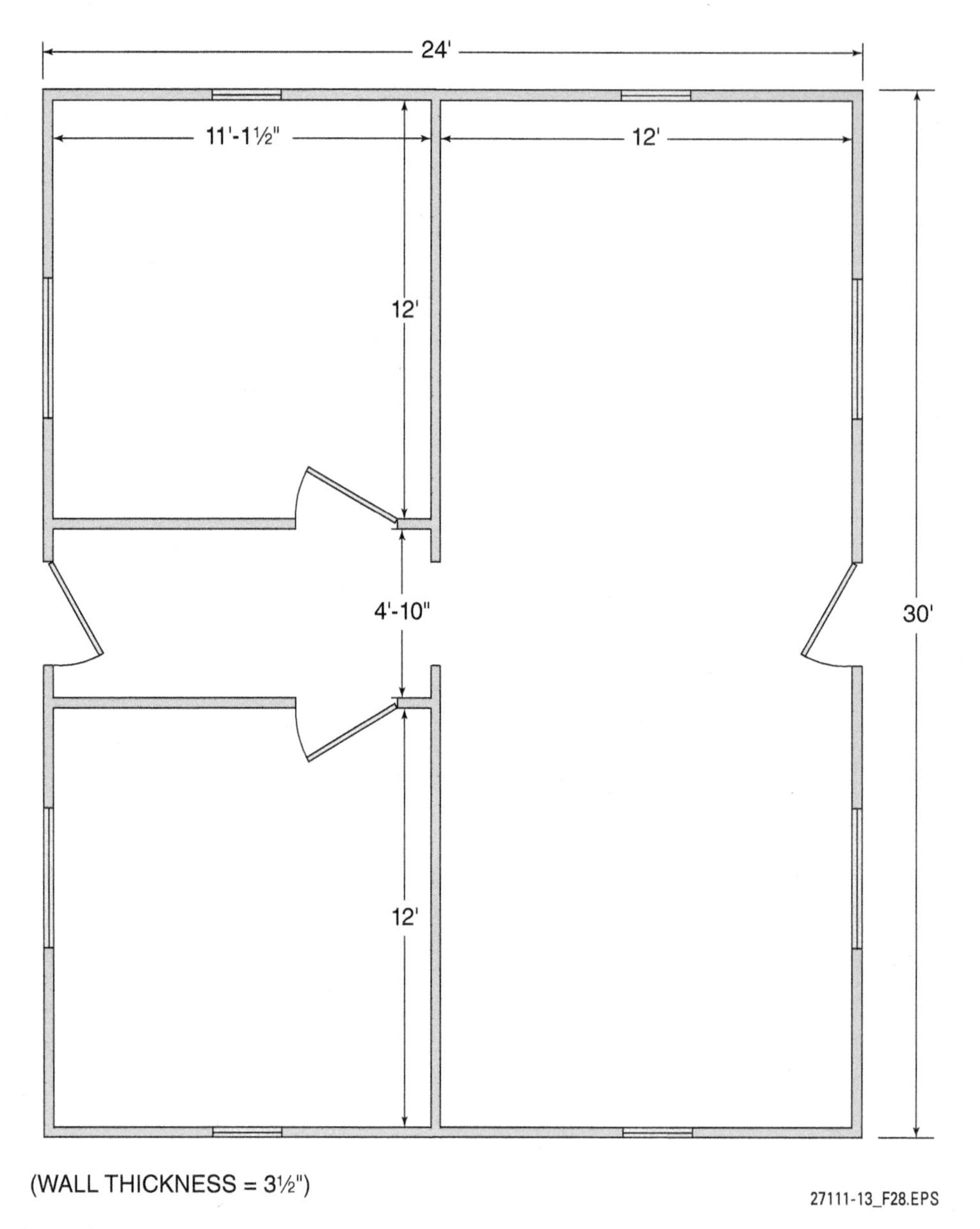

Figure 28 Sample floor plan.

How Accurate Is an Estimate?

How close does the estimate come to actual needs? Sometimes, it does not come close enough. In our example building, we estimated 162 studs based on one stud per lineal foot of wall length. However, if you calculate the number of studs based on the actual framing spaced 16" on center, it takes:

- 17 studs to span each of the two 24' walls
- 21 studs to span each of the two 30' walls
- 21 studs for the 28'-10" interior partition (excludes the width of the two interior and two exterior stud walls, which is 1'-2" total)
- 9 studs to span each 11'-1½" interior partition

So, that subtotal is:

$$17 + 17 + 21 + 21 + 21 + 9 + 9 = 115$$

But you need to add an extra full stud and trimmer on each side of each opening. There are 13 openings, so ...

$$13 \times 4 = 52$$

...add that to the previous subtotal ...

$$115 + 52 = 167$$

However, about 15 full studs will be eliminated where the openings will be, so subtract that from the subtotal ...

$$167 - 15 = 152$$

Then add two extra studs for each of the six partition Ts ...

$$152 + 12 = 164$$

...and two extra studs for each exterior corner ...

$$164 + 8 = 172$$

Of course, you'll need studs for cripples and for extra blocking here and there. Then, some of the delivered lumber will be unusable because it is twisted or warped. Plus, measuring and cutting mistakes will be made. So, you're well over the 162 pieces calculated by allowing one stud every 12". Be smart and add a safety factor. It might vary depending on the size and complexity of the job, but for our example here, 15 percent (for a total of about 186 studs) would have been about right. On bigger jobs, the safety factor might be smaller or even nil. That's because the total amount of lumber is greater, and the safety margin already built into the one-stud-per-foot will cover all needs.

5.0.0 Section Review

1. When estimating the amount of lumber needed for plates, to account for the soleplates and double top plate, multiply the wall length by _____.
   a. 2
   b. 3
   c. 4
   d. 6

2. Using a general rule of thumb, a 15' × 30' rectangular building will require _____.
   a. 45 studs
   b. 60 studs
   c. 90 studs
   d. 120 studs

3. For one header constructed with two 2 × 12s and spanning a 4'-0" rough opening, what length of 2 × 12 should you order?
   a. 4'
   b. 6'
   c. 8'
   d. 10'

4. Braces extend at a 45-degree angle from the soleplate to the _____.
   a. fire stop
   b. rough sill
   c. top plate
   d. trimmer stud

SECTION SIX

6.0.0 ALTERNATIVE WALL SYSTEMS

Objective

Identify alternative wall systems.

a. Describe how concrete walls are constructed.
b. Explain the difference between standard interior wall systems and alternative interior wall systems.

Trade Terms

Buck: Well-braced frame placed inside a concrete form to provide an opening for a door or window.

Monolithic: Concrete that is cast as a single unit; for example, a concrete floor and joists.

Sound transmission class (STC): Rating scale for the transmission of sound through a wall or ceiling of a building.

Tilt-up concrete construction: Concrete construction method in which the wall sections of a building are cast horizontally on the job site or precast off-site and are lifted into position using a crane.

The structural framework of large buildings such as office buildings, hospitals, apartment complexes, and hotels is usually made from concrete and/or structural steel. The exterior finish is often concrete panels that are either prefabricated and raised into place or concrete that is placed in forms at the site. Exterior walls (curtain walls) may also be made of glass in a metal or concrete framework. *Figure 29* shows a curtain wall section being installed. Before the concrete is placed, provisions must be made for electrical, plumbing, and data-cabling pathways.

In some buildings, the framework is made of structural steel and the floors of cast-in-place concrete. Wall panels that are fabricated off-site are lifted into place and bolted or welded to the steel (*Figure 30*).

Figure 31 shows the structure of a building in which all the structural framework is made of concrete and steel. Each component of the structure requires a different type of form. In this case, the floor and beams were made as a monolithic unit using integrated floor and beam forms, which were removed once the concrete hardened.

27111-13_F29.EPS

Figure 29 Installing a corner curtain wall panel.

27111-13_F30.EPS

Figure 30 Curtain wall under construction.

Tilt-up concrete construction is used for commercial applications. In tilt-up construction, the concrete for the wall panels is cast on the concrete floor slab or casting bed. The panels are then tilted into place on the footing using a crane (*Figure 32*) and temporarily braced. The panels are welded together to provide support.

Tilt-up concrete construction is most common in one- or two-story buildings with a slab at grade, and is popular for warehouses, low-rise offices, churches, and a variety of other commercial

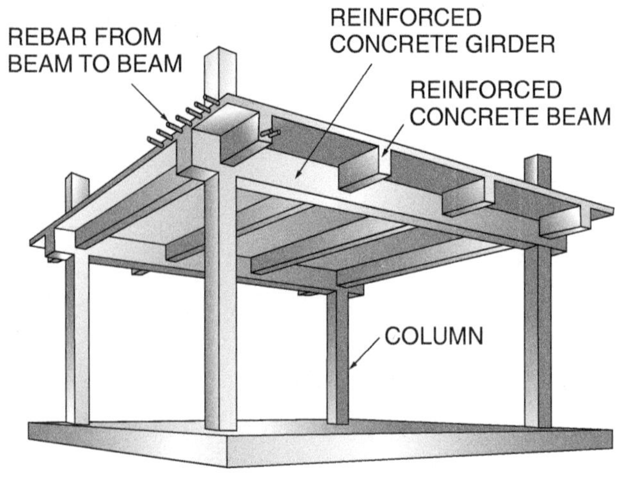

Figure 31 Concrete structure.

and multifamily residential applications. Tilt-up panels of 50' in height are not uncommon. They typically range from 5" to 8" thick, but thicker walls can be obtained when using lighter-weight concrete.

Tilt-up concrete construction involves lifting extremely heavy panel loads—often the heaviest lift on a job. Because of the large weight involved, always use caution. All lifting operations should be performed by qualified personnel.

6.1.0 Concrete Wall Openings

For concrete walls, openings for windows and doors are made by installing wood or metal bucks between the forms, as shown in *Figure 33*. The bucks are then removed after the concrete has set and the forms are stripped. Pipe and cable chases are formed using fiber, plastic, or metal tubes inserted into the form. The tubes typically remain in place.

Figure 32 Tilt-up panel being lifted into place.

6.2.0 Interior Walls and Partitions

Walls and partitions in commercial applications must meet the fire and soundproofing requirements specified in the local building code. In some cases, a frame wall with ½" gypsum board on either side is satisfactory. In extreme cases, such as the separation between offices and a manufacturing space in a factory, it may be necessary to combine a concrete masonry unit (CMU) wall with fire-resistant gypsum board, along with rigid and/or fiberglass insulation, as shown in *Figure 34*. This is especially true if there is an explosion or fire hazard.

While steel studs may be used in residential construction, they are the standard for framing walls and partitions for commercial construction. Once the studs are installed, one or more layers of gypsum board and insulation are applied. The type and thickness of the gypsum board and insulation depend on the fire rating and soundproofing requirements. Soundproofing needs vary from one use to another and are often based on the amount of privacy required for the intended use. For example, executive and physician offices may require more privacy than general offices.

Sound reduction and fire resistance requirements can significantly affect the wall thickness. For example, a wall with a low sound transmission class (STC) and fire resistance rating might use only 3½" thick (2½" steel studs with ½" gypsum board on each side), while walls with a high STC and fire resistance rating might have a total thickness of over 12" (see *Figure 35*).

As discussed previously, the fire rating specified by the applicable building code determines the types and amount of material used in a wall or partition. As shown in *Figure 35*, a one-hour-rated wall might be constructed using single sheets of ⅝" gypsum board on wood or 25-gauge steel studs. A two-hour-rated wall requires heavy-gauge steel studs and two layers of fire-resistant gypsum board.

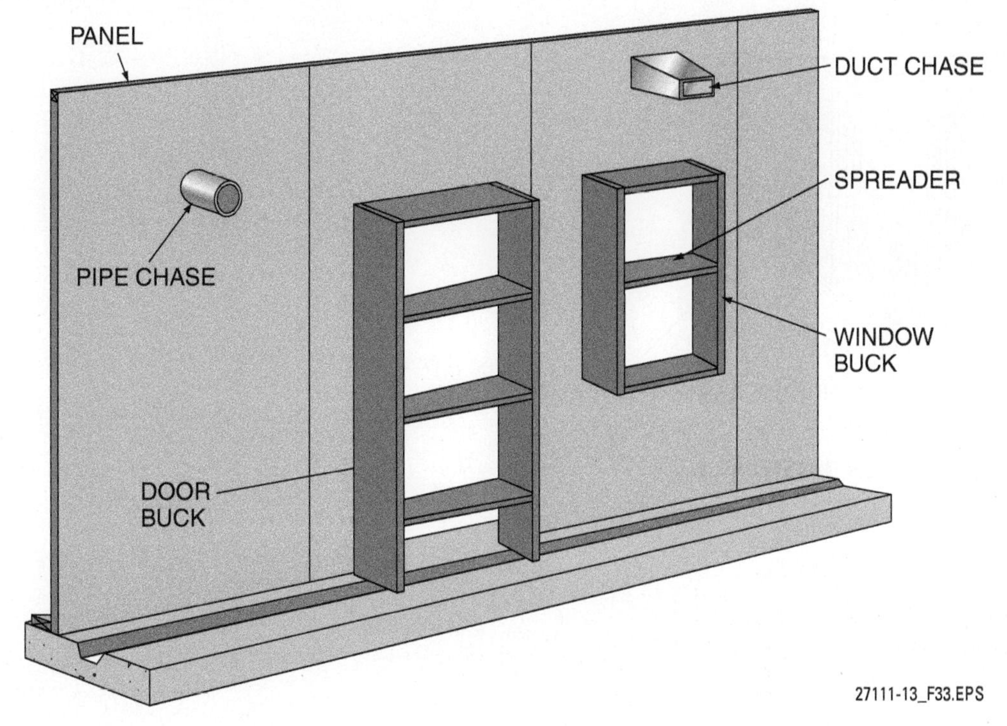

Figure 33 Framing openings in concrete walls.

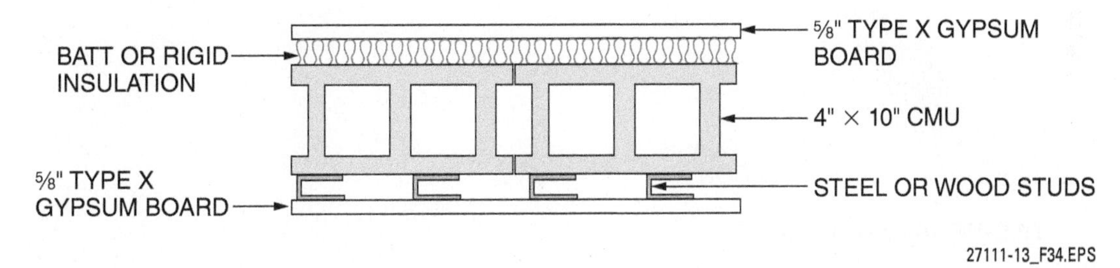

Figure 34 High fire-/noise-resistant partition.

Tilt-Up Records

The Tilt-Up Concrete Association (**www.tilt-up.org**) maintains records of tilt-up projects. As of 2010, several amazing statistics had been recorded, including the following:

- Heaviest tilt-up panel—339,000 pounds (153,768 kilograms)
- Largest tilt-up panel by area—2,950 square feet (274.1 square meters)
- Tallest tilt-up panel—96'-9" feet (29.49 meters)
- Widest tilt-up panel—81'-7½" (24.88 meters)
- Largest number of panels in a single building—1,310
- Largest tilt-up building (floor area)—3,420,000 square feet (317,718 square meters)

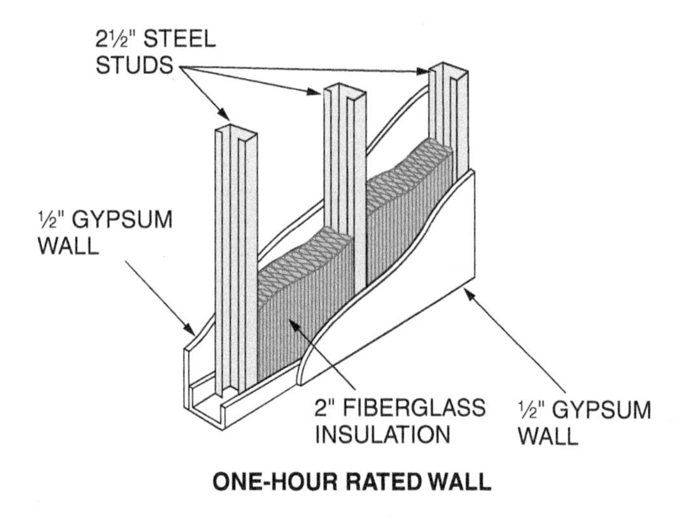

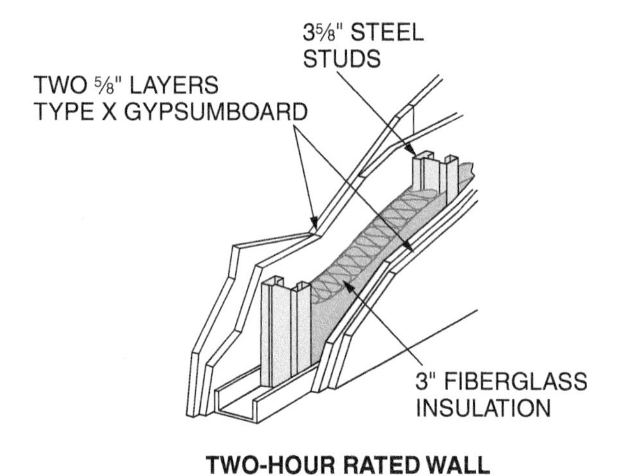

Figure 35 Partition wall examples.

6.0.0 Section Review

1. Exterior walls that are made of glass in a metal or concrete framework are referred to as _____.
   a. soffits
   b. curtain walls
   c. tilt-up walls
   d. monolithic construction

2. Which of the following will *not* increase the fire and soundproofing requirements of a wall?
   a. Use larger studs
   b. Install insulation in the wall cavity
   c. Increase the wall height
   d. Install two layers of gypsum board

SUMMARY

This module covered how to identify and use all the wall framing components; how to lay out interior and exterior walls; how to measure and cut lumber for framing members; how to assemble, erect, plumb, brace, and sheath walls; how to lay out and build doors, windows, and other wall openings; how to estimate wall framing materials; how to work with masonry walls; how to construct walls with steel studs; and much more.

The most important skill in wall and ceiling framing is accuracy. Walls must be straight, plumb, and square. Precise layout and measuring of studs and headers is also critical. Even a very small error can cause big problems. For example, over a span of 20', an error of only 1/16" at one end of the span could become 1¼" at the other end. The closer you get to the end of the wall, the more patching and fitting you will have to do to get the sheathing, windows, and exterior doors to fit.

In other words, saving a little time at the beginning can cost you a lot of time at the end. Remember this as you move forward in your training.

Review Questions

1. A short framing member that fills the space between the rough sill and the soleplate is a _____.
   a. spacer
   b. cripple stud
   c. trimmer stud
   d. top plate

2. The framing member that forms the side of a rough opening for a window or door is the _____.
   a. trimmer stud
   b. header
   c. soleplate
   d. cripple

3. A double top plate can be used to connect partitions that are perpendicular to a wall.
   a. True
   b. False

4. The stud is the main horizontal framing member in a wall.
   a. True
   b. False

5. The straightest, least defective studs are normally used for _____.
   a. window and door trimmers
   b. common studs
   c. corners
   d. partition Ts

6. The framing member that distributes the weight of the structure around a door or window opening is the _____.
   a. trimmer stud
   b. cripple stud
   c. top plate
   d. header

7. The width of a header is equal to the _____.
   a. width of the rough opening
   b. width of the opening plus the width of two trimmer studs plus the width of two common studs
   c. width of the opening plus the width of two trimmer studs
   d. width of the trimmer plus the width of two cripple studs

8. Fabricated headers normally use what material as spacers?
   a. Furring strips
   b. ½" plywood
   c. ¾" plywood
   d. 2 × 4 blocks

Maximum Span for Exterior Built-up Headers

Built-Up Header Size	Single-Story Load	Two-Story Load	Three-Story Load
2 × 4	3'-6"	2'-6"	2'
2 × 6	6'	5'	4'
2 × 8	8'	7'	6'
2 × 10	10'	8'	7'
2 × 12	12'	9'	8'

9. Refer to the *Review Question Table 1* to determine the maximum span of a 2 × 8 exterior header in a one-story building.
   a. 16'
   b. 4'
   c. 8'
   d. 20'

10. A header used when a load is heavy, or the span is very wide, is the _____.
    a. drywall header
    b. king header
    c. Blackmore header
    d. truss header

11. Steel studs are laid out to the _____.
    a. edge
    b. center line
    c. double top plate
    d. subfloor

12. The purpose of prepunched holes in steel studs is to _____.
    a. make them lighter
    b. save material
    c. provide runs for piping and wiring
    d. make it easier to attach drywall

13. Steel studs are attached to the bottom channel using _____.
    a. wood screws
    b. powder-actuated fasteners
    c. construction adhesive
    d. self-tapping sheet metal screws

14. If a 24" stud spacing is used for a two-story building, the first-floor framing must be constructed with _____.
    a. steel studs
    b. 2 × 6 lumber
    c. doubled 2 × 4s
    d. 2 × 8 lumber"

15. The common stud used for each side of a window opening is also called a _____.
    a. king stud
    b. cripple stud
    c. master stud
    d. trimmer stud

16. Door and window schedules can be found in the _____.
    a. legend
    b. architectural drawings
    c. details of the structural drawings
    d. manufacturer's directions

17. The length of a cripple stud above a window opening equals the length of a common stud less the height of the trimmer, combined with the thickness of the _____.
    a. soleplate
    b. rough sill
    c. top plate
    d. header

18. When calculating the length of a common stud for a wood frame floor, do *not* consider the thickness of the _____.
    a. soleplate
    b. subfloor
    c. underlayment
    d. double top plate

19. The preferred method of framing a wall is to _____.
    a. nail down the soleplate and built it in place
    b. assemble it on the ground
    c. lay out and assemble it on the subfloor
    d. have it built off-site

20. Pieces of 2 × 4 placed horizontally between each pair of studs are used to _____.
    a. brace the frame
    b. retard the spread of fire
    c. provide nailers for siding
    d. provide a place for carpenters to stand when installing ceiling joists

21. To prevent a wall section from sliding as it is being raised, you should install _____.
    a. cleats
    b. hold-down clamps
    c. retainers
    d. anchor bolts

22. A long 2 × 4 with a ¾" standoff block nailed to each end is used as a _____.
    a. wall brace
    b. straightedge
    c. gauge block
    d. trimmer stud

23. After all corner assemblies have been plumbed, the tops of the walls must be aligned using _____.
    a. a come-along
    b. adjustable jacks
    c. a builder's level
    d. temporary braces

24. When the sheathing acts as the finish surface, you should use material that is _____.
    a. ⅜" thick
    b. ½" thick
    c. ¾" thick
    d. 1" thick

25. When you are installing APA-rated sheathing material, space the nails at the panel edges _____.
    a. 6" apart
    b. 12" apart
    c. 16" apart
    d. 24" apart

26. Fasteners called masonry nails are used to_____.
    a. fasten masonry units together
    b. assemble concrete forms
    c. asten ceiling joists to CMUs
    d. attach furring strips to masonry surfaces

27. Standard industry practice when estimating the number of studs needed is to allow _____.
    a. 0.75 studs per foot of wall length
    b. one stud per foot of wall length
    c. 1.25 studs per foot of wall length
    d. one stud per 16" of wall length

28. Which of the following is *not* a typical use for tilt-up concrete construction?
    a. single-family residences
    b. warehouses
    c. churches
    d. multifamily residential buildings

29. Pressure-treated lumber must be used _____.
    a. below grade only
    b. when in contact with masonry
    c. for roof framing
    d. in low-humidity climates

30. The lowest horizontal member of a wall or partition, to which studs are nailed, is called the _____.
    a. floor flange
    b. rough sill
    c. soleplate
    d. footer

Trade Terms Quiz

Fill in the blank with the correct term that you learned from your study of this module.

1. The part of a door frame that comes in contact with the door is the _______.
2. _______ form window or door openings in concrete walls.
3. _______ is used to provide filler and support between framing members.
4. A(n) _______ is used to stiffen a wall and to connect splices, corners, and partitions that are at right angles to the wall.
5. A(n) _______ is a short framing member that fills the space between longer framing members.
6. A(n) _______ is a framing member that forms the sides of the rough opening for a door.
7. The top support member of a rough door or window opening is known as the _______.
8. The lower framing member of a rough door or window opening is known as the _______.
9. _______ provide(s) a nailing surface when a wall is attached to masonry.
10. The installation of sheathing, doors, and windows makes up the _______ stage of building construction.
11. Studs are nailed to the _______ at the upper end.
12. Sound transmission through a wall or ceiling surface is rated based on the _______.
13. When concrete for a floor and beam system is cast as a single unit, it is referred to as _______.
14. In _______, concrete wall panels are lifted into position using a crane.

Trade Terms

Blocking
Buck
Cripple stud
Double top plate
Drying-in
Furring strip
Header
Jamb
Monolithic
Rough sill
Sound transmission class (STC)
Tilt-up concrete construction
Top plate
Trimmer stud

Mark Robinson

Trades Instructor
Crossland Construction Company, Inc.
Columbus, Kansas

How did you get started in the construction industry?
I was always a shop-class type of person with wood and metal, but really got started in the industry right out of high school. I was working for a golf course that built a pro shop and storage building, and really fell in love with building things. At the golf course I learned how to operate heavy equipment through the construction of a new 18-hole course. I went into factory work for the next few years, but I stayed with construction trades on the side, learning and developing my skills until I got back into the industry full time.

Who or what inspired you to enter the industry?
An older craftsman that was in charge of building the pro shop took me under his wing as his right-hand guy. He tried to show me the right way of building from the ground up. Having him as a mentor was a major inspiration and a valuable resource to draw from. He not only taught me specifics on the trades, but also leadership methods and interpersonal skills.

What do you enjoy most about your career?
I enjoy seeing employees that are just starting in the industry, learning through the mentoring program that Crossland has in place, and watching them grow and develop a passion for working with their hands, and moving up in the company to run their own jobs.

Why do you think training and education are important in construction?
There is a big difference in the speed and cost of building now, compared to the past. You must have continual education to be able to compete in the industry. Every day, there are new construction technologies, methods, and materials to learn about.

Why do you think credentials are important in construction?
The benefit of credentials from an employee's perspective is that they define the employee's knowledge and skill level. It also shows their level of commitment and determination to better themselves. At Crossland, I've seen firsthand the dividends paid by our investment in our employees' education. Training and credentialing lead to a smarter workforce, which leads to better production, which leads to happier owners. It's a win-win for all involved.

How has training/construction impacted your life and career?
I started at Crossland as a laborer, and through hard work and education opportunities provided by the company, I've advanced into my current position as trades instructor. Now I get to share what I've learned with trainees and assist in seeing them through the certification programs.

Would you recommend construction as a career to others? Why?
I would definitely recommend a career in construction to anyone. I can't think of anything more rewarding than working on a building from the ground up, seeing the finished product, and knowing that you helped build it with your own hands.

What does craftsmanship mean to you?
To me, craftsmanship is the ability to build something with your hands that most people could not. It takes a craftsman to envision a finished building from a set of plans and understand what it will take to get it to that finished point.

Trade Terms Introduced in This Module

Blocking: A wood block used as a filler piece and a support between framing members.

Bucks: Well-braced frame placed inside a concrete form to provide an opening for a door or window.

Cripple stud: In wall framing, a short framing stud that fills the space between a header and a top plate or between the sill and the soleplate.

Double top plate: A plate made of two members to provide better stiffening of a wall. It is also used for connecting splices, corners, and partitions that are at right angles (perpendicular) to the wall.

Drying-in: Applying sheathing, windows, and exterior doors to a framed building.

Furring strip: Narrow wood strip nailed to a wall or ceiling as a nailing base for finish material.

Header: A horizontal structural member that supports the load over a window, door, or other wall openings.

Jamb: The top (head jamb) and side members of a door or window frame that come into contact with the door or window.

Monolithic: Concrete that is cast as a single unit; i.e., a concrete floor and joists.

Rough sill: The lower framing member attached to the top of the lower cripple studs to form the base of a rough opening for a window.

Sound transmission class (STC): Rating scale for the transmission of sound through a wall or ceiling of a building.

Tilt-up concrete construction: Concrete construction method in which the wall sections of a building are cast horizontally on the job site or precast off-site and are lifted into position using a crane.

Top plate: The upper horizontal framing member of a wall used to carry the roof trusses or rafters.

Trimmer stud: A vertical framing member that forms the sides of rough openings for doors and windows. It provides stiffening for the frame and supports the weight of the header.

Additional Resources

This module presents thorough resources for task training. The following reference material is suggested for further study.

Builder's Essentials: Advanced Framing Methods. Kingston, MA: R.S. Means Company.

Builder's Essentials: Framing & Rough Carpentry. Kingston, MA: R.S. Means Company.

Framing Floors, Walls and Ceilings. Newtown, CT: Taunton Press.

Framing Walls (DVD). Newtown, CT: Taunton Press.

Graphic Guide to Frame Construction. Newtown, CT: Taunton Press.

Precision Framing for Pros by Pros. Newtown, CT: Taunton Press.

The Proper Construction and Inspection of Ceiling Joists and Rafters (DVD and workbook). Falls Church, VA: International Code Council.

International Code Council. A membership organization dedicated to building safety and fire prevention through development of building codes, **www.iccsafe.org**

National Association of Home Builders. A trade association whose mission is to enhance the climate for housing and the building industry, **www.nahb.org**

Figure Credits

Benson Industries LLC, Figure 29

SkillsUSA, Module opener

Southern Forest Products Association, Figure SA01

Weyerhaeuser Wood Products, Figures SA01 and SA02

Tilt-Up Concrete Association, Figure 32

Section Review Answer Key

Answer	Section Reference	Objective Reference
Section One		
1. b	1.0.0	1
2. c	1.1.0	1a
3. a	1.2.0	1b
4. b	1.3.0	1c
5. c	1.4.1	1d
Section Two		
1. b	2.1.0	2a
2. c	2.2.0	2b
Section Three		
1. d	3.1.0	3a
2. c	3.2.0	3b
3. c	3.3.3	3c
Section Four		
1. c	4.0.0	4
Section Five		
1. b	5.1.0	5a
2. c	5.2.0	5b
3. d	5.3.0	5c
4. c	5.4.0	5d
Section Six		
1. b	6.0.0	6a
2. c	6.2.0	6b

NCCER CURRICULA — USER UPDATE

NCCER makes every effort to keep its textbooks up-to-date and free of technical errors. We appreciate your help in this process. If you find an error, a typographical mistake, or an inaccuracy in NCCER's curricula, please fill out this form (or a photocopy), or complete the online form at **www.nccer.org/olf**. Be sure to include the exact module ID number, page number, a detailed description, and your recommended correction. Your input will be brought to the attention of the Authoring Team. Thank you for your assistance.

Instructors – If you have an idea for improving this textbook, or have found that additional materials were necessary to teach this module effectively, please let us know so that we may present your suggestions to the Authoring Team.

NCCER Product Development and Revision
13614 Progress Blvd., Alachua, FL 32615
Email: curriculum@nccer.org
Online: www.nccer.org/olf

❑ Trainee Guide ❑ Lesson Plans ❑ Exam ❑ PowerPoints Other ____________

Craft / Level: Copyright Date:

Module ID Number / Title:

Section Number(s):

Description:

Recommended Correction:

Your Name:

Address:

Email: Phone:

27112-13

Ceiling Joist and Roof Framing

Overview

Ceilings and roofs must be framed prior to installing ceiling finish materials and exterior sheathing, respectively. In many cases, trusses are now installed in lieu of ceiling joists and roof rafters being installed individually. However, there are occasions when carpenters need to stick-frame all or part of a roof. In such cases, it is essential to be able to perform the length and angle calculations needed to correctly size and cut each type of rafter.

Module Seven

Trainees with successful module completions may be eligible for credentialing through the NCCER Registry. To learn more, go to **www.nccer.org** or contact us at **1.888.622.3720**. Our website has information on the latest product releases and training, as well as online versions of our *Cornerstone* magazine and Pearson's product catalog.

Your feedback is welcome. You may email your comments to **curriculum@nccer.org**, send general comments and inquiries to **info@nccer.org**, or fill in the User Update form at the back of this module.

This information is general in nature and intended for training purposes only. Actual performance of activities described in this manual requires compliance with all applicable operating, service, maintenance, and safety procedures under the direction of qualified personnel. References in this manual to patented or proprietary devices do not constitute a recommendation of their use.

27112-13
Ceiling Joist and Roof Framing

Objectives

When you have completed this module, you will be able to do the following:

1. Identify the components of ceiling framing.
   a. Describe the correct procedure for laying out ceiling joists.
   b. Describe how to cut and install ceiling joists on a wood frame building.
   c. Describe how to estimate the number of ceiling joists required for a building.
2. Identify common types of roofs used in residential construction.
3. Identify the components and define the terms associated with roof framing.
   a. Identify the two types of dormers.
   b. Describe how to use a framing square and a Speed Square™ for roof framing.
4. Describe the methods used to lay out a common rafter.
   a. Explain how to lay out rafter locations.
   b. Describe how to determine the length of a common rafter.
   c. Explain the correct procedure for laying out and cutting a common rafter.
5. Describe how to erect a gable roof.
   a. Describe how to install rafters.
6. Describe how to frame a basic gable end wall.
   a. Describe how to frame a gable overhang.
   b. Explain how to frame an opening in a roof.
7. Recognize the use of trusses in basic roof framing.
   a. Identify the various types and components of trusses.
   b. Identify the basics of truss installation.
   c. Identify the basics of truss bracing.
8. Describe the basics of roof sheathing installation.
9. Describe how to perform a material takeoff for a roof.
   a. Determine the materials needed for a gable roof.

Performance Tasks

Under the supervision of the instructor, you should be able to do the following:

1. Lay out ceiling joists.
2. Cut and install ceiling joists for a wood frame building.
3. Estimate the number of ceiling joists required for a building.
4. Lay out common roof rafters.
5. Cut and install roof rafters for a gable roof.
6. Frame a gable end wall.
7. Erect a gable roof using trusses.
8. Sheath a gable roof with an opening.
9. Perform a material takeoff for a roof.

Trade Terms

Barge rafter
Bottom chord
Dormer
False fascia
Gable
Hip roof
Lookout
Open web steel joist
Purlin
Ribband
Ridgeboard
Spreader bar
Stick-built framing
Strongback
Tag line
Tiedown
Unit rise

Industry-Recognized Credentials

If you're training through an NCCER-accredited sponsor you may be eligible for credentials from NCCER's Registry. The ID number for this module is 27112-13. Note that this module may have been used in other NCCER curricula and may apply to other level completions. Contact NCCER's Registry at 888.622.3720 or go to **www.nccer.org** for more information.

Contents

Topics to be presented in this module include:

Figures

Figures (Continued)

SECTION ONE

1.0.0 CEILING COMPONENTS

Objective

Identify the components of ceiling framing.

a. Describe the correct procedure for laying out ceiling joists.
b. Describe how to cut and install ceiling joists on a wood frame building.
c. Describe how to estimate the number of ceiling joists required for a building.

Performance Tasks 1 through 3

Lay out ceiling joists.

Cut and install ceiling joists for a wood frame building.

Estimate the number of ceiling joists required for a building.

Trade Terms

Bottom chord: The lower member of a truss.

Gable: The triangular wall enclosed by the sloping ends of a ridged roof.

Hip roof: A roof with four sides or slopes extending toward the center.

Ribband: A 1 × 4 nailed to the ceiling joists at the center of the span to prevent twisting and bowing of the joists.

Strongback: An L-shaped arrangement of lumber used to support ceiling joists and keep them in alignment.

After the walls are assembled and erected in a one-story building, carpenters install ceiling joists on top of the wall frames, spanning the narrow dimension of the building from top plate to top plate. Then, if the carpenters intend to build a common gable roof, they install rafters that extend from the ends of the ceiling joists to the peak, forming a triangle.

Modern roofs are commonly framed with roof trusses, which are installed rather than ceiling joists and roof rafters. The lowest member (bottom chord) of a truss serves the same purpose as the ceiling joist. The upper member serves the same purpose as roof rafters. If the building is taller than one story, floor joists or trusses are installed over the wall frame, and the platform and walls are framed and erected for the second story. The bottom chord of the floor joist or truss serves the same purpose as the ceiling joist.

Ceiling joists have two important purposes:

- Ceiling joists are the top of the six-sided box structure of a building. They prevent opposite walls from spreading apart.
- Ceiling joists provide a nailing surface for ceiling material, such as gypsum board, which is attached to the underside of joists.

1.1.0 Laying Out Ceiling Joists

As noted previously, ceiling joists extend all the way across the structure, from the outside edge of the top plate on one wall to the outside edge of the top plate of the opposite wall. Carpenters typically install ceiling joists across the narrow width of a building, usually at the same positions as the wall studs. This in-line framing allows roof loads to be transferred directly from the roof to the wall studs to the foundation.

If the spacing of the ceiling joists is the same as that of the wall studs, lay out the joists directly above the studs. This makes it easier to run ductwork, piping, and wiring above the ceiling.

Laying out ceiling joists for a gable roof is similar to laying out floor joists and wall studs. Measure along the top plate to a point 15¼" in from the end of the building. Mark it, square the line across the plate, and then use a measuring tape to mark every 16" (or 24", depending on the prints) along the length of the plate. To the right of each line, mark an *X* for the joist location. Then mark an *R* for each rafter on the left side of each mark (*Figure 1*). Repeat this procedure on the opposite wall and on any loadbearing partitions.

> **NOTE**
> If you are installing a hip roof, it is also necessary to mark the end double top plates for the locations of hip rafters. Start by marking the center of the end double top plate. Then measure and place marks for joists and rafters by measuring from each corner toward the center mark at 16" or 24" increments.

As discussed in the module *Floor Systems*, the actual allowable span for floor joists depends on the species, size, and grade of lumber, as well as the joist spacing and the load to be carried. The same is true for ceiling joists. If ceiling joists exceed the allowable span, two joist members must be spliced over a loadbearing wall or partition. *Figure 2* shows two ways to splice joists. The first method, *Figure 2A*, shows two joists that are over-

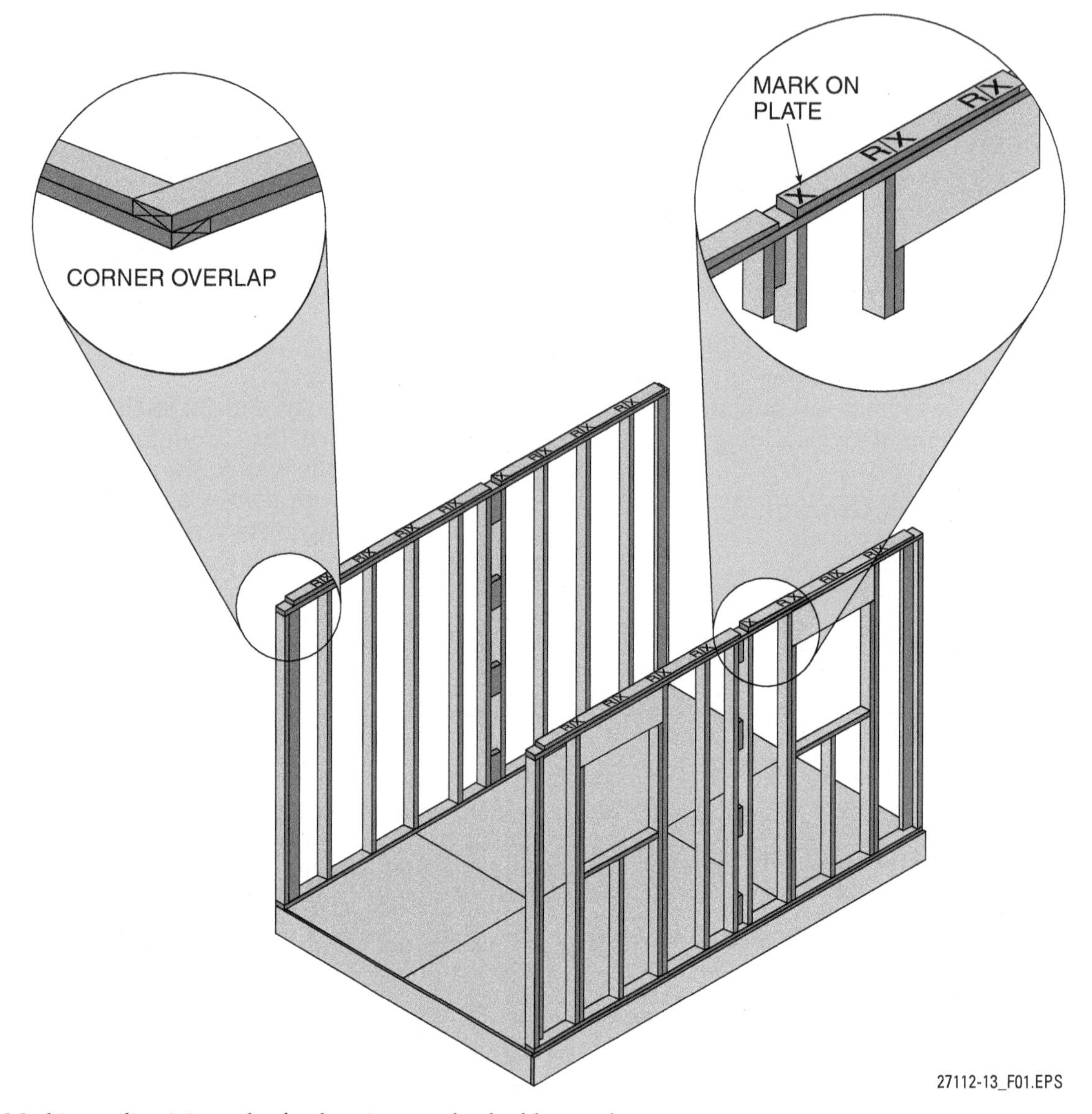

Figure 1 Marking ceiling joist and rafter locations on the double top plate.

lapped above a loadbearing partition. The overlap should be no less than 3".

Figure 2B shows another way to splice joists. Instead of overlapping the joist ends, the two joists butt together directly over the center of the loadbearing partition. A scab (shorter piece of joist material) is nailed to both joists to make a strong joint. Other materials used to reinforce the joist splice include plywood, 1× lumber, steel strapping, and special anchors. The architectural drawings specify the proper method for splicing the joists.

1.2.0 Cutting and Installing Ceiling Joists

The joists must be cut to the proper length, so the ends of the joists will be flush with the outside edge of the top plate. When determining joist length, ensure that enough extra length is allowed for any overlap of the joists above loadbearing partitions.

The upper edge of the outside ends of the joists must be cut at an angle to match the rafter pitch, as shown in *Figure 3*. This allows the roof sheathing to lie flush on the roof framing.

After you have cut the ceiling joists to the proper length, have made the angle cut, and have positioned them in the correct places, toenail the joists to the top plate. The architectural drawings may also specify the installation of metal anchors.

After installing the ceiling joists, nail a ribband or strongback across the joists to prevent twisting or bowing (*Figure 4*). Strongbacks are used for longer spans. In addition to preventing the ceiling joists from twisting or bowing, strongbacks provide support for the joists at the center of the span.

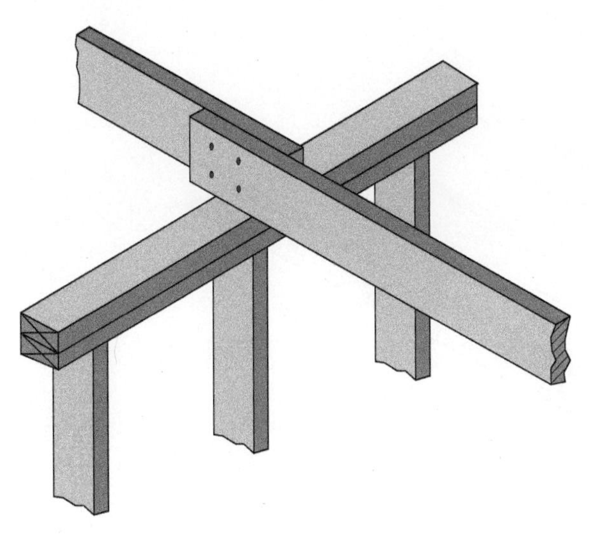

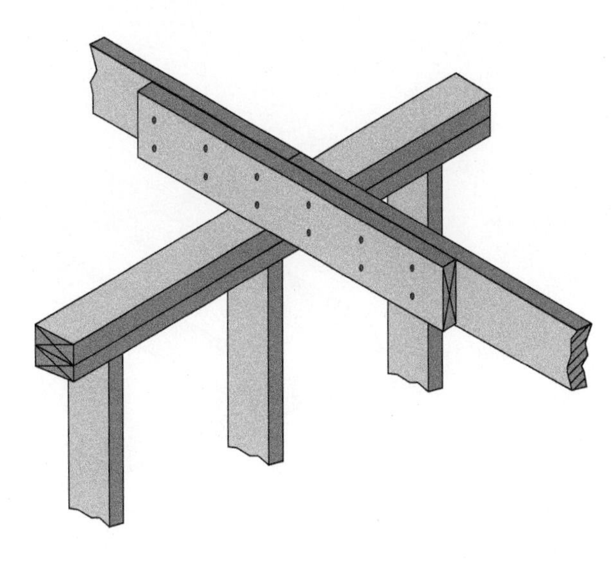

Figure 2 Splicing ceiling joists.

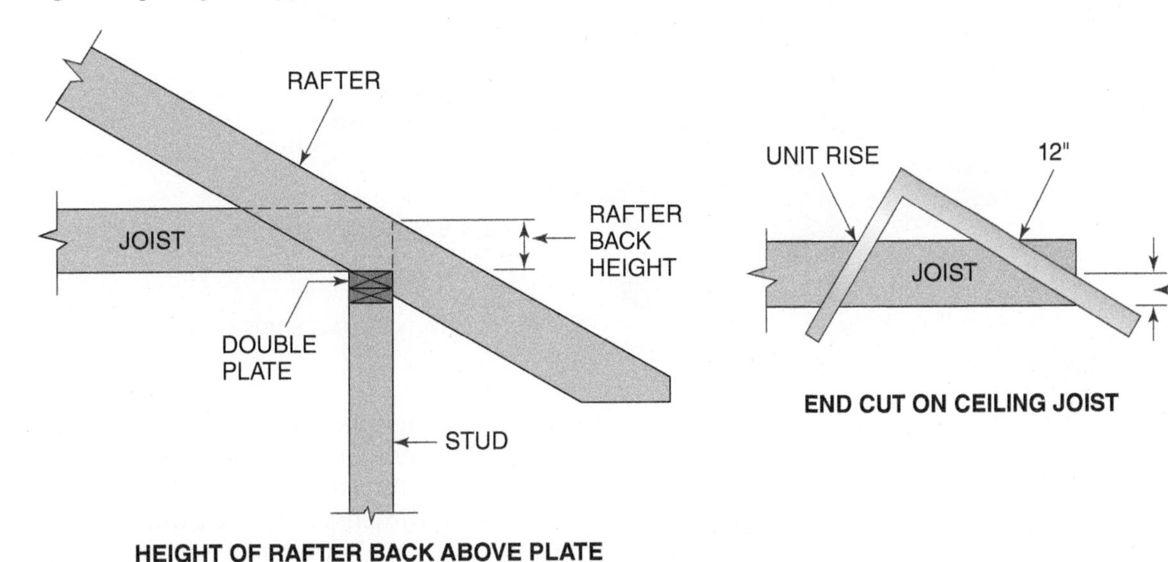

Figure 3 Cutting joist ends to match roof pitch.

Case History

Joist Installation

Two carpenters were working on the second floor of a house. There was no guardrail or cover over the floor opening for the stairway. While installing a ceiling joist, one of the carpenters fell through the stairway opening to the concrete basement below, receiving a fatal head injury.

The Bottom Line: Always install appropriate guardrails and cover floor openings. Use fall-protection equipment when working above floor level.

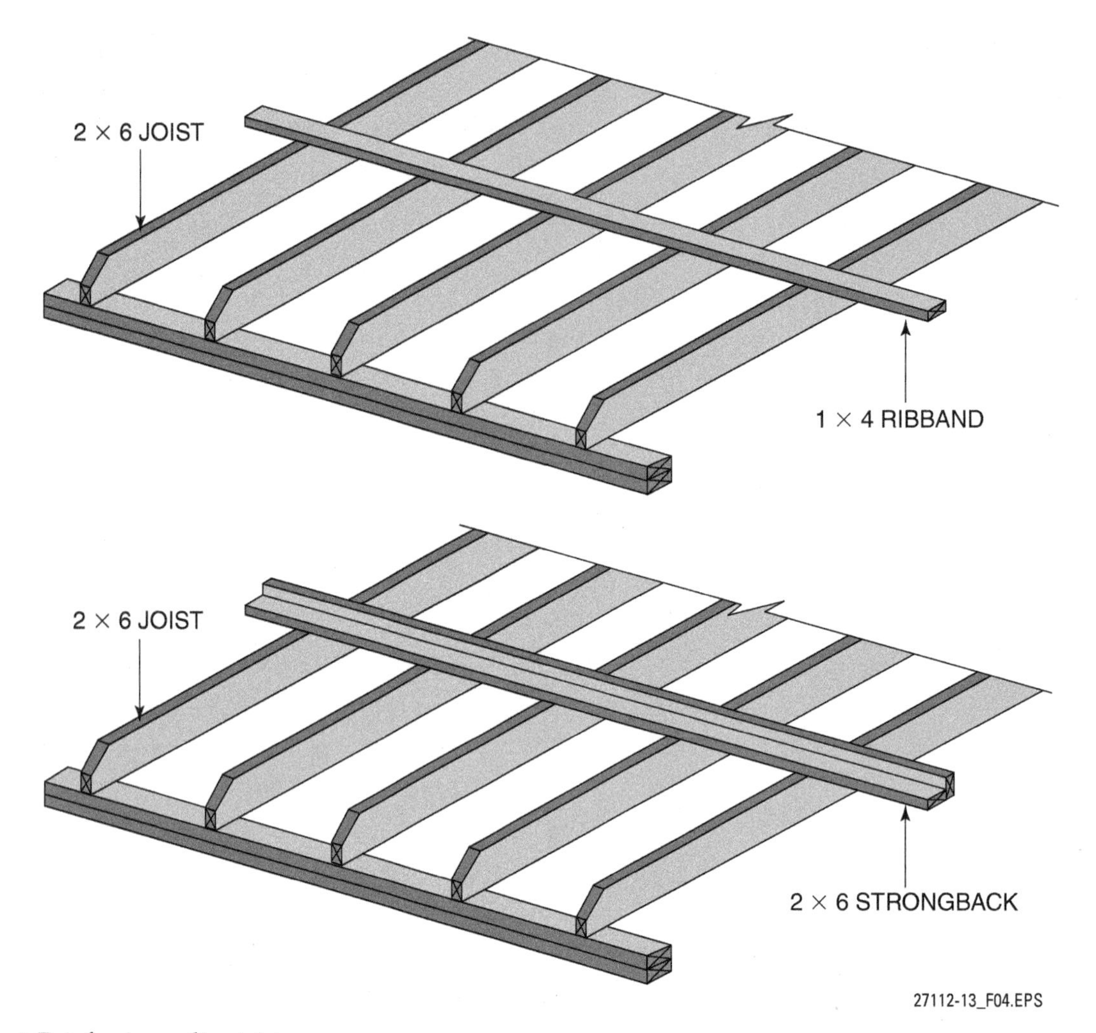

Figure 4 Reinforcing ceiling joists.

1.3.0 Estimating Ceiling Joists

As mentioned previously, it is more likely that carpenters will install roof trusses (which combine ceiling joists and roof rafters into one unit) instead of joists in a one-story building. Roof trusses and modern roof framing methods are covered later in this module. Here the focus is on framing with ceiling joists.

Step 1 Determine the span of the building.

Step 2 Determine the number of joists based on the spacing and then add one for the end joist.

Step 3 Multiply the span by the number of joists to determine the total length. Add 3" per joist per splice where joists will be spliced at loadbearing walls, partitions, and girders.

Example:

The example building is 24' × 30'. The ceiling joists will span across the short dimension (24'). For this example, there is a loadbearing partition down the long dimension of the building that breaks the 24' span into two 12' spans.

Divide the long dimension of the building (30') by 16" to determine the number of joist locations. First, convert 30' to inches:

$$30' \times 12'' \text{ per foot} = 360''$$

Divide that by 16":

$$360'' \div 16'' = 22\tfrac{1}{2} \text{ (round up to 23)}$$

Overlapped Splices

An overlapped splice is superior to a butted splice because a butted splice reduces by one-half the critical joist-bearing surface, or the part of the joist that is supported by the loadbearing wall beneath it. Also, a butted splice requires more materials and labor than the simpler, stronger overlapped splice.

Then add one for the end joist:

$$23 + 1 = 24$$

The 24' width is too big to span with one long joist, so you'll span it with two 12' joists. Multiply the last number by two:

$$24 \times 2 = 48 \text{ pieces}$$

Now calculate the total length, adding 3" for each joist at each splice.

$$48 \times 12' = 576'$$

$$48 \times 3'' = 144''$$

$$144'' \div 12 = 12'$$

$$576' + 12' = 588'$$

So how much lumber do you order? Twelve-foot lengths of 2 × 6 would be 3" short, so 14-foot lengths would be a better choice. You'd end up with 1'-9" of waste from each piece of lumber, but you can use the waste for bridging and blocking.

GOING GREEN

The Use of Wood in Construction

Wood is a naturally renewable resource. Since the 1940s, the forest growth in the United States has exceeded the harvest. Approximately one billion trees are planted a year. The manufacture of wood is very energy efficient when compared to the energy used to produce other building materials. It takes 5 times more energy to create one ton of cement, 24 times more energy to create one ton of steel, and 126 times more energy to create one ton of aluminum.

1.0.0 Section Review

1. When ceiling joists are to be spliced, the proper method of splicing is indicated in the _____.
   a. manufacturer's instructions
   b. architect's renderings
   c. electrical plan
   d. architectural drawings

2. Ribbands are used for longer spans of ceiling joists to prevent twisting and bowing.
   a. True
   b. False

3. How many joists are required for an 8' × 10' tool shed when the joists are spaced 16" on center (OC)?
   a. 6
   b. 7
   c. 8
   d. 9

Section Two

2.0.0 Common Roof Types

Objective

Identify common types of roofs used in residential construction.

Trade Terms

Hip roof: A roof with four sides or slopes extending toward the center.

Stick-built framing: Framing method in which framing members are installed one at a time (rather than as a prefabricated assembly)

Roof framing is the most demanding of the framing tasks. Floor and wall framing generally involves working with horizontal or vertical lines. Residential roofs are typically sloped to shed water from rain or melting snow. In areas where there is heavy snowfall, roofs must be constructed to bear the extra weight of the snow load. Since a roof is sloped, laying it out involves working with precise angles in addition to vertical and horizontal lines. In this section, common types of roofs used in residential construction are covered.

The most common types of roofs used for residential construction are shown in *Figure 5* and described below:

- *Gable roof* – A gable roof has two slopes that meet at the ridge (center) to form a gable at each end of the building. It is the most common type of roof because it is simple, economical, and can be used on any type of structure.
- **Hip roof** – A hip roof has four sides or slopes extending toward the ridge of the building. Rafters at the corners extend diagonally from the top plate to meet at the ridge. Additional rafters are framed into these rafters.
- *Mansard roof* – The mansard roof has four sloping sides, each of which has a double slope. As compared with a gable roof, this design provides more available space in the upper level of the building. The upper slope is typically not visible from the ground.
- *Gable-and-valley roof* – This roof consists of two intersecting gable roofs. The two roofs meet at a valley.
- *Hip-and-valley roof* – This roof consists of two intersecting hip roofs.
- *Gambrel roof* – The gambrel roof is a variation of the gable roof. The main difference is that each of the sloping sides of a gambrel roof has a double slope. The gambrel roof provides more available space in the upper level than a gable roof.
- *Shed roof* – Also known as a lean-to roof, the shed roof is a flat, sloped construction. It is common on high-ceiling contemporary construction and is often used on additions.

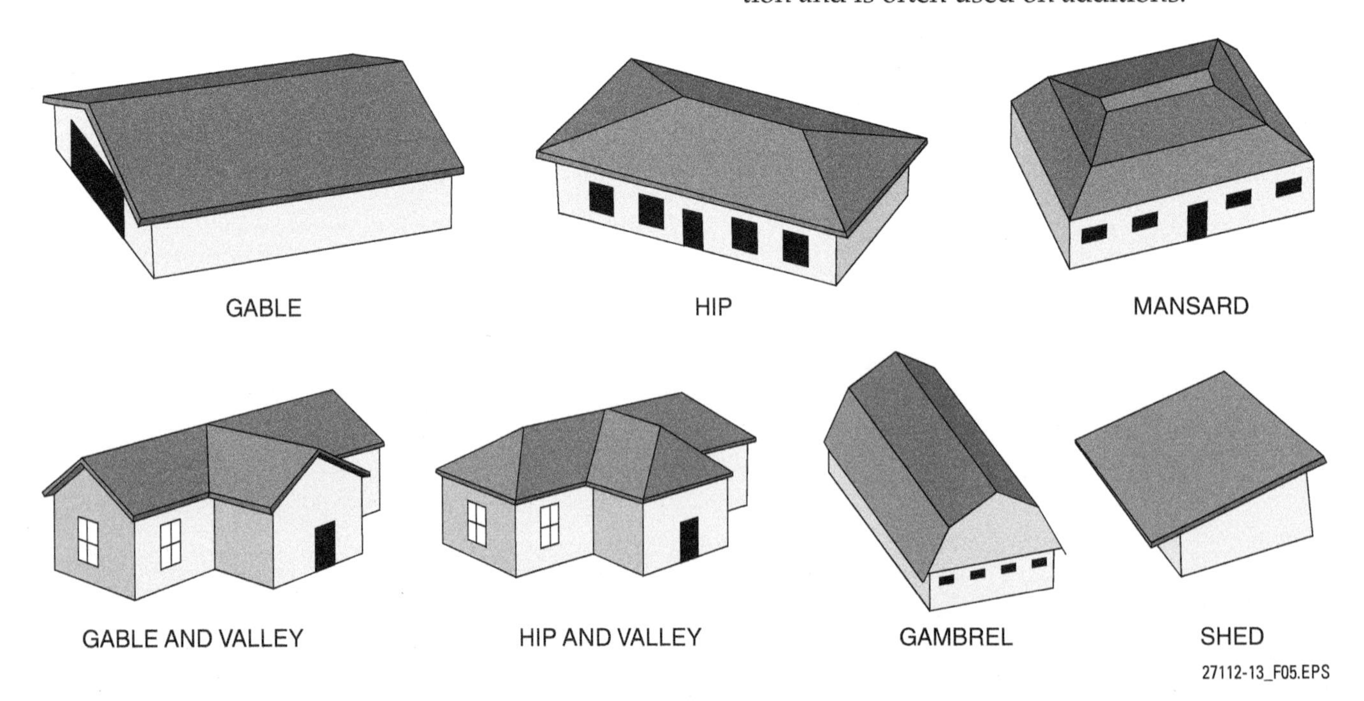

Figure 5 Types of roofs.

There are two basic roof framing systems: stick built and trusses. In **stick-built framing**, ceiling joists and rafters are laid out and cut by carpenters on-site and the frame is constructed one framing member (stick) at a time. In trussed roofs, the roof framework, consisting of both the ceiling joists and roof rafters, is prefabricated off-site. Trusses are then raised into position using cranes.

Hip Roofs

Hip roofs are an efficient roof design to use in areas with heavy wind loads, such as in areas that are prone to hurricanes. According to a study conducted by the Center for Building Science and Technology, when model homes with various roof designs were exposed to high wind conditions, homes with hip roofs "would be far more efficient under high winds and hurricane conditions than a typical structure." In addition, the study also concluded:

- Roofs with multiple slopes, such as hip roofs, perform better under wind conditions than gable roofs.
- Wind loads placed on a roof are uplift forces. This is the reason that stapled roofs were banned in Florida in 1993 following Hurricane Andrew.
- Structural failure is often progressive; in other words, the failure of one connection or component triggers the failure of another connection or component until total collapse of the structure.
- Certain areas of a roof, such as the ridge, eaves, and corners, are subject to higher wind pressures.

2.0.0 Section Review

1. A hip roof has _____.
   a. two sloping sides
   b. four sloping sides
   c. four sloping sides, each with a double slope
   d. two sloping sides, each with a double slope

2. A gable roof has _____.
   a. two sloping sides
   b. four sloping sides
   c. four sloping sides, each with a double slope
   d. two sloping sides, each with a double slope

SECTION THREE

3.0.0 ROOF FRAMING COMPONENTS

Objective

Identify the components and define the terms associated with roof framing.

a. Identify the two types of dormers.
b. Describe how to use a framing square and a Speed Square™ for roof framing.

Trade Terms

Dormer: Framework extending from the roof surface that provides additional light and ventilation to an attic space.

Ridgeboard: Horizontal board (typically a 2× member) placed at the ridge of the roof. Upper ends of the rafters are fastened to it.

Unit rise: The amount (in inches) that a rafter will rise per foot of run.

Rafters provide the main framework for roofs. The main components of a roof are shown in *Figure 6* and described here:

- **Ridgeboard** – The highest horizontal roof member. It helps to align the rafters and tie them together at the upper end. The ridgeboard is one size larger than the rafters.
- *Common rafter* – A structural member that extends from the top plate to the ridge in a direction perpendicular to the wall plate and ridge. Rafters often extend beyond the top plate to form the overhang (eaves) that protect the side of the building.
- *Hip rafter* – A roof member that extends diagonally from the outside corner of the top plate to the ridge.
- *Valley rafter* – A roof member that extends from the inside corner of the top plate to the ridge along the lines where two roofs intersect.
- *Jack rafter* – A roof member that does not extend the entire distance from the ridge to the top plate of a wall. The hip jack and valley jack rafters are shown in *Figure 6*. A rafter fitted between a hip rafter and a valley rafter is called a cripple jack rafter. It touches neither the ridge nor the plate.
- *Plate* – The wall framing member that rests on top of the wall studs. Commonly referred to as the top plate or double top plate. It is sometimes called the rafter plate because the rafters rest on it.

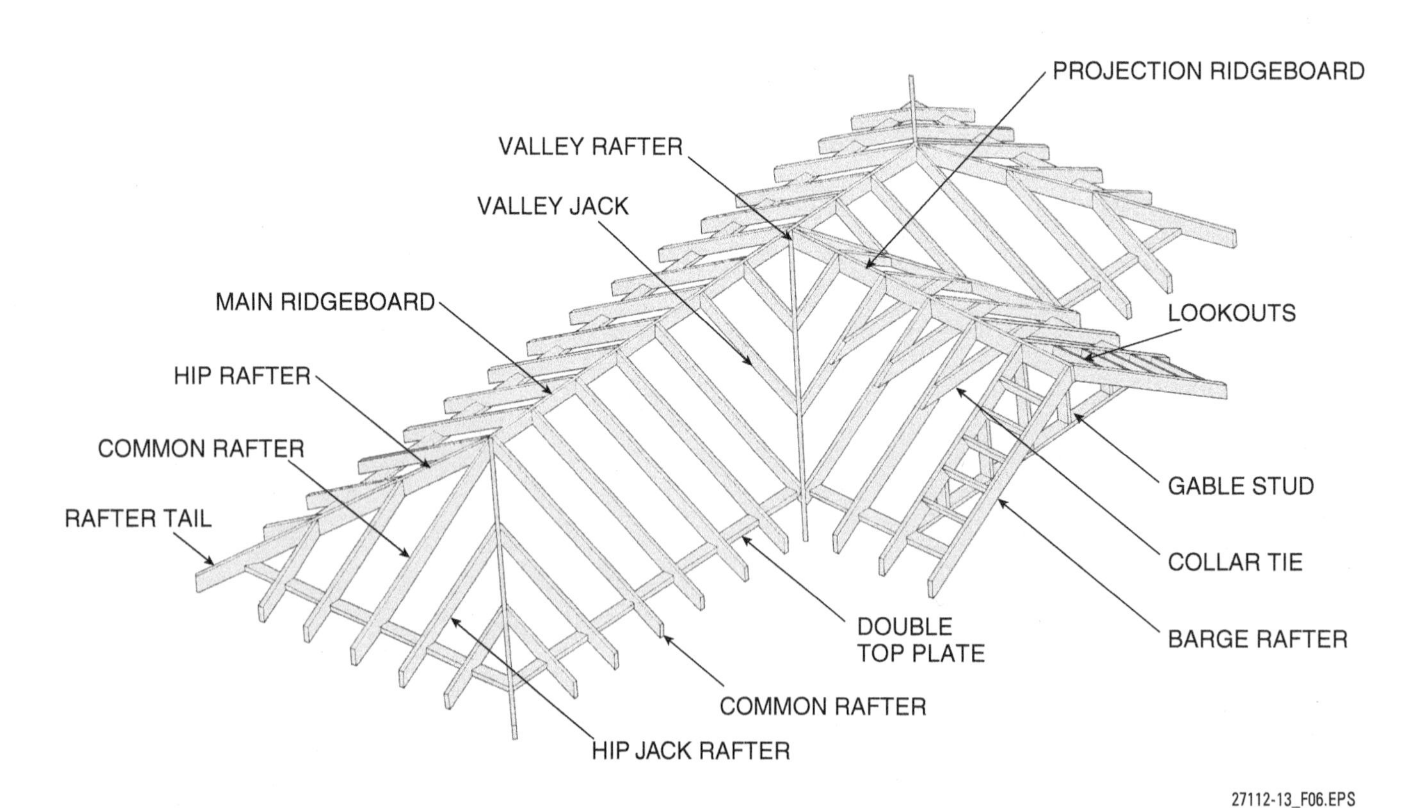

Figure 6 Roof framing members.

As you can see in *Figure 7*, on any pitched roof, rafters are at a downward angle from the ridgeboard. Therefore, the length of the rafter is greater than the horizontal distance from the plate to the ridge. In order to calculate the correct rafter length, a carpenter must factor in the slope of the roof.

Here are some additional terms related to rafter layout:

- *Span* – The horizontal distance from the outside of one exterior wall to the outside of the opposing exterior wall.
- *Run* – The horizontal distance from the outside of the top plate to the center line of the ridgeboard (usually equal to half of the span).
- *Rise* – The total height of the rafter from the top of the plate to the top of the ridge. This is stated in inches per foot of run.
- *Pitch* – The angle or degree of slope of the roof in relation to the span. Pitch is expressed as a fraction; for example, if the total rise is 6' and the span is 24', the pitch would be ¼ (6 over 24).
- *Slope* – The inclination of the roof surface expressed as the relationship of rise to run. It is stated as a unit of rise to so many horizontal units; for example, a roof that has a rise of 5" for each foot of run is said to have a 5-in-12 slope (*Figure 7*). The roof slope is sometimes referred to as the roof cut.

The first step in determining the correct length of a rafter is to find the unit rise, which is usually shown on the building's elevation drawing. The unit rise is the number of inches the rafter rises vertically for each foot of run. The greater the rise per foot of run, the greater the slope of the roof.

3.1.0 Dormers

A dormer is a framed structure that projects out from a sloped roof. A dormer provides additional space and is often used in Cape Cod–style homes, which are single-story dwellings in which the attics are often used for additional bedrooms. A shed dormer (*Figure 8*) is a good way to obtain a large amount of additional living space. If it is added to the rear of the house, it can be done without affecting the appearance of the house from the front.

A gable dormer (*Figure 9*) serves as an attractive addition to a house, in addition to providing a little extra space as well as some light and ven-

27112-13_F08.EPS

Figure 8 Shed dormer.

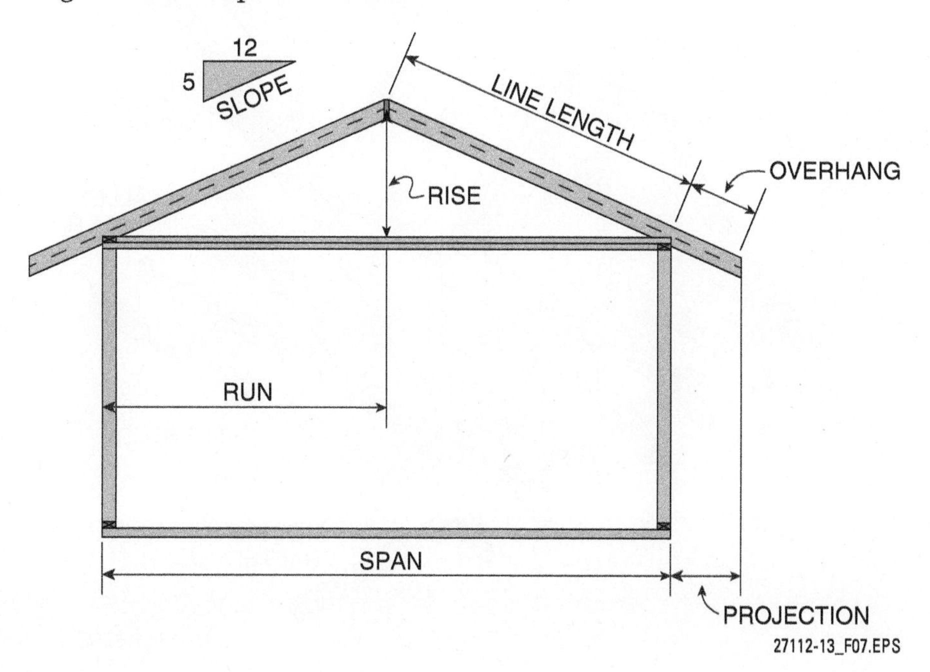

27112-13_F07.EPS

Figure 7 Roof layout terminology.

tilation. They are sometimes used over garages to provide a small living area or studio.

3.2.0 Rafter Framing Tools

There are several ways to calculate the length of a rafter. It can be done with a rafter square or Speed Square™, or it can be done using a calculator. Since a carpenter typically has a rafter square readily available, we will focus on this tool and the associated methods.

3.2.1 Rafter Square

A rafter square is a type of framing square that is calibrated to show the length per foot of run for each type of rafter (*Figure 10*). Note that the tongue is the short (16") section of the square, and the blade (or body) is the long (24") section. The corner is known as the heel. Rafter tables are provided on the back of the square. The rafter tables typically indicate the rafter dimensions in length per foot of run, but some give length per given run.

The rafter square is used to determine the rafter length and to measure and mark the cuts that must be made in the rafter (*Figure 11*). As you can see, you can relate the pitch and slope to the rise per foot of run. The rise per foot of run is always the same for a given pitch or slope. For example, a

Pitch and Slope

Carpenters may use the terms *pitch* and *slope* interchangeably on the job site, but the two terms actually refer to two different concepts. Slope is the amount of rise per foot of run and is always referred to as a number in 12. For example, a roof that rises 6" for every foot of run has a 6-in-12 slope (the 12 simply refers to the number of inches in a foot). Pitch, on the other hand, is the ratio of rise to the span of the roof and is expressed as a fraction. For example, a roof that rises 8' over a 32' span is said to have a pitch of 1/4 ($\frac{8}{32} = \frac{1}{4}$).

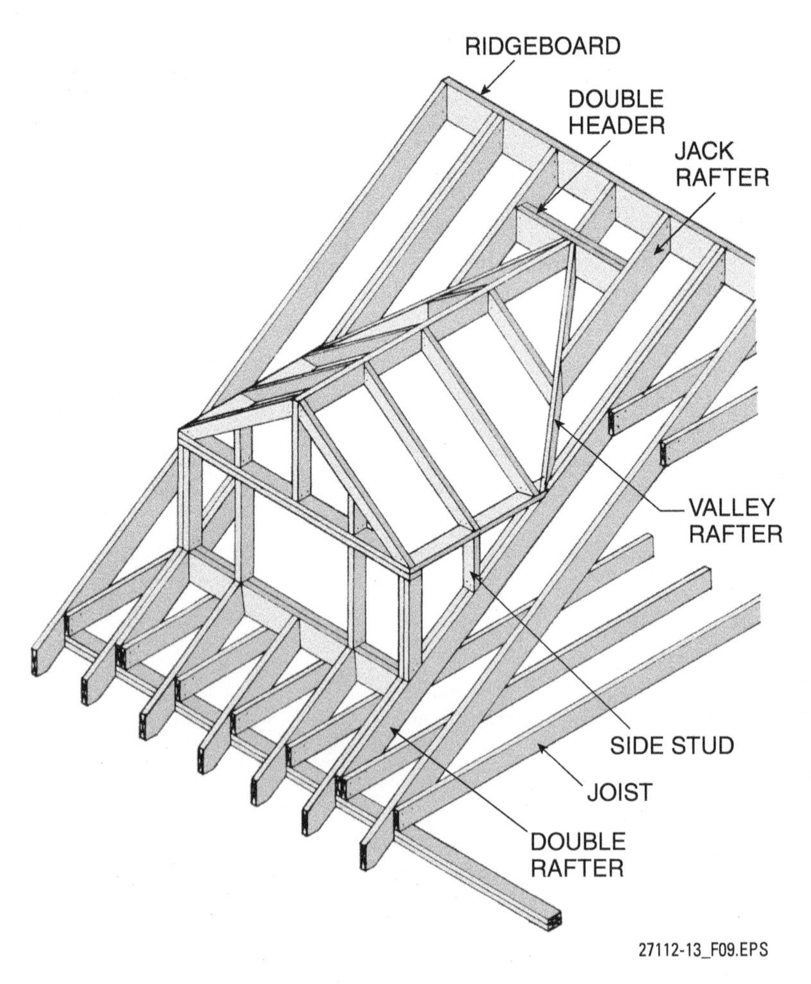

Figure 9 Gable dormer framing.

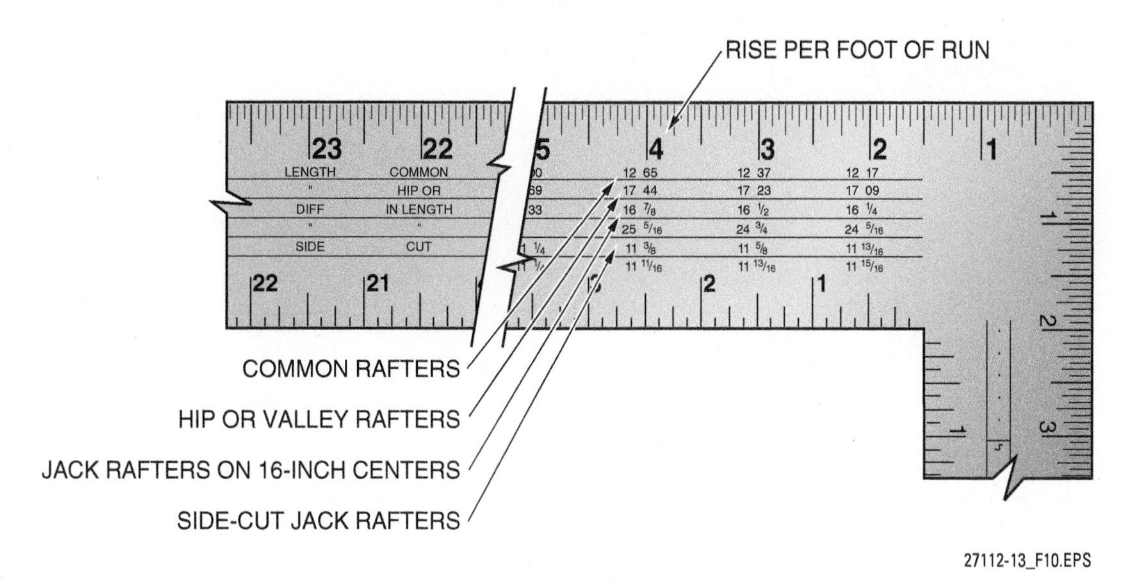

27112-13_F10.EPS

Figure 10 Rafter tables on a rafter square.

pitch of 1/2, which is the same as a 12-in-12 slope, equals 12" of rise per foot of run.

3.2.2 Speed Square™

The Speed Square™, commonly referred to as a Super Square™ or Quick Square™, is a combination tool consisting of a protractor and a miter square. A standard Speed Square™ is a 7" triangular tool with a large outer triangle and a smaller inner triangle (*Figure 12*). The large triangle has a 7" scale on one edge, a full 90-degree scale on another edge, and a T-bar on the third edge. The inner triangle has a 2" square on one side. The Speed Square™ is the same on both sides. A 12" Speed Square™ is used for stair layout.

The pitch of the roof must be known to use a Speed Square™. When you purchase a Speed Square™, an instruction booklet is usually packaged with the square. This booklet typically contains (among other information) tables that show the required rafter length for every pitch. Refer to the manufacturer's instructions for specific details regarding the use of the Speed Square™.

Gable Dormers

Dormers protrude from a sloped roof and create additional space in the building.

27112-13_SA01.EPS

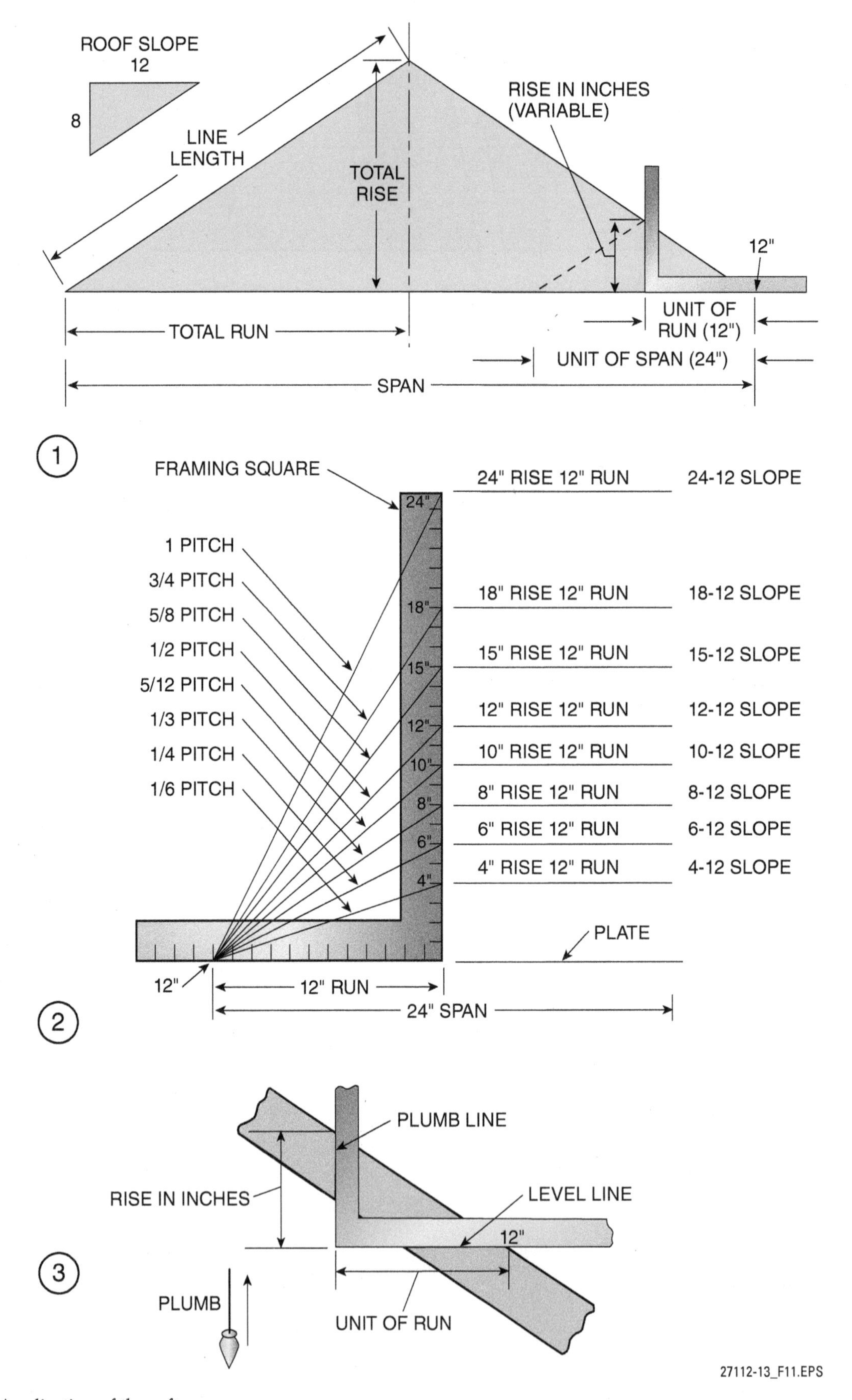

Figure 11 Application of the rafter square.

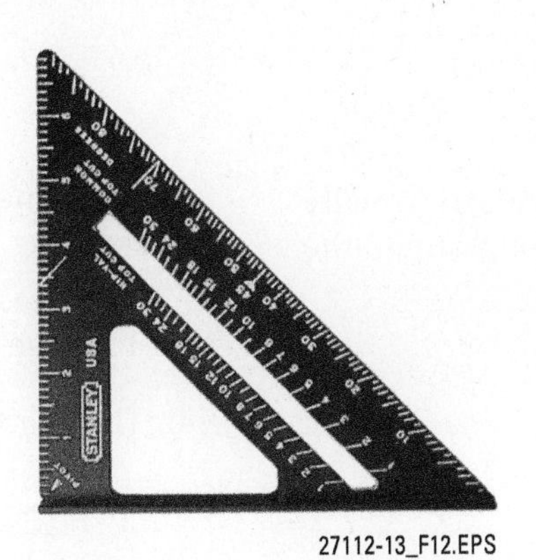

27112-13_F12.EPS

Figure 12 Speed Square™.

Calculating Rise and Pitch

Now that you are familiar with the concepts of pitch, rise per foot of run, and total rise, practice your skills by completing the following example problems.

Find the pitch for the following roofs:

Total Rise	Span	Pitch
8'	24'	
9'	36'	
6'	24'	
12'	36'	
8'	32'	

Find the total rise for the following roofs:

Pitch	Span	Rise per Foot of Run	Total Rise
1/2	24'	12"	
1/6	32'	4"	
1/3	24'	8"	
5/12	30'	10"	
3/4	24'	18"	

Find the rise per foot of run for the following roofs:

Span	Total Rise	Rise per Foot of Run
18'	6'	
24'	8'	
28'	7'	
16'	4'	
32'	12'	

3.0.0 Section Review

1. A roof framing member that extends diagonally from the outside corner of the top plate to the ridge is the _____.

   a. valley rafter
   b. jack rafter
   c. hip rafter
   d. ridgeboard

2. For a roof with a 36' span, the run is _____.

   a. 12'
   b. 18'
   c. 24'
   d. 36'

3. A dormer is framed into the gable end of a building.

   a. True
   b. False

4. The short leg of a rafter square is known as the _____.

   a. blade
   b. heel
   c. tongue
   d. body

SECTION FOUR

4.0.0 LAY OUT A COMMON RAFTER

Objective

Describe the methods used to lay out a common rafter.

a. Explain how to lay out rafter locations.
b. Describe how to determine the length of a common rafter.
c. Explain the correct procedure for laying out and cutting a common rafter.

Performance Task 4

Lay out common roof rafters.

Trade Terms

Barge rafter: A gable end roof member that extends beyond the gable to support a decorative end piece. Also referred to as a fly rafter.

False fascia: The board that is attached to the tails of the rafters to straighten and space the rafters and provide a nailing surface for the fascia. Also called subfascia and rough fascia.

Laying out the framing for a roof involves three main tasks:

1. Laying out rafter locations on the top plate
2. Determining the length of each rafter
3. Laying out and cutting the rafters

Rafters must be laid out and cut so that the ridge end will fit squarely on the ridgeboard and the tail end will present a square surface for the false fascia board. In addition, a bird's mouth (*Figure 13*) must be cut at the correct location and angle for the rafter to rest squarely on the plate.

4.1.0 Laying Out Rafter Locations

The following is a basic procedure for marking the rafter locations on the top plate (*Figure 14*) for 24" OC construction. Keep in mind that in most cases, the ceiling joists would be in place and the rafter locations would have already been marked.

Step 1 Locate the first rafter flush with one end of the top plate.

Step 2 Measure the thickness of a rafter. Mark and square a line the thickness of a rafter in from the end.

NOTE

The distance between the last two rafters may be less than 24", but not more than 24".

Step 3 Use a measuring tape to space off and mark the rafter locations every 24" all the way to the end of the top plate. Square the lines across the top of the plate.

Step 4 Repeat the process on the opposite top plate, starting from the same end as before.

Step 5 Cut the ridgeboard to length, allowing for a barge rafter at each end, if required.

Step 6 Place the ridgeboard on the top plate and mark it for correct position. Then measure and mark the rafter locations with an R on the ridgeboard.

To determine the number of rafters you need, count the marks on both sides of the ridge or top plates.

4.2.0 Determining the Length of a Common Rafter

Here is an easy way of determining the required length of a rafter:

Step 1 Start by measuring the building span. Then, divide the span by 2 to determine the run.

Step 2 Determine the rise. This can be done in either of the following ways:
- Calculate the total rise by multiplying the span by the pitch (e.g., 40' span × 1/4 pitch = 10' rise).
- Find it on the slope diagram on the roof plan, as discussed previously.

Cutting Sequence

Some carpenters prefer to make the tail end plumb cut while laying out the rafter. Others prefer to tail-cut all the rafters at once after the rafters are installed. One method isn't necessarily better than the other and the choice is simply a matter of preference.

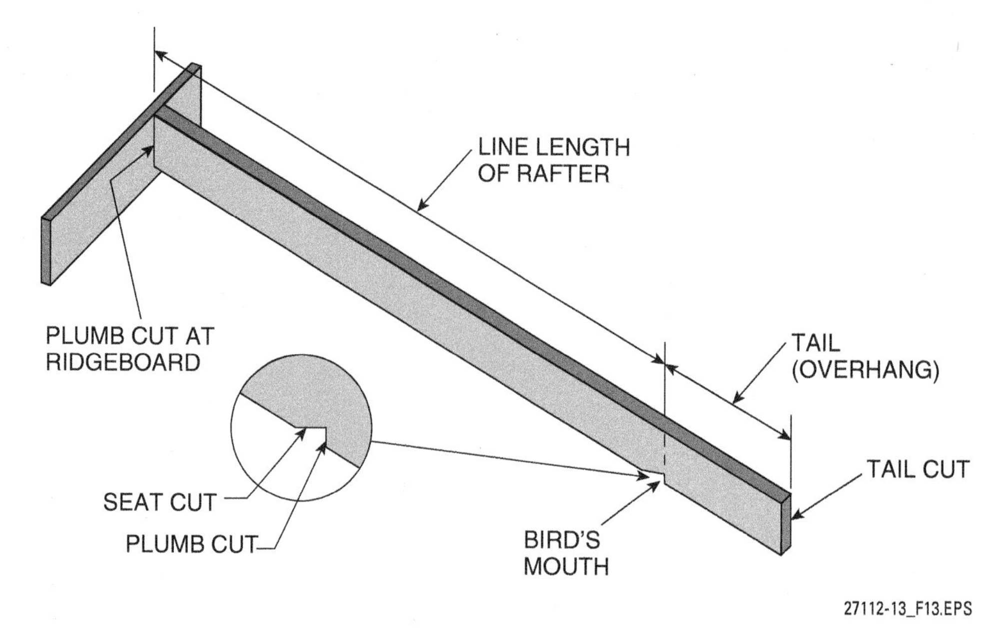

Figure 13 Parts of a rafter.

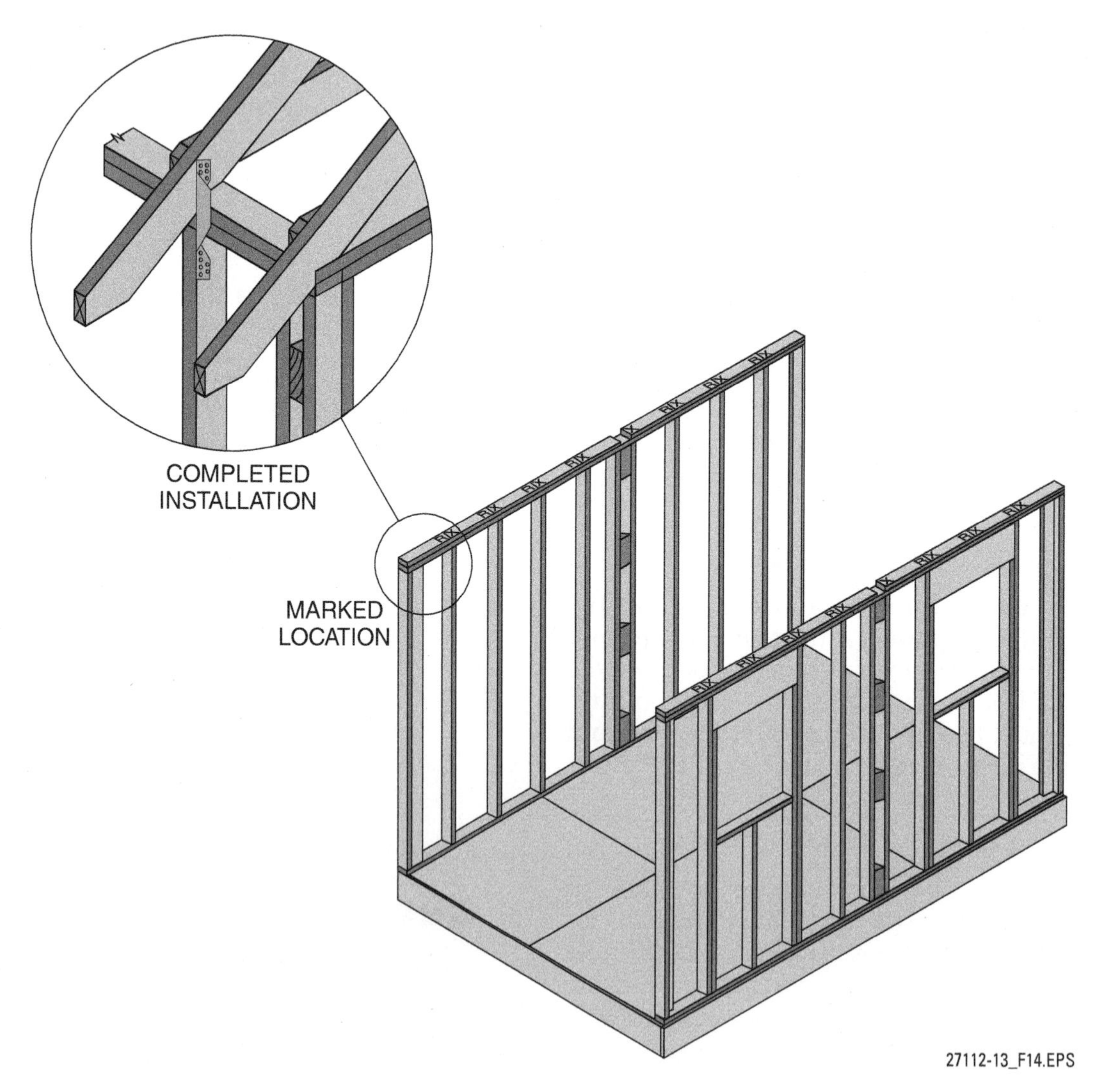

Figure 14 Marking rafter locations.

Step 3 Divide the total rise (in inches) by the run (in feet) to obtain the rise per foot of run.

Step 4 Look up the required length on the rafter tables on the rafter square.

For example, if the roof has a span of 20', the run would be 10'. Then, assume that the drawings show the rise per foot of run to be 8". Locate the 8 on the rafter square and note the value in the top row of the table. The correct rafter length is 14.42" ($14\frac{5}{12}$") per foot of run. Since there is 10' of run, the rafter length would be $144\frac{1}{6}$" (*Figure 15*). If an overhang is shown on the drawings, the overhang length must be added to the rafter length.

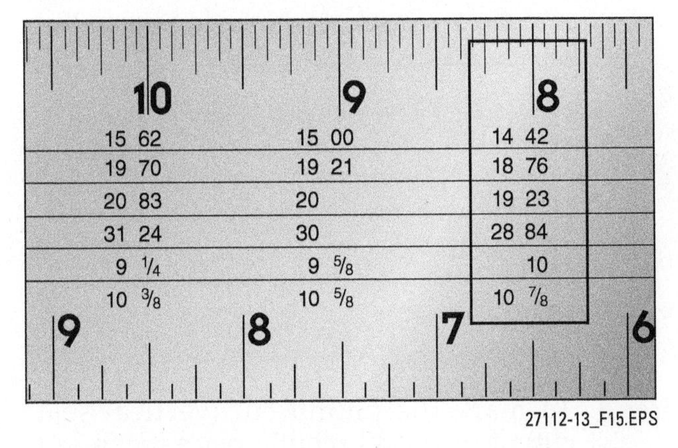

Figure 15 Determining rafter length with the rafter square table.

Another way to determine rafter length is to measure the distance between 8 on the rafter square blade and 12 on the tongue, as shown in *Figure 11*, and multiply this distance by the run.

Yet another method of determining the approximate length of a rafter is the rafter square step-off method shown in *Figure 16*. In this procedure, the ridge plumb cut is determined, then the rafter square is stepped once for every foot of run. The final step marks the plumb cut for the bird's mouth. This is not the preferred method because it is not as precise as other methods.

4.3.0 Laying Out and Cutting Common Rafters

The next two sections are overviews of two procedures for laying out and cutting common rafters.

4.3.1 Rafter Square Method

Step 1 Start with a piece of lumber a little longer than the required length of the rafter, including the tail. If the lumber has a crown or bow, it should be at the top of the rafter. Lay the rafter on sawhorses with the crown (if any) at the top.

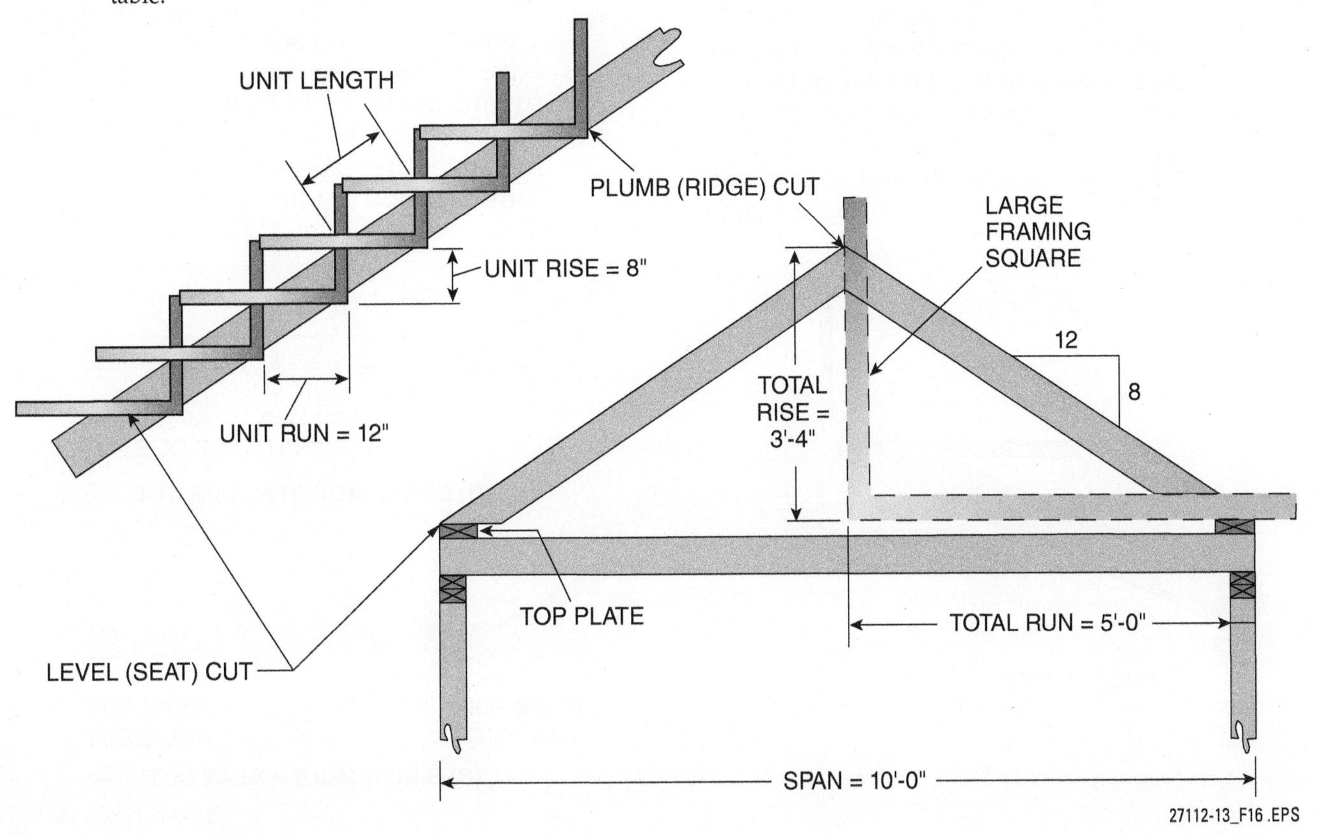

Figure 16 Rafter square step-off method.

Step 2 Start by marking the ridge plumb cut using the rafter square (*Figure 17A*). Be sure to subtract half the thickness of the ridgeboard. Make the cut.

Step 3 Measure the length of the rafter from the plumb cut mark to the end (excluding the tail) and mark the bird's mouth plumb cut (*Figure 17B*). Reposition the rafter square and mark the bird's mouth seat cut (*Figure 17C*).

Step 4 Make the end plumb cut, then cut out the bird's mouth. Cut the bird's mouth part of the way with a circular saw, then use a handsaw to finish the cuts.

Step 5 Check the first rafter to ensure the proper fit. Use the first rafter as a template for marking the remaining rafters. As the rafters are cut, stand them against the building at the joist locations.

4.3.2 Speed Square™ Method

Step 1 Choose a piece of lumber that is slightly longer than that needed for the rafter. (Remember, the eaves or overhang are not included in the measurements found in the Speed Square™ instruction manual. The length of the overhang is added after the length of the rafter is determined.) Place the lumber on a pair of sawhorses with the top edge of the rafter stock facing away from you. Be sure the crown is facing away from you (the crown, if any, will be the top edge of the rafter). The rafter will be cut for a building that is 24'-wide with a 4" rise per foot of run. Therefore, the length of the rafter from the center of the ridgeboard to the seat cut (bird's mouth) will be 12'-7¾". Any overhang must be added to this.

Step 2 To mark the top plumb cut, start at the right-hand end of the board. Holding the Speed Square™ on the pivot point with the 4 mark lined up with the top edge of your board, draw a line along the edge of the Speed Square™, as shown in *Figure 18A*. From the long point of this mark, use a tape measure along the top edge of the rafter to lay out the length required (12'-7¾") and make a mark, as shown in *Figure 18B*.

> **NOTE**
> When working with the Speed Square™, there is no measuring line. The length is measured on the top edge of the rafter. Assume that the right side of the lumber will be the top of the rafter. Place the Speed Square™ against the top edge of the lumber and set the square with the 4 on the common scale.

Step 3 To mark the plumb cut (vertical seat cut) of the bird's mouth, place the Speed Square™ with the pivot point against that mark, and move the square so the 4 on the common scale is even with the edge. Draw a line along the edge of the square, as shown in *Figure 18C*. This line represents the outside wall of your building and will establish the vertical seat cut (bird's mouth) or plumb cut.

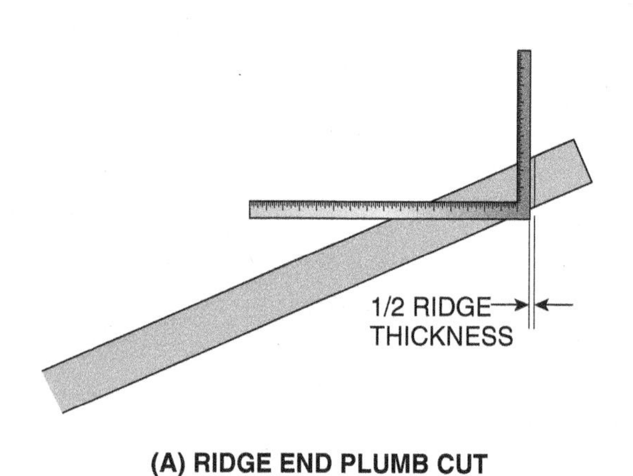

(A) RIDGE END PLUMB CUT

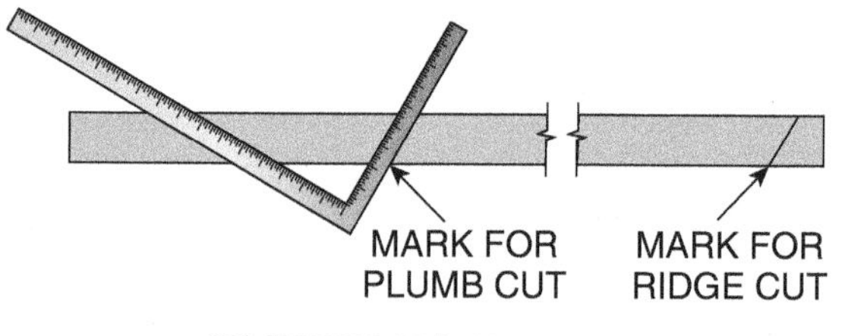

(B) BIRD'S MOUTH PLUMB CUT

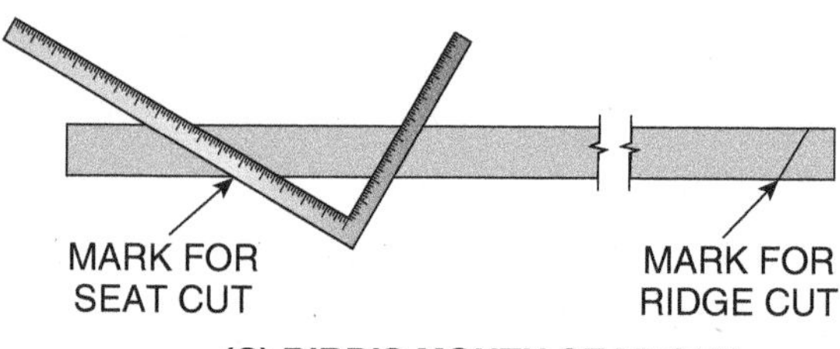

(C) BIRD'S MOUTH SEAT CUT

27112-13_F17.EPS

Figure 17 Marking the rafter cuts using a rafter square.

Step 4 To lay out the bird's mouth seat cut, reverse the Speed Square™ so that the short line at the edge of the square (above the 70-degree mark) is even with the line previously drawn. Draw the line so that the edge of the square is even with the lower edge of the rafter, as shown in *Figure 18D.*

Step 5 Draw a second line at a right angle to the plumb line. This line will establish the completed bird's mouth seat cut, as shown in *Figure 18E.*

Step 6 The top and bottom cuts have been established for the total length of the rafter. If a ridgeboard is being used, be sure to deduct one-half of the thickness of the ridgeboard from the top plumb cut. If any overhang is needed, it should be measured at right angles to the plumb cut of the bird's mouth seat cut, as shown in *Figure 18F.*

Step 7 Once the measurement has been established, place the square against the top edge of the rafter, lining up the 4 on the common scale and marking it, as shown in *Figure 18G.*

Step 8 If a bottom or vertical cut is required, follow the procedure for cutting the horizontal cut of the bird's mouth seat cut. (In the procedure previously described, a pattern for a common rafter has been created. Two pieces of scrap material [1 × 2 or 1 × 4] should be nailed to the rafter, one about 6" from the top or plumb cut and one at the bird's mouth seat cut. These scrap pieces will allow you to align the pattern to the next piece of stock.)

NOTE

The *Appendix* provides information on the laying out and erecting hips and valleys.

Bird's Mouth

When making the cut for the bird's mouth, the depth should not exceed ⅓ the width of the rafter.

Calculating Common Rafter Lengths

Now that you are familiar with the method of arriving at the length of a rafter using the square, find the lengths of common rafters for the following spans:

Span	Run	Rise per Foot of Run	Rafter Length
26'	13'	6"	
24'	12'	4"	
28'	14'	8"	
32'	16'	12"	
30'	15'	7"	

Note: Twelve is a factor used to obtain a value in feet. Be sure to reduce or convert to the lowest terms.

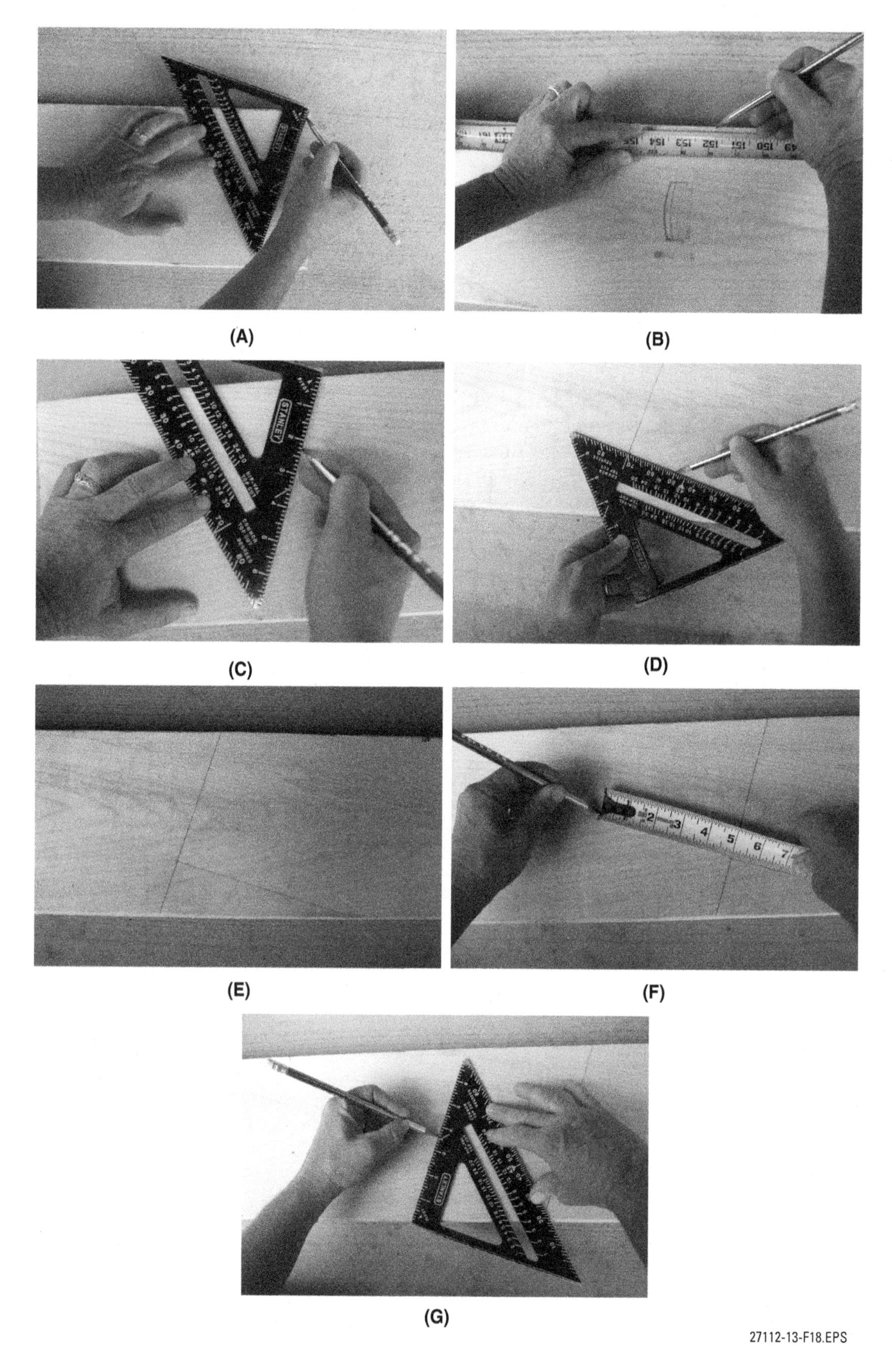

27112-13-F18.EPS

Figure 18 Laying out a common rafter using a Speed Square™.

4.0.0 Section Review

1. The part of a rafter that rests squarely on the top plate is the _____.
   a. heel
   b. tail end
   c. ridge end
   d. bird's mouth

2. The rafter locations should be laid out on the _____.
   a. top plate and ridgeboard
   b. soleplate and ridgeboard
   c. sill plate and top plate
   d. studs and ridgeboard

3. When the roof span is 36' and the pitch is 1/6, the total rise is _____.
   a. 3'
   b. 6'
   c. 8'
   d. 12'

4. When cutting a bird's mouth, a handsaw should be used for the finishing cuts.
   a. True
   b. False

Section Five

5.0.0 Erecting Gable Roofs

Objective

Describe how to erect a gable roof.

a. Describe how to install rafters.

Performance Task 5

Cut and install roof rafters for a gable roof.

Trade Terms

Purlin: A horizontal roof support member parallel to the plate and installed between the plate and the ridgeboard.

This section contains an overview of the procedure for erecting a gable roof. The layout and construction of hips and valleys is covered in the *Appendix*.

> **WARNING!** Be sure to follow applicable fall-protection procedures.

5.1.0 Installing Rafters

Rafters are installed using the following procedure:

> **NOTE** Do not mark a 2 × 4 with the total rise in advance.

Step 1 Start by placing boards or panels over the ceiling joists to walk on. Nail a rafter at each end of the ridgeboard. Then, lift the ridgeboard to a temporary position, secure it, and nail the bird's mouth of each rafter to the top plate (*Figure 19*). Nail the rafters in pairs.

Step 2 On the opposite side, start by nailing the bird's mouth to the joist, then toenail the plumb cut into the ridgeboard. Once this is done, use a temporary brace to hold the ridgeboard in place while installing the remaining rafters. Remember to keep the ridgeboard straight and the rafters plumb.

Step 3 Run a line and trim the rafter tails.

If the rafter span is long, additional support will be required. *Figure 20* shows the use of strongbacks, purlins, braces, and collar ties (collar beams) for this purpose. Two-by-six collar ties are installed at every second rafter. Two-by-four diagonal braces are notched into the purlins. Strongbacks are L-shaped members that run the length of the roof. They are used to straighten and strengthen the ceiling joists.

5.2.0 Framing an Opening in the Roof

It is sometimes necessary to make an opening in a roof for a chimney, skylight, or roof window (*Figure 21*). The following is a general procedure for framing such an opening:

Step 1 Lay out the opening on the floor beneath the opening and then use a plumb bob to transfer the layout to the roof. If you are framing a chimney, be sure to leave adequate clearance. If the opening is large, allow for double headers.

Step 2 Cut the rafters per the layout. Install the double headers and then install a double trimmer rafter on each side of the opening as shown in *Figure 21*.

> **NOTE** Install a brace across the rafters to be cut when creating an opening in the roof. It will temporarily hold the framing in place until the headers are installed.

Rafter Marks

When placing rafters into position on the top plate, ensure they are installed on the correct side of your marks. Start installing the rafters from one end of the roof and work toward the other end. Double-check each rafter position for consistency before nailing it into place. Roof framing mistakes are time consuming to fix and must be avoided.

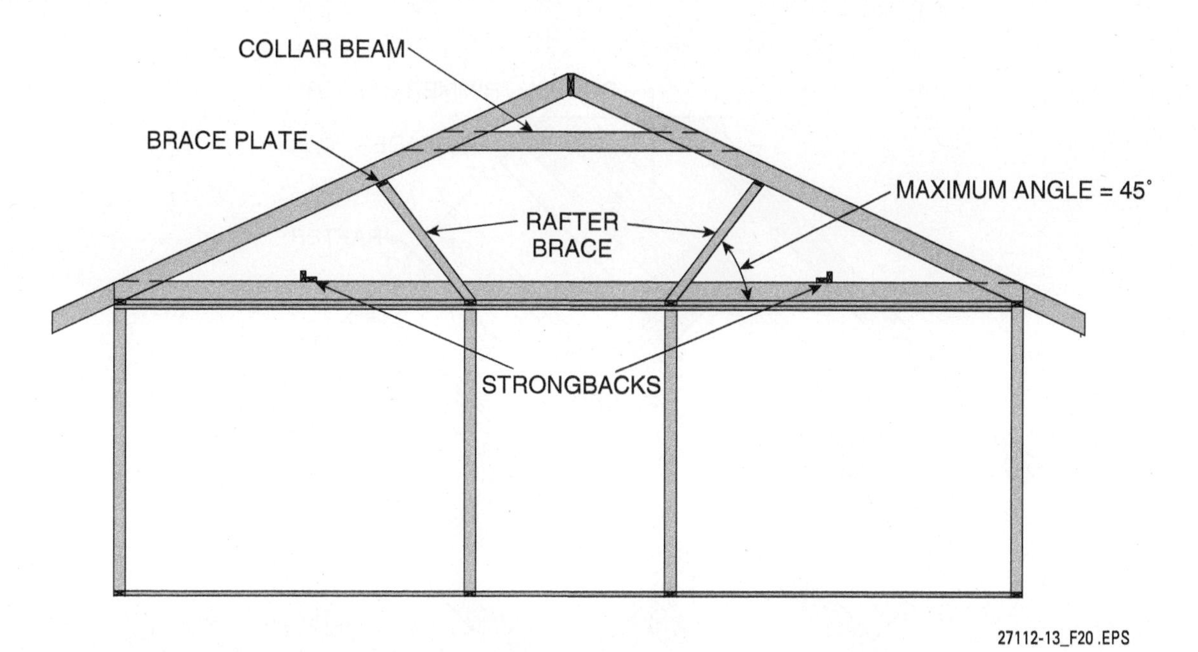

Figure 20 Bracing a long roof span.

Framing Roof Openings

Before cutting an opening in a roof, check with the site engineer to ensure you are proceeding according to specifications. Failure to check with the engineer may result in you cutting an incorrect rafter, and new rafters or trusses may be required. This type of error is very time consuming and expensive to fix. Also, always check local fire codes for the proper clearance around a chimney and other roof openings. If the framing for the opening is too close to a chimney, it could cause a fire.

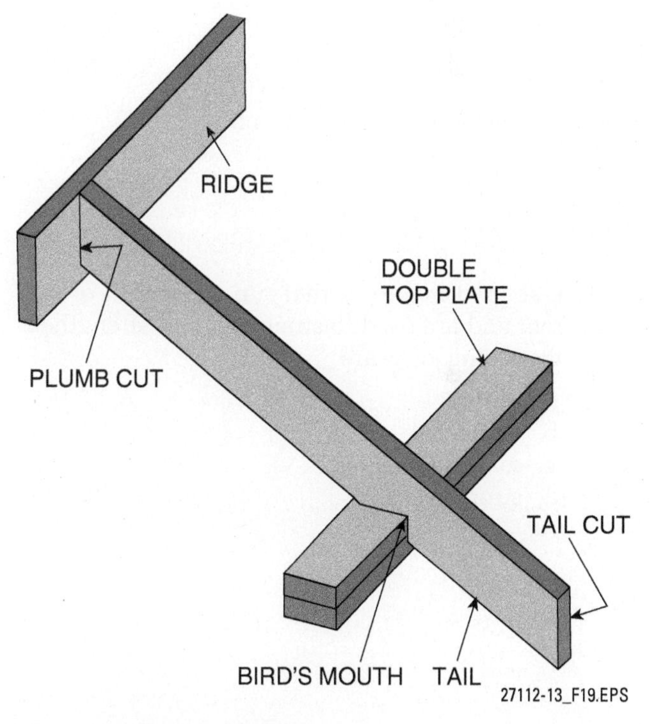

Figure 19 Rafter installation.

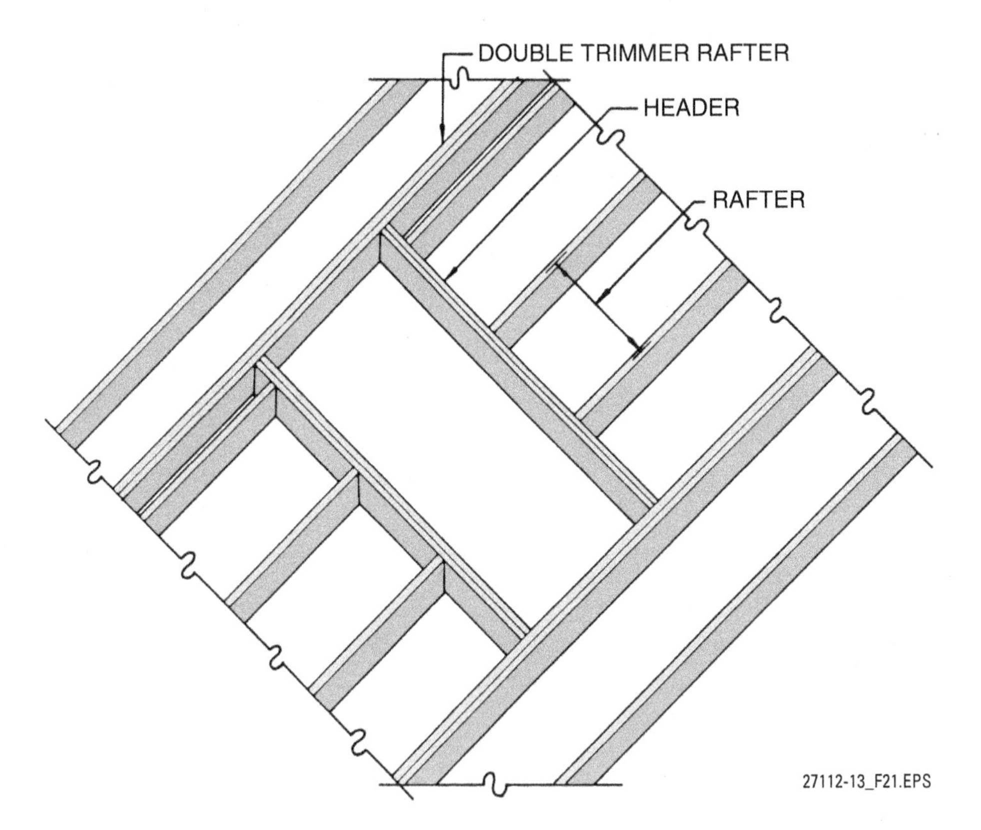

Figure 21 Framing an opening in a roof.

5.0.0 Section Review

1. L-shaped members that run the length of the roof and are used to straighten and strengthen the ceiling joists are _____.
   a. purlins
   b. collar ties
   c. diagonal bracing
   d. strongbacks

2. When laying out a roof opening, what tool is used to transfer the layout from the floor to the roof?
   a. Rafter square
   b. Plumb bob
   c. Carpenter's level
   d. Wall level

Section Six

6.0.0 Frame Gable End Walls

Objective

Describe how to frame a basic gable end wall.

a. Describe how to frame a gable overhang.
b. Explain how to frame an opening in a roof

Performance Task 6

Frame a gable end wall.

Trade Terms

Lookout: A structural member used to frame an overhang.

With the roof framed in, now the gable ends must be framed in and enclosed to provide additional stability and support for the roofing materials. The enclosed spaces form attics, which must be vented to allow heat that rises from the lower floors of the building to escape to the outdoors. There is also a need to vent moisture that accumulates in an attic due to condensation that occurs when rising heat from below meets the cooler air in the unheated attic space. Several methods are used to vent roofs. *Figure 22* shows two types of gable end vents.

Notice that the lengths of the studs decrease as they approach the sides of the building. Each pair of studs must be measured and cut to fit. One method to frame a gable end opening is as follows:

Step 1 Start by plumbing down from the center of the rafter to the top plate, then mark and square a line on the top plate at this point.

Step 2 Lay out the header and sill for the vent opening by measuring on either side of the plumb line. (For this vent, measure 8" on either side of the plumb line.)

Step 3 Mark the stud locations above the wall studs, then stand a stud upright at the first position, plumb it, and mark the diagonal cut for the top of the stud.

Step 4 Measure and mark the next stud in the sequence. The difference in length between the first and second studs is a common difference, which can be applied to all remaining studs.

Step 5 Cut and install the studs as shown in *Figure 23*. Toenail or straight-nail the gable stud to the top plate. The notch should be as deep as the rafter.

Step 6 When studs are in place, cut and install the header and sill for the vent opening. Then lay out, cut, and install cripple studs above and below the vent using the same method as wall openings.

NOTE

When creating an opening in the gable end, Install a brace across the studs to be cut. It will temporarily hold the framing in place until the headers are installed.

6.1.0 Framing a Gable Overhang

If an extended gable overhang (rake) is required, the framing must be done before the roof framing is complete and the sheathing is applied. *Figure 24* shows two methods of framing overhangs. In the view on the left, which can be used for a small overhang, a 2 × 4 barge rafter and short lookouts are used. In the view on the right, 2 × 4 lookouts are let into the rafters. Check the local building code prior to notching any framing members.

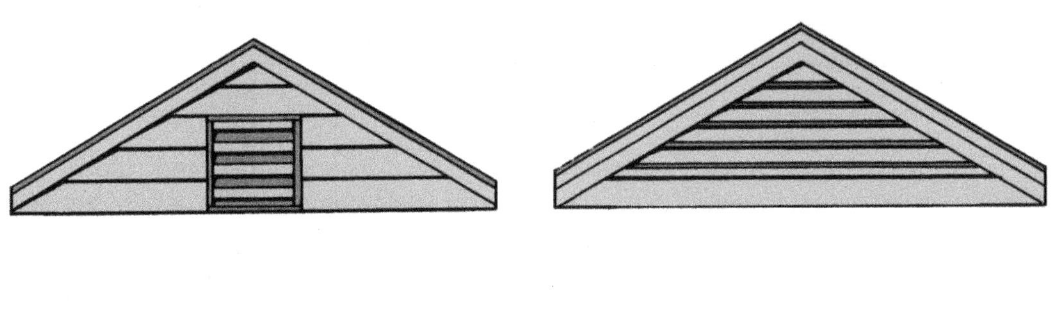

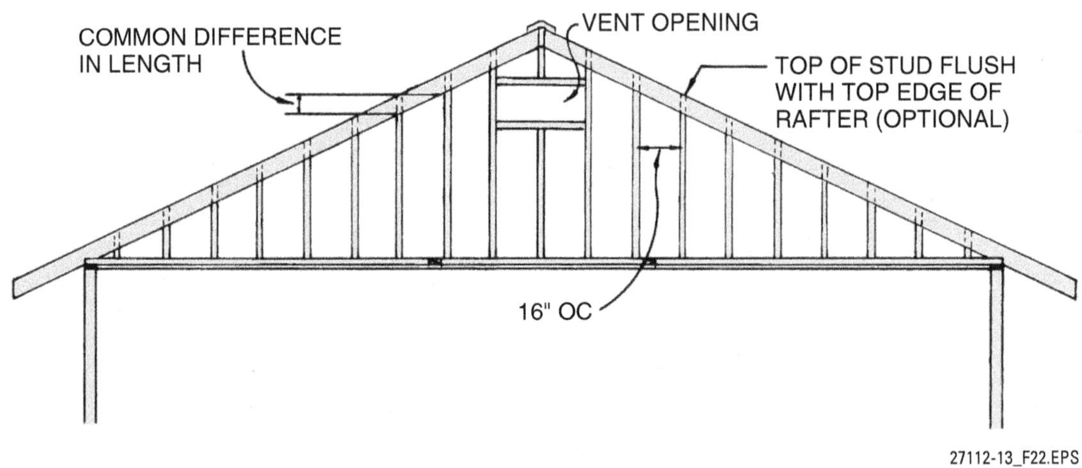

Figure 22 Gable end vents and frame.

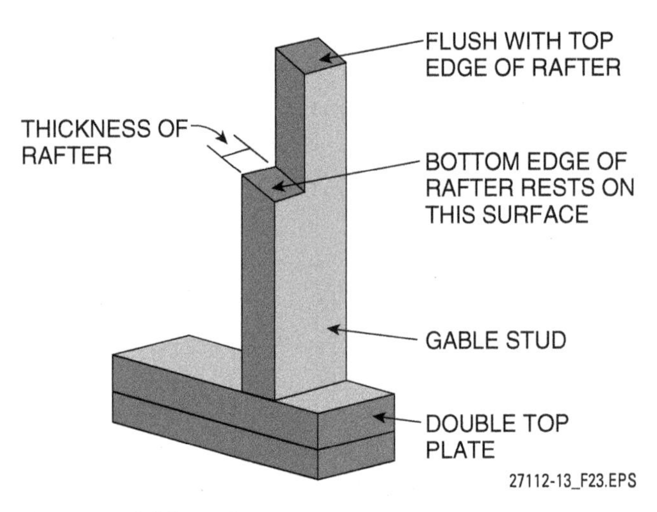

Figure 23 Gable stud.

Barge Rafter and Lookouts

A gable overhang consists of a barge (fly) rafter on the end and lookouts that connect the barge rafter to a common rafter. The lookouts provide structural support for the overhang and a solid foundation for a soffit or other decorative finish.

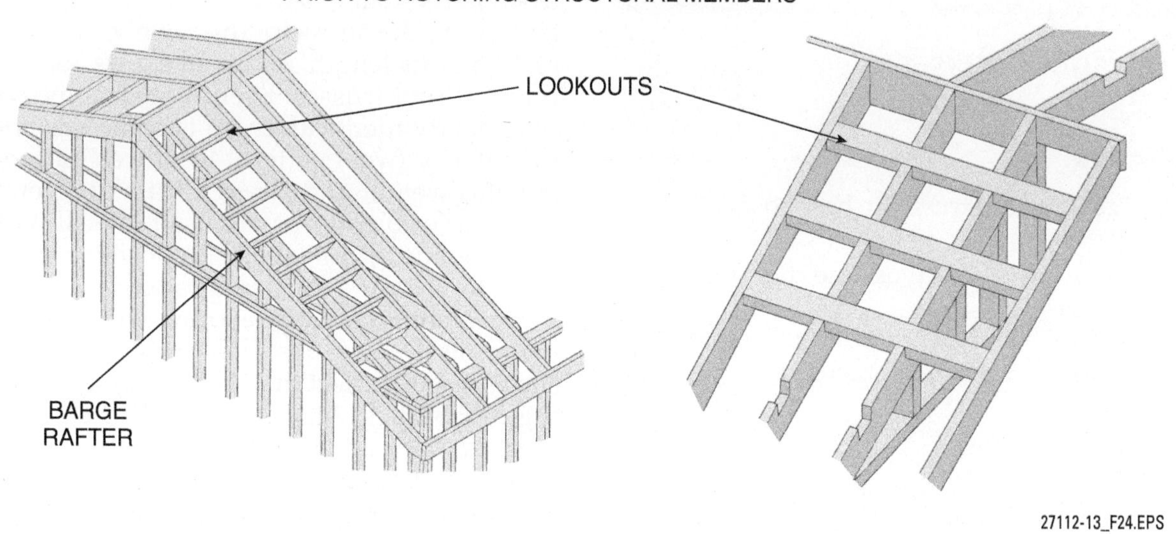

Figure 24 Two methods for framing a gable overhang.

6.0.0 Section Review

1. An attic space must be tightly sealed to prevent moisture from accumulating in it.
   a. True
   b. False

2. The barge rafter is connected to the common rafter using _____.
   a. joists
   b. headers
   c. diagonal bracing
   d. lookouts

SECTION SEVEN

7.0.0 ROOF TRUSSES

Objective

Recognize the use of trusses in basic roof framing.

a. Identify the various types and components of trusses.
b. Identify the basics of truss installation.
c. Identify the basics of truss bracing.

Performance Task 7

Erect a gable roof using trusses.

Trade Terms

Open web steel joist: Steel joist, also known as a bar joist, that is used to support roof loads.

Spreader bar: A rigging device used when lifting large members with a crane.

Tag line: Cord or rope attached to a structural member or other large unit that is being positioned by a crane, to assist in guiding the member into proper position.

Tiedown: Metal anchoring device used to fasten framing members to one another.

Roof trusses rely on a triangular arrangement of the chords and webs to transfer loads to the walls and foundation below. Trusses permit longer spans, which result in larger unobstructed areas within the building. Roof trusses are used in the majority of new residential and commercial construction.

7.1.0 Truss Construction

In most cases, it is much faster and more economical to use prefabricated trusses rather than installing individual rafters and joists. Even if a truss costs more to buy than the comparable framing lumber, truss erection requires significantly less labor than stick framing. Another advantage is that a truss will span a greater distance without a loadbearing wall than stick framing. Just about any type of roof can be framed with trusses.

Terminology associated with truss members is noted in *Figure 25*.

A truss is a framed or jointed structure that is designed so that when a load is applied at any intersection, the stress in any member is in the direction of its length. *Figure 26* shows some of the types of roof trusses. Even though some trusses look nearly identical, there is some variation in the interior (web) pattern. Each web pattern distributes weight and stress a little differently, so different web patterns are used to deal with different loads and spans. The decision of which truss to use for a particular application is made by the architect or engineer, and will be shown on the prints. Do not substitute or modify trusses onsite as it could affect their weight- and stress-bearing capabilities. Also, be careful when handling trusses; they are more delicate than standard lumber and cannot be thrown around or stored in a way that applies uneven stress to them.

Trusses are stored or carried on trucks either lying on their sides or upright in cradles, and protected from rain or other moisture. A crane should be used to lift the trusses from the truck. Some trucks are equipped with roller mats to unload the trusses. Unless the trusses are small and light, it is recommended that a crane be used to lift them into place on the building frame. *Figure 27* shows examples of erecting methods used for trusses. Note the tag lines, which are held by a worker on the ground to stabilize the truss while it is being lifted.

WARNING!

Installing trusses can be extremely dangerous. It is very important to follow the manufacturer's instructions for bracing and to follow all applicable safety procedures.

A single lift line can be used for trusses with a span of less than 20'. From 20' to 40', two chokers are needed. If the span exceeds 40', a spreader bar is required.

7.2.0 Installing Trusses

Truss installation poses several hazards for workers, with the greatest hazard being falls. These falls can be attributed to two primary reasons: truss installation occurs high above the ground and trusses are not stable until they are properly braced. The Occupational Safety and Health Administration (OSHA) requires fall-protection measures to be in place for activities that occur 6 feet or more above lower levels. Although personal fall arrest systems (PFAS) are widely used, they aren't appropriate safety equipment to use

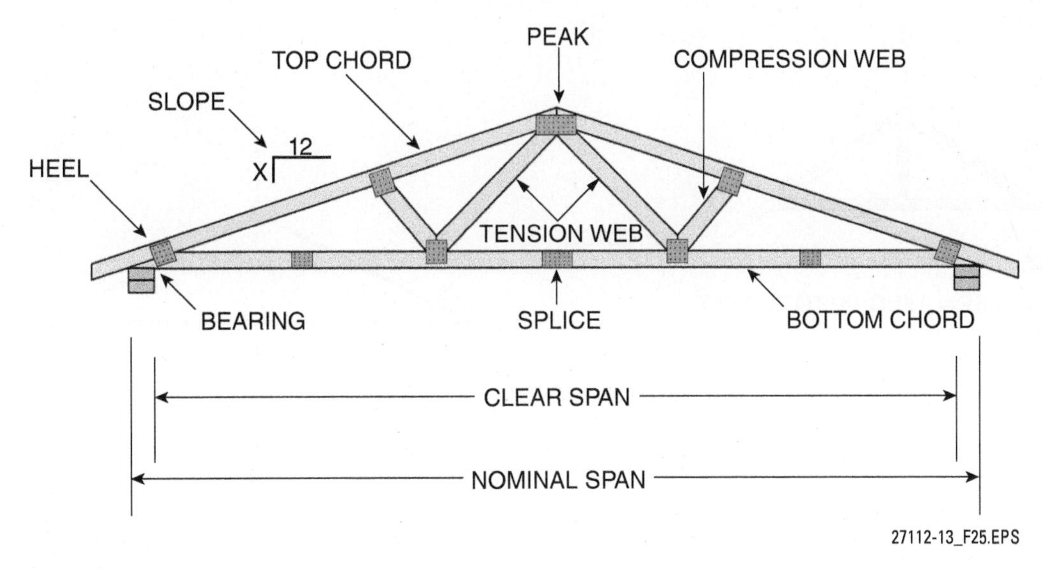

Figure 25 Truss components.

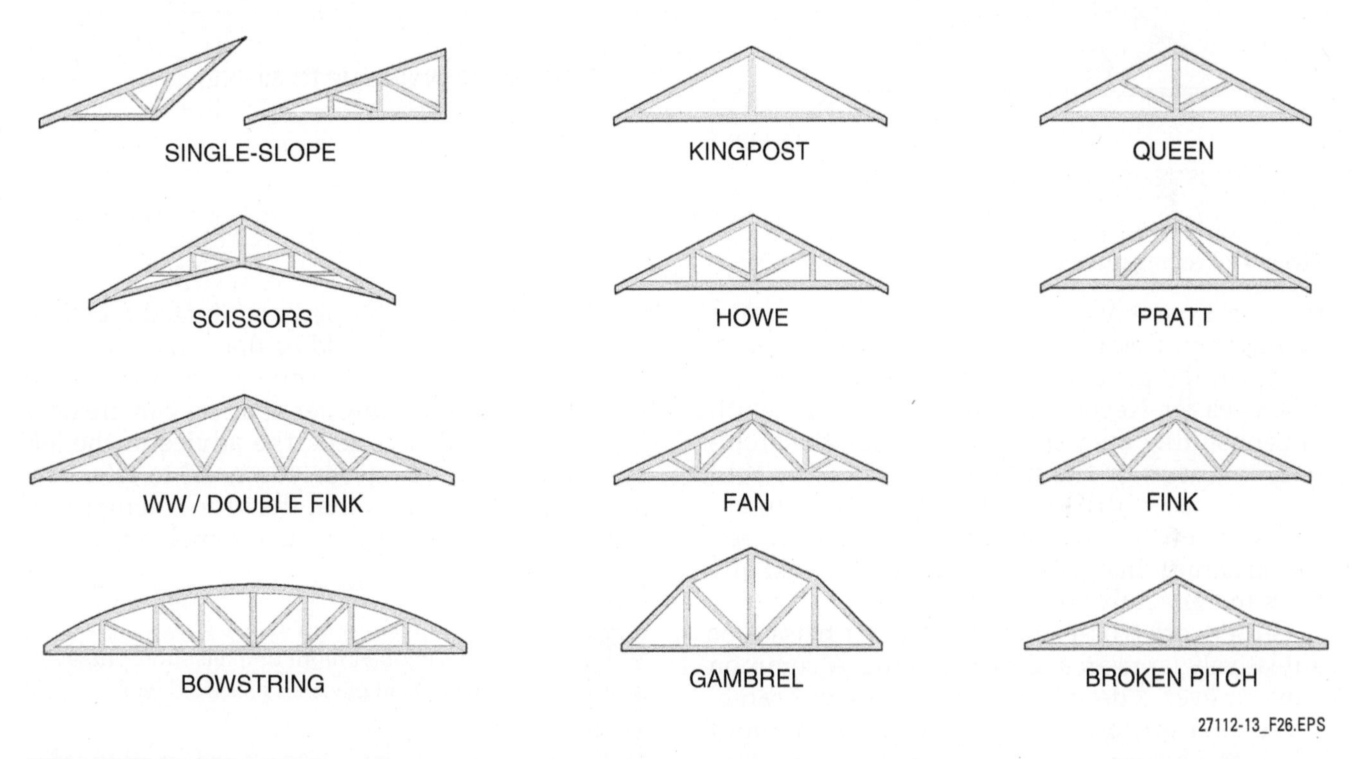

Figure 26 Types of trusses.

when starting truss installation. A single truss cannot support the weight of a worker.

Several means can be used to protect workers until a fully braced, multitruss section has been properly constructed. Truss sections can be assembled and sheathed on the ground and then lifted into place using a crane. Peak anchors and lifelines can be installed prior to the section being lifted into place. Lifts, scaffolds, and ladders can be used for the initial truss erection. Lifts provide a stable elevated platform, but workers must follow all safety procedures and conduct all work from inside the basket. Scaffolds can be used inside and outside of the building. Bracket scaffolds can be attached to the walls and provide a stable elevated walking surface. For some truss installations, platforms or stepladders are useful.

Once an assembled truss section has been installed, secured, and properly braced, it can be used as an attachment point for a PFAS.

Before installing roof trusses, refer to the framing plans for the proper truss locations. *Figure 28* is an example of a truss placement diagram. Truss placement diagrams are typically provided by a truss manufacturer to ensure the trusses are installed properly. If a truss is damaged before erec-

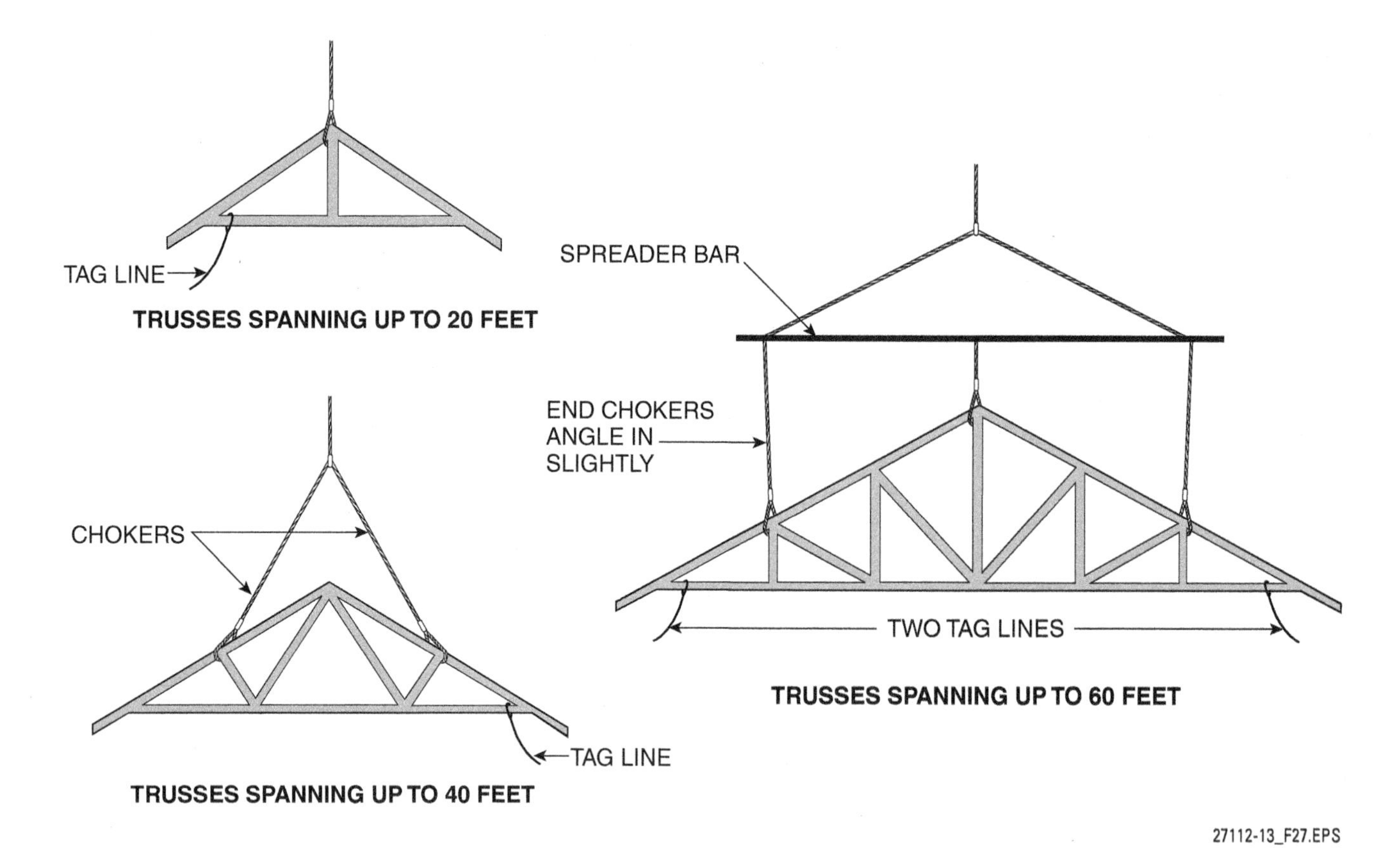

Figure 27 Erecting trusses.

tion, obtain a replacement or instructions from a qualified individual to repair the truss. Repairs made on the ground are usually better and are always easier. Never alter any part of a truss without consulting the job superintendent, the architect/engineer, or the manufacturer of the roof truss. Cutting, drilling, or notching any member without the proper approval could destroy the structural integrity of the truss and void any truss manufacturer warranties.

Girders are trusses that carry other trusses or a relatively large area of roof framing. A common truss or even a double common truss will rarely serve as a girder. If there is a question about whether a girder is needed, your job superintendent should consult with the truss manufacturer. Double girders are commonplace. Triple girders are sometimes required to ensure the proper load-carrying capacity. Always be sure that multiple-member girders are properly laminated together. Spacing the trusses should be done in accordance with the truss design. In some cases, very small deviations from the proper spacing can create a big problem. Always seek the advice of the job superintendent whenever you need to alter any spacing. Make certain that the proper temporary bracing is installed as the trusses are being set.

WARNING!

Never leave a job at night until all appropriate truss bracing is in place and secured well.

Light trusses (under 30' wide, for example) can be installed by lifting them up and anchoring them to the top plate, then pushed into place with

Truss Storage

Trusses should only be stored on-site for a short period of time. They are very large and take up a lot of storage area. In addition, they are more likely to get wet when stored on-site. When a truss is saturated with moisture, excess shrinkage will occur, causing structural damage to the webs and chords. If on-site storage of trusses is unavoidable, ensure they are well covered with a waterproof tarp and raised off the ground on pallets. During the planning phase of a project, truss delivery should be accounted for in the schedule so that they can be delivered and erected in a timely manner.

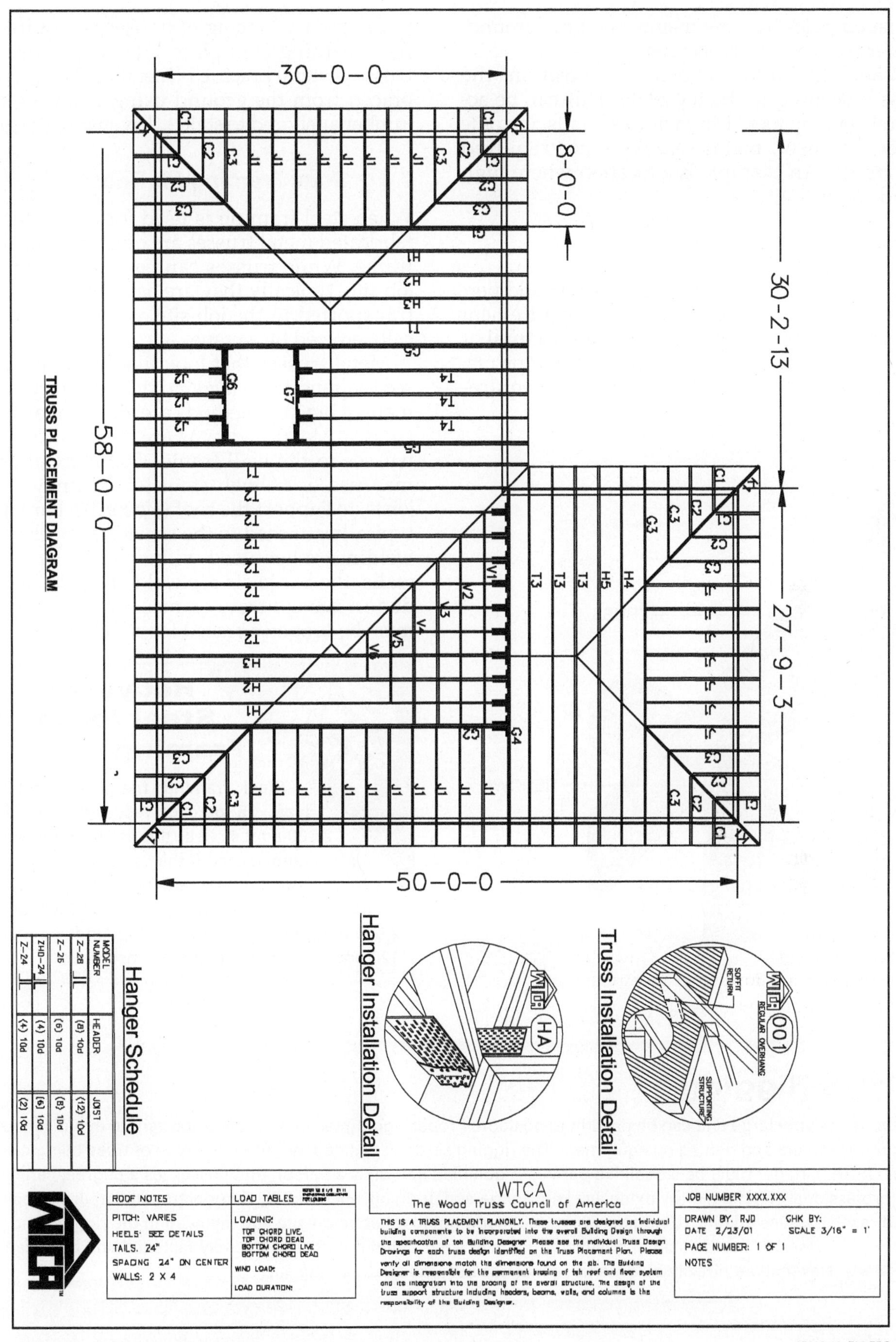

Figure 28 Example of a truss placement diagram.

Y-shaped poles by crew members on the ground. Larger trusses require a crane.

When the bottom chord is in position, the truss is secured to the top plate. This can be accomplished by toenailing with 10d nails. In some cases, however, metal tiedowns are required (*Figure 29*). An example is a location where high winds occur.

7.3.0 Bracing Roof Trusses

In most circumstances, it is necessary to add permanent bracing to the trusses. This requirement would be established by the certified structural or truss engineer and would appear on the prints. *Figure 30* shows an example of such a requirement.

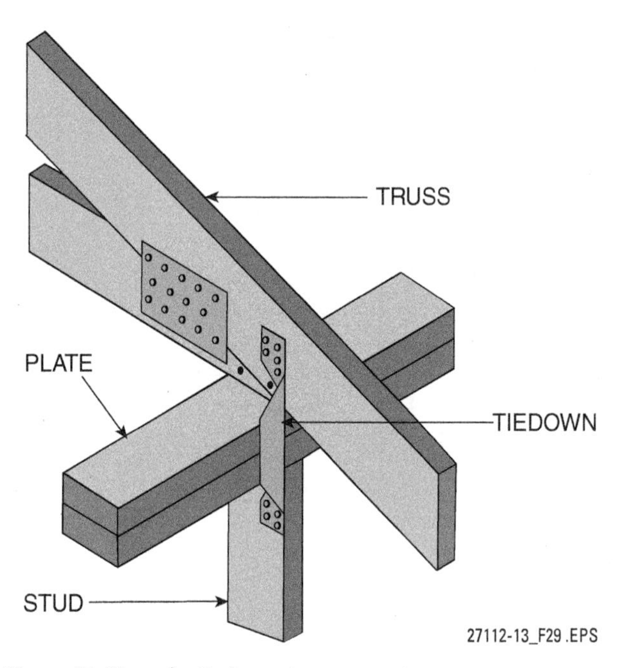

Figure 29 Use of a tiedown to secure a truss.

Temporary bracing of trusses is required until the sheathing is in place. Across the tops of the trusses, lateral bracing is installed. Gable ends are braced from the ground using 2 × 4s or similar lumber anchored to stakes driven into the ground.

7.4.0 Steel Framing for Roofs

When steel framing is used for a structure, prefabricated metal trusses are often used as the roof frame. While trusses can be constructed on the job site, typically they are assembled off-site and transported to the job site, where they are typically erected by a crane.

Metal trusses that look like wood are used in some applications (*Figure 31*). Temporary and permanent bracing methods are also similar to wood trusses. Metal tiedowns are used to connect the trusses to the wall frame. Commercial projects may use open web steel joists, known as bar joists, to support the roof (*Figure 32*). Bar joists are typically attached to the structural steel frame by welding. When used with concrete or masonry walls, the bar joists are welded to plates embedded in the concrete or masonry. Bar joists also must be properly braced for stability.

GOING GREEN

Recycling Steel for Steel Framing

Steel framing members contain at least 25 percent recycled steel. For a 2,000-sq-ft house, this is the equivalent of six scrapped cars. If the same house were built using wood, it would require approximately 40 to 50 trees. In addition, by recycling one ton of steel, 2,500 pounds of iron ore; 1,400 pounds of coal; and 120 pounds of limestone are conserved.

Truss Rigging

Trusses are very large and can be difficult to position. Proper rigging techniques are important to avoid seriously injuring workers and damaging equipment. The rigging setup varies depending on the type of truss being used. Per OSHA, a rigger must be a qualified person. A qualified person is a person that possesses a degree, certificate, or professional standing; or by extensive knowledge and, training, successfully demonstrate the ability to solve and resolve issues relating to rigging work. Qualified riggers must be on-site and involved in the rigging and lifting of trusses. Enlist the help of qualified riggers to ensure that the truss is securely fastened and balanced properly. Never stand directly below a truss that has been hoisted into the air.

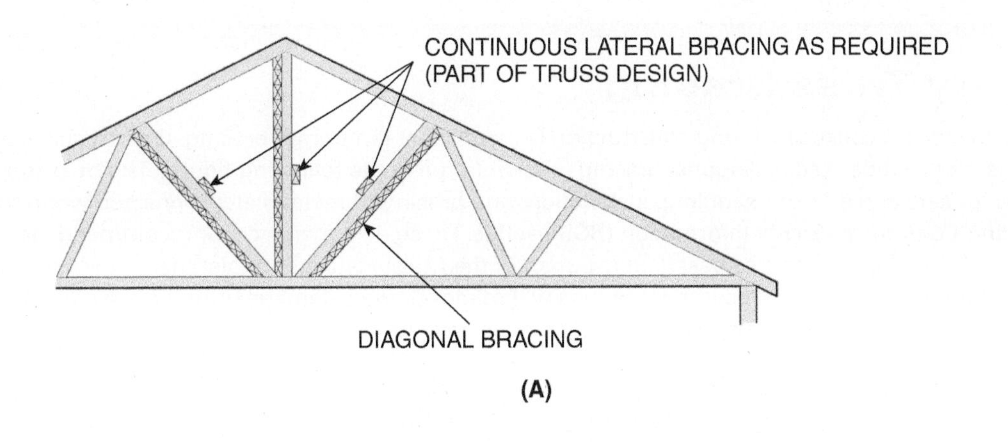

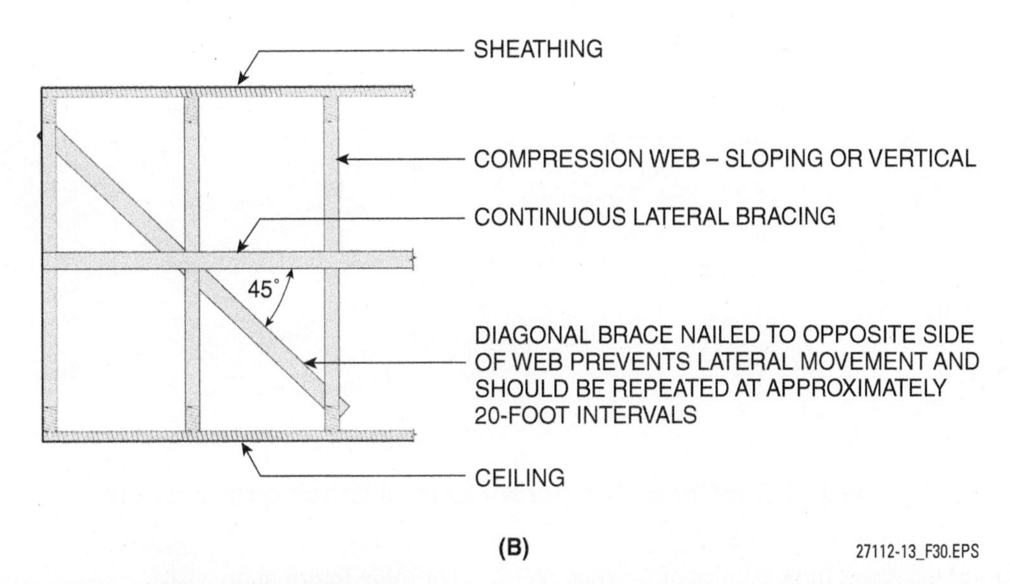

27112-13_F30.EPS

Figure 30 Example of permanent bracing specification.

27112-13_F31.EPS

Figure 31 Metal roof trusses.

27112-13_F32.EPS

Figure 32 Bar joists.

Temporary Truss Bracing

The leading cause of truss collapses during construction is insufficient temporary bracing. One crucial aspect of bracing that is often overlooked is diagonal bracing (shown in red in the following figure). The truss industry provides guides for safe and efficient handling, installation, and bracing of metal plate–connected wood trusses called the Building Component Safety Information (BCSI) series. These documents cover recommendations for correct hoisting and bracing procedures based on the span of the truss, installation tolerances, and limits on construction loading. Following BCSI recommendations will minimize truss damage during construction, lead to better long-term performance of the truss system, and create a safer work site.

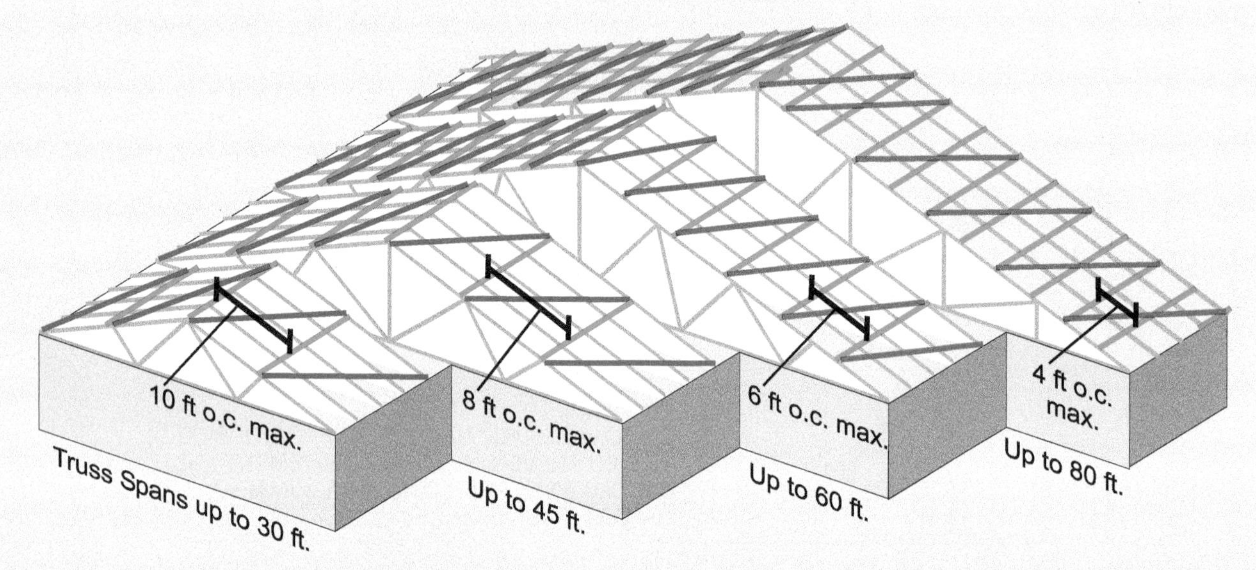

Spacing of Top Chord Temporary Lateral Bracing on Trusses

27112-13_SA02.EPS

Text courtesy of the Wood Truss Council of America (WTCA). For more information, visit **http://www.sbcindustry.com/**.

7.0.0 Section Review

1. A framed or jointed structure commonly used for roof frames is called a _____.
   a. king post
   b. corner assembly
   c. truss
   d. spreader

2. Trusses must be temporarily braced until _____.
   a. the roof finish material is installed
   b. the foundation is backfilled
   c. the owner takes occupancy of the structure
   d. roof sheathing is in place

3. In areas that are prone to heavy winds, what type of fastener or device may be required to attach trusses to the top plate?
   a. Screws
   b. Metal tiedowns
   c. Cleats
   d. Anchor bolts

4. Bar joists are fastened to the structural steel by _____.
   a. sheet metal screws
   b. toggle bolts
   c. anchor bolts
   d. welding

Section Eight

8.0.0 Roof Sheathing

Objective

Describe the basics of roof sheathing installation.

Performance Task 8

Sheath a gable roof with an opening.

Sheathing should be applied as soon as the roof framing is finished. The sheathing provides additional strength to the structure, and provides a base for the roofing material.

Some of the materials commonly used for sheathing are plywood, oriented strand board (OSB), waferboard, shiplap lumber, and common boards. When composition shingles are used, the sheathing must be solid so plywood or OSB is used. When solid sheathing is used, leave a ⅛" space between panels to allow for expansion. If wood shakes are used, the sheathing boards may be spaced.

The following is an overview of roof sheathing requirements using plywood or other 4 × 8 panel material.

Step 1 Start by measuring up 48¼" from where the finish fascia will be installed. Snap a chalkline at that point and lay the first panel down and nail it. Install H-clips midway between the rafters or trusses before starting the next course (*Figure 33*). H-clips eliminate the need for tongue-and-groove panels.

Step 2 Install the remaining panels. Stagger the panels by starting the next course with a half panel. Allow the edges to extend over the hip, ridge, and gable ends. Remove the excess sheathing off with a circular saw.

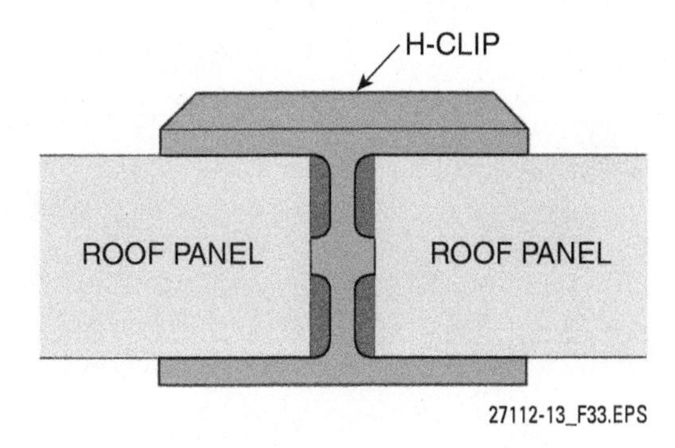

Figure 33 Use of H-clips.

Once the sheathing has been installed, an underlayment of asphalt-saturated felt or other specified material must be installed to prevent moisture from saturating the panels. For roofs with a slope of 4" or more, 15-pound roofer's felt is commonly used.

Material such as coated sheets or heavy felt that could act as a vapor barrier should not be used. They can allow moisture to accumulate between the sheathing and the underlayment. The underlayment is applied horizontally with a 2" top lap and a 4" side lap, as shown in *Figure 34A*. A 6" lap should be used on each side of the center line of hips and valleys. A metal drip edge is installed along the rakes and eaves to keep out wind-driven moisture.

In climates where snow accumulates, a waterproof underlayment, as shown in *Figure 34B*, should be used at roof edges and around chimneys, skylights, and vents. This underlayment has an adhesive backing that adheres to the sheathing. It protects against water damage that can result from melting ice and snow that backs up under the shingles.

Rafter Tails

Prior to installing the sheathing, check the rafter tails to ensure they form a straight line. If you made accurate measurements and cuts, the rafters should all be the same length. If they are not, shim or trim for alignment. This will avoid an unsightly, crooked roof edge and provide a solid nailing base for the false fascia.

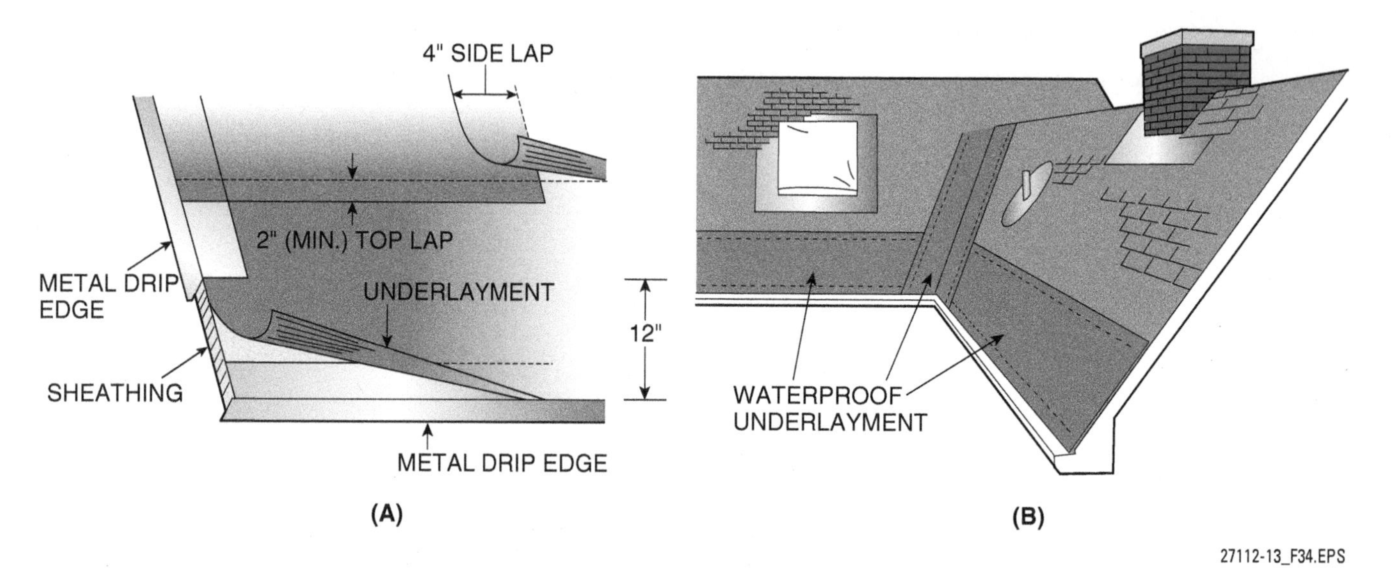

Figure 34 Underlayment installation.

Crane Delivery

If a crane is used to place a stack of sheathing on a roof, a special platform must be in place to provide a level surface. The platform must be constructed over a loadbearing wall so the weight of the sheathing doesn't cause structural damage to the framing.

Sheathing

If the trusses or rafters are on 24" centers, use ⅝" sheathing. With 16" centers, ½" sheathing may be used. This may vary depending on the local codes. Before starting construction, be sure to check the sheathing and fastener requirements in the local building code.

Case History

Roof Openings

A 21-year-old apprentice was installing roofing on a building with six unguarded skylights. During a break, he sat down on one of the skylights. The plastic dome shattered under his weight and he fell to a concrete floor 16 feet below, suffering fatal head injuries.

The Bottom Line: Never sit or lean on a skylight. Always provide appropriate guarding and fall protection for work around skylights and other roof openings.

Sheathing Safety

When installing sheathing on an inclined roof, always be sure to use a toeboard and safety harness. You can gain additional traction by wearing skid-resistant shoes. Falling from a roof can result in serious injury or death. Taking the time to follow proper procedures will minimize the chances of slipping or falling. The picture shows a safety harness anchor and toeboard.

27112-13_SA03.EPS

Felt Installation

The felt underlayment should be applied to the installed sheathing as soon as possible. It is important that both the sheathing and felt be dry and smooth at the time of installation. If the roof is damp or wet, wait a couple of days for it to dry completely before installing the felt. Moisture will cause long-term damage to both the sheathing and underlayment.

8.0.0 Section Review

1. The type of sheathing that should be installed if the roof finish is composition shingles is _____.
   a. plywood
   b. common board
   c. shiplap board
   d. particleboard

2. Roof underlayment should be lapped _____.
   a. 1" at the top and 2" along the sides
   b. 2" at the top and 4" along the sides
   c. 4" at the top and 6" along the sides
   d. 4" at the top and 8" along the sides

SECTION NINE

9.0.0 ESTIMATING ROOFING MATERIALS

Objective

Describe how to perform a material takeoff for a roof.

a. Determine the materials needed for a gable roof.

Performance Task 9

Perform a material takeoff for a roof.

Estimating the material you will need for a roof depends on several factors:

- Type of roof you are planning to construct
- Size of the roof
- Spacing of the framing members
- Load characteristics.

Local building codes address these factors and pertinent information is included on the prints. Lumber for conventional roof framing ranges from 2 × 4 to 2 × 10, depending on the span and the load. For example, 2 × 4 framing at 16" OC might support a 9' span. By comparison, 2 × 10 framing on 16" centers would support a span of more than 25'.

9.1.0 Determining Materials Needed for a Gable Roof

A material takeoff for a gable roof should include the rafters, ridgeboard, and sheathing.

9.1.1 Estimating Rafters

To determine how much lumber is needed for rafters on a gable roof:

Step 1 Determine the length of each common rafter (including the overhang) using the rafter square or another method.

Step 2 Calculate the number of rafters needed based on the rafter spacing (16", 24", etc.). Remember that one extra rafter will be needed for each gable end. You will also need barge rafters in the gable overhang. Note that these are usually one size smaller than the common rafters.

Step 3 Multiply the result by 2 to account for the two sides of the roof.

9.1.2 Estimating Ridgeboards

The ridgeboard is usually one dimension wider than the rafters. To determine how much lumber is needed for the ridgeboard on a gable roof:

Step 1 Determine the length of the plate on one side of the structure.

Step 2 Add more length as needed to account for gable overhang.

9.1.3 Estimating Sheathing

To determine the number of panels of sheathing required for a gable roof:

Step 1 Multiply the length of the roof (including overhangs) by the length of a common rafter. This yields half the area of the roof.

Step 2 Divide the roof area by 32 (the area of a 4 × 8 panel) to determine the approximate number of panels of sheathing needed. Round up if the division yields a partial sheet.

Step 3 Multiply by 2 to obtain the number of panels needed for the full roof area.

9.0.0 Section Review

1. Calculate the length of a ridgeboard for a gable roof with a 2' overhang on each end on a building measuring 24' × 40'.
   a. 24'
   b. 28'
   c. 40'
   d. 44'

2. Calculate the number of 4' × 8' OSB panels that are needed for a gable roof measuring 42' long with a common rafter length of 12'-6".
   a. 10
   b. 17
   c. 20
   d. 34

SUMMARY

The correct layout and framing of ceilings and roofs requires patience and skill. If the measurement, cutting, and installation work are not done carefully and precisely, the end result will never look right. Fortunately, there are many tools and reference tables that help to simplify the process. The important thing is to be careful and precise with the layout and cutting of the first ceiling joist or roof rafter of each type. These ceiling joists and rafters can be used as patterns for the others.

Review Questions

1. Ceiling joists span the narrow dimension of the building from _____.
   a. top plate to ridgeboard
   b. top plate to top plate
   c. header to header
   d. top chord to top chord

2. Which of the following statements about ceiling joists is *not* true?
   a. They should never be spliced between bearing points
   b. They usually span the narrow width of the building
   c. They are usually in line with the wall studs
   d. They provide a nailing surface for finishing material

3. The upper ends of the outside ends of ceiling joists must be cut _____.
   a. to match the rafter pitch
   b. flush with the top plate edge
   c. at a 45-degree angle
   d. with a bird's mouth

4. The type of roof that has four sides running toward the center of the building is the _____.
   a. gable roof
   b. hip roof
   c. shed roof
   d. gable and valley roof

5. The roof framing system in which carpenters assemble joists and rafters on-site is called _____.
   a. a site-assembled system
   b. a truss system
   c. a stick-built system
   d. a fabricated system

Refer to *Review Question Figure 1* to answer Questions 6–8.

6. Letter A indicates the _____.
   a. rise
   b. span
   c. run
   d. slope

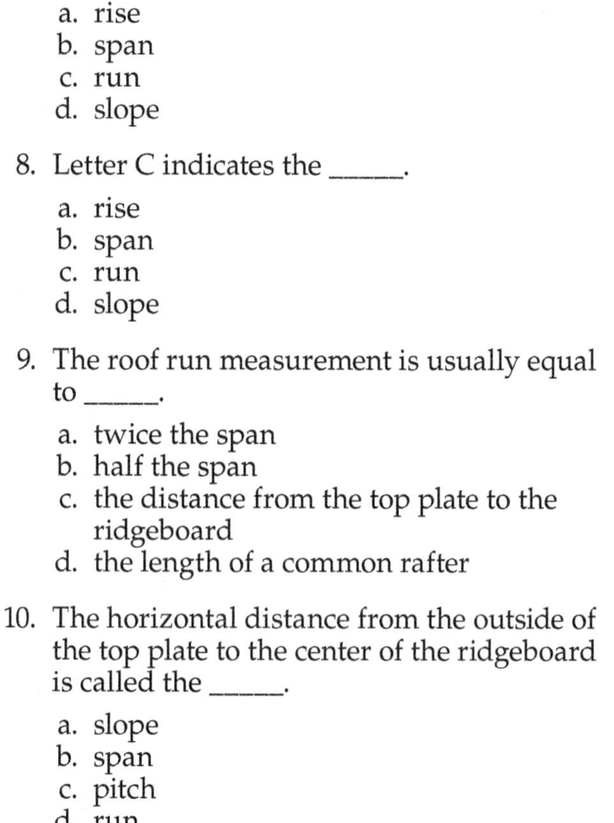

7. Letter B indicates the _____.
   a. rise
   b. span
   c. run
   d. slope

8. Letter C indicates the _____.
   a. rise
   b. span
   c. run
   d. slope

9. The roof run measurement is usually equal to _____.
   a. twice the span
   b. half the span
   c. the distance from the top plate to the ridgeboard
   d. the length of a common rafter

10. The horizontal distance from the outside of the top plate to the center of the ridgeboard is called the _____.
    a. slope
    b. span
    c. pitch
    d. run

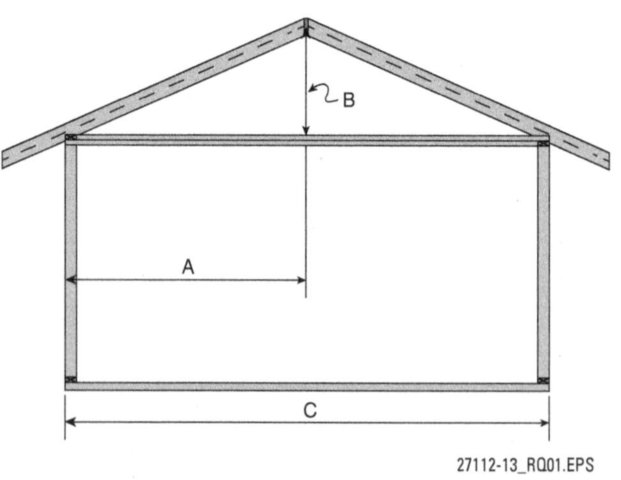

Figure 1

11. The distance a rafter rises vertically for each foot of run is known as the _____.
    a. rise
    b. slope
    c. unit rise
    d. projection

12. The pitch of a roof that has a rise of 8' and a run of 12' is _____.
    a. 1/3
    b. 2/3
    c. 2/4
    d. 63/4

13. Half the distance from the outside of one exterior wall to the outside of the opposite exterior wall is the _____.
    a. run
    b. span
    c. rise
    d. pitch

14. A roof with slope of 4 in 12 has a rise per foot of run of _____
    a. 4"
    b. 8"
    c. 12"
    d. 16"

15. A rafter that touches neither the ridge nor the plate is called a _____.
    a. cripple jack rafter
    b. hip rafter
    c. valley rafter
    d. rafter plate

16. The greater the unit rise of the rafter, _____.
    a. the stronger it must be
    b. the steeper the roof
    c. the flatter the roof
    d. the harder it is to cut accurately

17. A roof that rises 9" for each foot of run has a slope of _____.
    a. 6 in 12
    b. 9 in 12
    c. 12 in 9
    d. 12 in 12

Refer to *Review Question Figure 2* to answer Questions 18 and 19.

18. The length of a common rafter for a 20'-wide building with a 10" rise per foot of run is _____.
    a. 13'
    b. 14'-8"
    c. 19'-7"
    d. 20'

19. The rise per foot of run of a roof with a pitch of 1/2 is _____.
    a. 6"
    b. 12"
    c. 18"
    d. 24"

20. To determine the final measurement when laying out a common rafter, it is necessary to deduct _____.
    a. half the thickness of the ridgeboard
    b. half the thickness of the rafter
    c. half the thickness of the top plate
    d. nothing

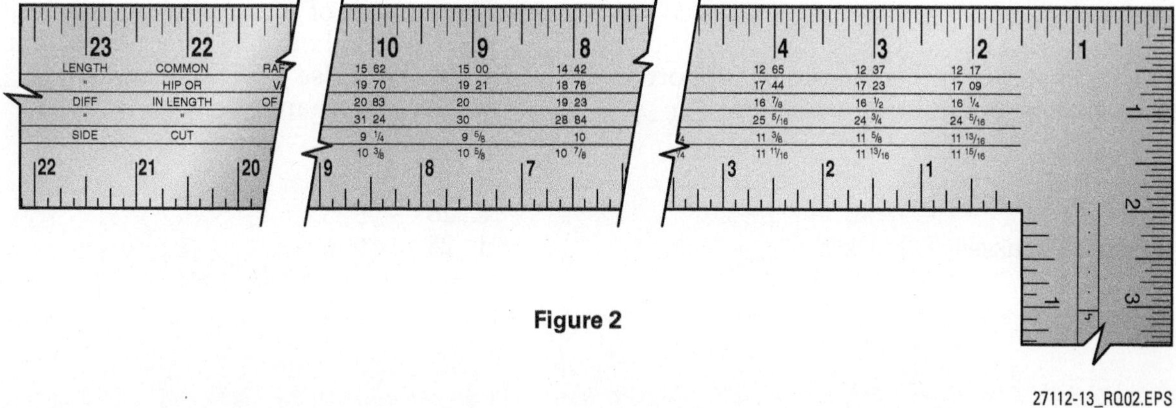

Figure 2

27112-13_RQ02.EPS

21. When you are using a Speed Square™, the length of the rafter is determined by_____.
   a. stepping the square
   b. reading it directly from the square
   c. using rafter tables in the instruction manual
   d. multiplying the run by the rise

22. A rafter support that parallels the top plate and the ridgeboard is known as a _____.
   a. purlin
   b. strongback
   c. collar tie
   d. ridgeboard

23. Double trimmers are used when framing _____.
   a. roof openings
   b. gable ends
   c. overhangs
   d. valleys

24. A strongback is installed to _____.
   a. frame a roof opening
   b. reinforce the ridgeboard
   c. straighten and strengthen the ceiling joists
   d. temporarily support rafters

25. When framing a gable end, it is necessary to measure and cut _____.
   a. diagonal braces
   b. each pair of studs
   c. each stud individually
   d. temporary supports

26. When rigging a truss with a span of 36', you must use a _____.
   a. spreader bar
   b. single choker
   c. double choker
   d. triple choker

27. Trusses that carry other trusses or support a large area of roof framing are called _____.
   a. beams
   b. girders
   c. bar joists
   d. double trusses

28. Positioning or installing trusses must be done under the supervision of a(n) _____.
   a. qualified rigger
   b. civil engineer
   c. architect
   d. experienced crane operator

29. When placing underlayment over sheathing for roofs with a slope of 4" or less, the commonly used roofer's felt is _____.
   a. 10 pound
   b. 15 pound
   c. 18 pound
   d. 22 pound

30. Immediately after nailing the plywood sheathing on a roof frame, you should install _____.
   a. a vapor barrier
   b. a felt underlayment
   c. a drip edge
   d. plywood clips

31. A waterproof underlayment should be used along roof edges _____.
   a. in areas prone to hurricanes
   b. in tropical climates
   c. in cold climates where snow accumulates
   d. in coastal locations

Refer to Figure 2 for questions 32 and 33.

32. The number of common rafters needed to frame a 20'-long gable roof for a house framed on 16" centers with no overhang is _____.
   a. 15
   b. 32
   c. 30
   d. 16

33. The number of sheets of plywood sheathing needed for a house 30' long with a span of 24' and with 8" rise per foot of run is _____. (Assume no overhang.)
   a. 14
   b. 18
   c. 26
   d. 28

34. Open web steel joists are typically attached to the structural steel frame _____.

a. using sheet metal screws
b. by welding
c. using lag bolts
d. with anchor bolts

35. Which type of roof is considered an efficient design in areas subject to hurricanes?

a. Shed roof
b. Gambrel roof
c. Gable roof
d. Hip roof

Trade Terms Quiz

Fill in the blank with the correct term that you learned from your study of this module.

1. A _______ roof has two slopes that meet at the center to form a point.
2. The board nailed to the rafter tails in order to space the rafters is the _______.
3. A(n) _______ is used when lifting large members with a crane.
4. The _______ extends beyond the gable and supports a decorative end piece.
5. An overhang is framed using short lengths of boards known as a(n) _______.
6. A(n) _______ should be attached to a truss that is being set by a crane.
7. The framework extending from the roof surface that provides additional light and ventilation is referred to as a(n) _______.
8. _______ are used to fasten framing members to one another.
9. A truss is fastened to a top plate by toenailing through the _______.
10. The upper ends of rafters are fastened to the _______.
11. A(n) _______ is an L-shaped arrangement of lumber that is fastened to the tops of long ceiling joists to keep them in alignment.
12. A roof with four sides extending toward the center is a(n) _______.
13. The amount of rise per 12" of run is referred to as the _______.
14. When framing members are assembled one piece at a time, it is referred to as _______.
15. A(n) _______ is a 1 × 4 member that is nailed to the ceiling joists to prevent twisting and bowing of the joists.
16. Commercial roofs are may be supported by _______, which are welded to the structural steel.
17. A(n) _______ is a roof support member that runs parallel to the plate and is installed between the plate and the ridgeboard.

Trade Terms

Barge rafter
Bottom chord
Dormer
Gable
Hip roof
Lookout
Open web steel joist
Purlin
Ribband
Ridgeboard
Spreader bar
Stick-built framing
Strongback
Tag line
Tiedown
Unit rise

Cornerstone of Craftsmanship

Vince Console

Head Instructor/Safety Officer
ABC Eastern Pennsylvania Chapter
Harleysville, Pennsylvania

How did you get started in the construction industry?
After working for some years in the restaurant industry, I decided it was time for a change. I had continued to develop the carpentry skills my grandfather had begun to teach me, and decided the construction industry was where I really wanted to be. So, I started working for myself in the residential construction industry.

Who or what inspired you to enter the industry?
My grandfather was a machinist by trade and a woodworker/furniture builder by avocation. As a child, I spent countless hours learning from him while helping him make furniture. Unfortunately, he passed away much too early and at that point had taught me enough to know that I needed to know much more.

What do you enjoy most about your career?
In construction, I enjoy the creativity in designing projects, and the instant gratification of seeing the results of your work at the end of a day and at the end of a project.
In the construction training industry, I enjoy developing and facilitating training programs that help students sharpen their hands-on craft skills, as well as the safety and communication skills needed to succeed in today's construction industry.

Why do you think training and education are important in construction?
Success in today's construction industry is dependent on a well-trained workforce. Workmanship, ethics, safety, efficiency, and teamwork are all components of a quality training program. The costs of bringing new employees into the field are daunting, so it is vitally important that the employee has a very short "learning curve" between being a new-hire and becoming a productive employee. Training and education, in a quality program, can help ensure a short learning curve.

Why do you think credentials are important in construction?
There are two reasons why credentials are important. First, credentials give the student a pathway to follow. A standardized learning program essentially creates a set of objectives and certifies the completion of the objectives. This is a source of self-pride in accomplishments and a motivation toward future accomplishments. Secondly, credentials give potential employers a proven method to assess a potential employee's job knowledge. As stated above, it helps in the selection of productive employees.

How has training/construction impacted your life and career?
I did not have the advantage of formal training in my craft, and so I know how much of a benefit training and education is to my students. Their careers are being positively impacted every single time they gain a new skill or add knowledge to an existing skill set. The result will be career advancement and everything that accompanies that, along with the additional pride in their achievement.

Would you recommend construction as a career to others? Why?
I would absolutely recommend construction as a career choice to others—if they are looking for a career that is physically and mentally challenging, and offers rewards for sincere efforts. The construction industry is a field in which you can progress as far as you want if you are willing to put in the investment effort. To those who meet that definition, the construction industry needs you. For those who are not motivated in those directions, the construction industry is not the place for you (it is not a career you should consider if it's not what you love to do).

What does craftsmanship mean to you?
Craftsmanship is the sum total of a number of components: pride, ethics, and professionalism in the work you do. A craftsman is a person who is detail oriented in the completion of their task and personally interested in the quality of work they produce. "Cutting corners" or "that's good enough" are not in a true craftsman's vocabulary.

Trade Terms Introduced in This Module

Barge rafter: A gable end roof member that extends beyond the gable to support a decorative end piece. Also referred to as a fly rafter.

Bottom chord: The lower member of a truss.

Dormer: Framework extending from the roof surface that provides additional light and ventilation to an attic space.

False fascia: The board that is attached to the tails of the rafters to straighten and space the rafters and provide a nailing surface for the fascia. Also called subfascia and rough fascia.

Gable: The triangular wall enclosed by the sloping ends of a ridged roof.

Hip roof: A roof with four sides or slopes extending toward the center.

Lookout: A structural member used to frame an overhang.

Open web steel joist: Steel joist, also known as a bar joist, that is used to support roof loads.

Purlin: A horizontal roof support member parallel to the plate and installed between the plate and the ridgeboard.

Ribband: A 1 × 4 nailed to the ceiling joists at the center of the span to prevent twisting and bowing of the joists.

Ridgeboard: Horizontal board (typically a 2× member) placed at the ridge of the roof. Upper ends of the rafters are fastened to it.

Spreader bar: A rigging device used when lifting large members with a crane.

Stick-built framing: Framing method in which framing members are installed one at a time (rather than as a prefabricated assembly).

Strongback: An L-shaped arrangement of lumber used to support ceiling joists and keep them in alignment.

Tag line: Cord or rope attached to a structural member or other large unit that is being positioned by a crane, to assist in guiding the member into proper position.

Tiedown: Metal anchoring device used to fasten framing members to one another.

Unit rise: The amount (in inches) that a common rafter will rise per foot of run.

Appendix

Laying Out and Erecting Hips and Valleys

An intersecting roof contains two or more sections sloping in different directions. Examples are the connection of two gable sections or a gable and hip combination such as that shown in *Figure A-1*. *Figure A-2* shows an overhead view of the same layout.

A valley occurs wherever two gable or hip roof sections intersect. Valley rafters run at a 45-degree angle to the outside walls of the building.

The material that follows provides an overview of the procedures for laying out hip and valley sections and the various types of rafters used in framing these sections. The first step is to lay out the rafter locations on the top plates (*Figure A-3*). The layout of the common rafters for a hip roof is the same as that for a gable roof. The next step is to lay out, cut, and install the common rafters and the ridgeboards, as described in the sections *Lay Out a Common Rafter* and *Installing Rafters*. At that point, you are ready to lay out the hip or valley section.

Hip Rafters

A hip rafter is the diagonal of a square formed by the walls and two common rafters (*Figure A-2*). Because they travel on a diagonal to reach the ridge, the hip rafters are longer than common rafters. The unit run is 17" (16.97" rounded up), which is the length of the diagonal of a 12" square. You can see this on the top line of the rafter tables on the rafter square. There are two hip rafters in every hip section.

For every hip roof of equal pitch, the hip rafter has a run of 17" for every foot of run of common rafter. This is a very important fact to remember. It can be said that the run of a hip rafter is 17 divided by 12 multiplied by the run of a common rafter. The total rise of a hip rafter is the same as that of a common rafter. For example, if a common rafter has an 8" rise per foot of run, the hip rafter would have a rise of 8" per 17" run. Therefore, the rise of a hip rafter would be the same as a common rafter at any given corresponding point.

To find the length of a hip rafter, first find the rise per foot of run of the roof, then locate that specific number on the inch scale line of the rafter square. Find the corresponding numbers on the second line of the rafter square. For example, assume a roof has a rise per foot of run of 4" (4 in

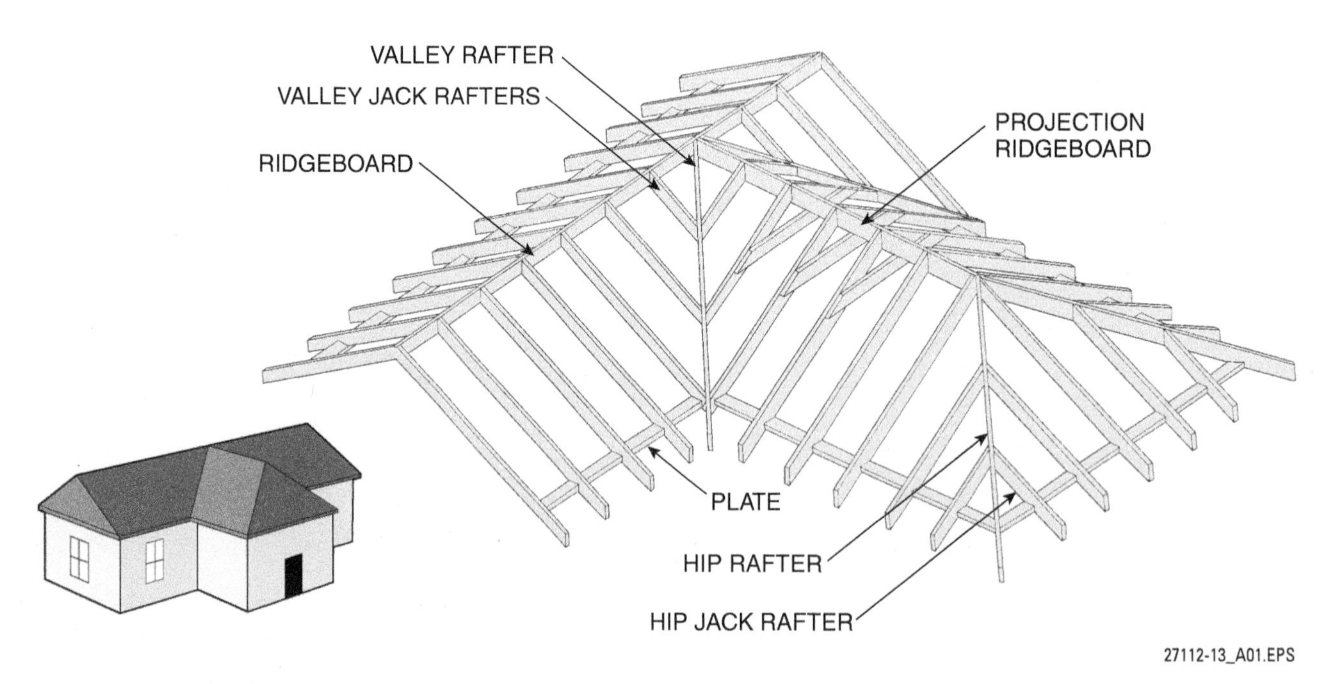

Figure A-1 Example of intersecting roof sections.

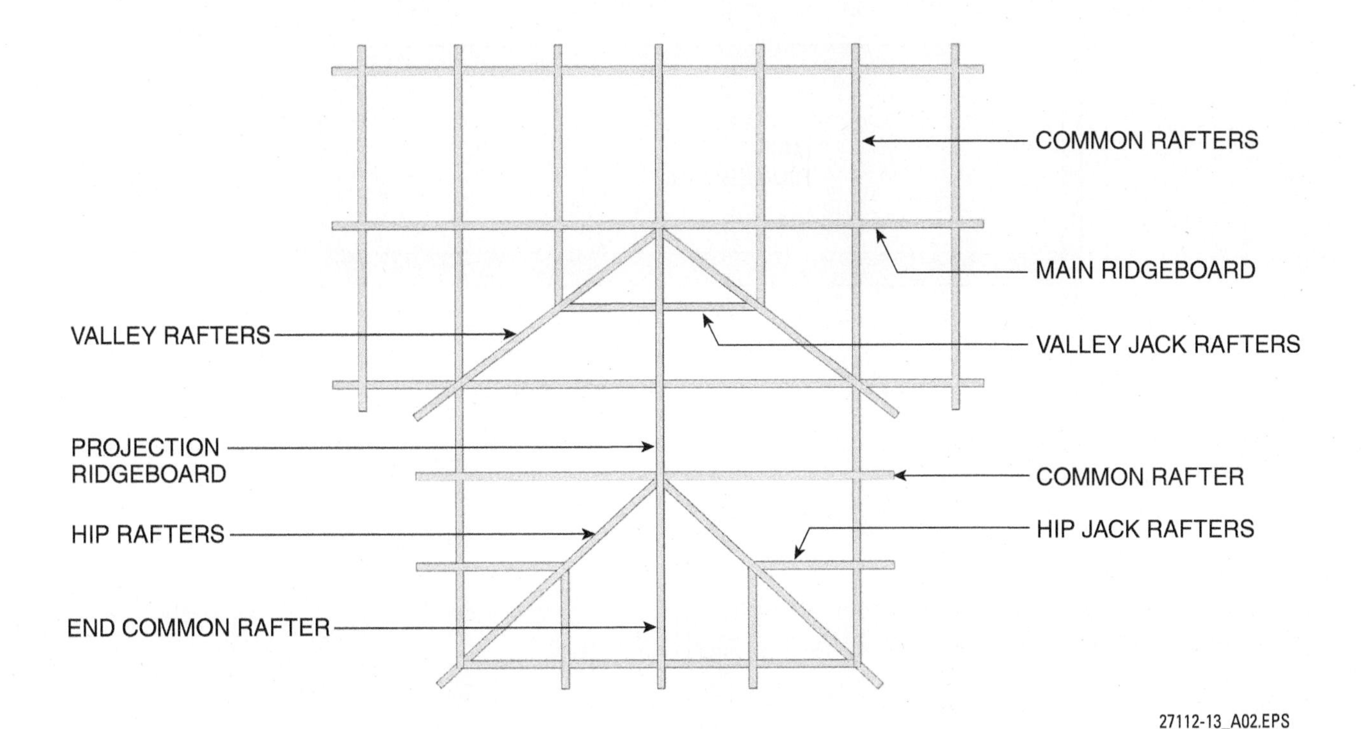

Figure A-2 Overhead view of a roof layout.

12). The span of the building is 10'. The second line under the 4 on the blade of the rafter (framing) square is 17.44. You have now established that the unit length of the hip rafter is 17.44" or 17$\frac{7}{16}$" for every foot of the common rafter. This unit length must be multiplied by the run of the structure, which is 5'. The sum of the two factors multiplied together must then be divided to find the length of the hip rafter. For example, 17.44" × 5 = 87.2", 87.2" ÷ 12 = 7.27' or 7'-3¼". Therefore, the total length of the hip rafter would be 7'-3 ¼".

No matter what the rise per foot of run is, the length of the common and hip rafters is determined in the same fashion. The method of laying out and marking a hip rafter is very similar to the layout of a common rafter, with one exception. A common rafter is laid out using 12 on the tongue and the rise per foot of run on the blade. A hip rafter is laid out by using 17 on the blade and the rise per foot of run on the tongue (*Figure A4*).

The basic procedure for laying out a hip rafter is as follows:

Step 1 Determine the length of the hip rafters using the rafter square (*Figure A-4*) or rafter tables, and add the overhang.

Step 2 Mark the plumb cuts on the hip rafters. The hip rafters must be shortened by half the diagonal thickness of the ridgeboard. Also, they must be cut on two sides to fit snugly between the common rafter and the ridgeboard (*Figure A-5*). These cuts are known as side cuts or cheek cuts.

Step 3 The bird's mouth plumb and seat cuts are determined in the same way as for a common rafter. However, the seat cut is dropped half the thickness of the rafter to align it with the top plane of the common rafter (distance *A*, *Figure A-6*). This drop is necessary because the corners of the rafters would otherwise be higher than the plane on which the roof surface is laid. Another way to accomplish this is to chamfer the top edges of the rafter. This procedure is known as backing. Most carpenters use the dropping method because it is faster.

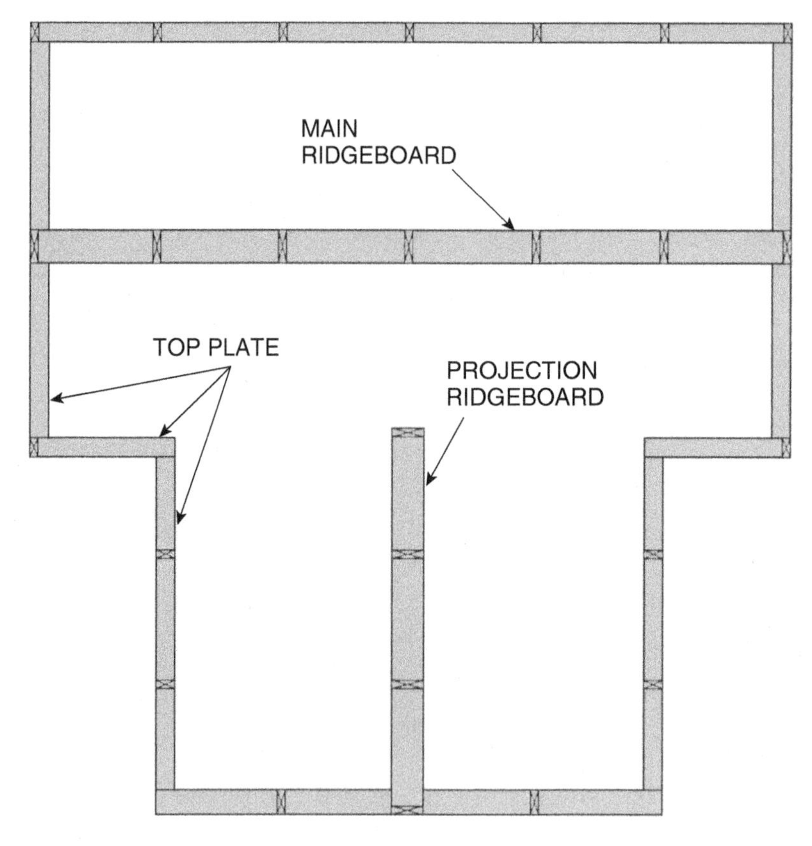

Figure A-3 Layout of rafter locations on the top plate.

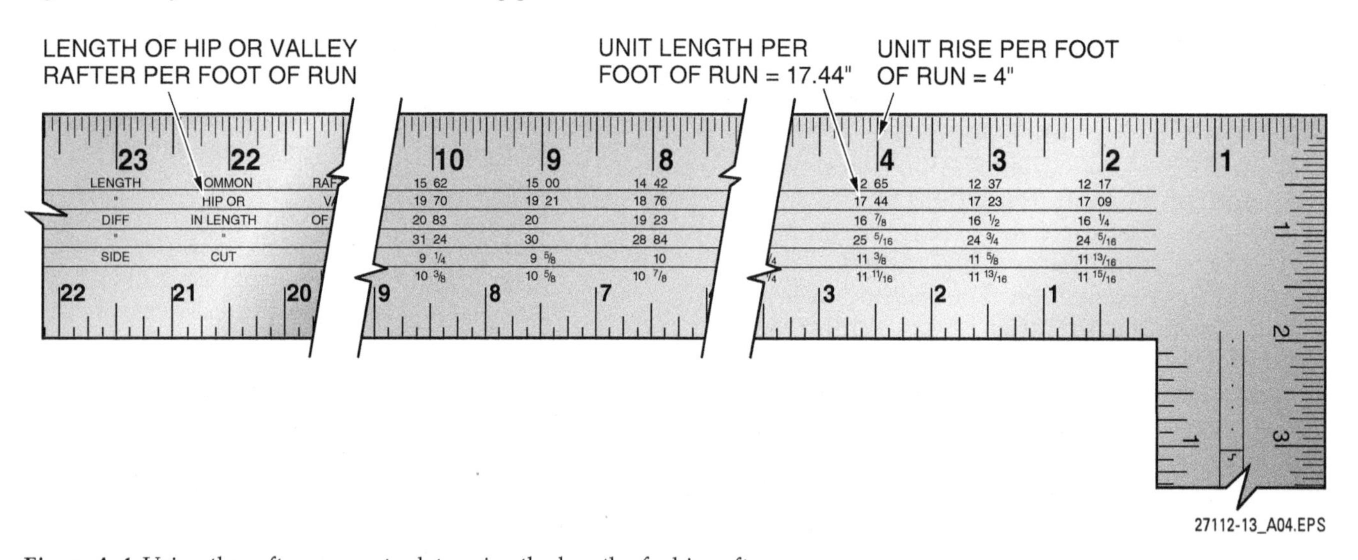

Figure A-4 Using the rafter square to determine the length of a hip rafter.

Valley Layout

Each valley requires a valley rafter and some number of valley jack rafters (*Figure A-7*). The layout of a valley rafter is basically the same as that of the hip rafter, with 17" used for the unit run. The only difference in layout between the hip and valley rafters is in the seat and tail cuts.

For a valley rafter, the bird's mouth plumb cut must be angled to allow the rafter to drop down into the inside corner of the building (*Figure A-8*). In addition, the tail end cuts must be made so that corner made by the valley will line up with the rest of the roof overhang. Like the hip rafter, the valley rafter must be aligned with the plane of the

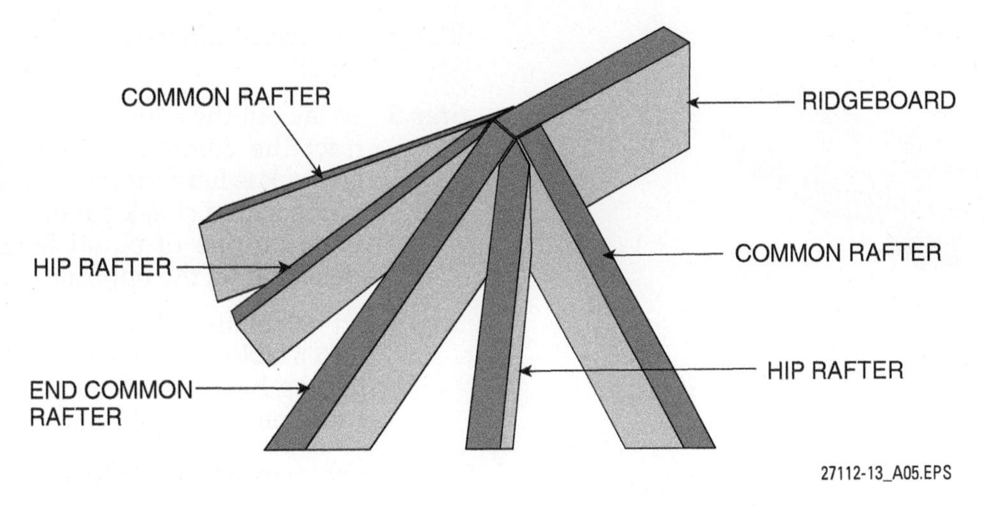

Figure A-5 Hip rafter position.

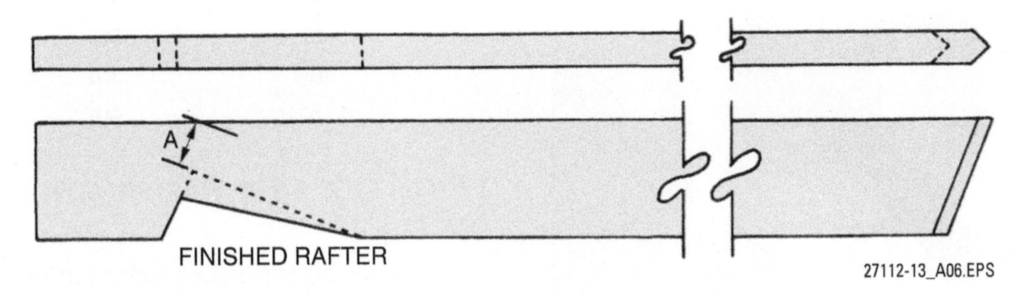

Figure A-6 Dropped bird's mouth cut.

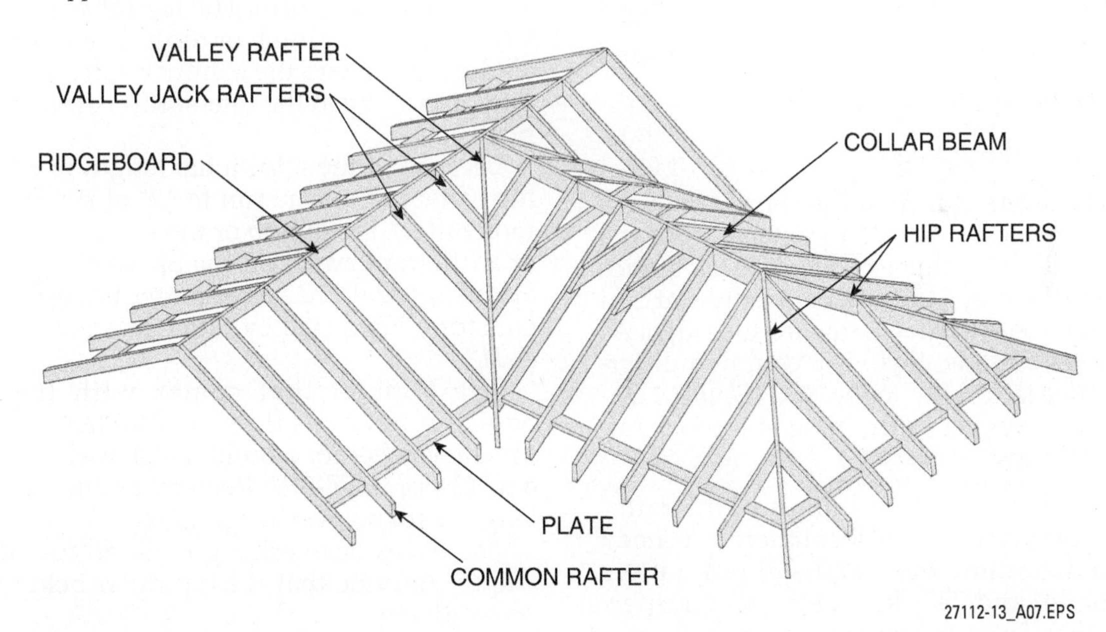

Figure A-7 Valley layout.

common rafter. Cheek cuts are also required on valley rafters to allow them to fit between the two ridgeboards.

The layout of valley jack rafters is the same as that for hip jack rafters, with the exception that the valley jack rafters usually run in the opposite direction; that is, toward the ridge.

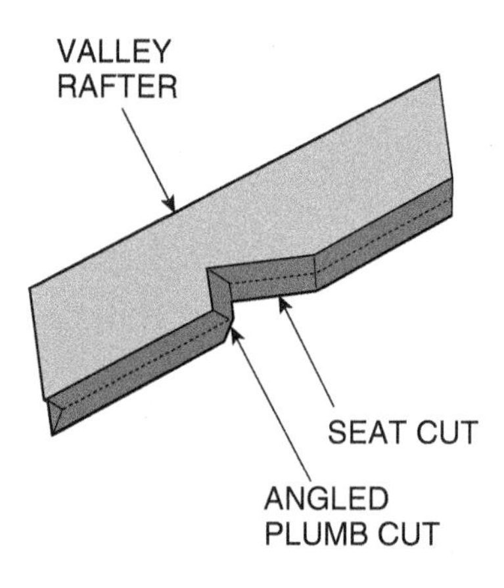

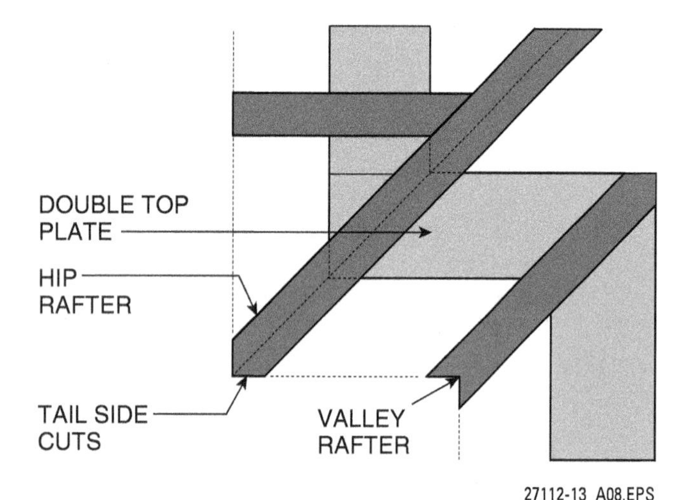

Figure A-8 Valley rafter layout.

Hip Jack Rafter Layout

Hip jack rafters run from the top plate to the hip rafters. Valley jack rafters extend from the valley rafter to the ridge (see *Figure A-1*). Notice that layout of valley jack rafters usually starts at the building line and moves toward the ridge. As with gable end studs, there is a common difference from one jack rafter to the next (*Figure A-9*).

Here is an overview of the hip and hip jack rafter layout process:

Step 1 The third and fourth lines of the rafter square have the information you need to determine the lengths of jack rafters for 16" and 24" OC construction, respectively. The number you read from the rafter square is the difference in calculated length between the common rafter and the jack rafter (*Figure A-10*). The longest jack rafter is referred to on the plans as the #1 jack rafter. Jack rafters must be cut and installed in pairs to prevent the hip or valley from bowing.

Step 2 Use the common rafter to mark the bird's mouth.

Step 3 To lay out the additional jack rafters, subtract the common difference from the last jack rafter you laid out. For each jack rafter with a cheek cut on one side, there must be one of equal length with the cheek cut on the opposite side.

Step 4 To calculate jack rafter lengths, use the rafter tables to find the lengths of the following hip jack rafters at 16" OC when a common rafter is 14'-6" long.

Rise per Foot of Run	Jack Rafter	Length
6"	2nd	
4"	4th	
9"	5th	
12"	3rd	
5"	4th	

Laying Out a Hip Rafter with a Speed Square™

Assume the hip rafter to be cut is on the same building as the common rafter (24' wide with a 4" rise per foot of run). The hip rafter must be cut with a different plumb or top cut because it is at a 45-degree angle to the common rafter (that is, sitting diagonally at the intersection of the two wall plates).

Because of the additional length of run in the hip rafter (17" of the run to 12" of run for a common rafter), the plumb or top cut must be figured at a different angle on the Speed Square™. The Speed Square™ has a set pattern to establish the cuts for a hip or valley rafter, as shown in *Figure A-11A*.

The booklet that comes with the Speed Square™ indicates that the total length of a hip or valley rafter for a building 24' wide and with a 4 in 12 rise is 17'-5¼". Remember, this is the total length with no overhang figured.

The eaves or overhang must be added to the length. Assume that the hip rafter being cut has no overhang.

Step 1 Choose a piece of lumber slightly longer than necessary. Place the lumber on a pair of sawhorses with the top edge and crown facing away from you. Assume that the right end of the lumber will be the top of the hip rafter. Place the Speed Square™ a few inches away from

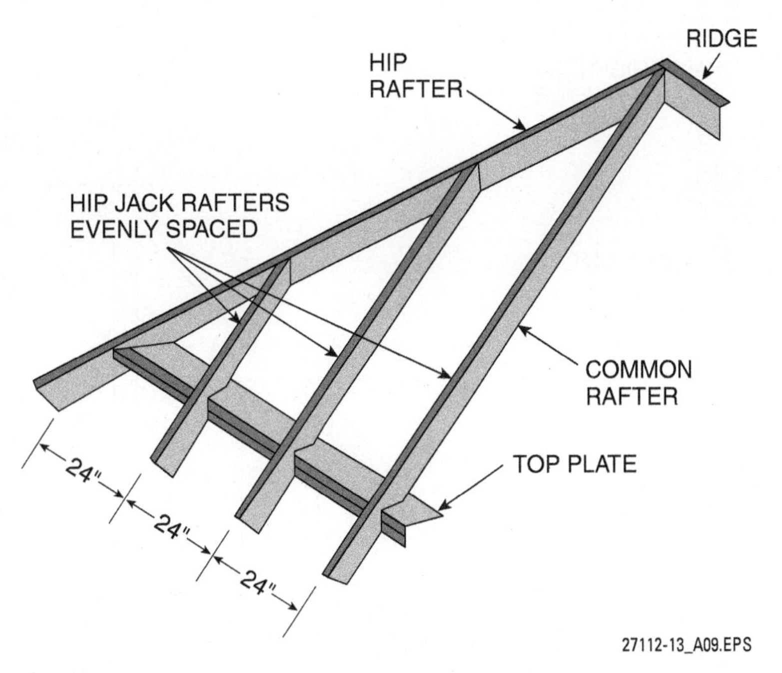

Figure A-9 Hip jack rafter locations.

the top. Move the Speed Square™ so that the pivot point and the 4 on the hip valley scale are even with the outside edge, and draw a line along the edge as shown in *Figure A-11B*. This will establish the top or plumb cut of the hip rafter.

Step 2 Lay out the length required for the hip rafter by measuring with a tape measure along the top edge of the rafter. Mark this point. Place the Speed Square™ with the pivot point on the edge of the mark and move the square until the 4 on the hip valley scale is even with the edge of the lumber; then draw a line along the edge of the square. This will establish the plumb cut of the seat cut (bird's mouth).

Step 3 To obtain the horizontal cut, reverse the square so that the small line at the pointed end of the Speed Square™ is covering the line previously drawn. Draw a line along the top edge of the Speed Square™. This will establish the bird's mouth seat cut, as shown in *Figure A-11C*.

Step 4 Place the lumber on edge and establish the center of the rafter. Draw a line on the center of the rafter so that it is over the ridge cut, as shown in *Figure A-11D*. This is done in order to establish the double 45-degree angle that must be cut so that the hip rafter will sit properly at the intersection of the two common rafters.

Step 5 Deduct ¾" plus half the diagonal measure of the ridge (1¹⁄₁₆"). Add the two together and mark that distance back from the original plumb cut (ridge cut).

If the ridge is 2" stock, the total measure deducted is 1³⁄₁₆".

Step 6 Using the Speed Square™, draw a 45-degree line from the edge of the second line drawn to the center line, and do the same on the opposite side of the center line, as shown in *Figure A-11E*.

Step 7 Make the ridge cuts and the seat cut. The hip rafter is now ready to be put in place. Be sure to check that the hip rafter is not up beyond the common rafter. The common rafter may need to be dropped before nailing the hip rafter in place. Use the hip rafter as a pattern to cut the other hip rafters.

The procedure outlined for making a hip rafter is the same procedure used to cut a valley rafter. There are two alternative methods of cutting the seat cut (bird's mouth) of the hip rafter. The first is to cut the rafter plate at the intersection of the two plates. The other alternative is to cut the seat cut at a double 45-degree angle.

The method of cutting a valley rafter is to cut a reverse 45-degree cut to fit into the intersecting corners.

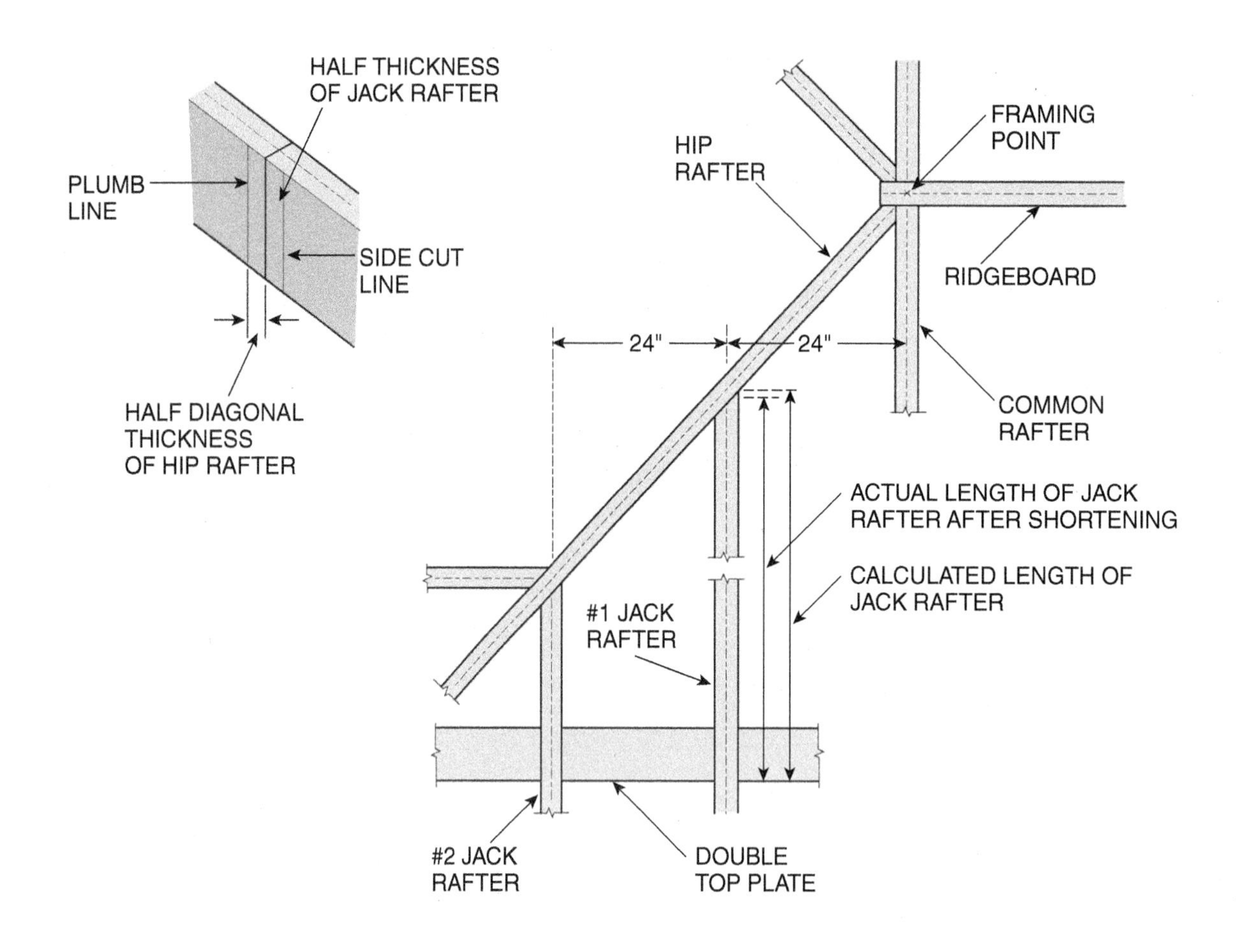

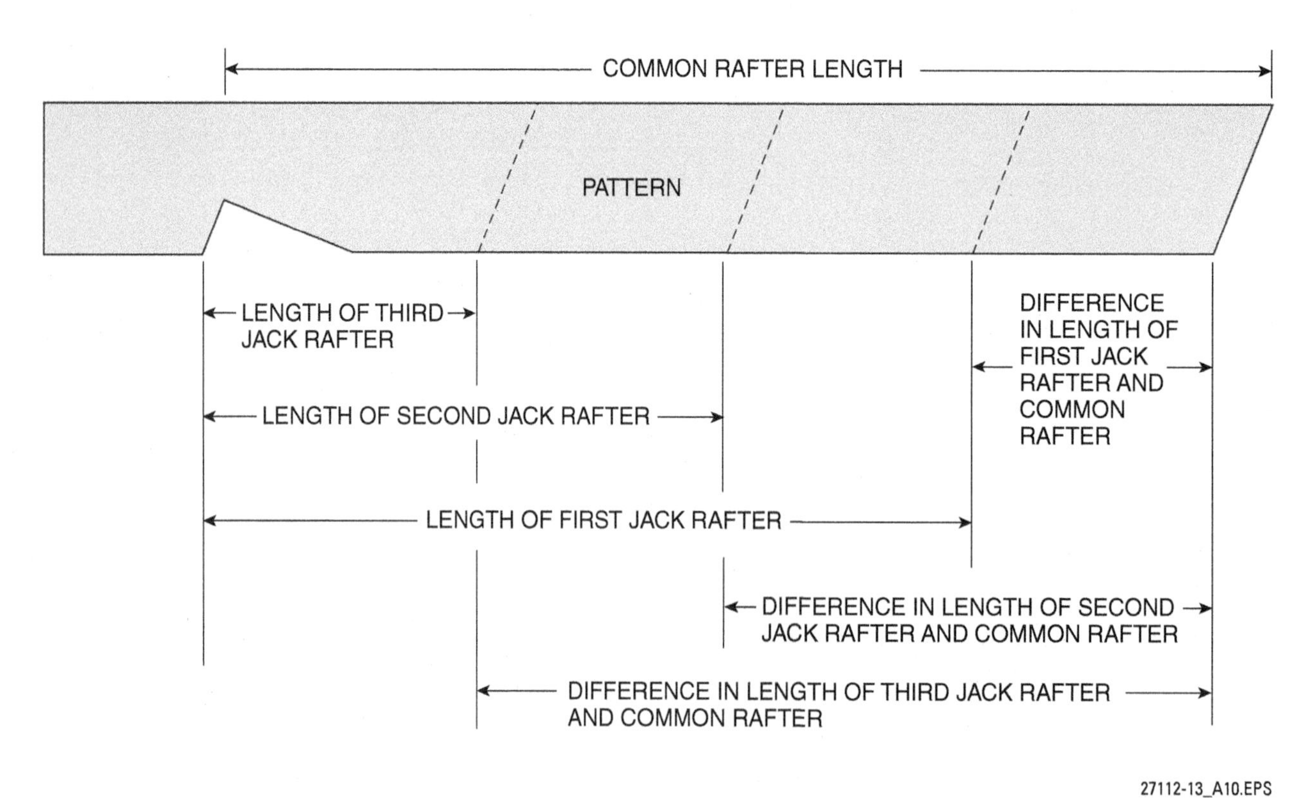

Figure A-10 Hip and hip jack rafter layout.

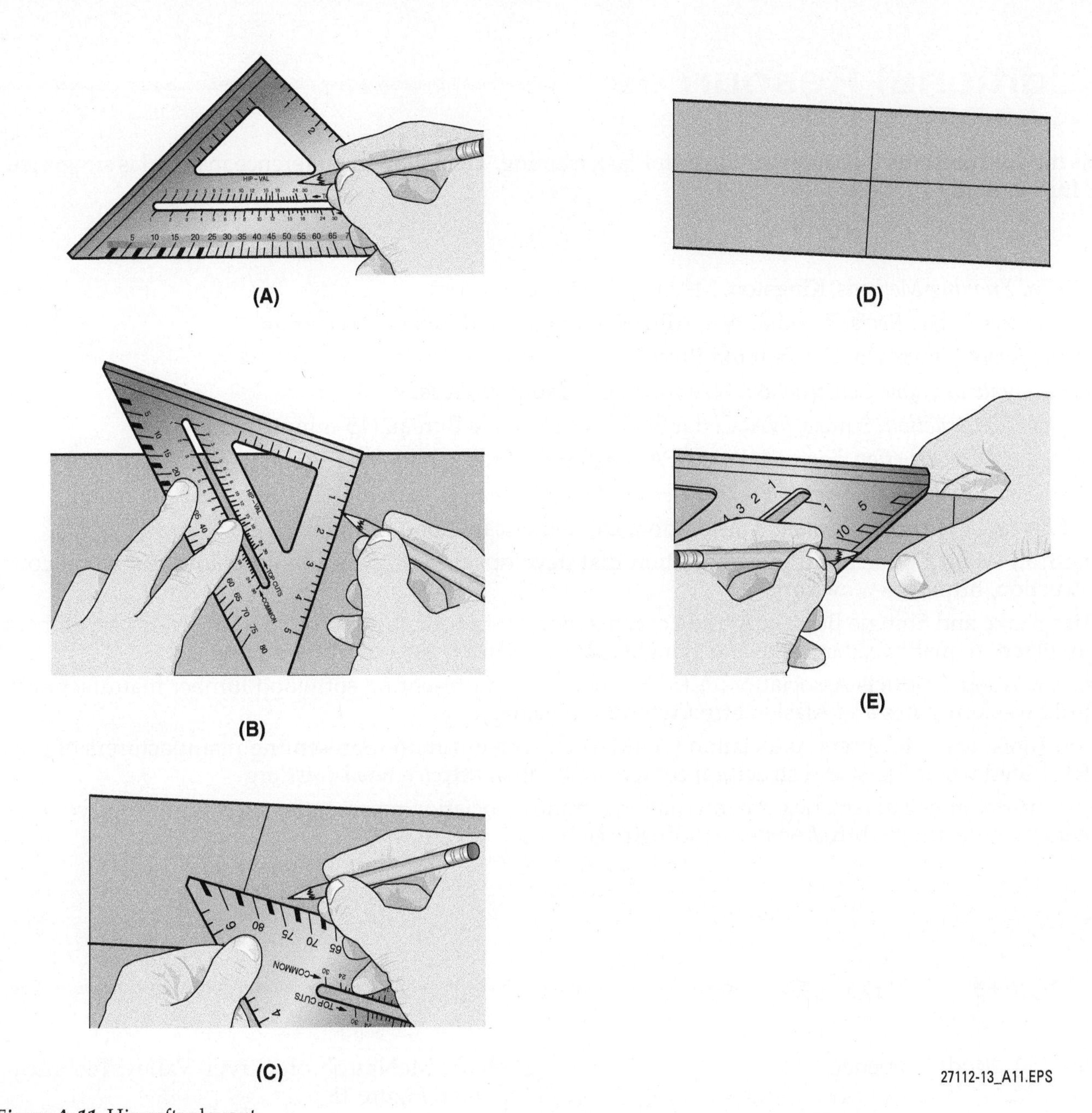

Figure A-11 Hip rafter layout.

Additional Resources

This module presents thorough resources for task training. The following reference material is suggested for further study.

Advanced Framing Methods. Kingston, MA: R.S. Means Company.

Build a Better Home: Roofs. Tacoma, WA: APA–The Engineered Wood Association.

Framing Roofs. Newtown, CT: Taunton Press.

Graphic Guide to Frame Construction. Newtown, CT: Taunton Press.

New Roof Construction. Sumas, WA: Cedar Shake and Shingle Bureau (15-minute video).

Quality Roof Construction. Tacoma, WA: APA–The Engineered Wood Association (15-minute video).

Roof Framer's Bible: The Complete Pocket Reference to Roof Framing. Jenkintown, PA: M.E.I. Publishing.

Wood Frame Construction Manual. Washington, DC: American Wood Council.

American Wood Council. A trade association that develops design tools and guidelines for wood construction, **http://www.awc.org**

Cedar Shake and Shingle Bureau. A trade organization that promotes the common interests of members involved in quality cedar shake and shingle roofing, **http://www.cedarbureau.org**

Western Wood Products Association. A trade association representing softwood lumber manufacturers in 12 western states and Alaska, **http://www2.wwpa.org**

Wood I-Joist Manufacturers Association (WIJMA). An organization representing manufacturers of prefabricated wood I-joist and structural composite lumber, **http://www.i-joist.org**

Wood Truss Council of America. An international trade association representing structural wood component manufacturers, **http://www.sbcindustry.com**

Figure Credits

SkillsUSA, Module opener

Zachary McNaughton, River Valley Technical Center, Figure 18

Section Review Answer Key

Answer	Section Reference	Objective Reference
Section One		
1. d	1.1.0	1a
2. a	1.1.0	1
3. d*	1.3.0	1c
Section Two		
1. b	2.0.0	2
2. a	2.0.0	2
Section Three		
1. c	3.0.0	3
2. b	3.0.0	3
3. b	3.1.0	3a
4. c	3.2.1	3b
Section Four		
1. d	4.0.0	4a
2. a	4.1.0	4a
3. b	4.2.0	4b
4. a	4.3.0	4c
Section Five		
1. d	5.1.0	5a
2. b	5.2.0	5
Section Six		
1. b	6.0.0	6a
2. d	6.1.0	6b
Section Seven		
1. c	7.1.0	7a
2. d	7.3.0	7c
3. b	7.2.0	7b
4. d	7.4.0	7
Section Eight		
1. a	8.0.0	8
2. b	8.0.0	8
Section Nine		
1. a*	9.1.2	9a
2. d*	9.1.3	9a

*Please see the following pages for further mathematical explanation of these answers.

Section Review Calculations

Section One

3. $10' \times 12"$ per foot $= 120"$

   $120" \div 16" = 7.5$ (round up to 8)

   $8 + 1 = 9$

Section Nine

1. $40' + 2' + 2' = 44'$

2. $42' \times 12.5' = 525$ sq ft

   525 sq ft $\div$ 32 sq ft $= 16.4$ (rounded up to 17 panels)

   $17 \times 2 = 34$ panels

NCCER CURRICULA — USER UPDATE

NCCER makes every effort to keep its textbooks up-to-date and free of technical errors. We appreciate your help in this process. If you find an error, a typographical mistake, or an inaccuracy in NCCER's curricula, please fill out this form (or a photocopy), or complete the online form at **www.nccer.org/olf**. Be sure to include the exact module ID number, page number, a detailed description, and your recommended correction. Your input will be brought to the attention of the Authoring Team. Thank you for your assistance.

Instructors – If you have an idea for improving this textbook, or have found that additional materials were necessary to teach this module effectively, please let us know so that we may present your suggestions to the Authoring Team.

NCCER Product Development and Revision

13614 Progress Blvd., Alachua, FL 32615

Email: curriculum@nccer.org
Online: www.nccer.org/olf

❑ Trainee Guide ❑ Lesson Plans ❑ Exam ❑ PowerPoints Other ____________________

Craft / Level: Copyright Date:

Module ID Number / Title:

Section Number(s):

Description:

Recommended Correction:

Your Name:

Address:

Email: Phone:

27109-13

Introduction to Building Envelope Systems

Overview

Carpenters are commonly called upon to install windows and exterior doors. The doors and windows encountered are typically be prehung. Many types of windows can be installed in a building: double hung, bay, casement, and jalousie, to name a few. There are also several types of exterior doors and door hardware, thresholds, and weather stripping. Window and door installation is a test of the framing quality. If a wall isn't square, installing windows and doors will be challenging.

Module Eight

ees with successful module completions may be eligible for credentialing through the NCCER Registry. To learn go to **www.nccer.org** or contact us at **1.888.622.3720**. Our website has information on the latest product releases ining, as well as online versions of our *Cornerstone* magazine and Pearson's product catalog.
edback is welcome. You may email your comments to **curriculum@nccer.org,** send general comments and **info@nccer.org**, or fill in the User Update form at the back of this module.
mation is general in nature and intended for training purposes only. Actual performance of activities his manual requires compliance with all applicable operating, service, maintenance, and safety procedures tion of qualified personnel. References in this manual to patented or proprietary devices do not constitute dation of their use.

y, Level One, Trainee Guide, Fifth Edition. NCCER.

27109-13
Introduction to Building Envelope Systems

Code Note

Codes vary among jurisdictions. Because of the variations in code, consult the applicable code whenever regulations are in question. Referring to an incorrect set of codes can cause as much trouble as failing to reference codes altogether. Obtain, review, and familiarize yourself with your local adopted code.

Objectives

When you have completed this module, you will be able to do the following:

1. Identify the components of the building envelope.
   a. Describe various ways that air infiltration can be minimized or prevented.
   b. Identify various types of fixed, sliding, and swinging windows.
   c. Identify the common types of exterior doors and explain how they are constructed
2. State the requirements for a proper window installation.
   a. Explain when jamb extensions are used.
   b. Identify common considerations when framing in glass blocks.
3. State the requirements for a proper door installation.
   a. Identify the differences between residential and commercial doors.
4. Identify the various types of locksets used on exterior doors and explain how they are installed.

Performance Tasks

Under the supervision of your instructor, you should be able to do the following:

1. Prepare a rough opening for proper window installation.
2. Prepare a rough opening for proper door installation.
3. Install a lockset.

Trade Terms

Building envelope
Building wrap
Casing
Conditioned space
Curb
Deadbolt
Expansion strip
Exterior insulation and finishing system (EIFS)
Flashing
Glazing
Hinge jamb
Hipped
Jamb
Light
Lockset
Muntin
Polyvinyl chloride (PVC)
Rail
R-value
Sash
Shim
Sill
Stile
Threshold
Window

Industry-Recognized Credentials

If you're training through an NCCER-accredited sponsor you may be eligible for credentials from NCCER's Registry. The ID number for this module is 27109-13. Note that this module may have been used in other NCCER curricula and may apply to other level completions. Contact NCCER's Registry at 888.622.3720 or go to **nccer.org** for more information.

Contents

Topics to be presented in this module include:

Figures

Section One

1.0.0 Building Envelope Components

Objective

Identify the components of the building envelope.

a. Describe various ways that air infiltration can be minimized or prevented.
b. Identify various types of fixed, sliding, and swinging windows.
c. Identify the common types of exterior doors and explain how they are constructed

Trade Terms

Building envelope: Consists of all building components, such as basement walls, exterior walls, floor, and roof, that separate conditioned space from unconditioned space or outdoor air.

Building wrap: Made of spun, high-density, polyethylene fibers randomly bonded into an extremely tough, durable sheet material and used to minimize air infiltration while allowing moisture to escape.

Conditioned space: An area of a building in which heating and/or cooling is deliberately supplied to it or is indirectly supplied through uninsulated surfaces of water or heating equipment or through uninsulated ducts.

Curb: A framework on which a skylight is mounted.

Exterior insulation and finishing system (EIFS): Method of exterior finish consisting of a coat of acrylic copolymers and portland cement applied over a base of exterior sheathing, insulation board, and reinforcing mesh.

Flashing: Sheet metal strips used to seal a roof or wall against leakage.

Glazing: Material such as glass or plastic used in windows, skylights, and doors.

Hipped: The external angle formed by the intersection of two adjacent sloping sides.

Jamb: The top and sides of a door or window frame that are in contact with the door or sash.

Light: Glass insert in a door.

Muntin: A thin framework used to secure panes of glass in a window or door.

Polyvinyl chloride (PVC): Plastic polymer used for electrical insulation and pipe.

Rail: Horizontal member of a window sash or panel door.

R-value: Measure of effectiveness of a material to insulate another material. A higher R-value indicates a higher insulating value.

Sash: The part of a window that holds the glass.

Sill: The lowest framing member of a window or exterior door frame.

Stile: Vertical member of a window sash or panel door.

Window: A wall opening that admits light and/or ventilation.

The final step in drying-in a house is the installation of windows and exterior doors. Since there are so many types and styles of windows and exterior doors, which are made by so many manufacturers, all the possibilities cannot be covered in this module. This module introduces the terms associated with doors and windows, and provides an overview of the various kinds of windows and exterior doors. In addition, installation practices related to windows and doors are covered.

The **building envelope** consists of all building components, such as basement walls, exterior walls, floors, and roof, that separate **conditioned space** from unconditioned space or outdoor air. For example, the walls and door between a kitchen and an unheated garage are part of the building envelope. However, walls that separate an unheated enclosed porch from the outside are not part of the building envelope. The energy code is mostly concerned with the building envelope.

The building envelope is affected by the types of construction materials used for the building and the installation practices that were used. For example, if premium construction materials are specified and used for a building, but they are installed poorly (such as loose-fitting joints), the quality of the building envelope will be significantly affected. The building is protected from air infiltration through the use of building wrap, windows, and doors. Thermal protection for the building envelope is provided by insulation.

1.1.0 Air-Infiltration Control

The exterior sheathing of a structure should be covered to prevent wind pressure from causing infiltration of outside air into the structure. To

achieve maximum energy efficiency in a structure, air infiltration must be strictly controlled.

Traditionally, structures have been covered with water-resistant building paper to help prevent water leakage through the primary barrier (siding) from reaching the sheathing or other structural components. In addition, the building paper had to be water permeable to allow moisture inside the walls of the structure to pass through and evaporate. (Moisture accumulating within the wall cavity can saturate the insulation, thus affecting the insulation qualities and may result in mold developing in the cavity.) To some extent, the paper reduced air infiltration of the structure, especially when board sheathing was used.

For a number of years, products called **building wrap** or house wrap have been used to replace building paper. Building wrap (*Figure 1*) is easier to apply and performs the same functions as building paper. When properly applied and sealed, the wrap provides a nearly airtight structure no matter what sheathing material is used. Most versions of these wraps are an excellent secondary barrier under all siding, including stucco and **exterior insulation and finishing systems (EIFS)**.

Most building wrap is made of spun, high-density polyethylene fibers randomly bonded into an extremely tough, durable sheet material. Building wrap is usually available in several versions and weights for residential and light commercial use. Special versions are available with vertical water channels permanently pressed into the material for stucco and EIFS. Building wrap is available in rolls ranging in widths from 18" to 10', and in lengths from 100' to 200'.

RESIDENTIAL APPLICATION

COMMERCIAL APPLICATION

27109-13_F01.EPS

Figure 1 Building wrap.

GOING GREEN

High-Performance Building Envelopes

Rinker Hall on the University of Florida's Gainesville campus earned Leadership In Energy and Environmental Design (LEED) Gold certification for its energy-efficient design. The architect was challenged to design a building that would fit in with the other brick buildings on campus, but not trap heat. The compromise was a freestanding masonry shade wall that serves as a second skin and helps match the design of the building to the site. The high-performance building envelope reflects heat and includes an air-infiltration barrier, a thermal break, and high-performance glass.

27109-13_SA01.EPS

Nails with large heads, nails or screws with plastic washers (*Figure 2*), or 1"-wide staples may be used to secure building wrap to the wall sheathing or exterior gypsum board. Screws and washers are used for steel construction. Special contractor's tape (*Figure 3*) or sealants compatible with the wrap are used to seal the edges and joints of the wrap.

> **WARNING!**
> Some building wraps are slippery and should not be used where they can be walked on. Because the surface is slippery, use pump jacks or scaffolds for exterior work above the lower floor. If ladders must be used, extra precautions must be taken to prevent the ladders from sliding on the wrap.

When installing building wrap, always refer to the manufacturer's instructions for specific installation information. In general terms, building wrap is installed as follows:

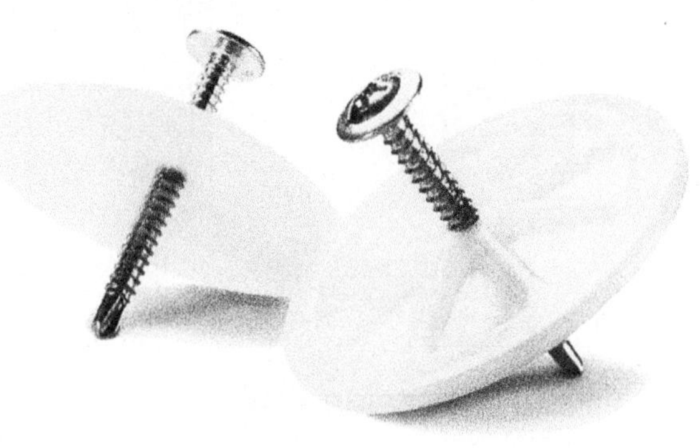

27109-13_F02.EPS

Figure 2 Screws with plastic washers, used to secure building wrap.

27109-13_F03.EPS

Figure 3 Contractor's tape.

Step 1 Using two people and beginning at a corner on one side of the structure, leave 6" to 12" of the wrap extended beyond the corner to be used as an overlap on the adjacent side of the structure (*Figure 4*). Align the roll vertically and unroll it for a short distance. Check that the stud marks on the wrap align with the studs of the structure. Also check that the bottom edge of the wrap extends over the foundation line by 1" and runs along the foundation line. Secure the wrap to the corner at 12" to 18" intervals.

Step 2 Unroll the wrap two or three more feet and ensure that it overlaps and runs along the foundation line. Secure the wrap vertically at 12" to 18" intervals on each stud using the stud marks as a fastening guide. Continue around the structure, covering all openings. If a new roll is started, overlap the end of the previous roll 6" to 12" to align the stud marks of the new roll with the studs of the structure.

Step 3 If the upper floors or parts of the structure require coverage, repeat Steps 1 and 2, starting above the existing wrap. Make sure that the bottom edge of the upper layer of wrap overlaps the top edge of the lower layer 6" to 12".

Step 4 At the top plate, ensure the wrap covers both members of the double top plate (*Figure 5*), but leave the flap loose for the time being.

Step 5 At each opening, use one of the following two methods to cut back the wrap:

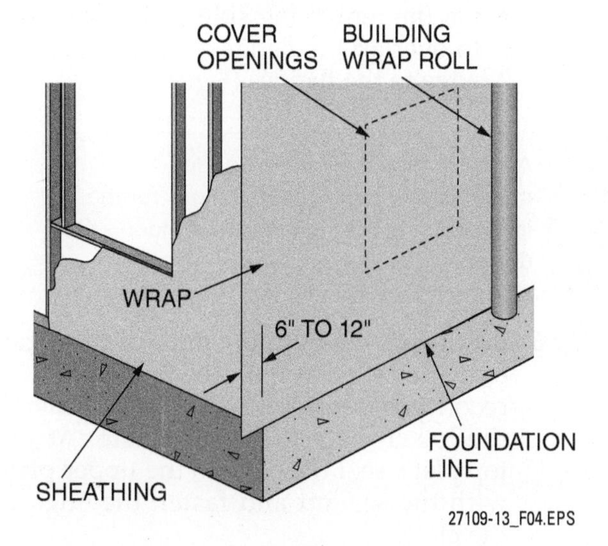

27109-13_F04.EPS

Figure 4 Starting a roll of building wrap.

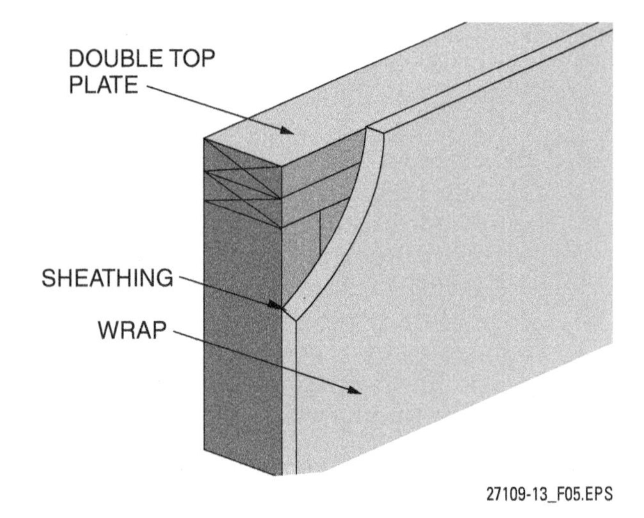

Figure 5 Top plate detail.

Method 1 – Uninstalled Windows/Doors:

- At the opening, cut the wrap as shown in *Figure 6*. Fold the three flaps around the sides and bottom of the opening and secure every 6". Trim off the excess.
- Along the outside, install 6" flashing at the bottom of the opening, then up the sides over the top of the wrap.
- Install head flashing at the top of the opening under the wrap and over the side flashing (*Figure 7*). Tape the flap ends to the head flashing using manufacturer-approved tape.

Method 2 – Installed Windows/Doors with Flanges:

- Create a top flap of the wrap. Insert head flashing under the flap and over the flange.
- Extend the flashing to the sides about 4" and tape the flap to the head flashing.
- On the remaining sides, trim the wrap to overlap the flange area and tape the edge to the flanges (*Figure 8*).

NOTE

Always follow the window or door manufacturer's recommendations for the installation of flashing around windows or doors.

Step 6 Secure all the bottom edges of the wrap to the foundation with the manufacturer-recommended joint sealant, then fasten the lower edge to the sill plate. At the top plate, seal the edge to the upper plate with the sealant and fasten the edge to the plate.

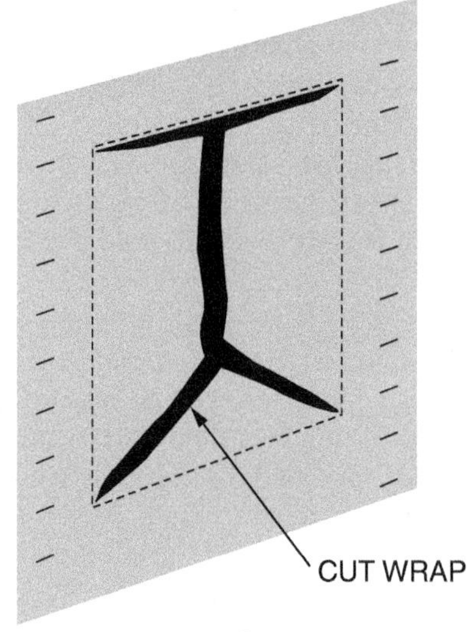

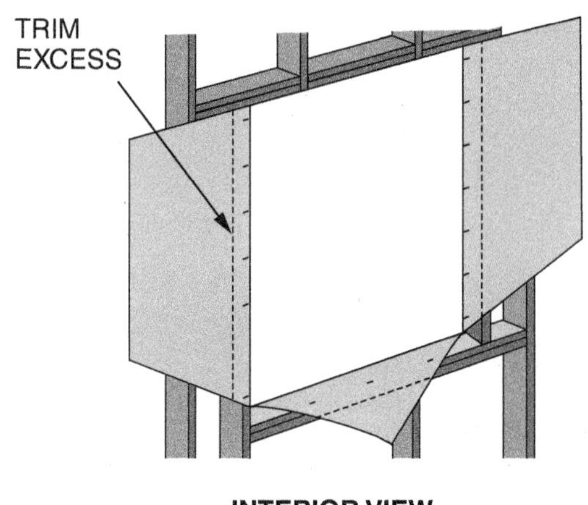

Figure 6 Cutting and folding building wrap at a rough opening.

Step 7 Seal all vertical and horizontal joints in the wrap with the recommended tape.

Step 8 Before applying the siding, repair any damage or tears in the wrap with tape or sealant.

1.2.0 Windows

A window refers to the entire window assembly, including the glass, sash, and frame. Window sashes can be made from wood, metal, or vinyl. Aluminum and steel are used in metal window construction. Wood windows must be protected with wood preservatives and/or paint to prevent decay. Ponderosa pine, carefully selected and kiln-dried to a moisture content of 6 percent to 12 percent, is commonly used in making wood-sash windows. Wood is often preferred over aluminum because it does not conduct heat as readily.

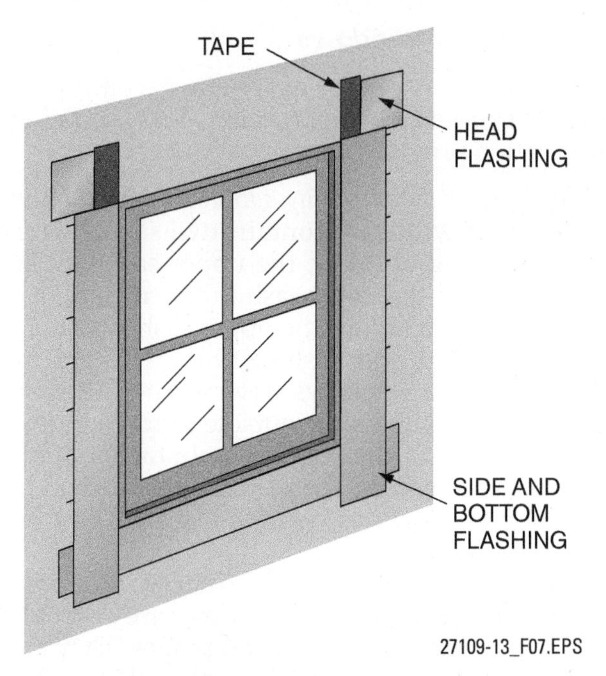

Figure 7 Installing flashing around an opening.

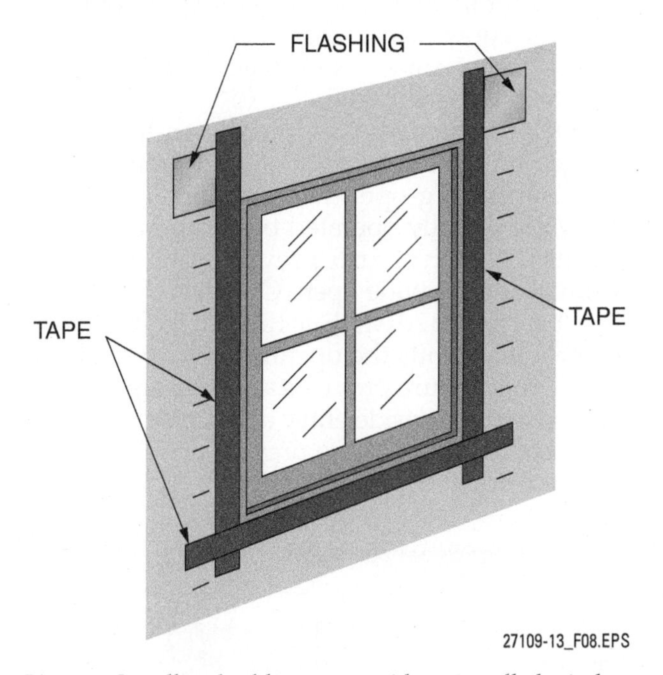

Figure 8 Installing building wrap with an installed window.

Metal window frames are usually filled with insulating material to make them more energy efficient. Although many sashes and frames are made of wood, it has become common for wood windows to be clad on the outside with aluminum or vinyl, providing the stability of wood and protecting the wood as well.

NOTE

An egress window is required in specific locations in a dwelling and is intended to provide an emergency means for exiting the dwelling. Egress windows must be installed in every room used for sleeping purposes, such as bedrooms, on any floor and in basements with habitable space. In new-home construction, the *International Residential Code®* (*IRC*) requires that you install an egress window in each bedroom. It also requires an egress window in the basement if habitable rooms will be finished in the basement.

Windows must meet specific *IRC* criteria to qualify as an egress window, as follows:

- Minimum opening width: 20"
- Minimum opening height: 24"
- Minimum net clear opening: 5.7 sq ft (5.0 sq ft for ground floor).
- Maximum sill height above floor: 44"

Net clear opening refers to the actual free and clear space available when the window is open; it is not the rough-opening size or the glass panel size.

An egress window must be operational from the inside without keys or tools. Bars, grilles, and grates can be installed over windows but must be operational without tools or keys and still allow the minimum clear opening.

Aluminum windows are much lighter, easier to handle, and more durable than wood windows. Aluminum windows are generally coated to prevent corrosion. However, they are not recommended for coastal areas where the corrosive effect of salt air can be extreme.

Steel windows are stronger than aluminum and wood, but are far more expensive. They are more common in commercial buildings than residential construction.

Vinyl window frames are made of impact-resistant **polyvinyl chloride (PVC)**. They are resistant to heat loss and condensation. Vinyl windows cannot be painted, so they must be purchased in a color that is compatible with the building. The sashes and frames of vinyl windows may distort when exposed to extreme temperatures. This can cause difficulty in operating the window, as well as increased air infiltration.

The basic component of a window assembly is the sash, which is the framework around the glass. *Figure 9* identifies the component parts of the sash. Note that the sections of glass are called lights. The sash may contain several lights or just one light with a false muntin known as a grille. The grille does not support the glass; it is simply installed on the face of the glass for decorative purposes.

The sash fits into, or is attached to, a window frame consisting of stiles and rails. The entire window assembly, including the frame and sash (or sashes), is shipped from the factory. The window frame consists of the head jamb, side jambs, and sill, as shown in *Figure 10.*

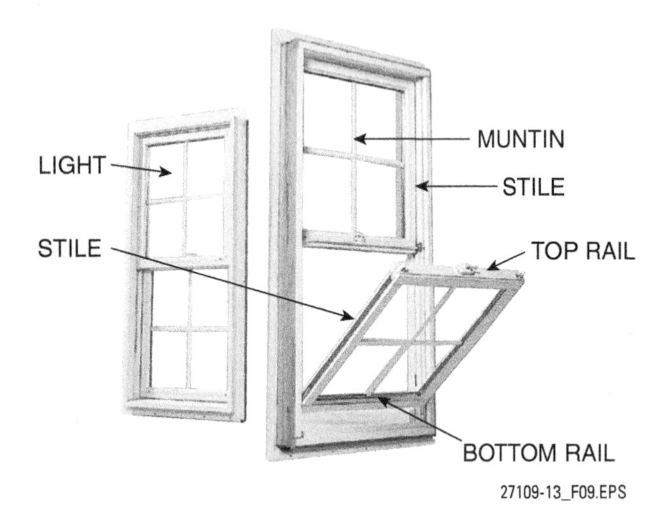

Figure 9 Parts of a sash.

1.2.1 Types of Windows

There are several types of windows commonly used in residential construction. The following information provides an overview of each type:

- *Single- and double-hung windows* (see *Figure 10*)–These windows contain two sashes. In a double-hung window, the upper sash can be lowered and the lower sash can be raised. In a single-hung window, the upper sash is stationary and the lower sash can be raised. Many double-hung windows have a tilt-and-wash feature that allows the outside of the window to be cleaned from the inside. In one application of this feature, the jamb can be pushed in enough to allow the top of the sash to pivot inward.
- *Fixed windows*–In a fixed window, the glass cannot be opened. Fixed windows are available in a wide variety of styles and shapes. They are often used to create an attractive, distinctive architectural appearance (*Figure 11*). Decorative glazing known as art glass (similar to stained glass) is sometimes used to achieve dramatic effects in these windows.
- *Casement windows*–Casement windows are hinged on the left or right of the frame so that they can swing open like a door (*Figure 12*). They are usually operated by a hand operator (crank). A swivel arm prevents the window from swinging wide open. Casement windows can be equipped with a limited ventilation control that limits the opening to a few inches. While this feature provides a certain amount of security from outside entry and accidental falls by occupants, it can also prevent people on the inside from getting out during an emergency.

Building Wrap

When building wrap is installed over a rough opening, don't cut the wrap diagonally across the entire opening. Instead, cut diagonally from the bottom corners to the center, then cut straight up from the center to the top of the opening, as shown below. Wrap the material over the sill and sides of the opening and secure it. Cut the wrap flush with the bottom of the header, but do not secure it. This will allow the nailing flange to slide under the wrap for a more secure installation.

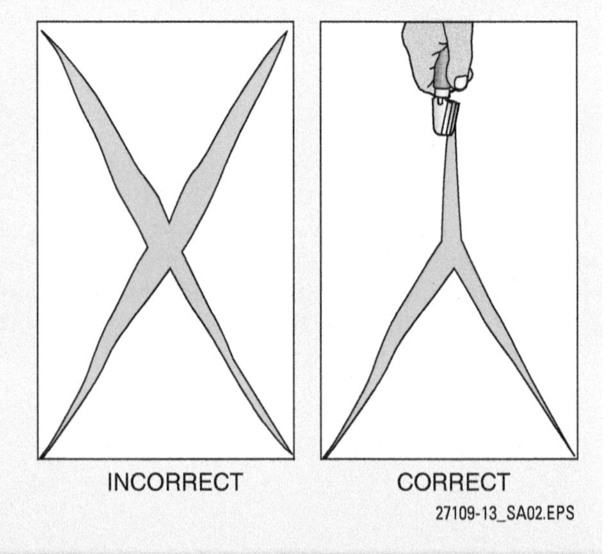

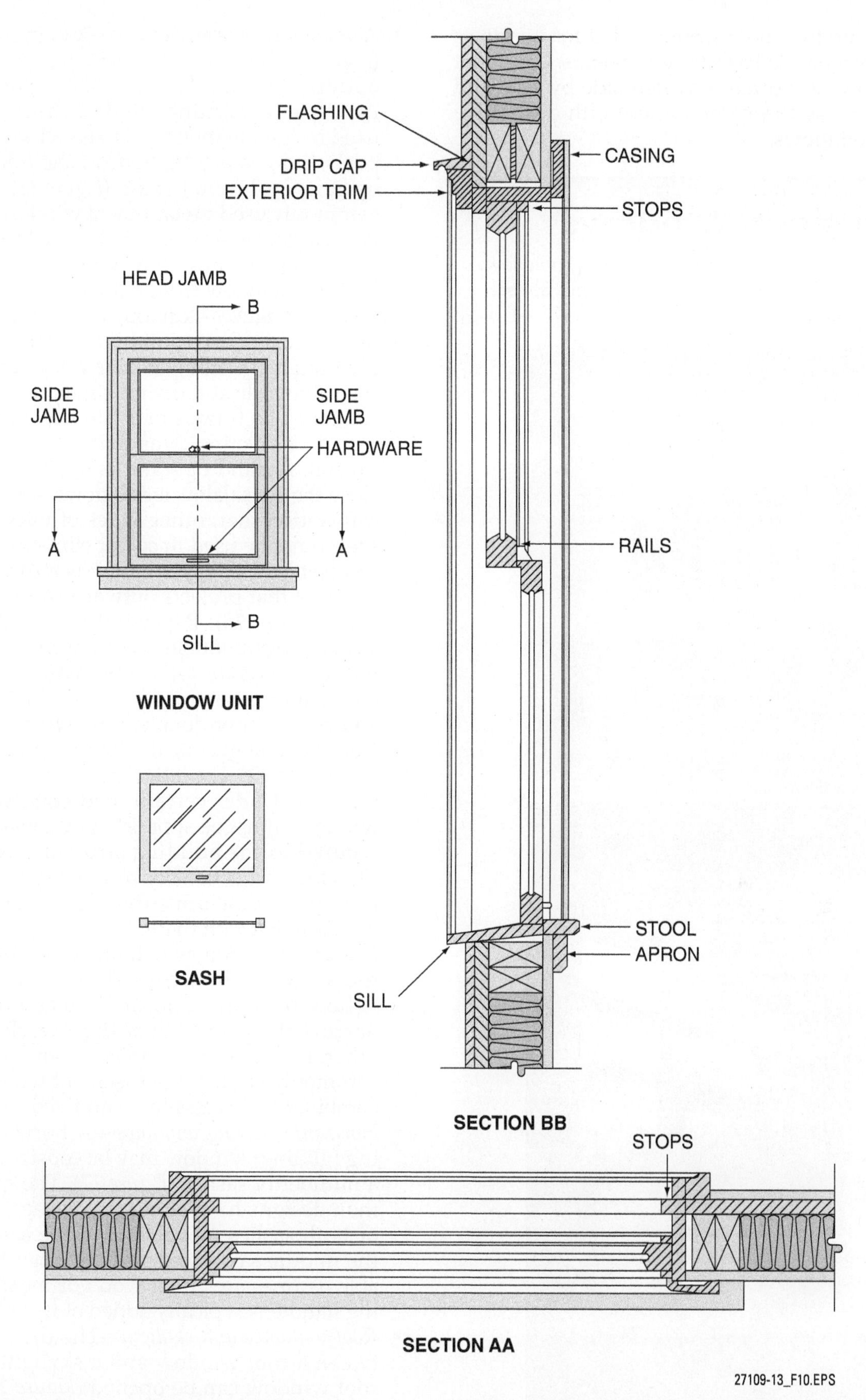

Figure 10 Parts of a double-hung wood window.

It is, therefore, not recommended for windows that may be used as emergency exits. Casement windows are often installed side by side and may be used in combination with decorative fixed windows.

Doors and Windows

The selection and placement of doors and windows can have a dramatic effect on the aesthetic appeal of a structure.

27109-13_SA03.EPS

- *Awning and hopper windows*–Awning windows (*Figure 12*) are hinged on the top and pushed outward to open. They are often operated by a hand crank. Awning windows are commonly used in combination with fixed windows. Hopper windows are hinged on the bottom and open inward from the top (*Figure 12*). They are commonly used for basement windows. A hopper window is equipped with a locking mechanism and also has pivot arms on the sides to keep the window from falling.
- *Jalousie windows*–A jalousie window consists of a series of horizontal glass slats, each in a pivoting metal frame (*Figure 12*). It can provide security while at the same time providing ventilation. The frames of a jalousie window are joined by pivoting arms so that they operate in unison. A hand operator is used to open and close the slats. Jalousie windows have more air infiltration than other types of windows and should not be used in colder climates.
- *Bay windows*–A bay window is a three-walled window that projects outward from the structure (*Figure 13*). Bay windows may be constructed from double-hung, casement, or fixed windows (*Figure 14*). Combinations are often used, such as a fixed window in the center and casement or double-hung windows on the sides. The angles of the bay window may be 30°, 45°, or 90°. The latter is known as a box bay window. Under certain load conditions, bay windows must be supported by cables that are secured to the building structure (*Figure 15*). This method is usually recommended for projecting window units that do not have a support wall beneath them.
- *Bow windows*–A bow window projects out from the structure in a curved radius (*Figure 16*). A bow window is normally made of several narrow, flat planes set at slight angles to each other. Like the bay window, a bow window is commonly made up of casement windows or a combination of casement and fixed windows.
- *Horizontal sliding windows*—A horizontal sliding (gliding) window may be constructed from a number of sashes (*Figure 17*). The most common designs have either two sashes, with one or both sashes movable, or three sashes with the middle sash fixed and the other two movable. Frames are either wood or metal. A locking handle is typically added at installation.
- *Roof windows and skylights*–The difference between a roof window and a skylight is that a roof window can be opened (*Figure 18*). There are many designs and shapes of skylights. Domed and hipped skylights are the most

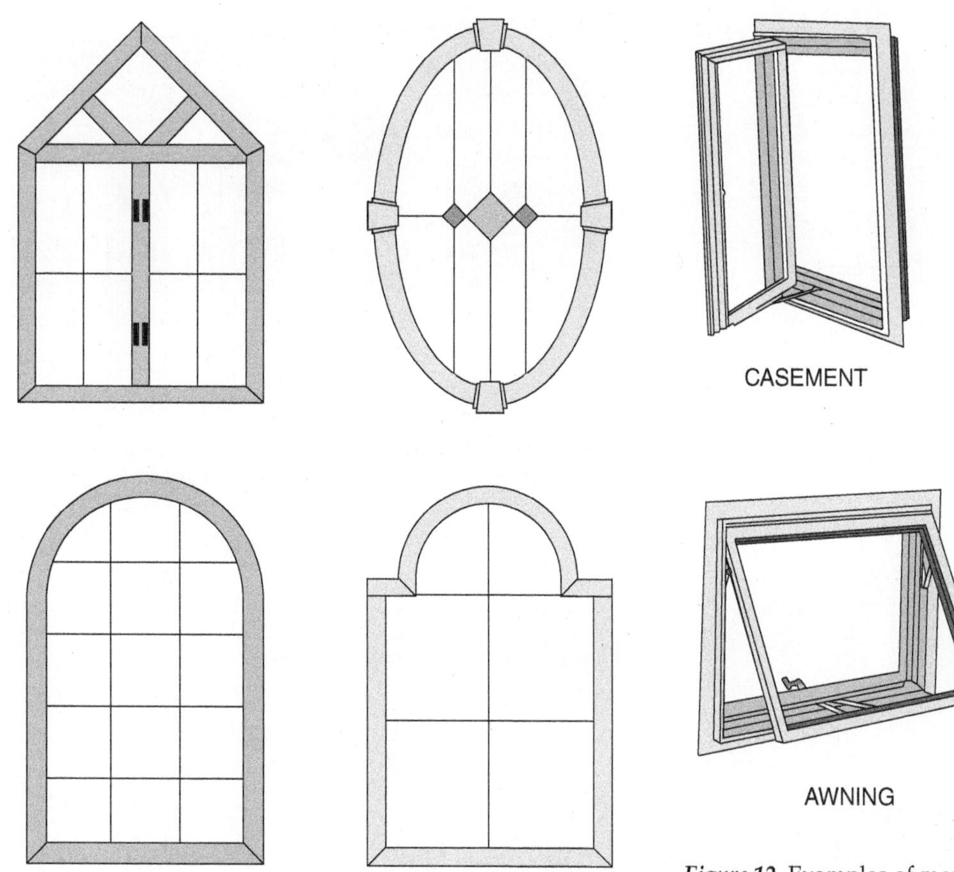

Figure 11 Examples of fixed windows.

common because they more readily shed rain and snow. Domed skylights are rounded, while hipped skylights have a peak. Skylights and roof windows are often installed using a built-up **curb** (*Figure 19*). Metal flashing is installed around the curb to prevent leakage. Another style of skylight is self-flashing. This type has an integral metal flange that is screwed into the roof and sealed on both sides with roofing cement during installation.

1.2.2 Types of Window Glass

There are several kinds of sheet glass made for use as glazing. Single-strength (SS) glass is about $\frac{3}{32}$" thick. It is used only for small lights of glass. Double-strength (DS) glass, which is about $\frac{1}{8}$" thick, can be used for larger lights. Heavy-duty glass ranges in thickness from $\frac{5}{16}$" to $\frac{7}{8}$".

Glass is a good conductor of heat. In any building, glass accounts for the majority of the heat loss in cold months and heat gain in warm months. When an engineer is sizing the heating/cooling system for a building, window glass normally accounts for 38 percent of the cooling load and 20 percent of the heating load (*Figure 20*). The term *load* refers to the amount of heat that must be added (heating mode) or removed (cooling mode) to keep the building comfortable for occupants.

Flashing

Metal flashing is essential to preventing moisture from entering the roof around the seams of roof windows, skylights, and chimneys. Always follow the manufacturer's instructions regarding proper installation procedures.

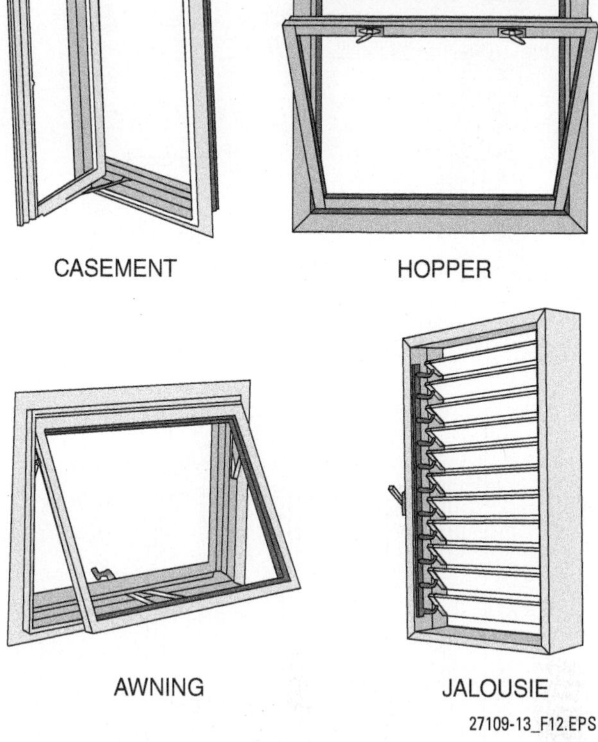

Figure 12 Examples of movable windows.

Figure 13 Bay window.

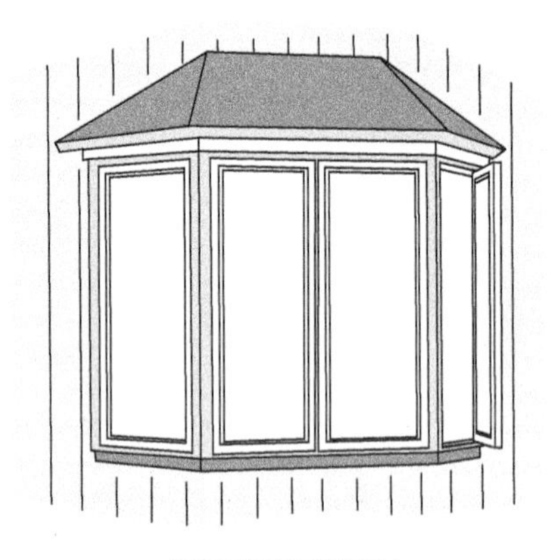

Figure 14 Examples of bay windows.

1.2.3 Energy-Efficient Windows

Money spent on energy-efficient windows is well spent. For example, the use of high-efficiency reflective glass instead of standard single-pane glass could reduce the cooling load by one-third. The load directly affects the size of the heating/cooling system required for the building, and therefore directly affects the original cost of the system, as well as the energy cost involved in operating the system.

Figure 15 Supported bay window.

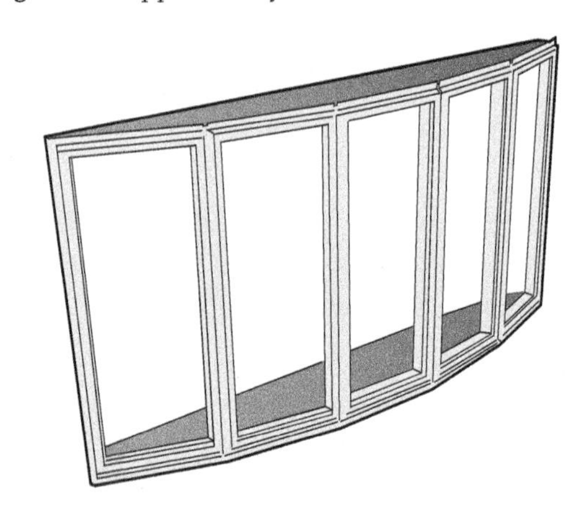

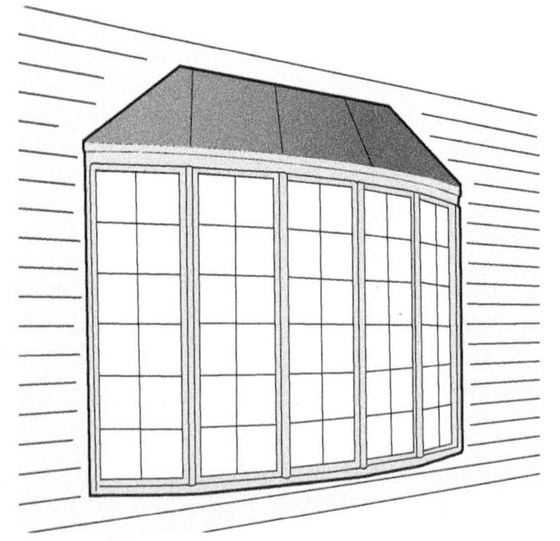

Figure 16 Bow windows.

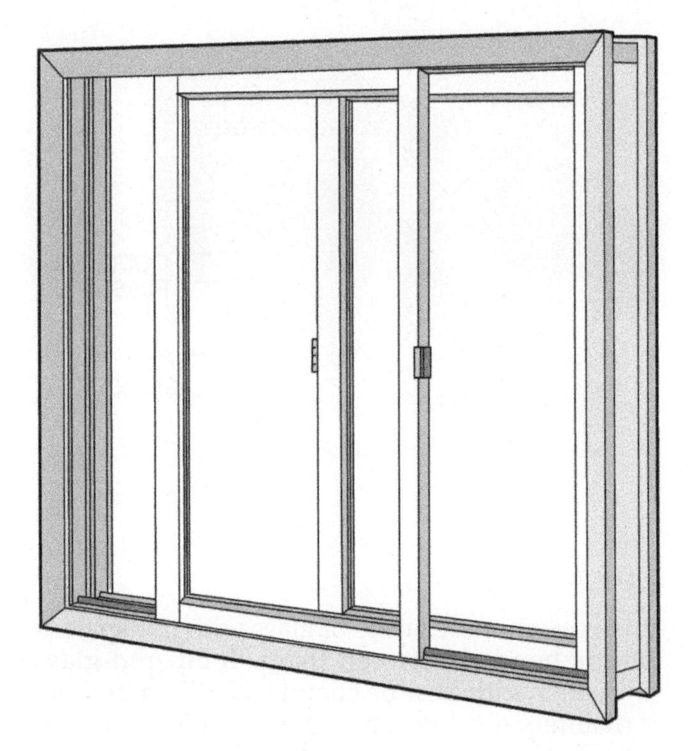
27109-13_F17.EPS

Figure 17 Sliding (gliding) window.

A single pane of glass provides very little insulation. It has an R-value (insulating value) of less than 1. (The greater the R-value, the greater the insulating value.) Adding another pane of glass with ½" of air space between them more than doubles the R-value. The air space between the panes of glass acts as an insulator. The larger the air space, the more insulation it provides. Windows are commonly designed with two or three layers of glass separated by 3⁄16" to 1" of air space in order to improve insulation quality (*Figure 21*). To obtain even more insulating value, the space between panes in some windows is filled with argon gas, which conducts heat at a lower rate than air. Where single-pane glass is used, it is common to add storm windows to the building.

A special type of glass, known as low-e (low-emissivity) glass, provides even greater insulating properties. Emissivity is the ability of a material to absorb or radiate heat. Low-e glass is coated with a very thin metallic substance on the inside of the inner pane of a double-pane window. In cold weather, radiated heat from walls, floors, furniture, etc., is reflected back into the room by the low-e coating instead of escaping through the windows. This reduces the heat loss, which in turn saves heating costs. In summer months, radi-

Transom Windows

Transom windows were common in the high-ceiling buildings erected in the early part of the twentieth century. Today, transom windows are becoming popular again because of the growing number of new high-ceiling homes. These decorative windows are usually shaped as a semicircle or rectangle, and may be either fixed windows or awning types. They are often used as decorative toppers for double-hung windows, as shown here.

27109-13_SA04.EPS

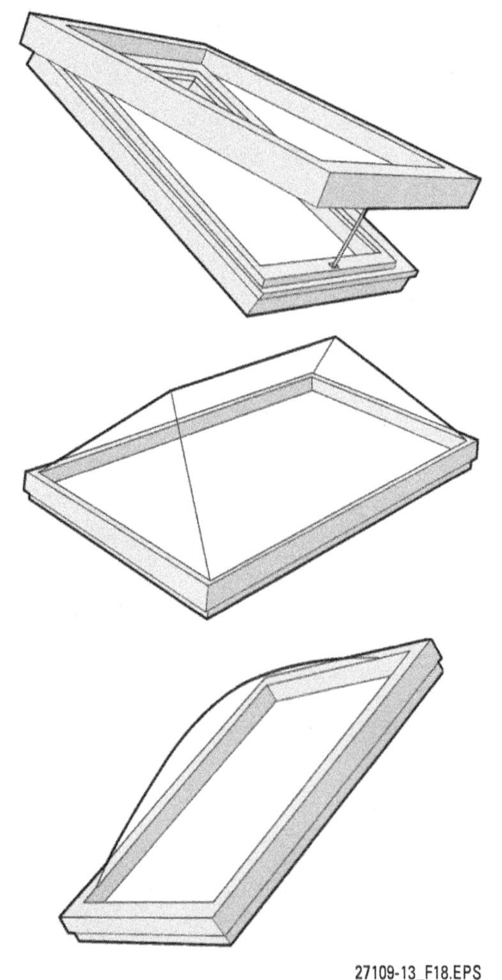

27109-13_F18.EPS

Figure 18 Roof windows and skylights.

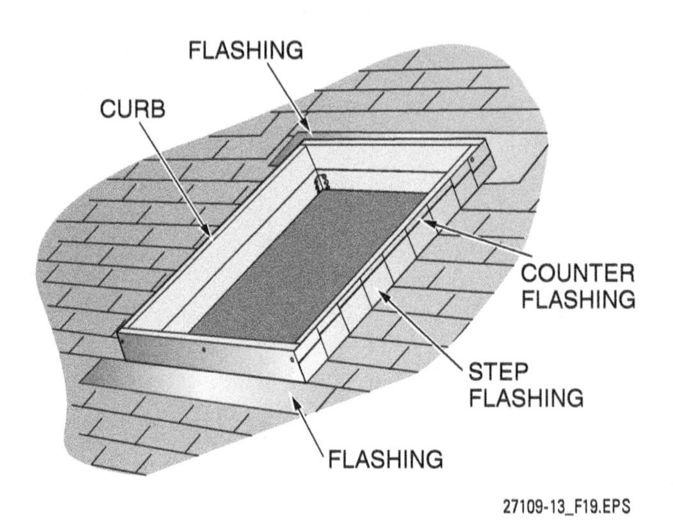

27109-13_F19.EPS

Figure 19 Skylight curb.

ated heat from outdoor sources such as the sun, roads, parking lots, etc., is reflected away from the building by the low-e coating. Although windows with low-e glass are considerably more expensive than standard windows, they usually pay for themselves in reduced heating and cooling costs within three or four years.

1.2.4 Safety Glass

Some local and state codes require the use of safety glass in windows with very low sills, and in those located in or near doors. Skylights and roof windows also require safety glass. There are several types of safety glass. Laminated glass contains two or more layers of glass with transparent plastic bonded between them. Tempered glass is treated with heat or chemicals. When broken, it disintegrates into tiny, harmless pieces. Wired glass has a layer of mesh sandwiched between the panes. The mesh keeps the glass from shattering. Transparent plastic (Plexiglas™) is also used for safety glazing applications, especially in skylights and doors.

1.3.0 Exterior Doors

Exterior doors are designed to provide security as well as insulation. Exterior doors, especially main entrance doors, are often designed to give a building an attractive appearance (*Figure 22*). Most residential exterior doors are prehung in frames with hinges and exterior casings applied. Once the door opening is framed in, the door simply needs to be installed and leveled. The first part of the following section covers residential doors.

Figure 23 shows the component parts of an exterior door installation. Like windows, doors have headers, side jambs, and sills. There are also several different kinds of exterior doors, as shown in *Figure 24*. A wide variety of styles and decorative designs for prehung exterior doors are available.

A panel door is the most common factory-built exterior door. Panel doors are made of vertical

Reflective Glass

Scientists working in the space program developed the first reflective glass. Space shuttles are subjected to high levels of friction when flying through Earth's atmosphere at great speeds. Reflective glass was developed in an attempt to reflect, transfer, and dissipate high temperatures. Reflective glass provides astronauts with a clear field of vision to navigate the shuttle, and also provides heat protection upon re-entry into the atmosphere.

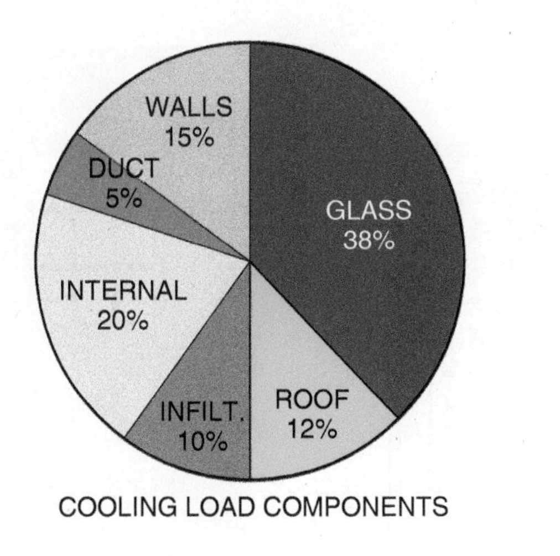

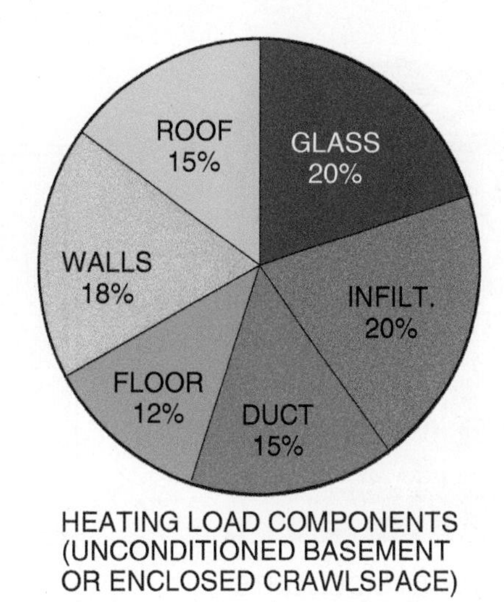

Figure 20 Cooling and heating load factors.

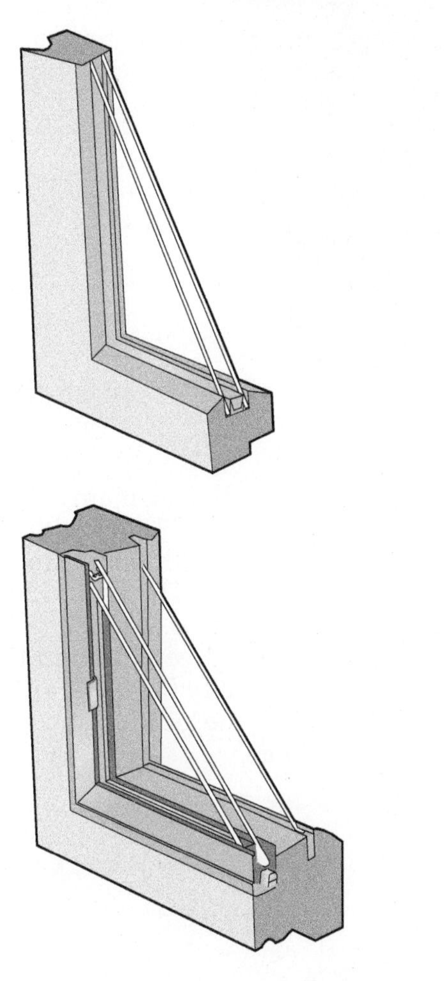

Figure 21 Double- and triple-pane windows.

members called stiles, cross members called rails, and filler panels. The panels are usually thinner than the stiles and rails and are rabbeted into them. Panel doors are more decorative than flush doors. Panel doors can be made of wood or metal.

An exterior flush door has a smooth surface made of wood veneer or metal and usually has a solid core of wood, composition board, or solidly packed foam. Hollow-core doors consist of a solid wood framework, metal honeycombs, or other materials that give the door rigidity. Hollow-core doors are not recommended for exterior use because they provide very little insulation and limited security. In addition, local building codes may prohibit their use in exterior entries. A flush door may contain a glass inset or it may have decorative moldings attached to the surface.

GOING GREEN

Smart Materials

Smart materials work by changing in response to environmental conditions. For example, smart windows may automatically increase their level of tint in response to higher levels of sunlight. Smart window blinds may automatically adjust to follow the sun's path. Some materials are under development that can change colors or other properties as well.

DOOR WITH SIDELIGHTS AND FANLIGHT

DOOR WITH SIDE PANEL (TRANSLUCENT)

DOOR WITH ORNAMENTAL TRIM

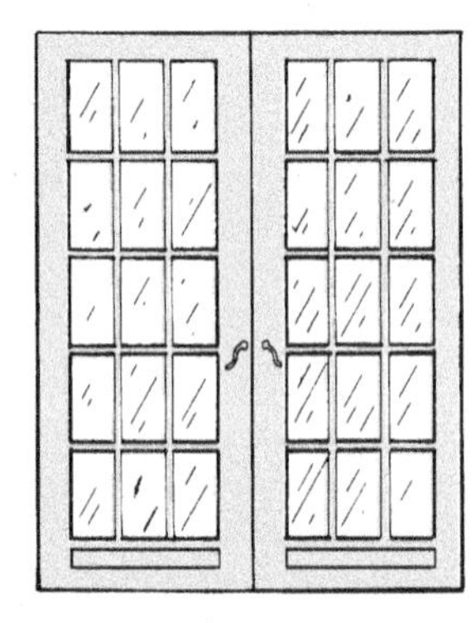

FRENCH DOORS

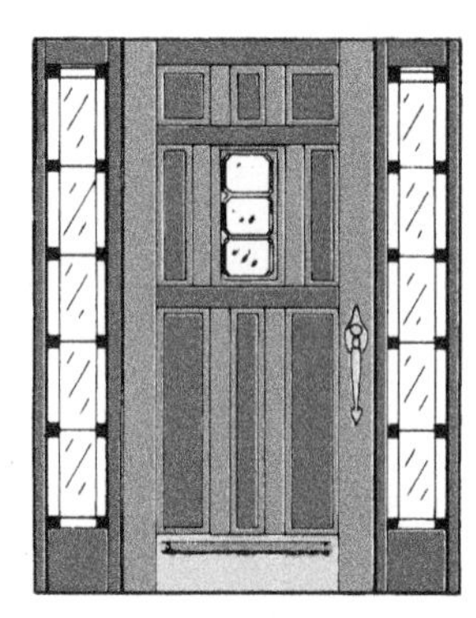

DOOR WITH SIDELIGHTS

27109-13_F22.EPS

Figure 22 Entry doors.

Sash doors may have a fixed or movable window. The window may be divided into lights. Various types of glass are used, including insulated, reinforced, and leaded. Louvered doors are popular as entry doors in warmer climates.

1.3.1 Exterior Door Sizes

Residential exterior doors are generally 1¾" thick and are available in a variety of widths and heights. A standard width for a main entry door is 36", although they are available in widths ranging from 18" to 40". Double-width doors are available in 60", 64", and 72" widths. The typical height for an exterior door is 80". Other available heights are 78", 79", 84", 90", and 96". Keep in mind that local building codes may specify the types and sizes of entry doors, as well as the location and minimum number of entry doors.

GOING GREEN

Energy-Efficient Glass

Low-e glass is coated with a thin metallic substance and is an effective way to control radiant heat transfer. Special heat-absorbing glass is also available. It contains tints that can absorb approximately 45 percent of incoming solar energy. This energy (heat) is then transferred from the window to the building structure. If these new types of glass aren't available, there are other alternatives that will increase energy efficiency in existing windows. For example, reflective film can be applied to windows to help contain heat loss in the winter and reflect sunlight in the summer. Caulk, weather stripping, and storm windows are also simple approaches to energy conservation.

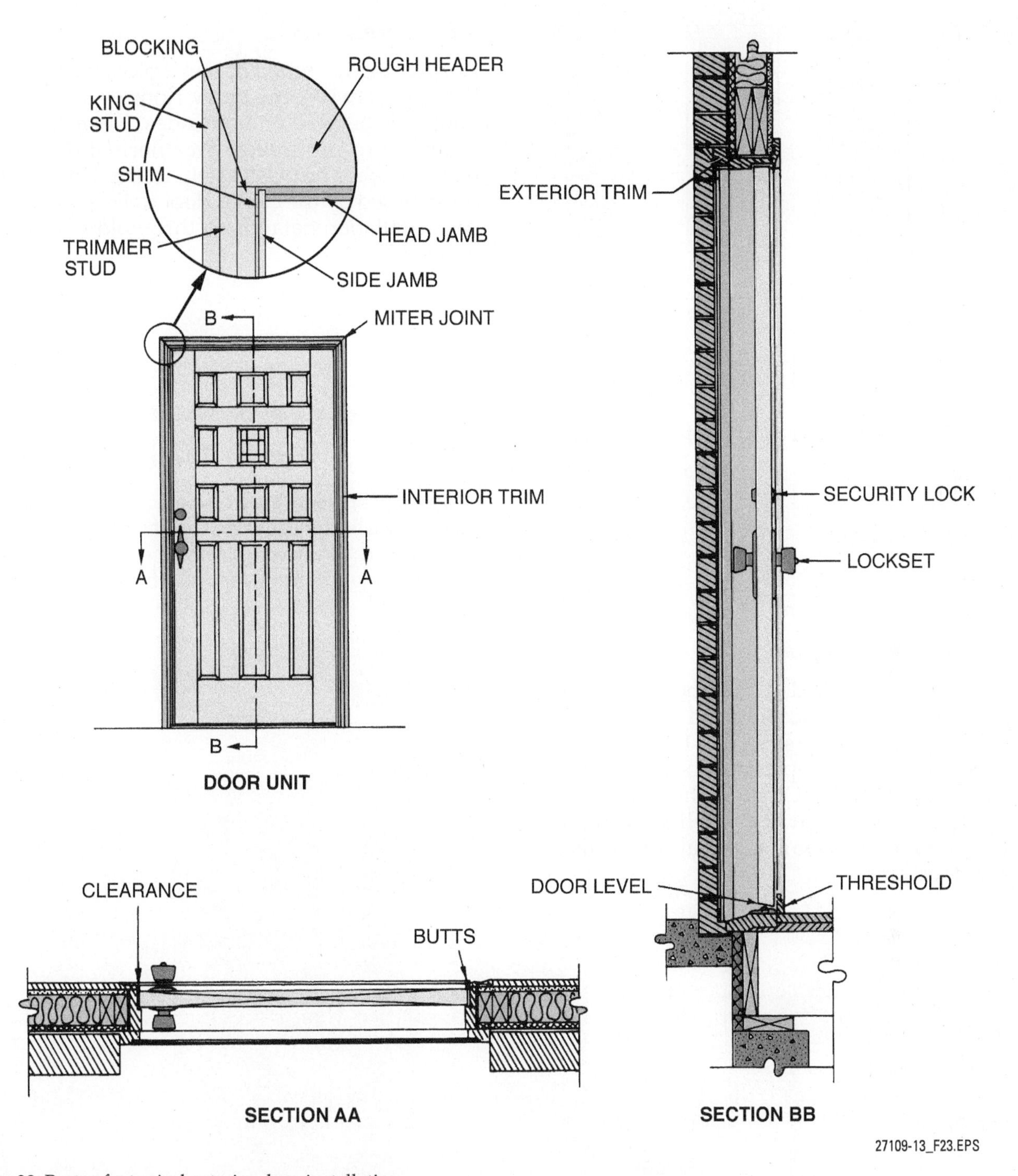

Figure 23 Parts of a typical exterior door installation.

Trimming Door Length

When trimming a veneer door to length, lay a straightedge along the length marks and score through the layer of veneer with a utility knife. This will prevent the veneer from splintering when the finish cut is made with a saw. Before making the cut, place a piece of tape on top of the veneer so the saw baseplate doesn't scratch the surface.

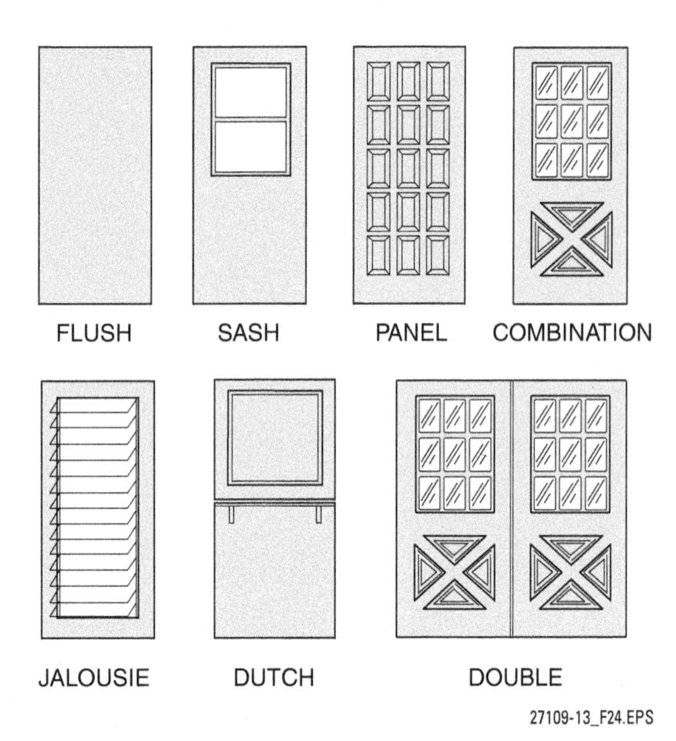

Figure 24 Types of entry doors.

1.3.2 Weather Stripping

Weather stripping is added to the bottom of a door to prevent heat from escaping and moisture from entering. One weather-stripping technique uses a rubber or vinyl sweep attached to the bottom of the door (*Figure 25*). Other methods are shown in *Figure 26*. A wide variety of self-adhesive and tack-on weather-stripping materials are available to seal door jambs.

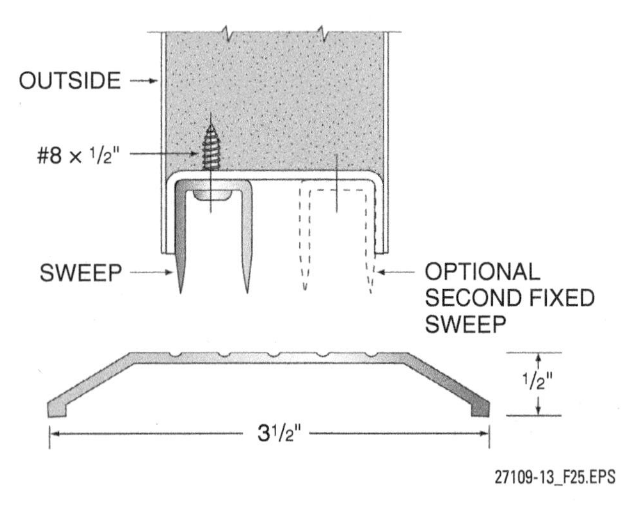

Figure 25 Fixed-bottom sweep.

1.3.3 Thresholds

A threshold is a wood or metal piece used to close the gap between the entry floor or sill and the door. It is beveled on both sides. *Figure 27* shows examples of thresholds. Weatherseal thresholds have a rubber or plastic strip in the center that is compressed by the closed door to keep out drafts. An installation detail for a threshold is shown in *Figure 27*.

Sliding Patio Doors

Sliding patio doors are a type of exterior door typically used for patios and porches. The door shown is solid oak with a painted aluminum exterior and a transom for use in high-ceiling rooms.

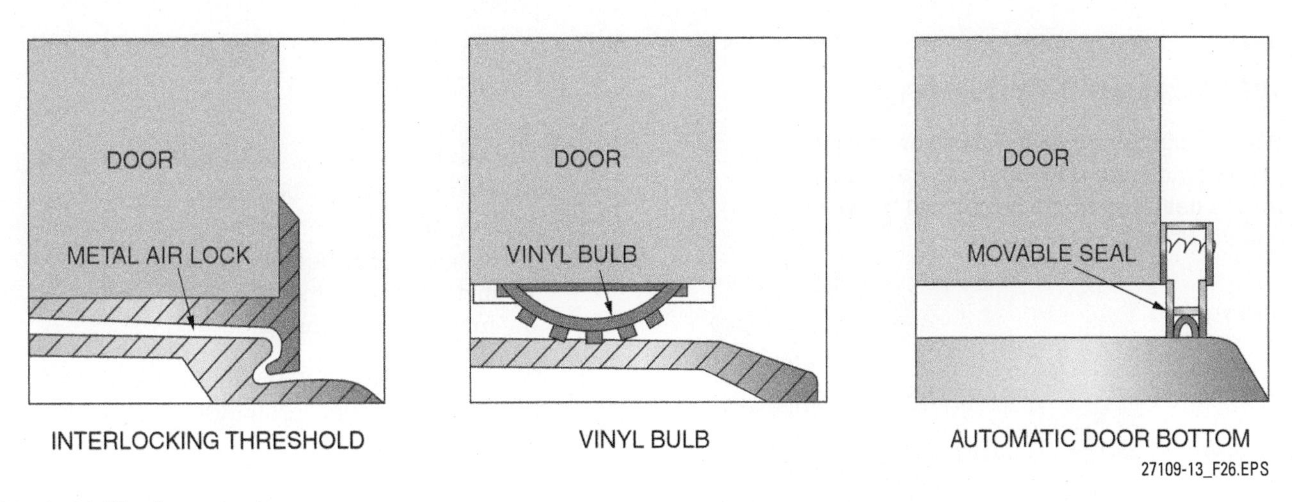

Figure 26 Weather stripping.

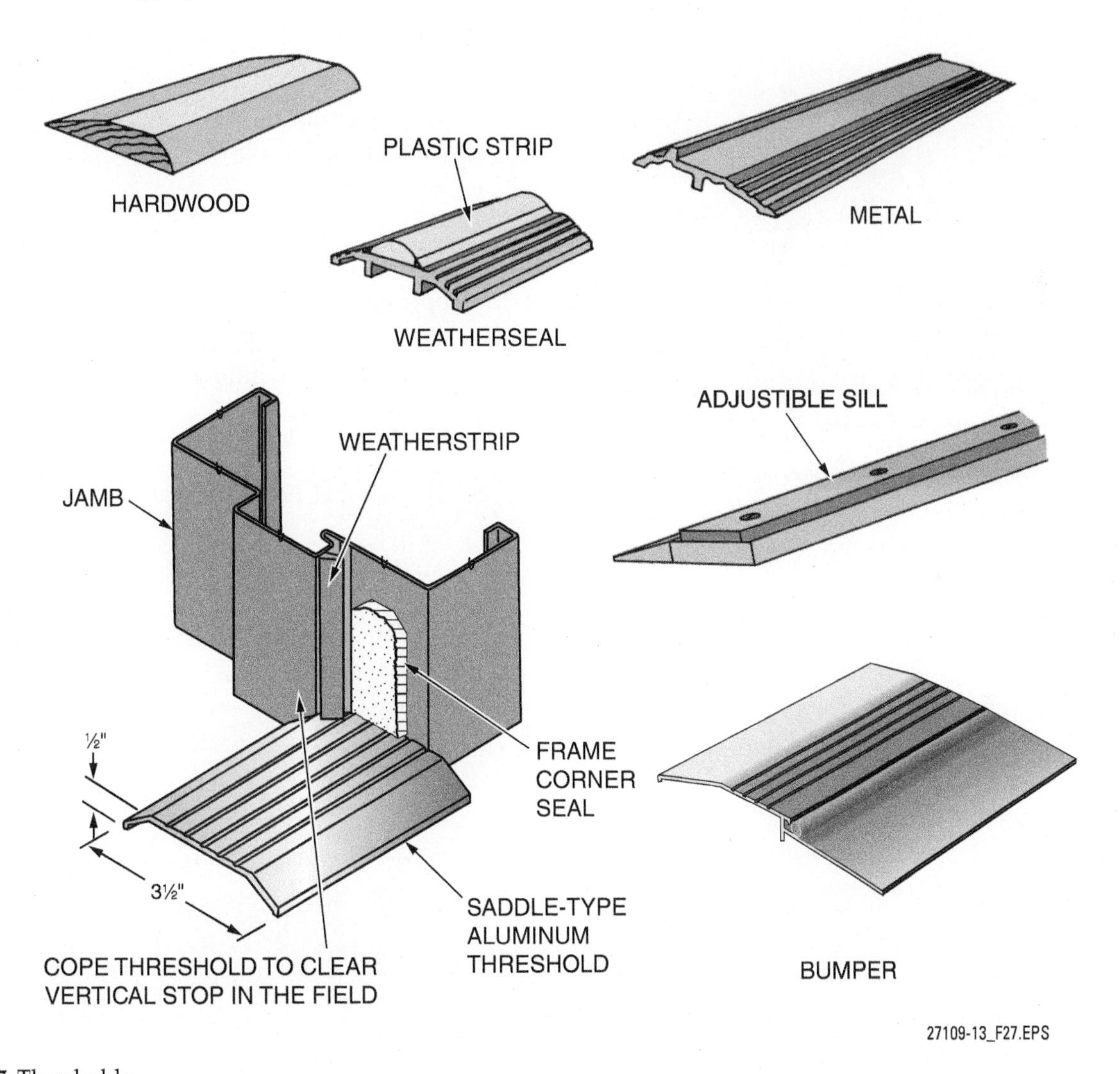

Figure 27 Thresholds.

Weather Stripping

Weather-stripping systems, such as thresholds, prevent moisture from entering under a door and reduce heat loss by creating a seal between a door and the floor.

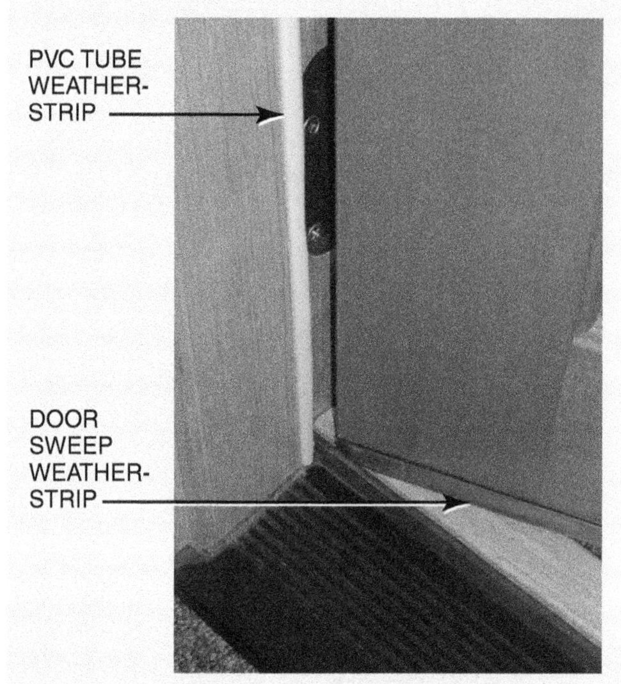

27109-13_SA06.EPS

1.0.0 Section Review

1. The edges and joints of the building wrap should be sealed using _____.
   a. contact cement
   b. waterproof adhesive
   c. metal furring channels
   d. special contractor's tape

2. When applying building wrap, all wall openings, such as doors and windows, should be covered.
   a. True
   b. False

3. Windows that swing open like a door are _____.
   a. fixed windows
   b. awning windows
   c. hopper windows
   d. casement windows

4. Which of the following windows *cannot* be opened?
   a. Awning windows
   b. Casement windows
   c. Skylights
   d. Roof windows

5. The typical height for an exterior door is _____.
   a. 64"
   b. 70"
   c. 76"
   d. 80"

SECTION TWO

2.0.0 WINDOWS

Objective

State the requirements for proper window installation.

a. Explain when jamb extensions are used.
b. Identify common considerations when framing in glass blocks.

Performance Task 1

Prepare a rough opening for proper window installation.

Trade Terms

Casing: Trim around a window or door.

Expansion strip: Resilient material added to a joint to allow expansion and contraction of adjacent materials while minimizing cracking.

Shim: Narrow, tapered piece of wood or composite material used to plumb or level a door or window.

There are many types of windows and many types of installations. *Figure 28* shows a cutaway of a vinyl-clad window.

The best approach is to follow the manufacturer-recommended procedure using the recommended tools, materials, and fasteners. The basic procedure for installing a prehung window is as follows:

Step 1 Ensure the window is shut before starting the installation. Also, ensure that the opening is plumb, level, and square, and is large enough to accommodate the window. To check the window opening for squareness, use a framing square. If the opening is not square, make the necessary adjustments and check again. Do not proceed until the opening is square and level. Allow approximately ¼" to ½" space between the rough header and the window head jamb to account for settling.

Step 2 Remove the blocks used to protect the window during shipment. Also remove any horns, which are side-jamb extensions that are also used for protection during shipment. If there are diagonal braces on the window, leave them in place temporarily.

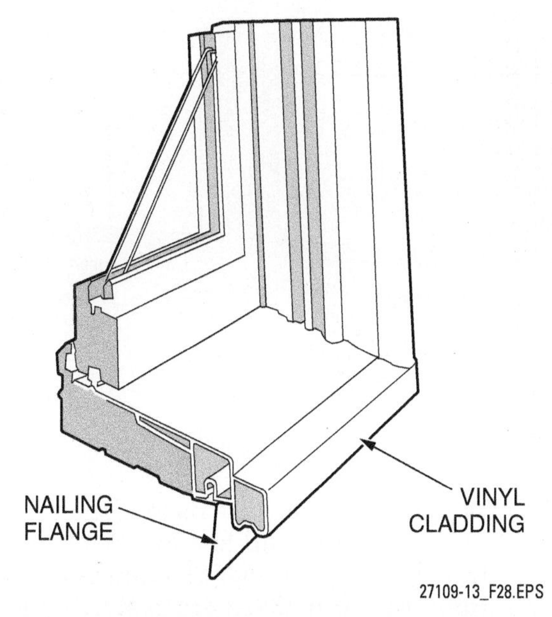

Figure 28 Cutaway of a vinyl-clad window.

Step 3 Install the window in the rough opening. Ensure there is enough assistance to avoid injuring yourself or damaging the window.

Step 4 If necessary, insert **shims**, wedges, or wood shingle tips under the window to raise it to the correct height. Temporarily nail one corner of the window through the flange or wood frame.

> **NOTE**
> Vinyl and metal windows, as well as vinyl- or metal-clad wood windows, are equipped with nailing flanges (fins) around the outside (see *Figure 28*). Use screws or roofing nails to secure the window to the framing members.

Step 5 Check the level of the sill. If it is not level, use shims to correct it. For windows with long sills, use shims at intermediate points to ensure that the sill is level with no sag (*Figure 29*). As each side is leveled, secure it with a nail or screw as recommended by the manufacturer.

Step 6 Plumb the side jambs using shims (*Figure 30*), but be sure not to shim too tight. When each jamb is plumb, secure that side.

Step 7 Verify the unit is plumb, level, and square. Operate the sash to make sure it works smoothly and does not bind. When it is satisfactory, finish securing the unit.

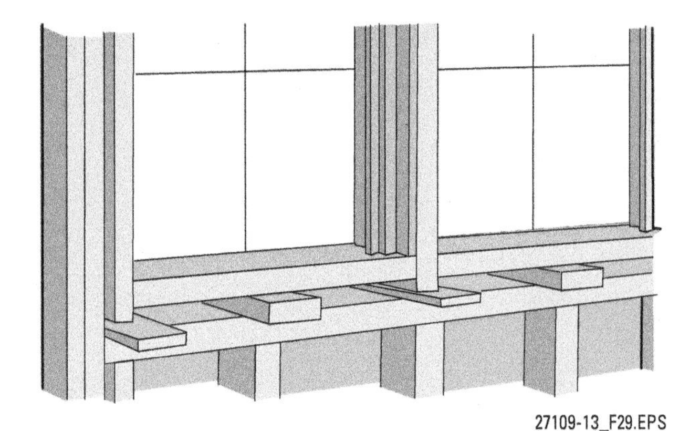

Figure 29 Leveling a window with shims.

When casing nails are used, use a nail set to drive in the nail to avoid denting the frame. Check the manufacturer's instructions to see if there are restrictions on nailing into the header.

> **NOTE**
> Casing nails should be used to secure a window unit to the building framing members. Don't drive the nails all the way in yet. Use a nail set to finish the job and prevent denting the frame. Keep the shims permanently in place by nailing through them during this step. Once the basic unit is installed, finish nails should be used on the trim.

Step 8 Pack insulation or apply expanding foam in the gaps between the trimmer studs and the jambs (*Figure 31*). Check the window manufacturer's recommendations before using expanding foam, as some types of foams apply pressure to the window components and thereby pinch the window.

2.1.0 Jamb Extensions

Modern windows are made for thinner walls than the walls found in older buildings. When replacing windows in older buildings, the jamb will likely need to be built out so it is flush with the wall (*Figure 32*). Jamb extensions are available as optional accessories from the window manufacturer or they can be ripped from a length of 1× lumber that is wide enough to fill the gap. Also, for 2 × 6 wall construction, most manufacturers supply extension jambs at 4⁹⁄₁₆", 5⅛", 6⁹⁄₁₆", etc., as options.

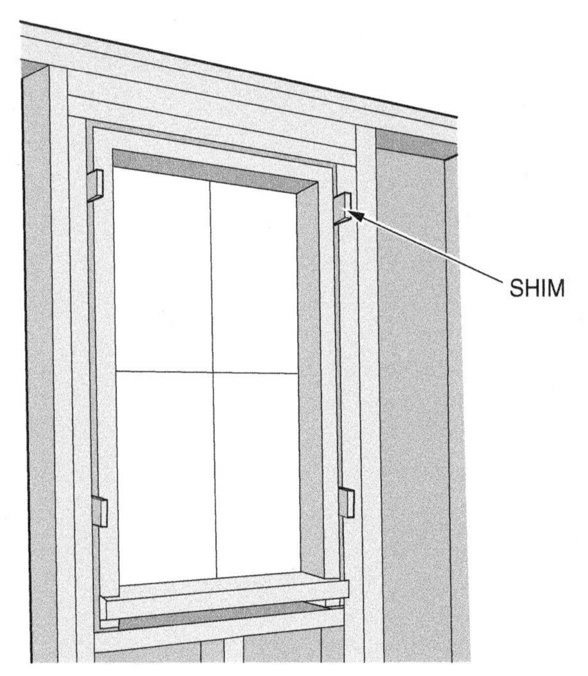

Figure 30 Plumb the side jambs.

Figure 31 Insulating between the jambs and trimmer studs.

2.2.0 Glass Blocks

Glass blocks may be used in place of standard fixed and movable windows to provide light while at the same time providing privacy. Glass blocks are generally 3⅞" thick and are available

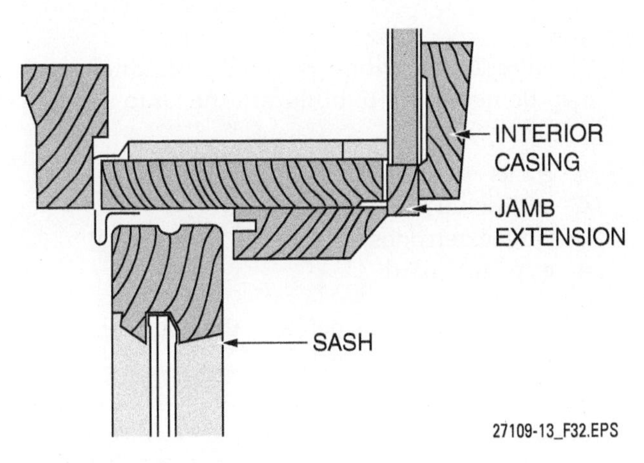

Figure 32 Jamb extension.

in three common nominal sizes: 6 × 6, 8 × 8, and 12 × 12. Actual block dimensions are ¾" less than nominal size to account for mortar joints. Glass blocks are made of two formed pieces of glass fused together in a way that leaves an air space in between. The air space provides excellent insulating qualities. Glass blocks are available individually or they are available in pre-assembled panels in a range of sizes.

Glass blocks are installed using masonry tools and white mortar or silicone. Prefabricated glass block window panels are available in a variety of sizes for direct installation. Some are available with a built-in vent for air circulation. Glass block windows can also be constructed from individual blocks, as shown in *Figure 33*. Glass block window systems include spacers, panel anchors, expansion strips, and glass blocks. If the window is larger than 25 square feet, panel reinforcing is also required.

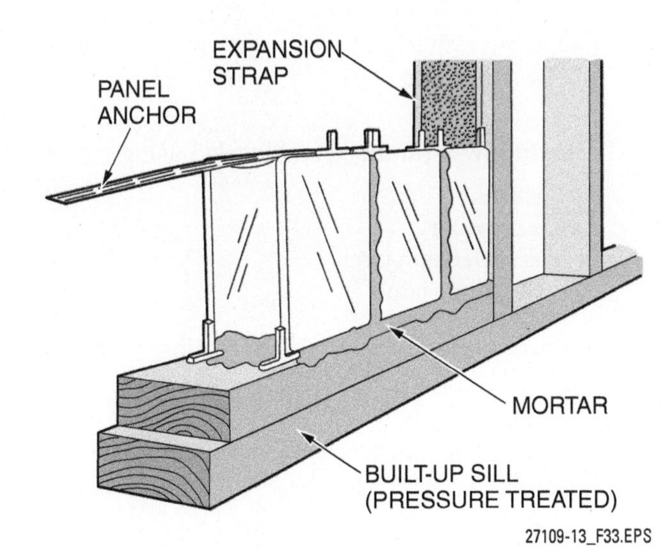

Figure 33 Glass block installation.

The following considerations are important when framing-in glass block windows. To determine the size of the opening required, multiply the nominal block size by the number of blocks and add ⅜" for expansion strips at the header and jambs. When the size of the opening is less than 25 square feet, the height should be no greater than 7' and the width no greater than 5'. When the opening is greater than 25 square feet, the panels should never be more than 10' wide or 10' high. A built-up sill should be used to protect the bottom row of blocks and the sill from damage. The built-up sill should be constructed of pressure-treated lumber since the sill makes direct contact with the mortar.

Glass Blocks

Fully assembled glass block window panels are available. The panels are faster to install than individual blocks, but are generally more expensive. Regardless of how the blocks are installed, glass blocks are approximately four times heavier than other windows of equal size. Check the instructions that come with the window kit for the weight specifications. Extra framing members may be required to support the window.

2.0.0 Section Review

1. A window is plumbed and leveled using _____.
   a. shims
   b. a crowbar
   c. an extension jamb
   d. the hinge jamb

2. Secure a window unit to the framing members using _____.
   a. duplex nails
   b. casing nails
   c. brads
   d. finishing nails

3. When replacing windows in older buildings, it may be necessary to build out the jamb using _____.
   a. 2 × 4s
   b. additional framing members
   c. jamb extensions
   d. gypsum board

4. The difference in size between the nominal and actual sizes of glass block is _____.
   a. ⅛"
   b. ¼"
   c. ⅜"
   d. ¾"

Section Three

3.0.0 Doors

Objective

State the requirements for proper door installation.

a. Identify the differences between residential and commercial doors.

Performance Task 2

Prepare a rough opening for proper door installation.

Trade Terms

Hinge jamb: The side of the door jamb into which the hinges are installed.

Lockset: The entire lock unit, including locks, strike plate, and trim pieces.

Threshold: A piece of wood or metal that is set between the door jamb and the bottom of a door opening.

Exterior doors are installed after the framing is complete, sheathing is installed, and building wrap is applied. Prehung door kits include the frame, threshold, exterior casing, and the door itself. A lockset is often included with the door kit.

Doors can be either left-hand or right-hand swing (*Figure 34*). Manufacturers use different methods of describing the door. The direction of swing is determined when facing the door from the outside of the building. An example of a manufacturer's product sheet is provided in the *Appendix*.

The following is a typical procedure for installing a prehung exterior door:

Step 1 As soon as the door is removed from the packaging, inspect it to make sure there is no damage. Remove the nails if the door was nailed shut for shipment.

Step 2 Ensure the rough opening is the correct size.

Step 3 Position the door unit and mark the inside of the threshold. Remove the door unit and apply a double bead of caulk to the bottom of the opening inside the line just marked.

Step 4 Center the door unit in the opening. Leave the factory-installed spacer shims between the door and the frame in place until the frame is securely attached to the rough opening.

Step 5 Ensure that the sill is level and the hinge jamb is plumb. Use shims to correct, if necessary.

Step 6 Adjust for the correct door clearance by inserting shims between the side jambs and rough opening at the bottom, top, and middle at the hinge butt locations. Additional shims may be required on larger doors.

Step 7 Nail the jamb through the shims using 16d finishing nails. Drive the nails home with a nail set to avoid marring the finish.

Step 8 Adjust the threshold so it makes smooth contact with the door without binding or leaving a space. Ensure the integrity of your original caulk bead. If broken, apply another caulk bead.

Step 9 Remove the top inner screws on the hinge and install 2½" to 3" wood screws through the door jamb and into the trimmer stud.

Step 10 Cut off all shims flush with the edge of the door casing and pack any gaps between the trimmer stud and the door casing with insulation.

3.1.0 Commercial Exterior Doors

Commercial exterior doors are selected by the architect to provide a secure and convenient means of entering and leaving a building. There is a wide variety of exterior doors available for commercial applications; the type and style selected for a specific location is based on its purpose. Doors used exclusively by employees are usually no-frills doors, while those used on loading docks are strictly utilitarian. The doors used by the public are selected to project a particular image and to make their use convenient. A trendy retail store

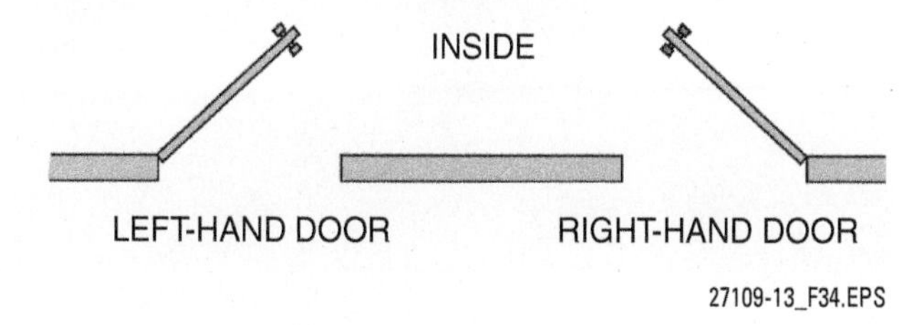

Figure 34 Door swing.

will use a very different style door than a supermarket, and an upscale hotel in New York City will have a different style door than a budget motel located near an interstate highway.

Figure 35 shows a revolving door on the Wrigley Building in Chicago. This beautiful entrance is perfect for a building that is headquarters for the Wm. Wrigley Jr. Company, as well as investment management companies, advertising agencies, marketing firms, a bank, and foreign consulates.

Figure 36 shows a thermally efficient door on a loading dock. This type of door is used when temperature control of the interior of the building is critical.

The installation of commercial doors is typically more complicated than residential doors. Due to their complexity, commercial exterior doors are often installed by teams that have been specially trained by the door manufacturer. In these cases, prepare the installation site according to the building plans and specifications. If you are required to install a commercial exterior door, obtain the door manufacturer's installation instructions and follow them very carefully.

27109-13_F35.EPS

Figure 35 Revolving door with ornate facade.

27109-13_F36.EPS

Figure 36 Thermally efficient door on a loading dock.

When a door is delivered to the job site, carefully inspect the door for damage before accepting it. Note any damage to the door packaging on the delivery invoice. Do not accept delivery of any severely damaged doors, and report the damage to your supervisor. Don't remove any packaging if there is suspicion that the product is damaged. The packaging material helps to protect the door during storage.

Once the door has been accepted, store it in a secure location. The door should be stored in the unopened shipping container to protect it. Store the door in an area that permits good air circulation. Doors should never be stored in an area where they are exposed to the elements. When doors are stored outside, be certain that they are off the ground and protected. If the door must be covered with plastic, ensure that there is enough of an air gap to prevent humidity buildup. Protect metal doors from corrosives such as lime, mortar, and cement.

CAUTION

When installing a commercial door, perform only the work you are trained and qualified to perform. Other work, such as electrical installations, must be completed by a qualified worker.

To minimize problems during door installation, ensure that wall construction is done according to the drawings and specifications. The wall adjacent to the opening must be plumb and the opening must be square and level. Although there are a number of adjustments the installation team can use to compensate for tolerance variations, it is much better to ensure that the installation site is prepared properly.

3.0.0 Section Review

1. The direction of door swing is determined when _____.
   a. facing the door from the inside of the building
   b. the architect develops the plans
   c. facing the door from the outside of the building
   d. the door is installed in the rough opening

2. Commercial exterior doors are often installed by teams that have been specially trained by the door manufacturer.
   a. True
   b. False

SECTION FOUR

4.0.0 LOCKSETS

Objective

Identify the various types of locksets used on exterior doors and explain how they are installed.

Performance Task 3

Install a lockset.

Trade Term

Deadbolt: A square-head bolt in a door lock that requires a key to move it in either direction.

Locksets of various designs are typically installed on residential and commercial exterior doors. *Figure 37* shows examples of locksets that might be used on residential exterior doors. Mortise locksets (*Figure 38*) are more secure and are therefore considerably more expensive than most other types of locksets. For that reason, they are more common in commercial buildings than in residential construction. Tubular locksets (*Figure 39*) are less secure than cylindrical locksets (*Figure 40*). Neither is an excellent choice where security is a major concern.

> **NOTE**
> When mortising for a doorplate strike or hinge, score around them, then score every ¼" down the mortise before using a chisel. This method allows small distances to be chiseled out, producing a clean, accurate mounting area on the jamb for the strike or hinge.

In many cases, homeowners desire a security lock or deadbolt lock in addition to the locking mechanism built into the doorknob assembly. Deadbolts are either single- or double-cylinder types. A double-cylinder deadbolt lockset requires a key on both sides and would be used where there is glass close to the lock. If the door is solid, a security lock or single-cylinder deadbolt (requiring a key on the outside only) will suffice.

Lock manufacturers include installation instructions and drilling templates (*Figure 41*) in the packaging for locksets and other locking devices. Professionals generally prefer to use a boring jig and boring bit, along with a mortise marker instead of a template, because this method is faster and more accurate.

Locksets are usually installed at a height of 36" to center from the floor. Correct measuring is extremely important because the bolt on the lockset must fit into a strike plate installed on the door jamb (*Figure 42*). Check the manufacturer's instructions to determine if the backset is 2⅜" or 2¾".

Always keep plenty of shims (*Figure 43*) on hand when installing windows and doors. Shims are easy to cut by simply scoring them with a utility knife, then snapping them off along the score.

Figure 37 Examples of exterior door locksets and security locks.

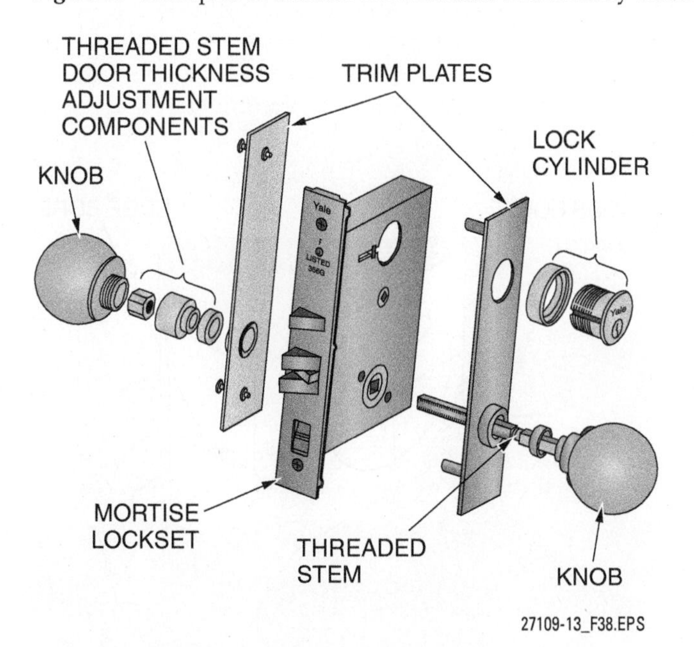

Figure 38 Exploded view of a mortise lock.

Measuring the Rough Opening

Doors can be unwieldy to carry and position, so always check the rough opening size before attempting to move the door into place.

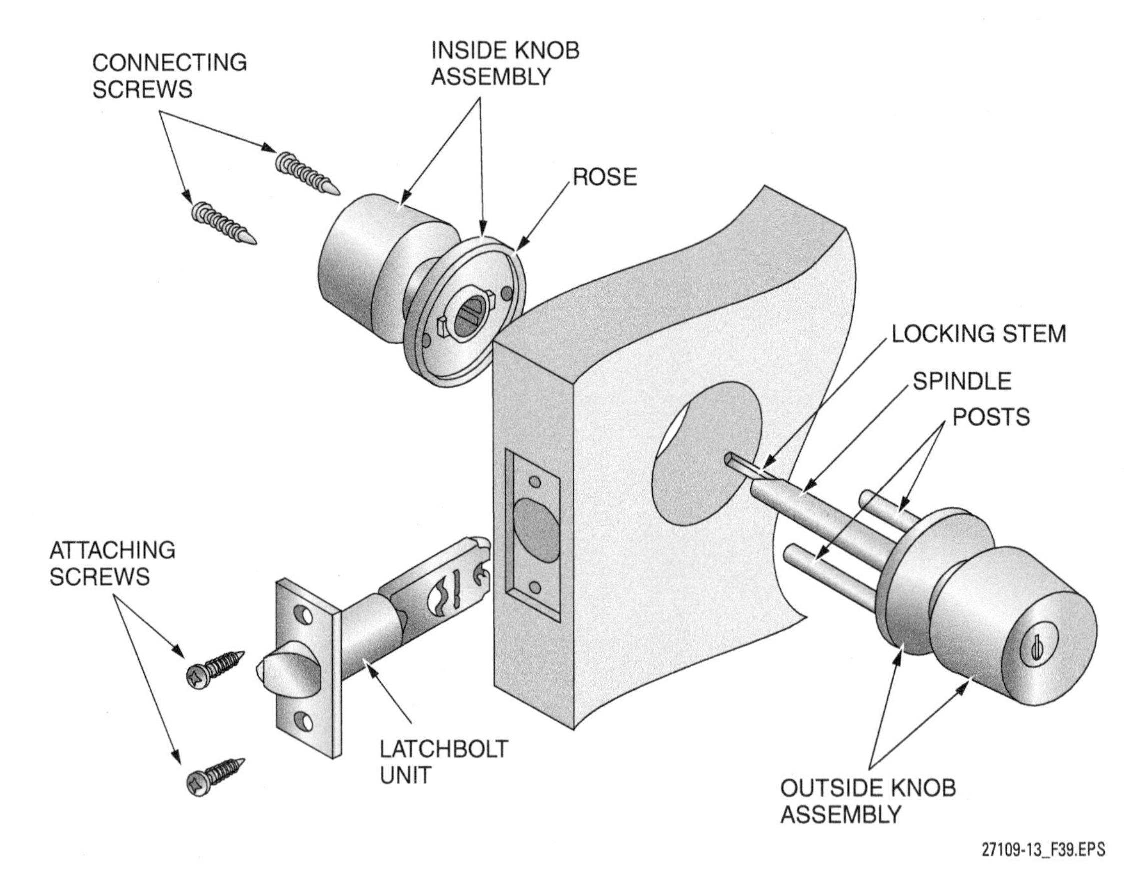

Figure 39 Exploded view of a tubular lockset.

Figure 40 Heavy-duty cylindrical locksets.

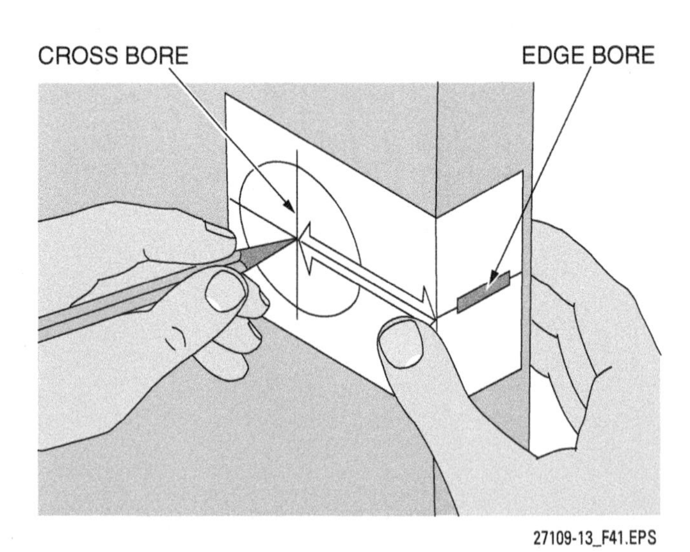

Figure 41 Using an installation template.

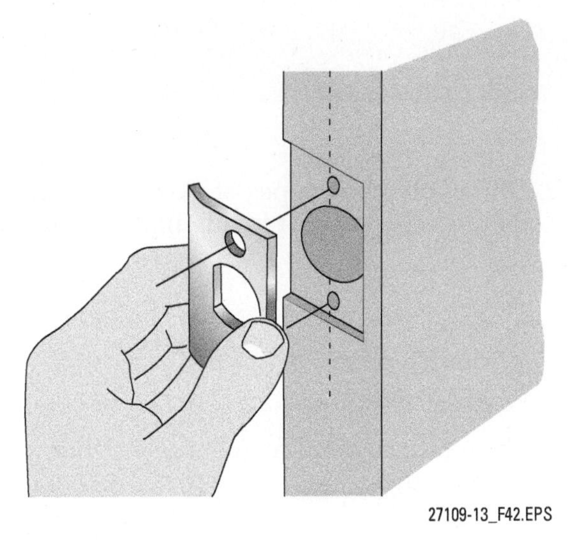

Figure 42 Installing a door strike.

Figure 43 Shims.

4.0.0 Section Review

1. For locksets, the typical center-to-floor distance is _____.
   a. 24"
   b. 32"
   c. 36"
   d. 42"

SUMMARY

There are many different styles and types of windows and exterior doors, made by many different manufacturers. The overwhelming majority of the windows and exterior doors encountered by carpenters are the prehung type. While the basic designs and installation requirements are essentially the same, the specifics vary from one manufacturer to another. Selection and installation may also depend to some extent on local building codes. The important thing to remember is to follow the instructions provided by the manufacturer for each window and become familiar with local building codes.

Review Questions

1. The framework around the glass in a window is the _____.
   a. schenker
   b. sash
   c. casing
   d. header

2. A double-hung window contains _____.
   a. a single, fixed sash
   b. two fixed sashes
   c. one fixed sash and one movable sash
   d. two movable sashes

Refer to *Review Question Figure 1* to answer Questions 3 and 4.

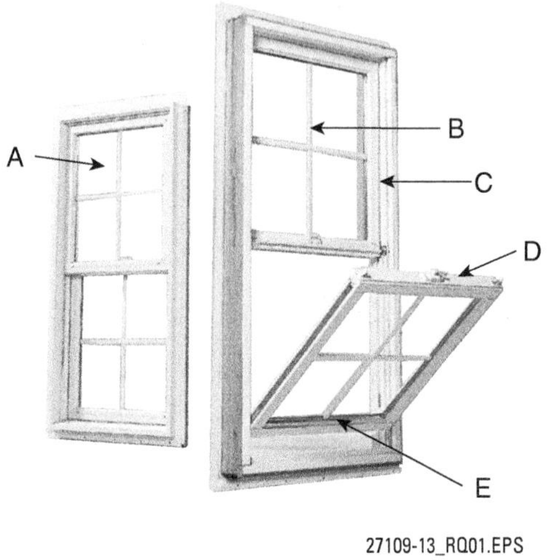

27109-13_RQ01.EPS

Figure 1

3. Letter B is pointing to a _____.
   a. muntin
   b. rail
   c. sash
   d. light

4. Letter C is pointing to a _____.
   a. sash
   b. jamb
   c. stile
   d. muntin

5. Of the following types of glazing, the one that has the greatest insulating value is _____.
   a. DS
   b. low-e
   c. double-pane
   d. argon-filled

6. The top horizontal member of a window frame is the _____.
   a. header
   b. stile
   c. trimmer
   d. head jamb

7. The most common thickness of a residential exterior door is _____.
   a. 1¼"
   b. 1½
   c. 1¾"
   d. 2"

8. When using casing nails to install a window, what tool should be used to drive the head of the nail below the surface without denting the frame?
   a. Plumb bob
   b. Casing tool
   c. Nail set
   d. Jab saw

9. To determine the size opening needed for a glass block window, add an allowance for expansion strips at the header and at the jamb of _____.
   a. ⅜"
   b. ½"
   c. ⅝"
   d. ¾"

10. Once a window is installed, trim should be fastened in place using _____.
    a. casing nails
    b. panel adhesive
    c. staples
    d. finish nails

11. The horizontal member at the bottom of a door opening is the _____.
    a. threshold
    b. stile
    c. footer
    d. rail

12. When installing an exterior door, the top inner screw on each hinge should be replaced with a _____.
    a. lag bolt
    b. 2½" or 3" wood screw
    c. 1½" drive screw
    d. 16d nail

13. Shims can be most easily cut to size by _____.
    a. scoring with a knife and snapping
    b. using tin snips
    c. sawing
    d. using a chisel and mallet

14. Which of the following types of locksets is the most secure?
    a. Cylindrical
    b. Mortise
    c. Deadbolt
    d. Tubular

15. Building wrap is made from _____.
    a. polyethylene fiber
    b. fiberglass
    c. polyurethane strands
    d. recycled paper

Trade Terms Quiz

Fill in the blank with the correct term that you learned from your study of this module.

1. The exterior walls, doors, windows, and roof are all part of the _______.
2. Wood windows commonly use aluminum or _______ for cladding.
3. A door with a light uses _______ to secure the panes of glass.
4. The glass inserts in a door are called _______.
5. Locks on traditional double-hung windows would most likely be attached to the top _______ of the windows.
6. The lock on a traditional panel door would most likely be attached to the _______ opposite the hinges.
7. A skylight that forms a peak is called a(n) _______ skylight.
8. A(n) _______ is the part of a window that holds the glass.
9. A layer of _______ is installed over the wall sheathing to minimize air infiltration.
10. A skylight is often mounted on a raised frame known as a(n) _______.
11. When you walk through a door you step over the _______.
12. The trim around a window or door is known as the _______.
13. A(n) _______ requires a key to move in in either direction.
14. The material used as glass in doors or windows is referred to as _______.
15. The efficiency of insulation is known as its _______.
16. The living room, bedrooms, kitchen, and other rooms to which heated or cooled air is delivered are the _______.
17. A(n) _______ allows passage of light and may be used for ventilation.
18. _______ is an acrylic/portland cement exterior finish often used on commercial buildings.
19. The lockset is installed in the _______ of a panel door.
20. _______ allow materials to change dimension as a result of heating or cooling.
21. _______ is installed around a wall or roof opening to seal it against leakage.
22. Use _______ to level or plumb a window or door in the rough opening.
23. A(n) _______ must be installed to provide security to a building's inhabitants.
24. The side of a door frame opposite to the side where the strike plate is installed is the _______.
25. The window stiles butt into the _______ on the lower end.

Trade Terms

Building envelope
Building wrap
Casing
Conditioned space
Curb
Deadbolt
Expansion strip
Exterior insulation and finishing system (EIFS)
Flashing
Glazing
Hinge jamb
Hipped
Jamb
Light
Lockset
Muntins
Polyvinyl chloride (PVC)
Rail
R-value
Sash
Shim
Sill
Stile
Threshold
Window

John Ambrosia

Consultant and Master Trainer
JJA Construction & General Industry Training Consultants
Lawrenceville, Georgia

How did you get started in the construction industry?
I came from a family of craftsmen (electricians, stone masons, carpenters, etc.) and started out as a laborer at an early age. After getting out of the US Navy, I went through a formal apprenticeship program, became a journeyman carpenter, and worked my way up to being a general superintendent for the top 20 Engineering News Record (ENR) contractors. My construction tenure also includes owning a company, working as architect representative, owner representative, etc.

Who or what inspired you to enter the industry?
My inspiration was my uncles, who were journeymen, and recognizing the honor of working with your hands. Another inspiration was the knowledge that I wasn't limited to being a craftsman but could move into positions of leadership within the construction industry as long as I maintained my reputation and integrity.

What do you enjoy most about your career?
After over 30 years in the construction arena, it is great to know that I can continue to share and educate our future workforce. We all start out by working at a job, and some of us decide this is a career path we can pursue. The lucky ones find they have a vocation.

I receive a lot of communication and correspondence from many associates that I have worked with, coached, and mentored throughout my career. Knowing I had some small contribution in their present success is rewarding and validating.

Why do you think training and education are important in construction?
All training and education is good but I question what some of our educational entities actually provide. Not just a degree, our credentials provide a chance for success at the career the student chose. Is this position going to be there when they are ready? Training in construction is just not for the jobs we have now, but more so in the future. We must listen to the client/company needs for qualified trained people and adjust our training efforts to accommodate this issue. This is a path for future craftsmen and leaders and ensures we are setting them up for success.

Why do you think credentials are important in construction?
I always preached that apprenticeship craft training is the other four-year program for continuing education, before it became popular. I believe that just because we get mud on our boots, sweat on our brow, and don't work in offices, we are still entitled to a professional credential recognizing the qualifications of our skills and our efforts to attain and continually practice these skills. I have been involved with NCCER since 1999 as a master trainer, craft trainer, master safety instructor, and subject matter expert (SME) for different curricula and will put our curricula, training, and validation up against anything out there.

Don't take it as my opinion, but numerous clients, companies, owners, and government departments make credentials a standard of qualification for their projects. After 30 years in the business, I have seen the demand for verifiable training programs that meet the client's requirements increase over the past 12 years, and it is getting more and more difficult to provide qualified manpower that meets the client's requirements.

How has training/construction impacted your life and career?
I had the luck to have many great coaches as I progressed along my career. I learned not just my craft from some great journeymen, but also the pride of a job well done. No matter what type of training—formal, classroom, e-training, etc.—it still takes time in the saddle and practice on the job. Things change so much all the time in the construction arena— materials, means and methods, design—that we must always pursue the path of knowledge and keep learning and asking questions. My wife and I were able to provide for our six children and let them pursue their goals, and my career in construction allowed that success, too.

Would you recommend construction as a career to others? Why?
Yes, but it is not for everyone. If you have a desire to be successful, the drive to eliminate self-ignorance, continually learn at your craft or position, and not be hesitant to question what you are learning, you can be successful in the construction industry.

Most people have communication breakdowns because we might understand the when and the how but don't understand the why things should be done a certain way. Ask for the why first and you can be successful sooner.

Where else can you go starting out as a laborer, own your own company, and work your way to supervising and managing multibillion-dollar projects. If I can do it, anyone can.

What does craftsmanship mean to you?
Integrity and reputation. No matter who you are or what title you might have, integrity and reputation is all we can take away with us at the end of the day.

Trade Terms Introduced in This Module

Building envelope: Consists of all building components, such as basement walls, exterior walls, floor, and roof, that separate conditioned space from unconditioned space or outdoor air.

Building wrap: Made of spun, high-density polyethylene fibers randomly bonded into an extremely tough, durable sheet material and used to minimize air infiltration while allowing moisture to escape.

Casing: Trim around a window or door.

Conditioned space: An area of a building in which heating and/or cooling is deliberately supplied to it or is indirectly supplied through uninsulated surfaces of water or heating equipment or through uninsulated ducts.

Curb: A framework on which a skylight is mounted.

Deadbolt: A square-head bolt in a door lock that requires a key to move it in either direction.

Expansion strip: Resilient material added to a joint to allow expansion and contraction of adjacent materials while minimizing cracking.

Exterior insulation and finishing system (EIFS): Method of exterior finish consisting of a coat of acrylic copolymers and portland cement applied over a base of exterior sheathing, insulation board, and reinforcing mesh.

Flashing: Sheet metal strips used to seal a roof or wall against leakage.

Glazing: Material such as glass or plastic used in windows, skylights, and doors.

Hinge jamb: The side of the door jamb into which the hinges are installed.

Hipped: The external angle formed by the intersection of two adjacent sloping sides.

Jamb: The top and sides of a door or window frame that are in contact with the door or sash.

Light: Glass insert in a door.

Lockset: The entire lock unit, including locks, strike plate, and trim pieces.

Muntin: A thin framework used to secure panes of glass in a window or door.

Polyvinyl chloride (PVC): Plastic polymer used for electrical insulation and pipe.

Rail: Horizontal member of a window sash or panel door.

R-value: Measure of effectiveness of a material to insulate another material. A higher R value indicates a higher insulating value.

Sash: The part of a window that holds the glass.

Shim: Narrow, tapered piece of wood or composite material used to plumb or level a door or window.

Sill: The lowest framing member of a window or exterior door frame.

Stile: Vertical member of a window sash or panel door.

Threshold: A piece of wood or metal that is set between the door jamb and the bottom of a door opening.

Window: A wall opening that admits light and/or ventilation.

Appendix

Door Manufacturer's Product Data Sheet

HANDING DIAGRAMS

INSIDE

KS LH — KS RH — KS RHA — KS LHA — RRA KS — LRA KS — LHR KS — RHR KS

"KS" = KEY SIDE

OUTSIDE

SUFFIX "A" = ACTIVE LEAF OF PAIRS

LOCATION

COL.1: OPENING NUMBER

COL.2: LOCATION

COL.3: FROM or TO
F = FROM
T = TO

COL.4: LOCATION

GENERAL

COL.5: QUANTITY REQUIRED

COL.6: NOMINAL DOOR OPENING DIMENSION

COL.7: ACTUAL HAND & SWING

FRAME DATA

COL.8: SERIES/PROFILE
SU = STANDARD PROFILE UNEQUAL RABBET FRAME
SQ = STANDARD PROFILE EQUAL RABBET FRAME
DU = DRYWALL PROFILE UNEQUAL RABBET FRAME
DQ = DRYWALL PROFILE EQUAL RABBET FRAME
BQ = BEFORE DRYWALL PROFILE EQUAL RABBET FRAME (APPLIED ANCHORS)
BU = BEFORE DRYWALL PROFILE UNEQUAL RABBET FRAME (APPLIED ANCHORS)
SE1,2 = DOUBLE EGRESS PROFILE FRAME
SQT = STANDARD PROFILE EQUAL RABBET THERMAL BREAK FRAME
SQW = STANDARD PROFILE EQUAL RABBET WEATHERSTRIPPED FRAME
DQW = DRYWALL PROFILE EQUAL RABBET WEATHERSTRIPPED FRAME
XXX = SPECIAL PROFILE, SPECIAL RABBETS

COL.9: GAGE, MATERIAL AND FINISH
_CRS = _GA. COLD ROLLED STEEL
_A60 = _GA. A60 GALVANNEAL STEEL
_G90 = _GA. G90 GALVANNEAL STEEL
_M40 = _GA. A40 GALVANNEAL STEEL, MADERA
FGFRP = FIBERGLASS, FIBER REINFORCED PLASTIC
_S01 = _STAINLESS STEEL 304-4
_S02 = _STAINLESS STEEL 316-4
_S03 = _STAINLESS STEEL 304-6
_S04 = _STAINLESS STEEL 316-6
_S05 = _STAINLESS STEEL 304-8
_S06 = _STAINLESS STEEL 316-8
"_" represents steel gage (12,14,16,18)
Common codes shown; additional gages, material and finishes available.

COL.10: FRAME DEPTH
300 = 3"
434 = 4-3/4"
578 = 5-7/8"

COL.11: FRAME THROAT
200 = 2"
334 = 3-3/4"
478 = 4-7/8"

COL.12: STRIKE
S = ANSI A115.1 & 2 (4-7/8")
C = CYLINDRICAL A115.2/(3) (2-3/4")
D = DOUBLE DOOR
P = PLAIN (NO PREP)

COL.13: LABEL

FIRE PROTECTION RATING		AGENCY
A = 3 HR		U = UNDERWRITERS LABORATORY
B = 1-1/2 HR		
C = 3/4 HR		F = FACTORY MUTUAL
D = 1-1/2 HR		W = WARNOCK HERSEY
E = 3/4 HR		
T = 1/3 HR		

COL.14: TYPE
BL = BORROWED LITE
SL = SIDELITE
TF = TRANSOM
DF = DUTCH
DA = DOUBLE ACTING
CM = COMMUNICATING

COL.15 ANCHORS
WMA = WIRE MASONRY ANCHOR
WS = WOOD STUD
MS = METAL STUD
FFA = FLOOR
EO = EXISTING OPENING
See anchor sheet for more

COL.16: FRAME ELEVATION

DOOR DATA

COL.17: DOOR SERIES AND CORE
RI = REGENT (IMPREGNATED HONEYCOMB CORE, HANDED)
RN = REGENT (NON-IMPREGNATED HONEYCOMB CORE, HANDED)
IU = IMPERIAL (POLYURETHANE FOAM CORE, HANDED)
OI = OMEGA (IMPREGNATED HONEYCOMB CORE, NON-HANDED)
OT = OMEGA DOOR W/250° TEMP. RISE CORE, NON-HANDED)
VU = VERSADOOR (POLYURETHANE FOAM CORE, NON-HANDED)
FT = FUEGO (250° TEMP. RISE CORE, HANDED)
MS = MEDALLION (VERTICALLY STIFFENED CORE, HANDED)
MJ = MEDALLION (VERTICALLY STIFFENED CORE, MARINE TECH, HANDED)
TX = THRULITE (STILE & RAIL DESIGN, HANDED)
AP = ARMORSHIELD (LEVEL III, BULLET RESISTANT POLYSTYRENE CORE, HANDED)
AS = ARMORSHIELD (LEVEL III, BULLET RESISTANT STEEL-STIFFENED CORE, HANDED)
A4 = ARMORSHIELD (LEVEL IV, BULLET RESISTANT CORE, HANDED)
SA = SOUNDTECH (ACOUSTICAL CORE, HANDED)
KU = KHEMPRO (FRP, HANDED)
BI = BETA (IMPREGNATED HONEYCOMB CORE, NON-HANDED, LIGHT DUTY)
GU = GAMMA (POLYURETHANE FOAM CORE, NON-HANDED, LIGHT DUTY)

COL.18: GAGE, MATERIAL AND FINISH
_CRS = _GA. COLD ROLLED STEEL
_A60 = _GA. GALVANNEAL STEEL
20T60 = 20 GA. A60 GALVANNEAL TEXTURED STEEL
_G90 = _GA. G90 GALVANNEAL STEEL
_M40 = _GA GALVANNEAL STEEL, MADERA
FGFRP = FIBERGLASS, FIBER REINFORCED PLASTIC
_S01 = _GA. STAINLESS STEEL 304-4
_S02 = _GA. STAINLESS STEEL 316-4
_S03 = _GA. STAINLESS STEEL 304-6
_S04 = _GA. STAINLESS STEEL 316-6
_S05 = _GA. STAINLESS STEEL 304-8
_S06 = _GA. STAINLESS STEEL 316-8
"_" represents steel gage (12,14,16,18,20)
Common codes shown; additional gages, material and finishes available.

COL.19: DESIGN
F = FLUSH
G = HALF GLASS
G3 = HALF GLASS W/3 LITES
FG = FULL GLASS
N660 = 6 x 60 NARROW LITE
V = 10 x 10 VISION LITE

EMBOSSED PANEL DESIGNS
E601 = 6 PANEL
E801 = 8 PANEL
EC05 = CROSSBUCK W/9 LITES
See door elevation sheet for more

COL.20: LOCK PREP
LC1 = CYLINDRICAL (Gov. 160/161)
LM1 = MORTISE (Gov.86-4)
LMO = BLANK (MORTISE) (Gov. 86-5)
LCO = PLAIN (NO LOCK PREP)

COL.21: LABEL

COL.22: DOOR ELEVATION

COL.23: REMARKS
ASTRAGALS, LOUVER, VISION LITE
See Door Accessories sheet
P___ = COLORSTYLE OR SPECIAL FINISH (NOTE FINISH CODE)
SG___ = MADERA STAIN FINISH (NOTE FINISH CODE)

LEGEND COMPUTER INDEX

(FOR USE WITH SCHEDULE SHEET)

SYMBOLS

N.B.C. = NOT BY CECO N.I.C. = NOT IN CONTRACT

DETAIL NUMBER / SHEET NUMBER

QUANTITY OF ANCHORS / TYPE OF ANCHORS

Ceco Door Products

DRAWN BY ______ DATE ______ CHK'D BY ______ DATE ______		
SHT ___ OF ___ CONTRACT NO.	ISSUE DATE	REVISIONS	BY

An **ASSA ABLOY** Group company

FORM 1090-100A 7/15/03

ASSA ABLOY

27109-13_A01.EPS

Additional Resources

This module presents thorough resources for task training. The following reference material is suggested for further study.

Window and Door Manufacturers Association. A trade organization representing 145 US window and door manufacturers, **www.wdma.com**

The National Fenestration Rating Council (NFRC). The nation's recognized authority for measuring and evaluating window energy performance, **www.nfrc.org**

Window & Door magazine. An information source for manufacturers, distributors, and dealers of windows and doors, **www.windowanddoor.com**

Figure Credits

SkillsUSA, Module opener

Pella® Windows and Doors, Figures SA03 and 13

Andersen Windows, Inc., Figures 9 and RQ01

Section Review Answer Key

Answer	Section Reference	Objective Reference
Section One		
1. d	1.1.0	1a
2. a	1.1.0	1a
3. d	1.2.1	1b
4. c	1.2.1	1b
5. d	1.3.1	1c
Section Two		
1. a	2.0.0	2
2. b	2.0.0	2a
3. c	2.1.0	2a
4. d	2.2.0	2b
Section Three		
1. c	3.0.0	3
2. a	3.1.0	3a
Section Four		
1. c	4.0.0	4

NCCER CURRICULA — USER UPDATE

NCCER makes every effort to keep its textbooks up-to-date and free of technical errors. We appreciate your help in this process. If you find an error, a typographical mistake, or an inaccuracy in NCCER's curricula, please fill out this form (or a photocopy), or complete the online form at **www.nccer.org/olf**. Be sure to include the exact module ID number, page number, a detailed description, and your recommended correction. Your input will be brought to the attention of the Authoring Team. Thank you for your assistance.

Instructors – If you have an idea for improving this textbook, or have found that additional materials were necessary to teach this module effectively, please let us know so that we may present your suggestions to the Authoring Team.

NCCER Product Development and Revision

13614 Progress Blvd., Alachua, FL 32615

Email: curriculum@nccer.org
Online: www.nccer.org/olf

❑ Trainee Guide ❑ Lesson Plans ❑ Exam ❑ PowerPoints Other ____________

Craft / Level: Copyright Date:

Module ID Number / Title:

Section Number(s):

Description:

Recommended Correction:

Your Name:

Address:

Email: Phone:

27110-13

Basic Stair Layout

OVERVIEW

Although prefabricated stairways are available in a variety of designs, job-built stairways or forms for concrete stairs may be required. Laying out and cutting stair stringers requires precise measuring and cutting, and the ability to perform math calculations. In addition, stairway construction is more code-driven than most other construction tasks because of the potential tripping and falling hazards.

Module Nine

Trainees with successful module completions may be eligible for credentialing through the NCCER Registry. To learn more, go to **www.nccer.org** or contact us at **1.888.622.3720**. Our website has information on the latest product releases and training, as well as online versions of our *Cornerstone* magazine and Pearson's product catalog.

Your feedback is welcome. You may email your comments to **curriculum@nccer.org,** send general comments and inquiries to **info@nccer.org**, or fill in the User Update form at the back of this module.

This information is general in nature and intended for training purposes only. Actual performance of activities described in this manual requires compliance with all applicable operating, service, maintenance, and safety procedures under the direction of qualified personnel. References in this manual to patented or proprietary devices do not constitute a recommendation of their use.

27110-13
Basic Stair Layout

Code Note

Codes vary among jurisdictions. Because of the variations in code, consult the applicable code whenever regulations are in question. Referring to an incorrect set of codes can cause as much trouble as failing to reference codes altogether. Obtain, review, and familiarize yourself with your local adopted code.

Objectives

When you have completed this module, you will be able to do the following:

1. Identify the types of stairways.
   a. Identify how residential and commercial stairways differ.
2. Identify the various components associated with stairs.
3. Identify terms associated with stair framing.
   a. Define *headroom*.
   b. Define *stringer* and explain when more than two stringers are used.
   c. Define *treads* and *risers* and explain the importance of uniform tread depths and riser heights.
   d. List the minimum stairway width requirements for residential and commercial structures.
   e. Describe the difference between handrails and guards.
   f. Identify situations that carpenters may be confronted with when framing stairwells.
4. Describe the procedure used to determine the total rise, number and size of risers, and number and size of treads required for a stairway.
   a. Explain how to calculate the riser height, tread depth, and total run for a stairway.
   b. Describe how to calculate stairwell opening sizes.
5. Describe the procedure to lay out and cut stringers, risers, and treads.
   a. Explain how to lay out and cut a stringer.
   b. Describe how to properly reinforce a stringer.
   c. Summarize how concrete stairways are formed.

Performance Tasks

Under the supervision of your instructor, you should be able to do the following:

1. Calculate the total rise, number and size of risers, and number and size of treads required for a stairway.
2. Lay out and cut a stringer.

Trade Terms

Baluster
Balustrade
Closed stairway
Geometrical stairways
Guard
Handrail
Headroom
Housed stringer
Landing
Newel post
Nosing
Open stairway
Pitch board
Rise and run
Riser
Skirtboard
Stairway
Stairwell
Stringer
Total rise
Total run
Tread
Unit rise
Unit run

Industry-Recognized Credentials

If you're training through an NCCER-accredited sponsor you may be eligible for credentials from NCCER's Registry. The ID number for this module is 27110-13. Note that this module may have been used in other NCCER curricula and may apply to other level completions. Contact NCCER's Registry at 888.622.3720 or go to **nccer.org** for more information.

Contents

Topics to be presented in this module include:

Figures

SECTION ONE

1.0.0 TYPES OF STAIRWAYS

Objective

Identify the types of stairways.

a. Identify how residential and commercial stairways differ.

Trade Terms

Closed stairway: A stairway that has solid walls on each side.

Geometrical stairway: A winding stairway with an elliptical or circular design.

Landing: A horizontal area at the end of a flight of stairs or between two flights of stairs.

Open stairway: A stairway that is open on at least one side.

Riser: A vertical board under the tread of a stairway.

Stairway: The complete set of stairs extending from one level of a structure to another level.

Tread: The horizontal member of a stairway upon which a person steps.

Safety is the most important concern in the design of any stairway. Most building codes include detailed requirements for stairway construction. The drawings for a building provide information on laying out and constructing stairways. This information is typically given on the plan and elevation views, as well as the section and details.

Don't assume anything when constructing stairways. Building codes for stairways vary from state to state. In addition, code requirements for commercial and residential stairways can be significantly different.

It's important to stay current on the building code in effect in your jurisdiction. They may change periodically, and stairs are one area in the code that is subject to frequent change.

Stairways include the stair framing, risers, and treads. A variety of standard stairway designs are available, or stairways can be custom built. There are many ways to classify stairways. One way to classify stairways is based on whether they are open, closed, or a combination of open and closed (*Figure 1*). An open stairway is open on at least one side. A closed stairway has solid walls on each side. Combination open/closed stairways are essentially closed stairways with part of the stairway exposed. Building codes may have special requirements for closed stairways. Refer to your local building code before construction of any stairway.

Another classification of stairways relates to the shape. Straight-run stairways (*Figure 1*) are the simplest type of stairway. They extend from one floor level to another without interruption. L-shaped stairways extend along two perpendicular (90 degree) walls, and include a landing where the stairway changes direction (*Figure 2*).

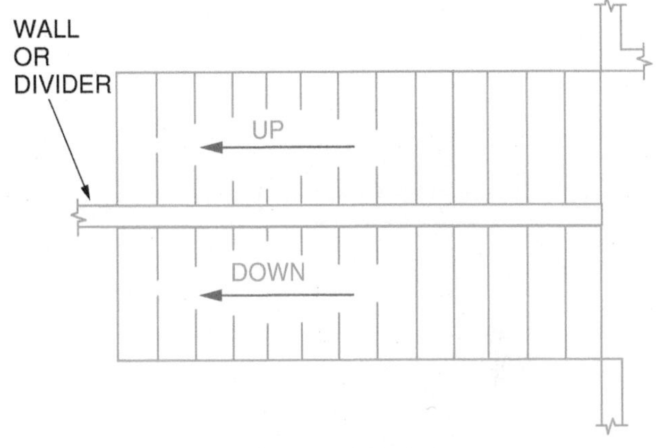

OPEN STAIRWAY

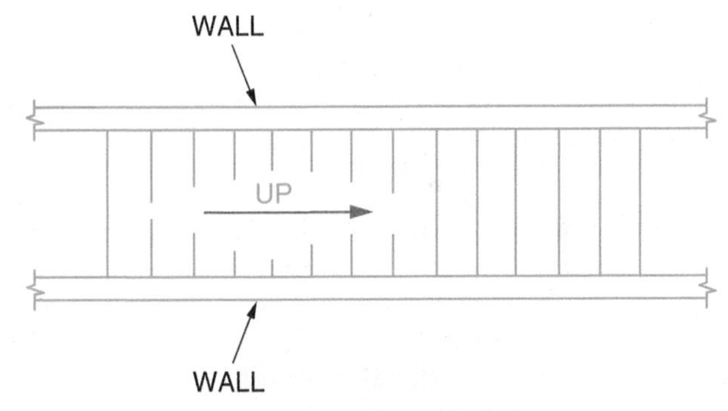

CLOSED STAIRWAY

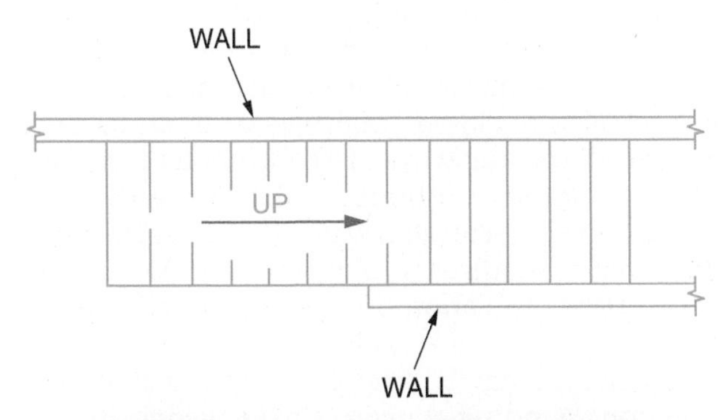

COMBINATION OPEN/CLOSED STAIRWAY

27110-13_F01.EPS

Figure 1 Open, closed, and combination open/closed stairways.

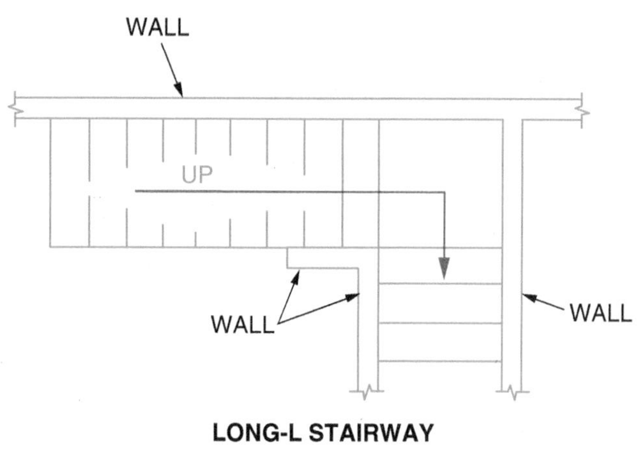

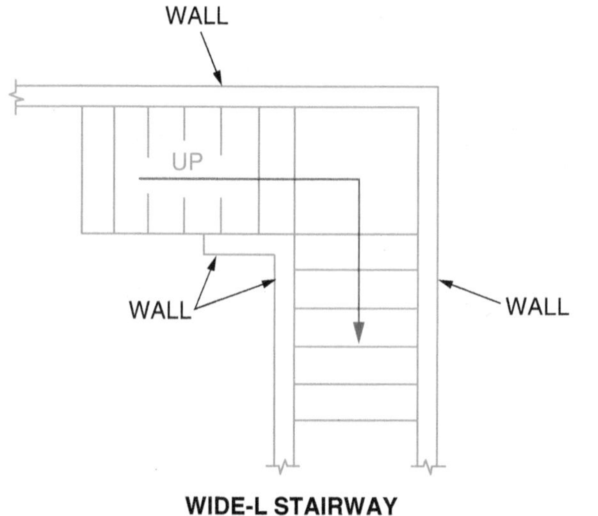

27110-13_F02.EPS

Figure 2 Long-L and wide-L stairways.

U-shaped stairways have two sets of stairs that run parallel to one another (*Figure 3*). Winding stairways are a type of circular stairway (*Figure 4*). Winding stairways are difficult to construct and are commonly prefabricated off-site before being erected in their final location.

When a series of steps are straight and continuous without breaks (such as those formed by a landing), they are referred to as a flight of stairs, staircase, or stairway. Stairway arrangements can also be used to classify stairways. Straight-run or straight-flight stairways are continuous from one level to another, without breaks in the flight of stairs. A straight-run stairway is the easiest to build. However, it requires a long stairwell, which can cause space utilization problems in smaller structures. Platform stairways have landings at regular intervals, and may be straight or make a change in direction. Winding stairways have wedge-shaped treads at the change in directions and may not be permitted by the local building code.

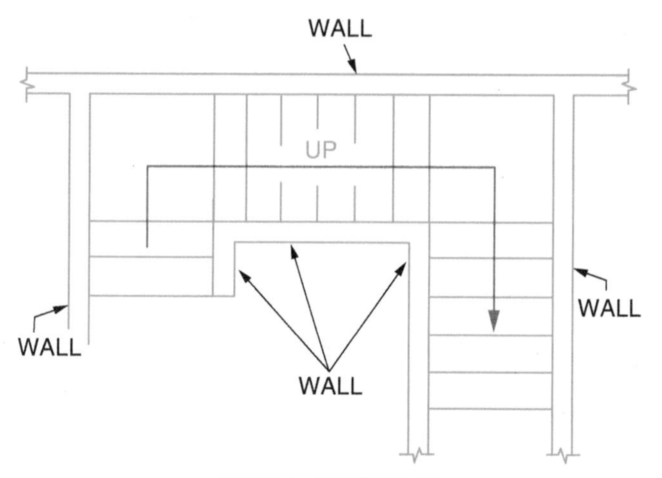

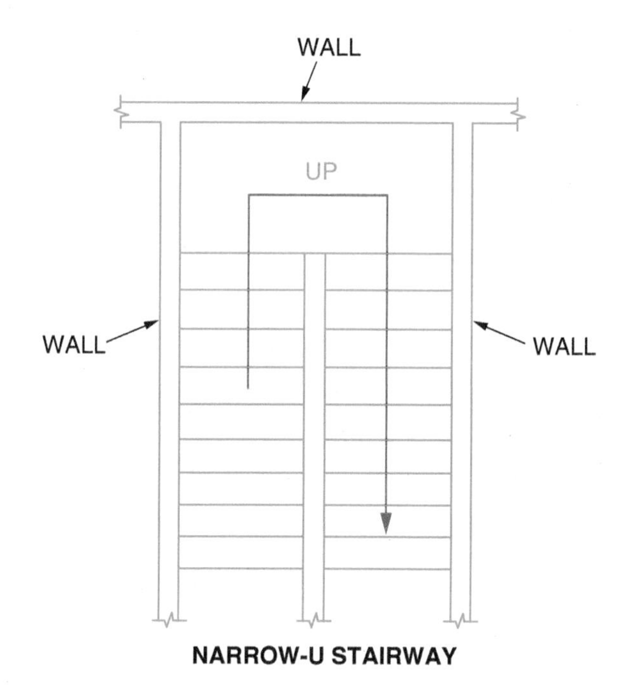

27110-13_F03.EPS

Figure 3 Wide-U and narrow-U stairways.

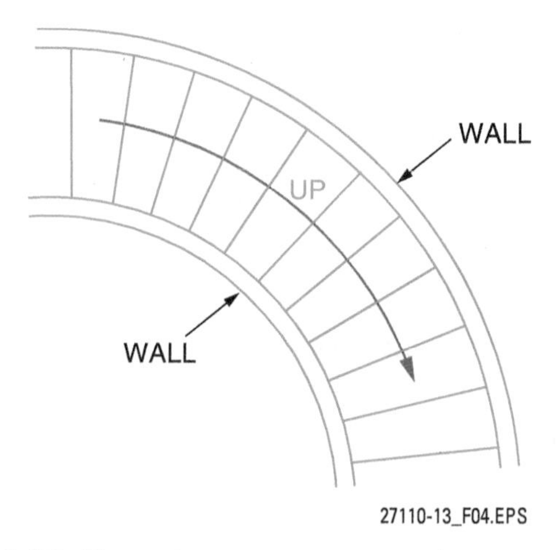

27110-13_F04.EPS

Figure 4 Winding stairway.

Geometrical stairways are circular, elliptical, or spiral in shape and gradually change direction in their ascent from one level to another. The geometric change in this type of stairway can be anywhere from 90 to 360 degrees from the starting point. Geometrical stairways and winding stairways include landing stairs where the direction of the stairway changes. Landing stairs are wedge-shaped stairs, which are designed to take up a limited amount of space. Due to the wedge-shaped design, geometrical stairs can be dangerous if not properly constructed. They should be used only where no other method is feasible.

1.1.0 Commercial Stairways

Most commercial structures of two stories will only require main stairs and service stairs. Structures of two or more stories may also incorporate escalators and/or elevators. In some jurisdictions, the service stairs are classified as fire stairs. This means the stairway is not used unless there is no other exit from a building.

There are a variety of ways to construct main stairs and service stairs. Depending on the design criteria, main stairs can be constructed from a vast array of products, ranging from relatively inexpensive laminated lumber to elaborately designed wrought iron.

The World's Longest Stairway

The service stairway for the Niesenbahn funicular railway near Spiez, Switzerland, is the world's longest stairway. Measuring 5,476 feet long, this stairway, which is used only by employees, has 11,674 steps.

Source: Guinness World Records

Elegant Stairs

Elaborate custom-built stairs are commonly installed in expensive residences. These stairs are usually prefabricated in a shop from expensive hardwoods and are assembled on site. In rare cases, including some restoration work, they are custom-fabricated on site using specialized tools.

27110-13_SA01.EPS

1.0.0 Section Review

1. The type of stairway that extends from one level to another without interruption is a(n) _____.

   a. U-shaped stairway
   b. straight-run stairway
   c. L-shaped stairway
   d. landing

SECTION TWO

2.0.0 STAIR COMPONENTS

Objective

Identify the various components associated with stairs.

Trade Terms

Baluster: A supporting column or member; a support for a railing, particularly one of the upright columns of a balustrade.

Balustrade: A stair rail assembly consisting of a handrail, balusters, and posts.

Guard: A rail secured to uprights and erected along the exposed sides and ends of platform stairs, etc.

Handrail: A member supported on brackets from a wall or partition to furnish a handhold.

Headroom: The vertical and clear space in height between a stair tread and the ceiling or stairs above.

Housed stringer: A stair stringer with horizontal and vertical grooves cut (mortised) on the inside to receive the ends of the risers and treads. Wedges covered with glue are often used to hold the risers and treads in place in the grooves.

Newel post: An upright post supporting the handrail at the top and bottom of a stairway or at the turn of a landing; also, the main post about which a geometrical staircase winds.

Nosing: The portion of a stair tread that extends beyond the face of the riser.

Skirtboard: A baseboard or finishing board at the junction of the interior wall and floor. Also called a finished stringer.

Stringer: The inclined member that supports the treads and risers of a stairway.

The main components of a flight of stairs are defined in the following paragraphs and shown in *Figure 5*.

- *Tread* – The horizontal surface of a step.
- *Riser* – The piece forming the vertical face of the step. Depending on the application or jurisdiction, commercial stair codes may require a solid riser. Some codes may allow the risers to be omitted (open risers).
- *Cutout stringer* – A cutout **stringer** provides the main support for the stairway. Center cutout stringers may be required by building code when the width of a wood stairway exceeds 30", or multiples of 30", or if tread material thinner than 1½" is used on stairs that are 36" wide. Other types of stringers used as side supports for stairs are the dadoed or **housed stringers** (see *Figure 6*). On a dadoed stringer the treads fit into slots cut in the stringer. Dadoed stringers are typically used for open-riser designs. A housed stringer is similar to a dadoed stringer, except that tapered slots are provided for the treads and risers. The treads and risers are secured with wedges and glue. Housed stringers are commonly prefabricated in a shop and delivered to the job site.

The terminology used for other components of most interior stairways is defined as follows and shown in *Figure 7*:

- *Ending newel post* – The ending **newel post**, if used, is the uppermost supporting post for the **handrail** or **guard**. When the upper end of the stairway is terminated at a wall, the newel post may be omitted or half of a newel post is used.
- *Landing newel post* – The main post supporting the handrail at the landing.
- *Gooseneck* – A bent or curved section in the handrail.
- *Spindle* – A spindle or **baluster** is the upright piece that runs between the handrail or guard and the treads or a closed stringer. The balusters, handrail or guard, and newel posts make up the **balustrade**.
- *Handrail* – A handrail is used for support when ascending or descending a stairway. It may be mounted on the walls or on a guard for open stairways. *Figure 8* shows most of the handrail requirements for commercial structures. For residential applications, handrails without extensions at lower ranges of height are usually permitted. In some cases, additional lower handrails are permitted for children.
- *Guard* – Per the *International Building Code®*, a guard is a building component located at or near open sides of elevated walking surfaces that minimizes the chance for falling to a lower level. On the open side of commercial stairways or mezzanines, a guard that is 42" high is typically required. Guards used in residential applications, such as for lofts or balconies, are usually permitted at heights of 36".
- *Starting newel post* – The main post supporting a handrail at the bottom of the stairway.
- *Headroom clearance* – **Headroom** clearance is the closest distance between any portion of a step of the stairway and any overhead structure such as a ceiling.

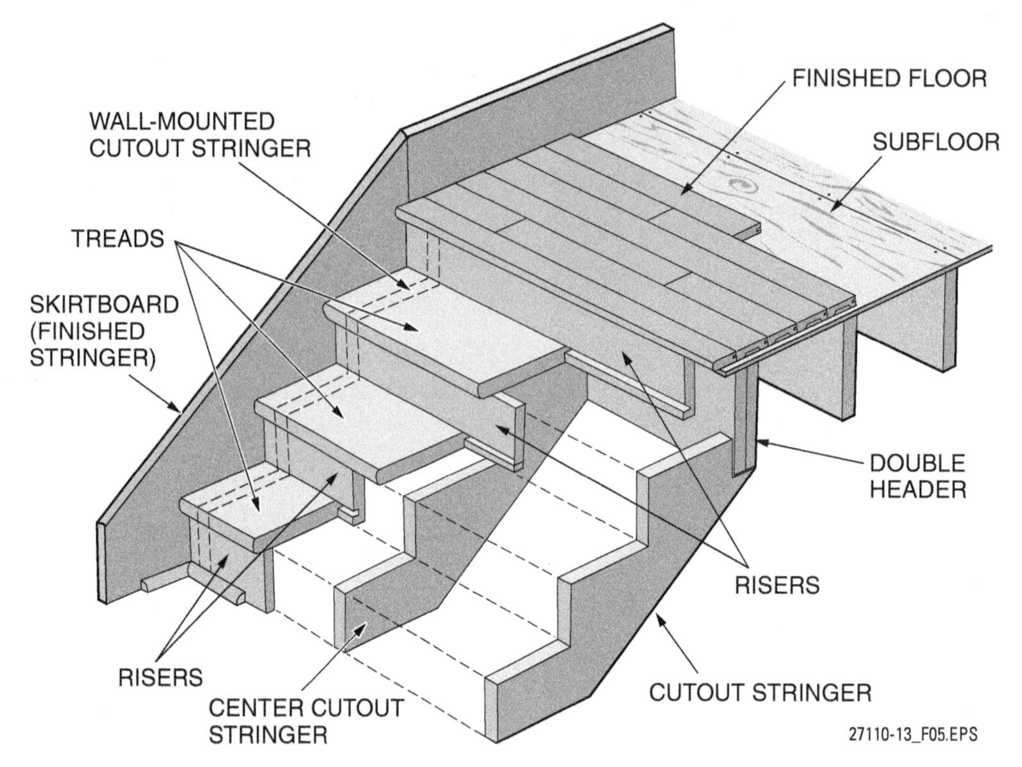

Figure 5 Main stair components.

- *Skirtboard* – A skirtboard (finished stringer) is a finished board fastened to the wall side of the stairway. The top of the board is parallel to the slope of the stairway, and the ends terminate horizontally with the wall base molding. A housed stringer supporting the treads may be used instead of a skirtboard.
- *Nosing* – Nosing is the projection of the tread beyond the face of the riser.
- *Closed finished stringer* – A closed finished stringer is a finished piece fastened to the stair stringer that is also the mounting base for balusters. A housed stringer supporting the treads may be used instead of a closed stringer.
- *Landing (platform)* – A flat section that breaks the stairway into two sections between floors or the floor at the top or bottom of a stairway. Landings must be used when the vertical distance (between floors) from the top of a stairway to the bottom exceeds 12' or as specified by code. A platform must be as wide and as deep as the stairway width for each section of the stairs abutting it.
- *Open finished stringer* – A finished piece that matches the stair stringer supporting the stair treads.
- *Nosing return (end nosing)* – On a stairway with an open stringer, the nosing return is the projection over the face of the tread at the end of the tread.

The *International Building Code*® (IBC) specifies minimum and/or maximum requirements for commercial stairways and handrails. The following are examples of the IBC requirements:

- *Stairway width* – 44 inches minimum (Exception: 36 inches minimum for occupancies serving fewer than 50 people)
- *Stair tread depth* – 11 inches minimum
- *Stair riser height* – 7 inches maximum, 4 inches minimum
- *Headroom* – 80 inches minimum
- *Nosing* – 1¼ inches maximum
- *Vertical rise* – 12 feet maximum between floor levels or landings
- *Handrail height* – 34 inches minimum, 38 inches maximum from top of rail to nosing

Safe Stairway Design and Construction

Stairway incidents result in roughly 4,000 deaths and a million injuries requiring hospital treatment each year; therefore, stairway design and construction is strictly controlled by building codes and regulations.

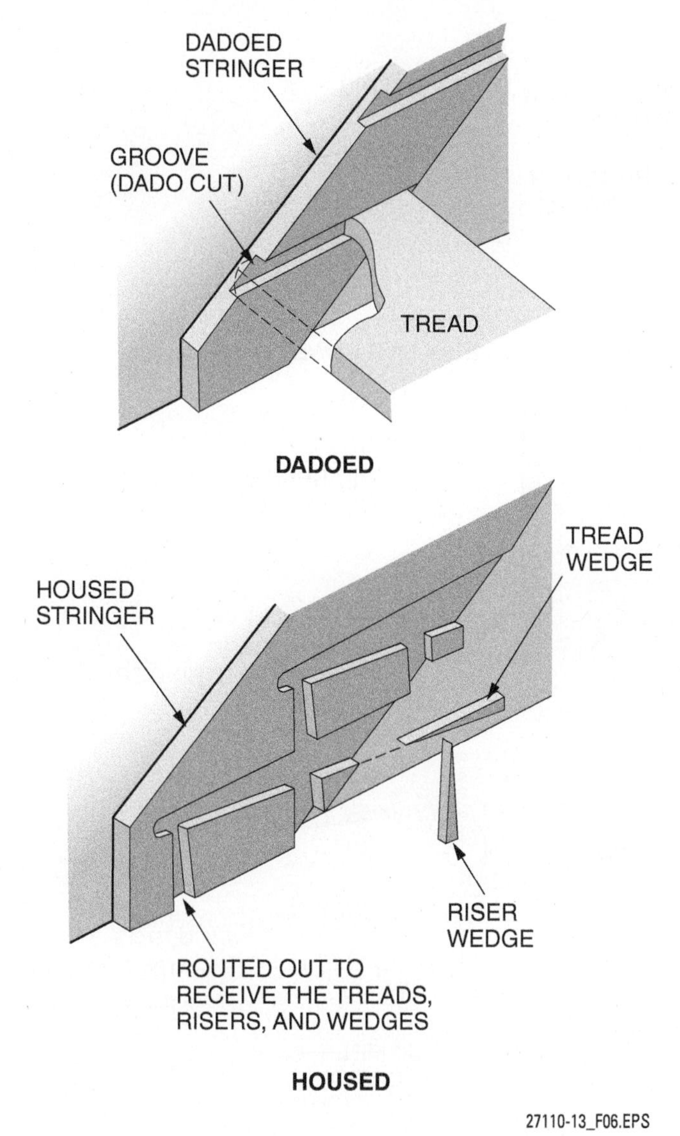

Figure 6 Dadoed and housed stringers.

Guard vs. Guardrail

Terminology cited by different sources may be similar, but the definitions may have drastically different meanings. As noted previously, the *International Building Code®* defines a guard as a building component located at or near open sides of elevated walking surfaces that minimizes the chance for falling to a lower level. A similar term—guardrail—is defined and cited by OSHA in *Subpart M*. OSHA defines a guardrail as a barrier erected to prevent employees from falling to lower levels. The top-edge height of top rails, or equivalent guardrail system members, must be 42 inches plus or minus 3 inches above the walking/working level. When conditions warrant, the height of the top edge may exceed the 45-inch height.

The *International Residential Code®* (IRC) specifies minimum and/or maximum requirements for residential stairways and handrails. The following are examples of the IRC requirements:

- *Stairway width* – 36 inches minimum
- *Stair tread depth* – 10 inches minimum
- *Stair riser height* – 7¾ inches maximum
- *Headroom* – 80 inches minimum
- *Nosing* – ¾ inch minimum, 1¼ inches maximum
- *Vertical rise* – 12 feet maximum between floor levels or landings
- *Handrail height* – 34 inches minimum, 38 inches maximum from top of rail to nose

Always refer to the local building code for stairway requirements for the particular type of work you are performing in your area.

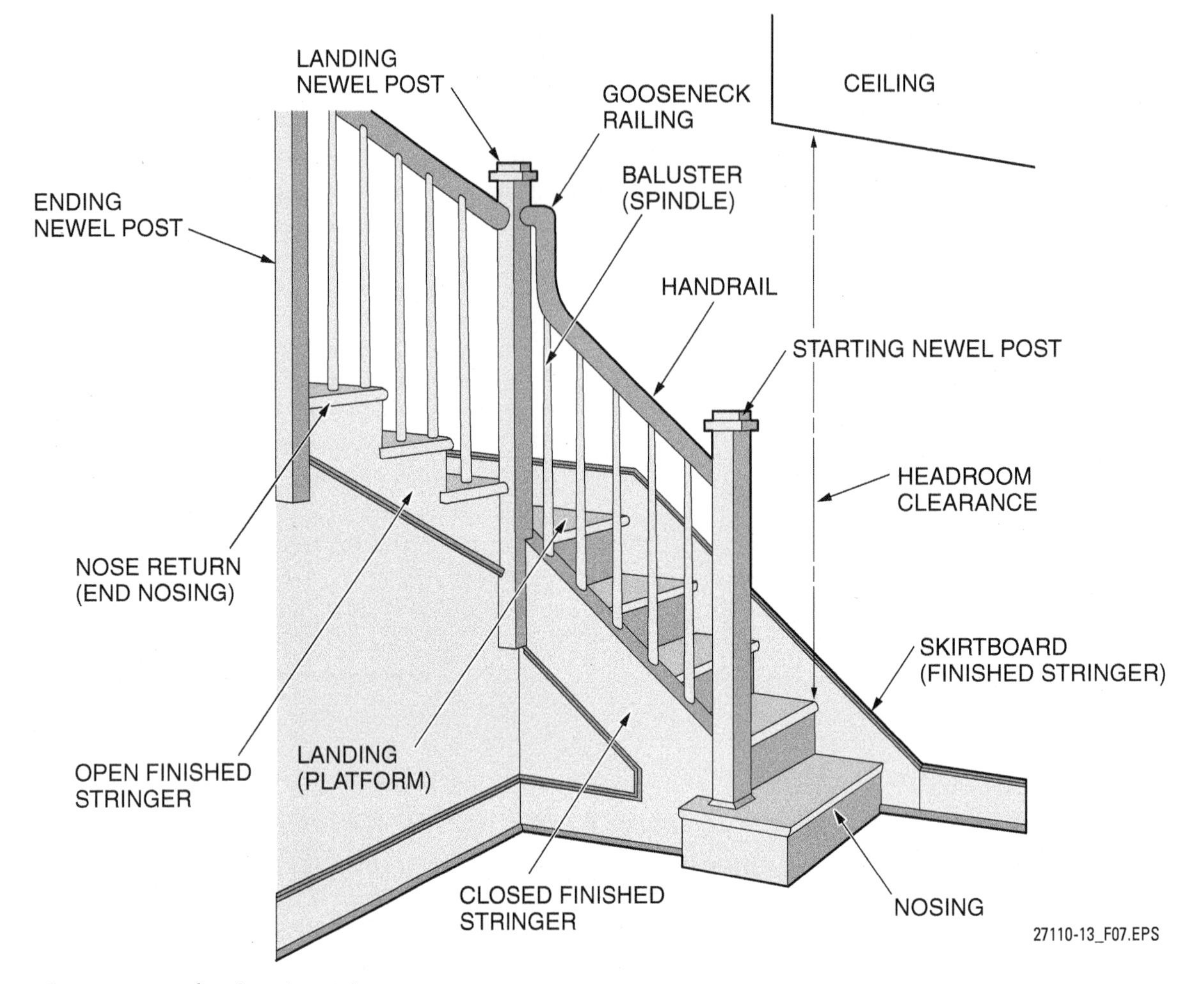

Figure 7 Components of an interior stairway.

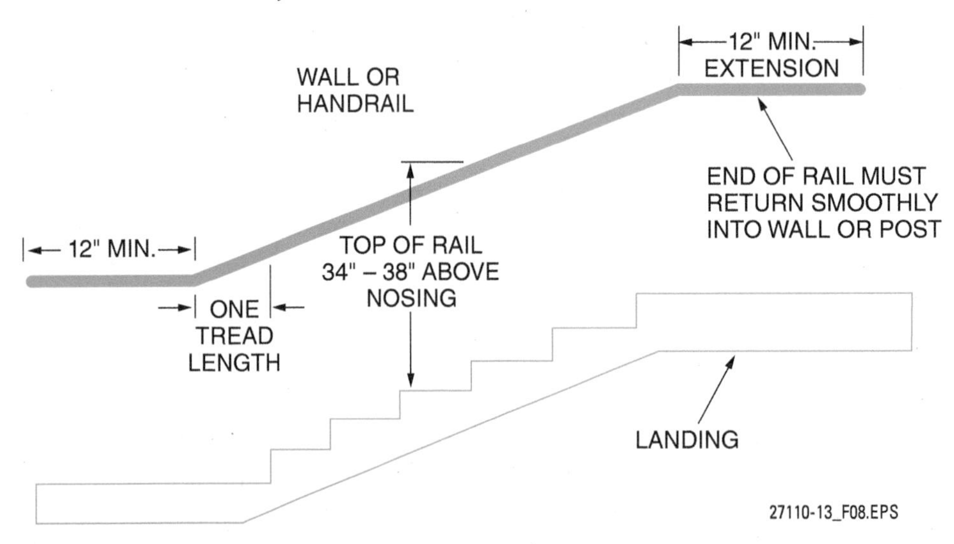

Figure 8 Commercial handrail requirements.

2.0.0 Section Review

1. Landings must be used when the vertical distance (between floors) from the top of a stairway to the bottom exceeds _____.
   a. 4'
   b. 6'
   c. 12'
   d. 18'

Section Three

3.0.0 Stair Framing Terminology

Objective

Identify terms associated with stair framing.

a. Define *headroom.*
b. Define *stringer* and explain when more than two stringers are used.
c. Define *treads* and *risers* and explain the importance of uniform tread depths and riser heights.
d. List the minimum stairway width requirements for residential and commercial structures.
e. Describe the difference between handrails and guards.
f. Identify situations that carpenters may be confronted with when framing stairwells.

Trade Terms

Rise and run: A term used to indicate the degree of stairway incline.

Stairwell: A compartment extending vertically through a building into which a stairway is placed.

The procedure for building stairways will vary from one locality to another due to differences in building codes. Rough framing of stairs is typically done during the framing of the building. Regardless of procedure, the rough opening in the floor in combination with the height difference from floor to floor determine the length and width of a stairway. Check the local building code for stairway requirements in your area. Keep in mind that code requirements for residential construction may vary from commercial requirements.

3.1.0 Headroom

Headroom is the vertical distance between a stair tread nosing and any structure or ceiling above the stairway. The minimum headroom required for residential and commercial stairways is 6'-8". However, 7' headroom is desirable, if at all possible.

3.2.0 Stringers

Stringers must be constructed in accordance with the local building code. The size of the stock used for stringers is noted on the drawings. Stock used for stringers should be selected at one time and the crowns should be matched. Typically, center cutout stringers are required for wood stairways wider than 30", in multiples of 30", or if tread material thinner than 1½" is used on stairs that are at least 36" wide.

> **NOTE**
>
> Many states and localities specify minimum and maximum requirements for stairways that may differ from the *International Building Code®* (IBC) and *International Residential Code®* (IRC). Before starting any stair construction, be aware of the IBC/IRC requirements and any superseding state or local codes that govern stairway construction. Furthermore, if construction drawings pertain to the project, they must be followed for the construction of the stairs. If the drawings are incomplete or conflict with the building codes, a supervisor should be notified to obtain or clarify the needed information.
>
> As construction of a structure proceeds, the stairway (with temporary stair treads) should be installed as soon as possible to allow workers to access the upper levels more easily with minimum delay and maximum safety. Stairways are not finished until all danger of damage from workers and materials is eliminated.

3.3.0 Treads and Risers

For optimum safety and comfort, it is important that the dimensions of stair treads and risers be uniform within any stairway. As a person walks down a flight of stairs, the stride is uniform. Therefore, the system to support the person (the risers and treads) should also be uniform. If a tread and/or riser is constructed using material that is a different size than the others, or if the tread depth or riser height is not uniform, it may create an unsafe condition and result in a fall. Per the IRC, the greatest tread depth in a stairway should not exceed the smallest tread depth by more than ⅜" (9.5 mm), and the greatest riser height should not exceed the smallest riser height by more than ⅜".

3.4.0 Width Requirement

The minimum stairway width required is also specified in the building code. The minimum stairway width for a residential structure is 36", while the minimum for a commercial structure is 44" (see *Figure 9*).

3.5.0 Handrails and Guards

A handrail is used on stairways to assist people when ascending or descending a stairway by furnishing a continuous rail for support along the side(s). A handrail differs from a guard in that a guard is erected on the exposed sides and ends of stairs and platforms, and may incorporate a handrail. Open stairways have a low partition or banister along the open edge of the stairway. The handrail in a closed stairway is called a wall rail because it is attached to the wall with special brackets. As a rule, standard-width stairways have a handrail only along one side. However, stairways wider than 44" usually require handrails on both sides. Always refer to the local code for specific handrail and guard requirements.

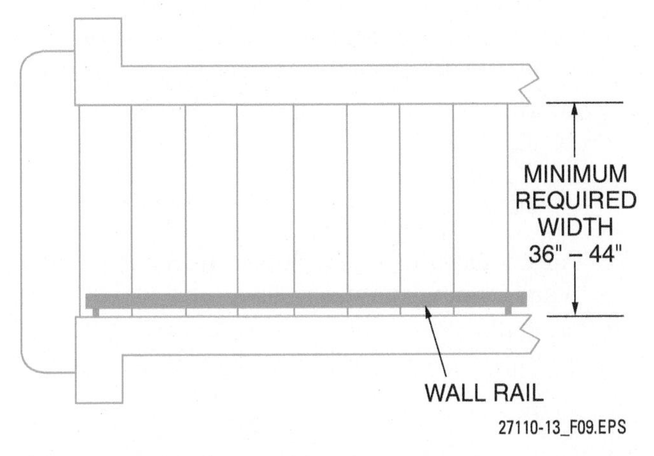

Figure 9 Minimum stairway width.

3.6.0 Stairwells

Stairwells must be constructed so they are wide enough and long enough to provide the code-required width and headroom for the stairway. If the drawings do not specifically detail the stairs, carpenters can be confronted with several situations at a job site.

In one situation, the stairwells may not have been constructed and the carpenter must lay out and frame any stairwell openings as well as the stairs. In another circumstance, someone else may have already framed the stairwell opening and the task will be to frame the stairs to provide as much headroom as possible with an optimum rise and run for the stairway.

In a one-story structure, the stairway would be a set of service stairs to the basement. The stairwell opening should be constructed to allow code-required headroom. When framing a stairwell, the construction must include double trimmer joists and headers to support weight placed on or around the stairwell opening (see *Figure 10*).

The size of the stairwell opening will vary with the type of stairway being installed. As stated earlier, building codes have requirements for minimum headroom, riser height, and tread width. Always check the local building code for specific stairway requirements.

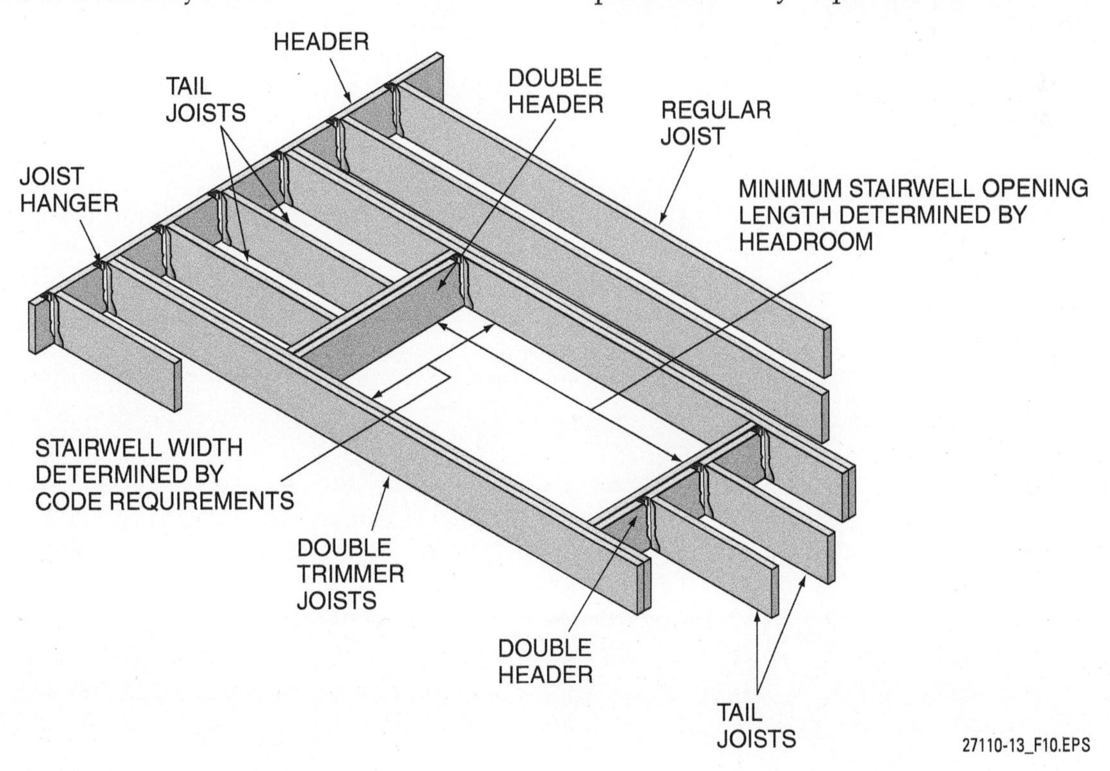

Figure 10 Typical framed stairwell opening.

Building Code Variations

Local building codes for stairways vary widely from one location to another. Before beginning stairway construction, always check the building code for requirements regarding the type of stairways permitted, minimum headroom, proper tread depth/riser height relationship, and handrail quantity and height.

3.0.0 Section Review

1. The minimum headroom for residential and commercial stairways is _____.
   a. 6'-6"
   b. 6'-8"
   c. 7'-0"
   d. 7'-8"

2. How many center cutout stringers will be needed for a stairway measuring 90" wide?
   a. 1
   b. 2
   c. 4
   d. 6

3. A stairway containing riser heights of 7½", 7⅝", and 7¾" is within tolerance per the *International Residential Code®*?
   a. True
   b. False

4. The minimum width for a residential stairway is _____.
   a. 28"
   b. 32"
   c. 36"
   d. 44"

5. Depending on local code, handrails may be required on both sides of a stairway measuring a minimum of _____.
   a. 28" wide
   b. 32" wide
   c. 36" wide
   d. 44" wide

6. Additional structural members are *not* required when framing in a stairwell.
   a. True
   b. False

Section Four

4.0.0 Stair Calculations

Objective

Describe the procedure used to determine the total rise, number and size of risers, and number and size of treads required for a stairway.

a. Explain how to calculate the riser height, tread depth, and total run for a stairway.
b. Describe how to calculate stairwell opening sizes.

Performance Task 1

Calculate the total rise, number and size of risers, and number and size of treads required for a stairway.

Trade Terms

Total rise: The vertical distance of a stairway one floor to the floor above.

Total run: Horizontal length of a stairway.

Unit rise: The vertical distance from the top of one stair tread to the top of the next one above it; also called stair rise.

Unit run: The horizontal distance from the face of one riser to the face of the next riser.

If architectural plans are being used for a project, the **total rise** and **total run** of the stairways in the building are commonly specified by the architect, in addition to the **unit rise** (riser height), **unit run** (tread width), number of risers and treads, and the size of the stairwell opening (see *Figure 11*). In these cases, the stairwell openings are framed first according to the drawings. Then, the carpenter lays out and constructs the rough stairway. If stairway details are not available, a carpenter may be required to design and construct the stairways and stairwells.

4.1.0 Stairway Design

An important element in stair design is the mathematical relationship between the riser height (unit rise) and tread depth (unit run) dimensions. The ratio of these two elements (or similarly, the ratio of the total rise and total run of the stairway) determines the slope of the stairway. One of three simple, generally accepted rules is commonly applied when determining the riser-to-tread (unit rise-to-run) ratio:

- *Rule 1* – Tread width + riser height = 17" to 18"
- *Rule 2* – Tread width + (2 × riser height) = 24" to 25"
- *Rule 3* – Tread width × riser height = 70" to 75"

Check the local building code to determine which of the rules is applicable in your area. Rule 1 is the most common rule used on the job site. For residential stairways, Rule 1 allows a maximum rise of 7¾" for the code-minimum tread depth of 10". For residential stairways, Rules 2 and 3 allow a maximum rise of 7½" for the code-minimum tread depth of 10".

Given the unit rise of the stairway, the applicable rule can be used to determine the longest and shortest recommended unit run (tread) length. Using any recommended tread length equal to or greater than the code minimum (usually 10"), the total run length of the stairway can then be determined. The stairway riser height (unit rise) is typically determined first.

4.1.1 Determining Riser Height

The preferred slope of a stairway is typically between 30 and 35 degrees for maximum ease and safety in ascent and descent. The *International Residential Code®* limits the tread depth to a minimum of 10" and the riser height to a maximum of 7¾", which, if used, can result in a slope of 38 degrees. Therefore, a greater tread depth or smaller riser height should be used to achieve the preferred angle of 30 to 35 degrees.

Whenever possible, a unit rise between 7" and 7½" and a unit run between 10½" and 12" is recommended; this will result in slopes of 30 to 35 degrees. The maximum riser height and minimum tread depth permitted by code are often used in buildings to minimize stairwell space.

> **NOTE**
> OSHA requires that guardrails be placed around any stairwell opening during its construction. In addition, the normal access point to the stairs must be roped off or barricaded until the temporary or permanent stairs are installed.

The riser height (unit rise) can be determined as outlined in the following procedure:

Step 1 From the drawings, determine the final thickness, in inches, of the floor assembly above the stairway. Include the thickness of the anticipated or actual finished floor, subfloor, floor joists, any furring, and the

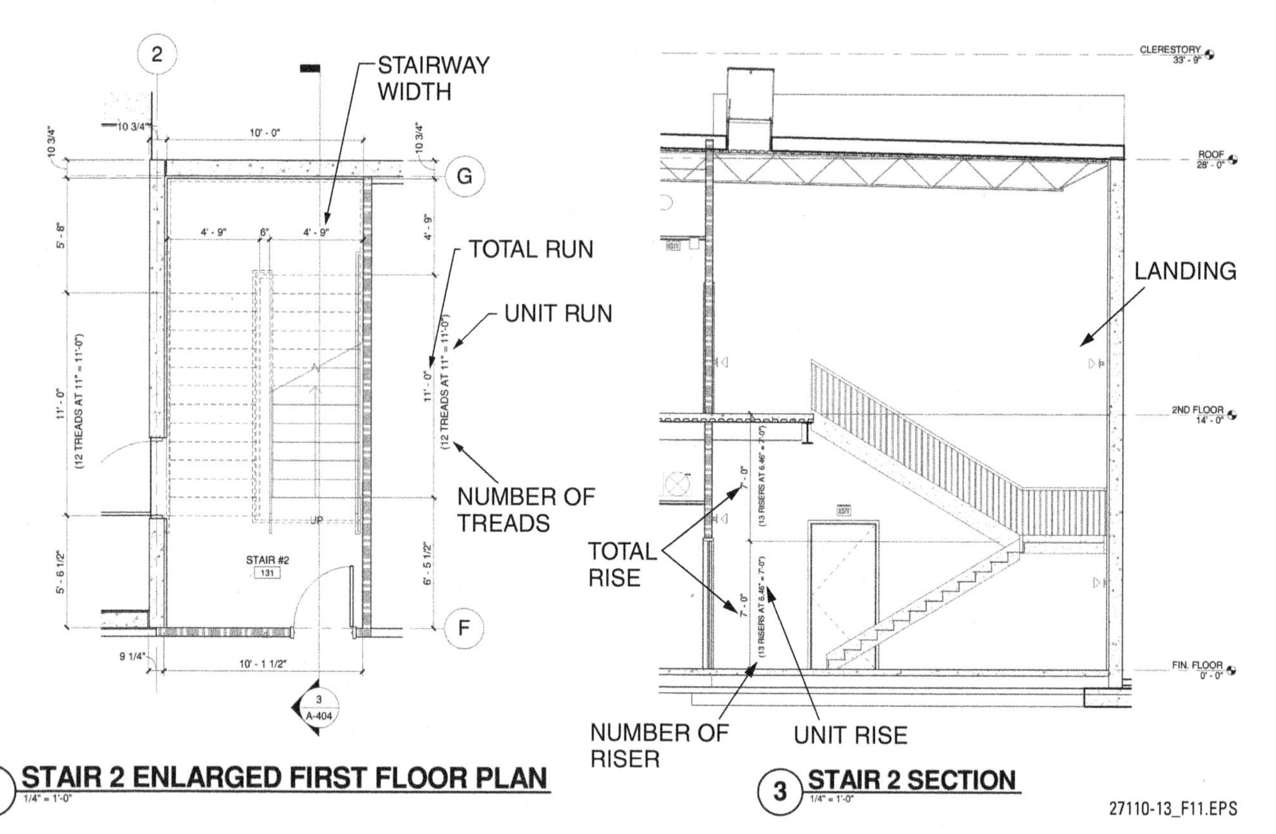

Figure 11 Stairway information noted on prints.

anticipated or actual finished ceiling. As shown in *Figure 12*, the floor assembly thickness for our example stairway is 8 ½".

Step 2 Add the floor assembly thickness (Step 1) to the vertical distance, in inches, from the anticipated or actual finished ceiling to the anticipated or actual finished floor used at the base of the stairs. This sum is the total rise of the stairway. For our example:

Total rise = 8½" + 8'-1½"

= 8½" + (8' × 12") + 1½"
= 106"

> **NOTE**
> The total rise of a straight stairway normally must never exceed 12' (144") or as specified by applicable building codes. If the limit is exceeded, an intermediate landing must be used.

Step 3 Divide the total rise of the stairway determined in Step 2 by 7" (the optimum riser height). The result is usually a whole number with a decimal remainder. The whole number normally represents the optimum number of risers desired.

106" ÷ 7" = 15.14 or 15 risers

Riser/Tread Combinations

To construct a staircase having the proper angle for comfortable climbing and descending, a riser height of between 7" and 7½" and a tread width of between 10" and 11" is recommended. In addition, a rule of thumb states that the sum of one riser and one tread should equal between 17" and 18". Some examples include:

Riser Height	Tread Width	Total
7"	11"	18"
7¼"	10"	17¼"
7⅜"	10½"	17⅞"
7½"	10½"	18"

Step 4 Divide the total rise by the number of risers determined in Step 3. This results in the unit rise (riser height) in inches and, usually, a decimal remainder. Examine the result to determine if it falls within the desired range of 7" to 7.5" or is equal to or below the maximum riser-height limitation for the applicable code. If not, or if a different height is required, use the next higher or lower number of risers and repeat Step 4. Increasing the number of risers decreases the riser height. Conversely, decreasing the number of risers increases the riser height.

$$\text{Unit rise} = 106" \div 15 = 7.066"$$
$$= 7" + \text{decimal remainder of } 0.066"$$

In this case, the riser height falls within the desired range of 7" to 7.5". However, if the number of risers is decreased to 14, the riser height would be 7.57", which is only a fraction of an inch more than the desired range of 7" to 7.5". Although 14 risers is within code, 15 risers would yield a rise closer to 7", which would be more comfortable for use by the occupants.

Step 5 Convert the decimal remainder to a fractional inch value (32nds) by multiplying the decimal remainder by 32 and rounding up or down to the nearest whole number. This whole number represents the number of 32nds of an inch equal to the decimal remainder. Reduce the fraction to 16ths or 8ths if possible and combine with the whole number of inches of the unit rise determined in Step 4.

$$\text{Unit rise} = 7" + 0.066"$$
$$= 7" + (0.066" \times 32 = 2.11 = \tfrac{2}{32}" = \tfrac{1}{16}")$$
$$= 7\tfrac{1}{16}"$$

4.1.2 Determining Tread Depth (Unit Run) and Total Run

The tread depth (unit run) is measured from the face of one riser to the face of the next and does not include any nosing (*Figure 13*).

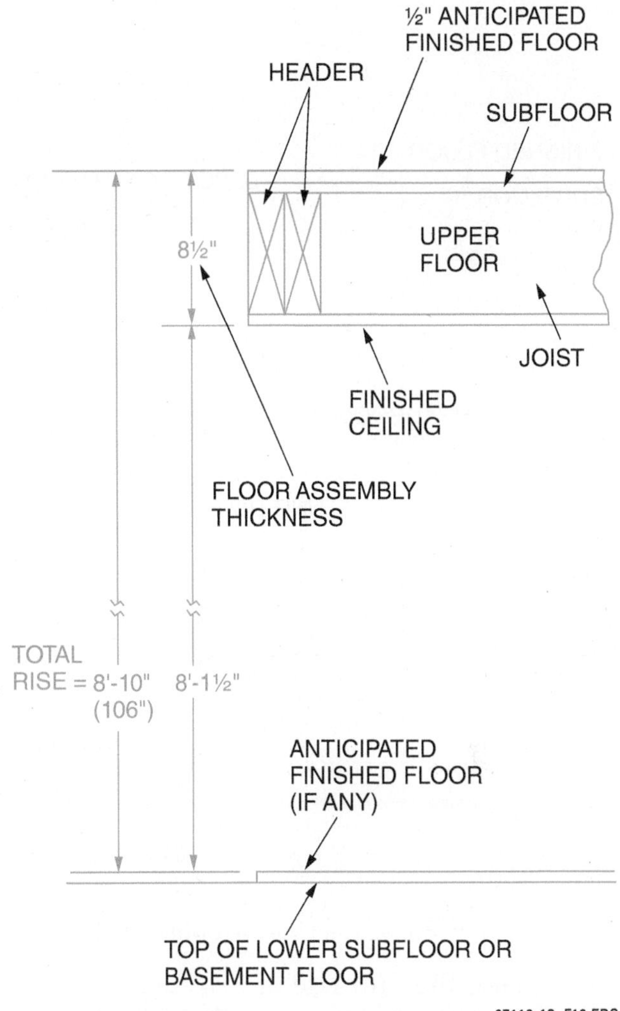

Figure 12 Method of determining total stair rise and unit rise.

To determine the unit run, use the unit rise and apply one of the rules (Rule 1, 2, or 3) as defined by the applicable code. Typically, the minimum permissible unit run is used in the calculation to conserve the space required for the stairwell opening.

To determine the total run if the stringer uses the stairwell header as the last riser, subtract one from the number of risers (one less tread required) and multiply the result by the unit run.

If the stringer is mounted flush with the top of the floor, as shown in the alternate configuration in the figure, multiply the number of risers by the unit run (same number of treads as risers). For this example, refer to *Figure 13*.

Optimizing Design

Be aware that because of space limitations in residential and light commercial structures, most prints or specifications will detail or specify the steepest stairs (maximum unit rise) with the minimum amount of headroom (minimum unit run) permitted by the applicable building code for the area.

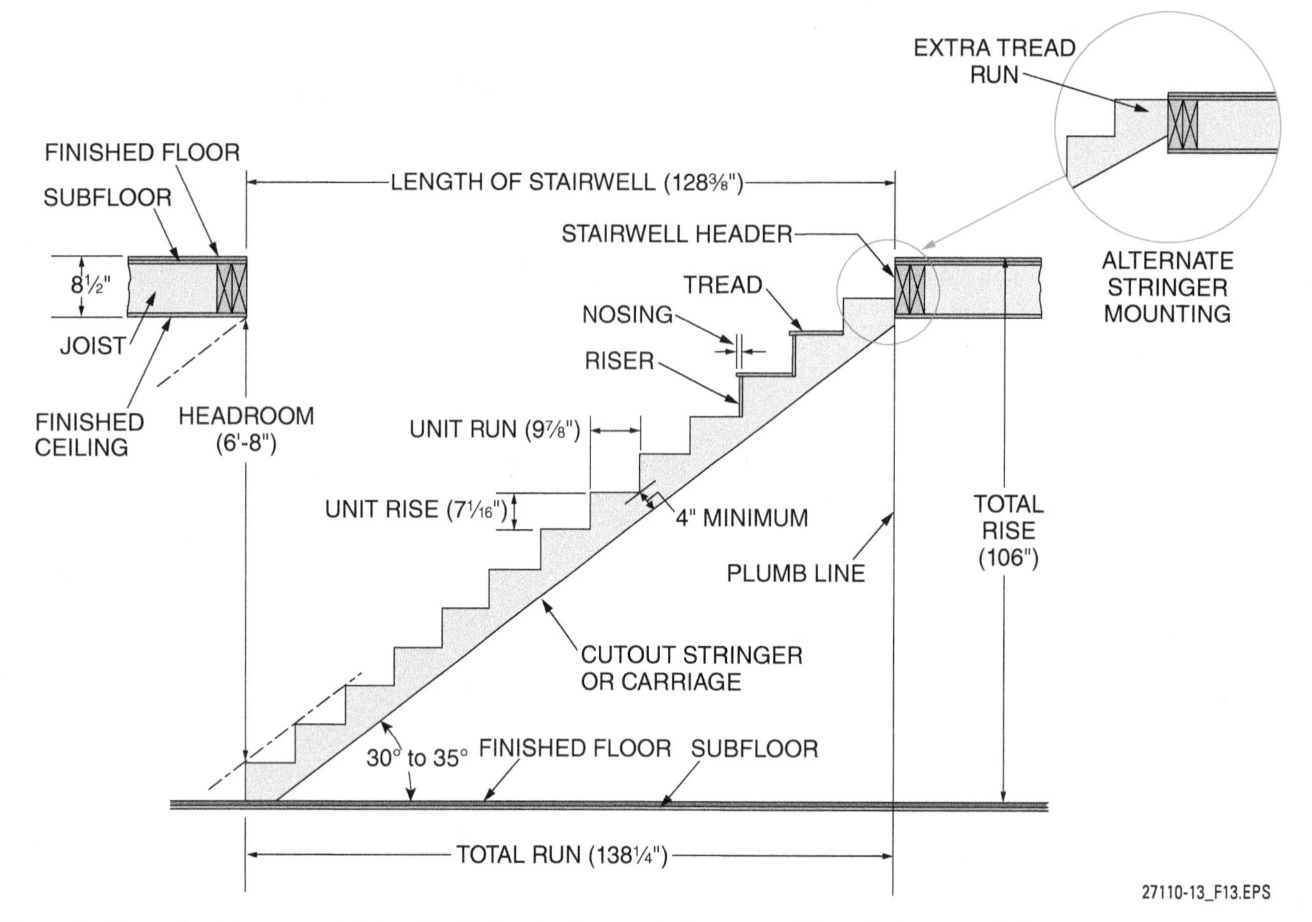

Figure 13 Example stairway and stairwell with terminology and dimensions (floor thickness varies).

Step 1 Using the unit rise of $7\frac{7}{16}$" determined from the previous example and the relationship defined in Rule 2, solve for the unit run as follows:

$$\text{Maximum unit run} = 25" - (2 \times \text{unit rise or } 7\tfrac{7}{16}")$$

$$\text{Minimum unit run} = 24" - (2 \times \text{unit rise or } 7\tfrac{7}{16}")$$

The maximum and minimum allowable unit runs are:

$$\text{Maximum unit run} = 25" - 14\tfrac{7}{8}" = 10\tfrac{7}{8}"$$

$$\text{Maximum unit run} = 24" - 14\tfrac{7}{8}" = 9\tfrac{7}{8}"$$

Step 2 For this example, a minimum unit run of $9\frac{7}{8}$" will be used. If the stairwell header is used as the last riser, it requires one less tread than the number of risers and the total run will be:

$$\text{Total run} = (\text{number of risers} - 1) \times \text{unit run}$$

$$\text{Total run} = (15 - 1) \times 9\tfrac{7}{8}" = 138\tfrac{1}{4}"$$

If the top end of the stringer is to be flush with the top of the floor as depicted in the alternate mounting shown in *Figure 13*, then the same number of treads as risers is required, and the total run would be:

$$\text{Total run} = \text{number of risers} \times \text{unit run}$$

$$\text{Total run} = 15 \times 9\tfrac{7}{8}" = 148\tfrac{1}{8}"$$

4.2.0 Stairwells

If a stairwell has been framed but the unit rise and run of the stairway are not specified, they must be determined. If the stairwell hasn't been framed, the stairwell opening should be determined as follows for the desired headroom.

4.2.1 Determining Stairwell Openings

The width of a stairwell opening for a straight-run stairway is determined by the desired width of the stairway plus the thickness of any skirtboard(s). The stairwell width for an open U-shaped stairway could be as wide as both stairways, plus the skirtboards, plus the amount of space required to turn the U-shaped handrails.

The length of the stairwell depends on the slope and total rise of the stairway. Stairs with a low angle and/or low total rise require a longer stairwell to provide adequate headroom. To find the required length of a stairwell, use the previously determined unit rise and run for the stairway along with the desired headroom and the following procedure.

Step 1 Determine the final thickness of the floor assembly above the stairs in inches. Include the thickness of the anticipated finished floor, subfloor, floor joists, any furring, and the anticipated finished ceiling (from *Figure 13* and the previous examples).

Floor thickness = 8½" = 8.5"

Headroom required = 6'-8"
= (6' × 12") + 8" = 80"

Unit rise = 7$\frac{1}{16}$" = 7.06"

Unit run = 9⅞" = 9.875"

Step 2 Add the floor thickness (Step 1) to the code-required or desired headroom (in inches) and divide the sum by the riser height (unit rise) in inches. Round the answer up to the next whole number. This number represents the number of risers from the top of the stairs down to, or slightly below, the headroom clearance point.

Number of risers required =
(8.5" + 80") ÷ 7.06"
= 12.54 (round up to 13)

Step 3 Multiply the tread depth (unit run) by the whole number obtained in Step 2 to obtain the stairwell length in inches. This length will be correct if the header at the top of the stairway will be the top riser. If the stringers will be framed flush with the top of the header, add one additional tread depth (a unit run) to the overall length of the stairwell. In this example, assume that the stairwell header is used as the last riser and the addition of one unit run to the length will not be required.

Stairwell length =
unit run × number of risers

Stairwell length =
9⅞" × 13 = 128.375" = 128⅜"

In stairways involving higher floor-to-ceiling heights, the stairwell length will be less than the total run.

4.2.2 Determining Stairway Unit Rise and Run Used for a Preframed Stairwell

In some instances, the stairwell openings may be framed but the unit rise, run, and headroom dimensions are not available. In these cases, the unit rise and run, total run (based on the length of the stairwell opening), total rise, and minimum headroom allowed by the applicable code must be determined. This can be accomplished using the following process:

Step 1 Determine the optimum unit rise (riser height) in inches, as previously described. Using the values shown in *Figure 13* as an example, the total rise of the stairway is determined to be 106" and the optimum unit rise is 7$\frac{1}{16}$".

Step 2 Determine the final thickness of the floor assembly above the stairs in inches. Include the thickness of the anticipated finished floor, subfloor, floor joists, any furring, and the anticipated finished ceiling. Referring to *Figure 13*, the floor thick-

Commercial Stairways

Per the *International Building Code*®, commercial stairways require a minimum tread depth of 11" and a maximum riser height of 7". These dimensions are generally regarded as the most comfortable for customer use; however, always check the local building code for specific stairway requirements.

Determining Stairway Rise

Instead of using the drawings to determine the total rise of a stairway, an alternate method is to measure the vertical distance from the anticipated or actual finished floor at the top of the stairs to the anticipated or actual finished floor at the base of the stairs.

ness is 8½" and the minimum headroom is 6'-8".

Step 3 Add the floor assembly thickness (Step 2) to the code-required headroom (in inches) and divide the sum by the unit rise determined in Step 1. Round the answer up to the next whole number. This number represents the number of risers of optimum height from the top of the stairs down to, or slightly below, the headroom clearance point.

Number of risers required = (8.5" + 6'-8") ÷ 7¹⁄₁₆" = 88.5" ÷ 7.06"

= 12.54 (round up to 13)

Step 4 Divide the stairwell length by the number of risers calculated in Step 3 to obtain the unit run (tread width) that may have been used to determine the length of the stairwell.

Unit run = 128⅜" ÷ 13 = 128.375" ÷ 13 = 9.875" or 9⅞"

Step 5 Check the unit run-to-rise ratio using the applicable rule to determine if the calculated tread width meets the minimum requirement. If it does not, the maximum allowable unit rise may have been used in the original calculations for the stairs. In that case, repeat all the steps using the maximum allowable unit rise. In this example, the unit run (and consequently, the unit rise) appears to be valid because it will satisfy Rule 1 for the minimum tread width.

Apparent Run-Length Errors

If the unit run appears to be too large and the resultant total run appears to be too long for the space allowed, the stairwell may have been calculated using more than the minimum required headroom, or it may have been lengthened by an extra tread width to accommodate for flush-mounting the stringers with the floor above. In that case, repeat the required calculations using one extra riser or greater headroom clearance (7' or more, if possible) to determine if either method provides the correct total run.

4.0.0 Section Review

1. According to Rule 1, when determining riser-to-tread ratios, a stairway with a 10½" depth width and 7" riser height is an acceptable combination.
   a. True
   b. False

2. When there are 19 units of rise, the stairwell header is used for the last riser, and a 10¼" tread depth is specified, the total run for the stairway is _____.
   a. 9'-8½"
   b. 12'-2"
   c. 14'-3⅞"
   d. 15'-4½"

Section Five

5.0.0 Stair Construction

Objective

Describe the procedure to lay out and cut stringers, risers, and treads.

a. Explain how to lay out and cut a stringer.
b. Describe how to properly reinforce a stringer.
c. Summarize how concrete stairways are formed.

Performance Task 2

Lay out and cut a stringer.

Trade Term

Pitch board: A board that serves as a pattern for marking cuts for stairs. The shortest side is the height of the riser cut; the perpendicular side is the tread width. Mainly used when there is great repetition such as in production housing.

5.1.0 Laying Out and Cutting a Stringer

Marking a cutout stringer is a relatively simple task. It can be done with a pitch board, framing square, or Speed Square™ (see *Figure 14*). When using a framing square, the blade represents the unit run and the tongue represents the unit rise. To make the stair layout go faster, a set of stair gauges (*Figure 15*) can be used to set the unit rise and run measurements on the framing square.

To determine the approximate length of the stringer, the principles of the right triangle can be used. The rise and run are known; therefore, they form two sides of the right triangle. The stringer is the hypotenuse, or third side, of the right triangle. Using the outside edge on the back of the framing square, locate the total run on the blade and the total rise on the tongue. Each inch increment on the outside edge of the framing square represents 1' and each small increment represents 1". Mark the points on paper. Measure the hypotenuse (diagonal) between the two points using the same edge of the blade. This will be the approximate length of the stringer (see *Figure 16*).

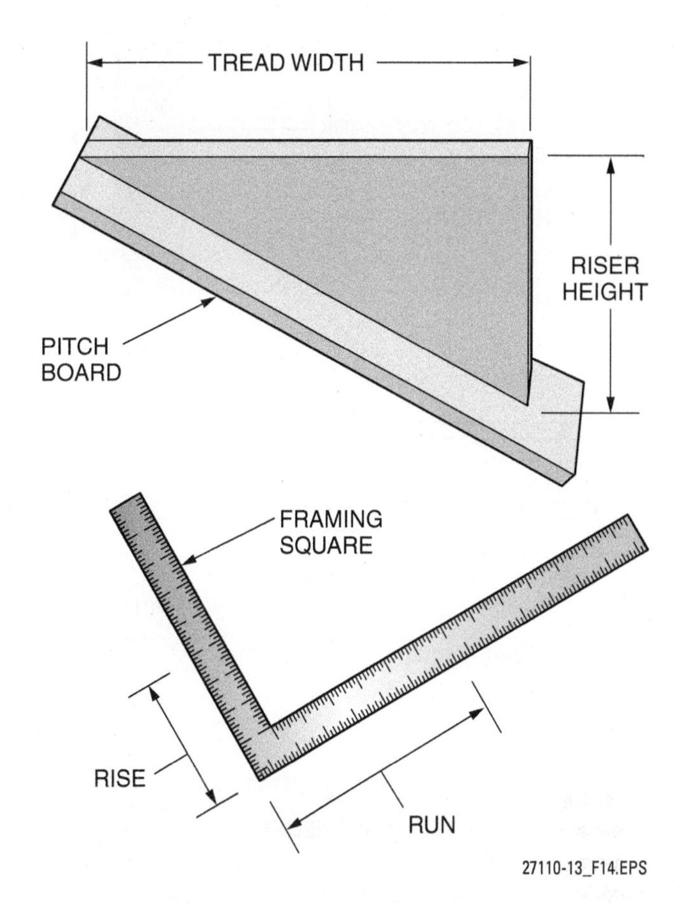

Figure 14 Pitch board and framing square.

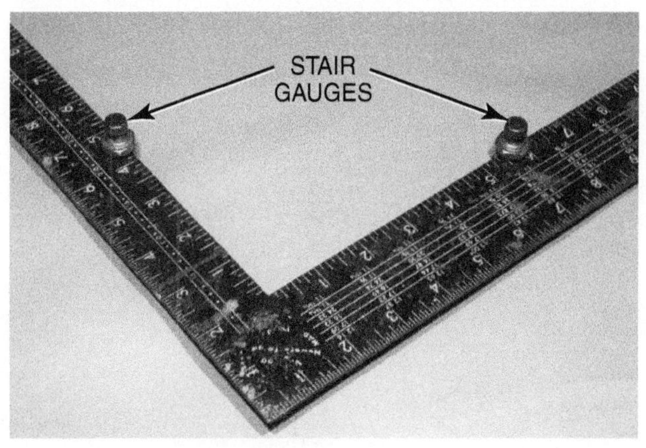

Figure 15 Stair gauges.

For example:

Total rise = 9'-3"

Total run = 12'-6"

Stringer = 15'-7" (use 16')

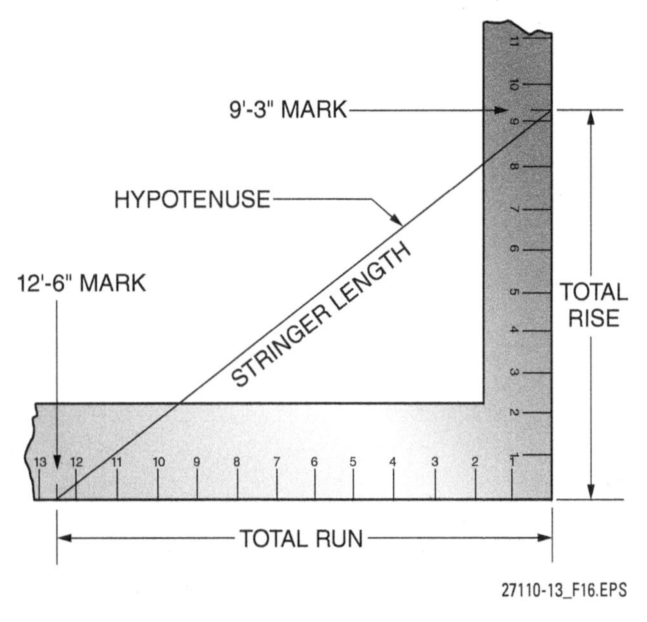

Figure 16 Framing square.

This measurement can also be calculated mathematically by applying the Pythagorean theorem: $(a^2 + b^2 = c^2)$. In this case: $(9'\text{-}3")^2 + (12'\text{-}6")^2 = (15'\text{-}7")^2$ or $(16')^2$.

The cutting and installation of stringers must be done with care so the carpenter finishing the stairs will not encounter difficulty in fitting the risers and treads. When laying out a stringer, all lines drawn must be thin and accurate. Cutting the risers and treads from the stringer can be done with a handsaw, or with a circular saw and finished with a handsaw. A handsaw must be used when finishing the cut so the cut does not extend farther than the riser line. A stringer must be cut accurately. If the stringer is too long for the allotted space, the treads will slant backwards; if the stringer is too short, the treads will slant forward. Laying out a stringer can be accomplished using one of the following two methods.

Using Two Framing Squares to Determine Stringer Length

Instead of marking the rise and run on a piece of paper and measuring between the points, the side of a second framing square, marked in twelfths, can be positioned over the rise and run points on the first square to form the hypotenuse representing the stringer. Then, the length of the stringer can be read directly on the second square.

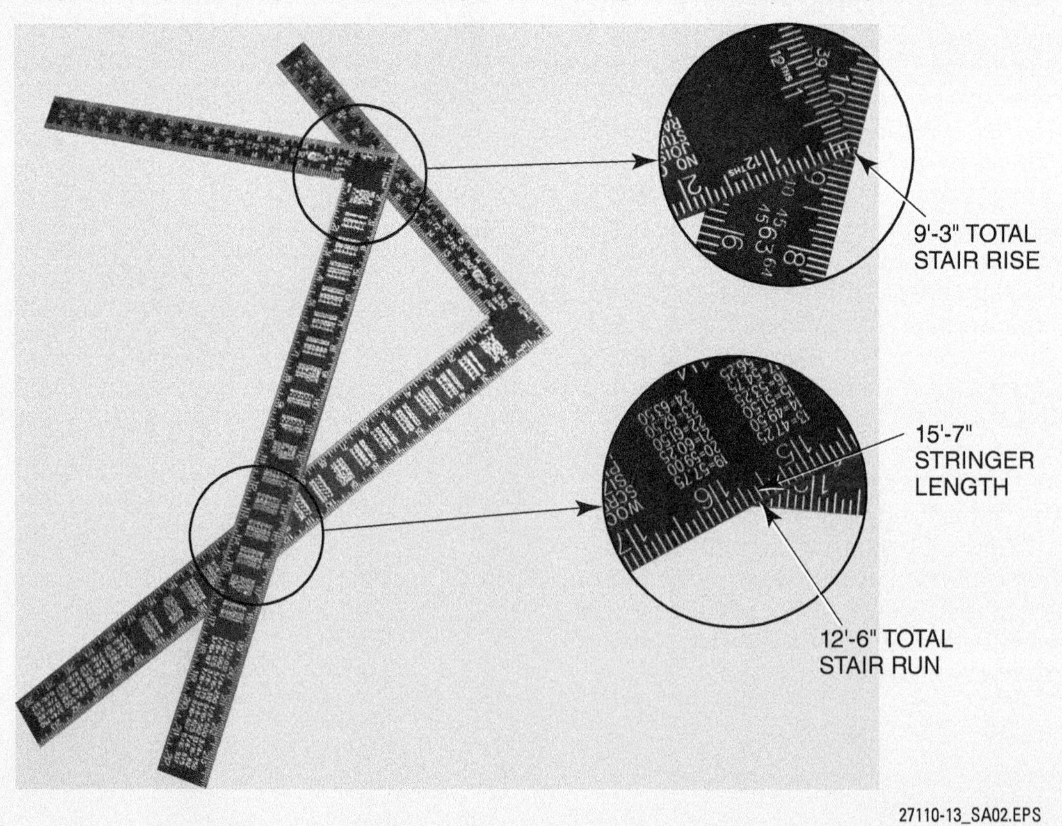

5.1.1 Method One

Step 1 Use 2 × 10s, 2 × 12s, or some other structurally sound size of lumber that is of a species and grade approved by the local code. The stock must be as straight as possible and installed with the crown up.

Step 2 Lay the stock on sawhorses or a work table. Place the top edge (good edge) away from you. All layout work is done from the top edge (see *Figure 17*).

Every system and technology used in a project comes with overhead. This overhead includes extra packaging, transportation, and other costs that do not add value to the product itself. When one product serves as multiple systems, the overhead for that product is a fraction of the overhead associated with the multiple systems it replaces.

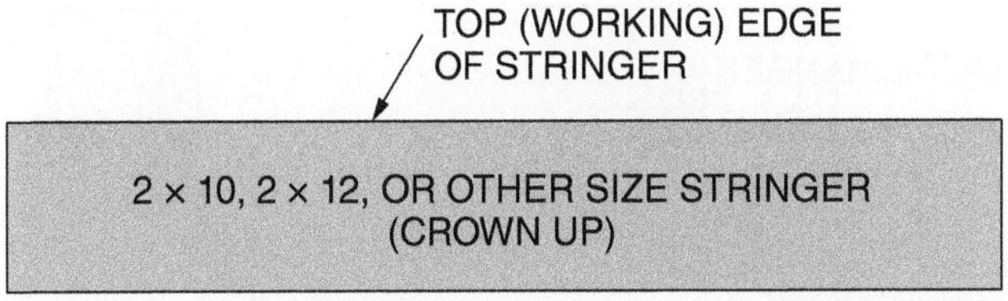

Step 3 Place the framing square at the end of the stringer as shown in *Figure 18A*. Mark the unit rise and unit run on the stringer using the inside edge of the square. Remove the square and check that there is a minimum of 4" between the junction of the rise/run and the bottom of the stringer as shown in *Figure 18B*. If not, a wider stringer must be used. Place the square back on the stringer, as shown in *Figure 18C*, and extend the unit run mark to the bottom of the stringer. This will be the portion of the stringer that rests on the floor. Identify the extended unit-run mark as a cut line.

> **WARNING!**
>
> Overcutting at the junctions of the rise/run can significantly weaken the stringer.

Step 4 Move the framing square to the next tread and mark the rise and run. Proceed to mark the rise and run for all the other steps. If the riser is to be installed as shown in *Figure 19*, then the stringer will be cut as it was laid out because the header will be the top riser (one more riser than the number of treads). If the stringer is to be placed so the top tread is at the level of the finished floor, the stringer will be cut with one more tread (see *Figure 20*).

Step 5 Starting at the left side, cut along the first line that goes from the top to the bottom.

TOP (WORKING) EDGE OF STRINGER

2 × 10, 2 × 12, OR OTHER SIZE STRINGER (CROWN UP)

27110-13_F17.EPS

Figure 17 Positioning of a stringer for layout.

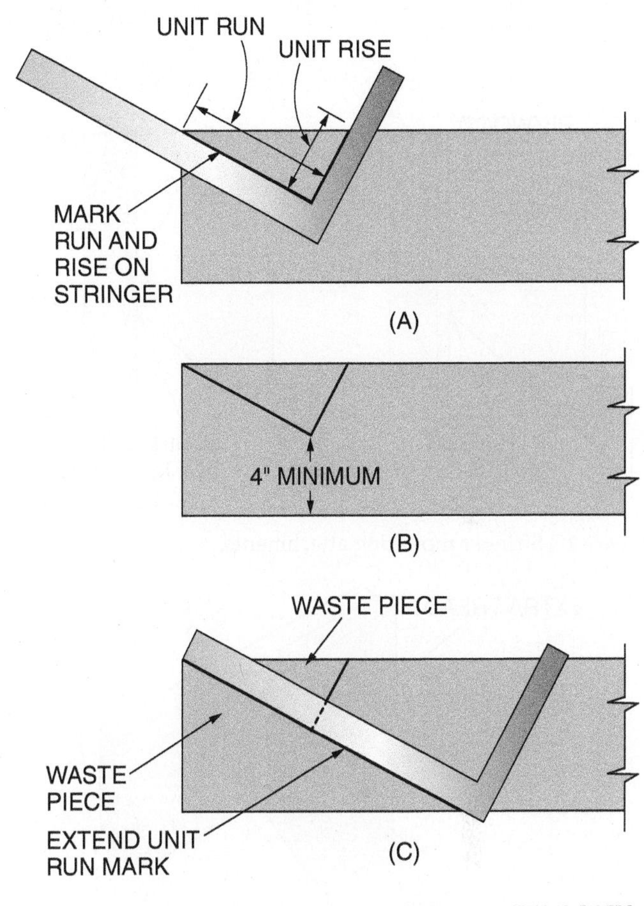

Figure 18 Marking a stringer for the floor-level cut line.

> **GOING GREEN**
>
> **Using Materials More Efficiently**
>
> One strategy for greening materials is to use materials more efficiently, which means getting more benefit from the materials you do use and/or using fewer materials to achieve the same result. One way to do this is to use multifunction materials. Multifunction materials do more than one thing as part of a building. They can speed up construction, save building materials, and reduce waste. For example, lumber cutoffs can be saved and used as blocking between studs or joists or as fire-stopping.

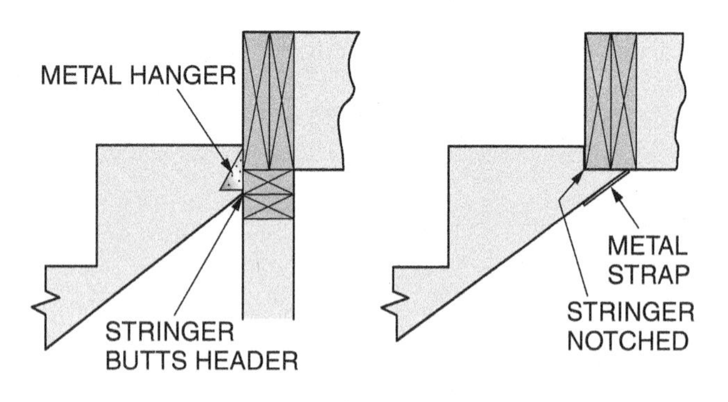

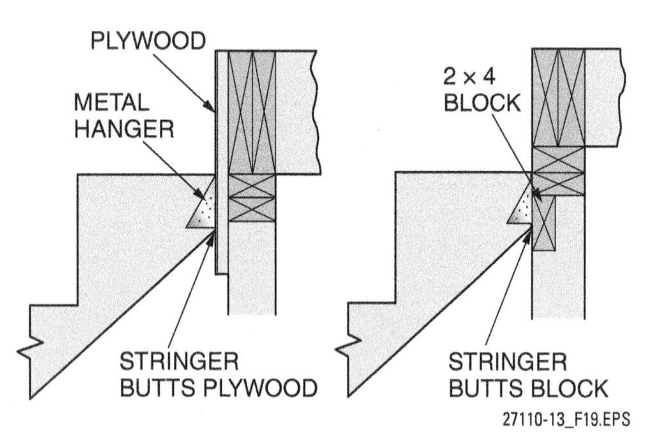

Figure 19 Stringer mounting attachments.

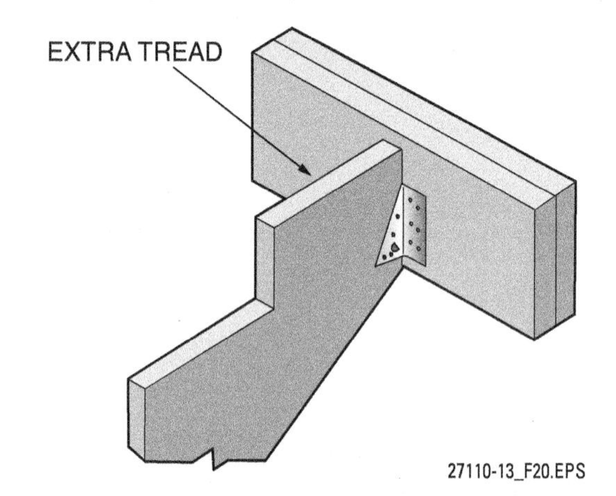

Figure 20 Flush stringer attachment.

All cuts must be done cautiously. Be sure not to overcut (see *Figure 21*).

Step 6 Cut the rest of the treads and risers.

Step 7 If the stringer is to sit on top of a finished floor, the stringer must be dropped (cut) the thickness of a tread so the top and bottom step of the stairs will have the same riser height (see *Figure 22*). For an irregular floor, the bottom of the stringer may need to be scribed. If the floor that the stringers will rest on will have some type of finished floor added later, the bottom of the stringer must be cut to compensate. The amount to be cut would be the difference between the finished floor thickness and the thickness of the tread (see *Figure 23*).

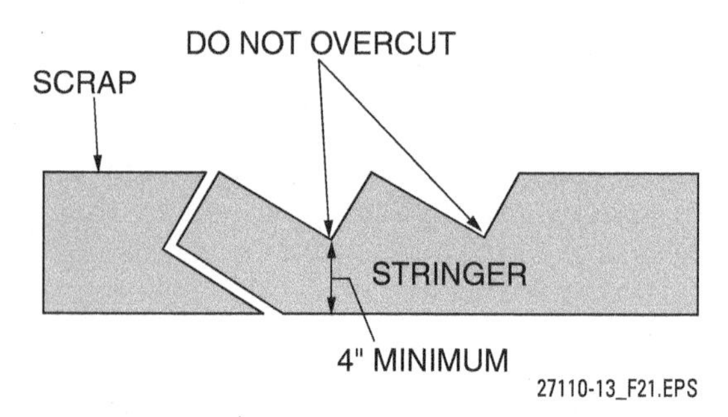

Figure 21 Stringer cuts.

Step 8 Use the completed stringer as a pattern to lay out the remaining stringer(s). Lay out the opposite wall stringer so that any bow faces in toward the stairway.

Step 9 If a stairway is 30" or less in width, only two stringers are required. For added strength and stability of the stairway, a third stringer may be centered between the two stringers. A stairway that is 30" or wider typically requires three stringers. Check the local building code for requirements.

Step 10 Install the skirtboards, if required. Then install the stringers, one on each side of the stairway. Nail the stringers to the wall studs with 16d common nails. Complete any other related construction activities as shown on the prints.

Step 11 Install rough treads of 2" nominal stock. These treads will stay in place until the finished risers and treads are ready to be installed. After removal, the rough treads can be reused on another project.

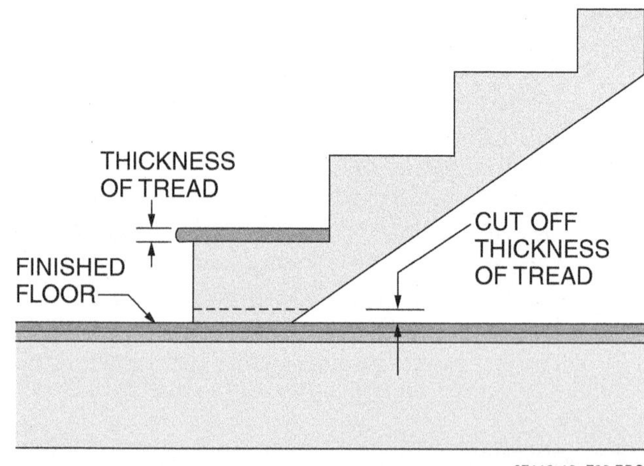

Figure 22 Dropping the stringer for tread thickness.

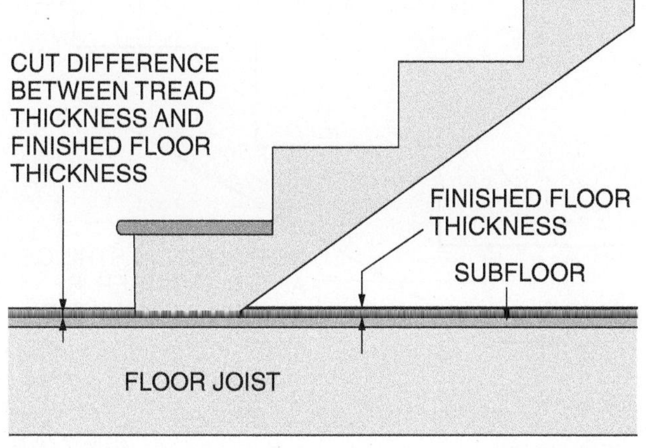

Figure 23 Dropping the stringer for tread thickness minus the finished floor thickness.

5.1.2 Method Two

Laying out stringers with a pitch board is similar to laying out stringers using a framing square. A pitch board is made by marking a piece of 1" stock with the correct rise and run for the stairway being constructed. After cutting the stock, a piece of 1 × 2 stock is fastened to the cut piece, enabling the pitch board to slide along the edge of the stringer (*Figure 24*). *Figure 25* illustrates the use of a pitch board.

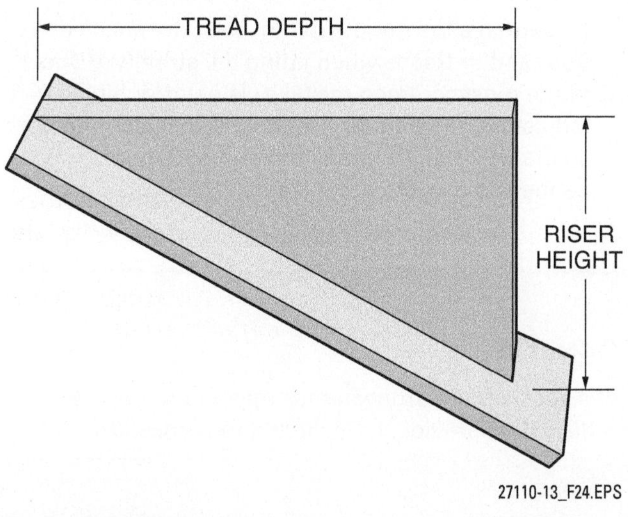

Figure 24 Pitch board.

Stringers that are cut and installed correctly will not cause problems for the carpenter installing the finished risers and treads. Service and main-stair stringers can be prepared in several ways:

- Stringers can be cut to receive finished risers and treads or finished treads only.
- Stringers can be dadoed to receive treads only (*Figure 26*).

5.2.0 Reinforcing Cutout Stringers

A 2 × 4 or 2 × 6 reinforcement known as a strongback may be fastened to longer cutout stair stringers that are supported only at the header and floor and do not receive the support of an adjacent wall. A strongback is secured to one side of the stringer below the cut notches to add strength and rigidity (see *Figure 27*).

It is also a common practice to secure a 2 × 4 ledger to the floor and/or header to add strength to the stair assembly. The stair stringers are notched to fit over the ledger (see *Figure 28*).

> **NOTE**
> If the ledger is fastened to a concrete floor, treated lumber must be used.

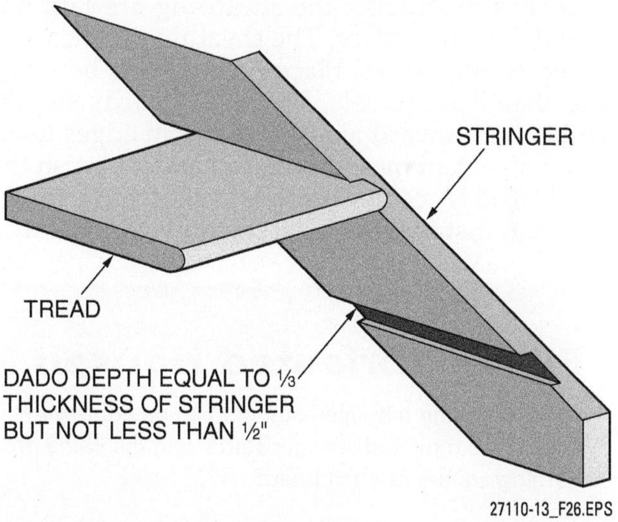

STRINGER
TREAD
DADO DEPTH EQUAL TO ⅓ THICKNESS OF STRINGER BUT NOT LESS THAN ½"
27110-13_F26.EPS

Figure 26 Dadoed stringer.

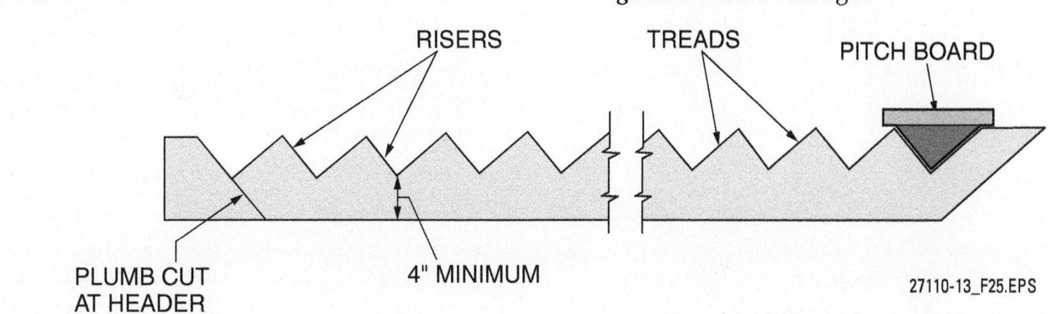

Figure 25 Using a pitch board.

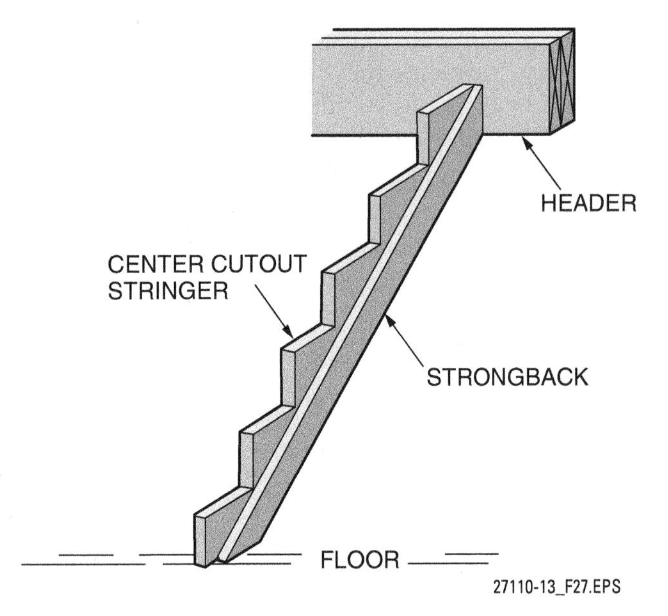

Figure 27 Reinforced cutout stringer.

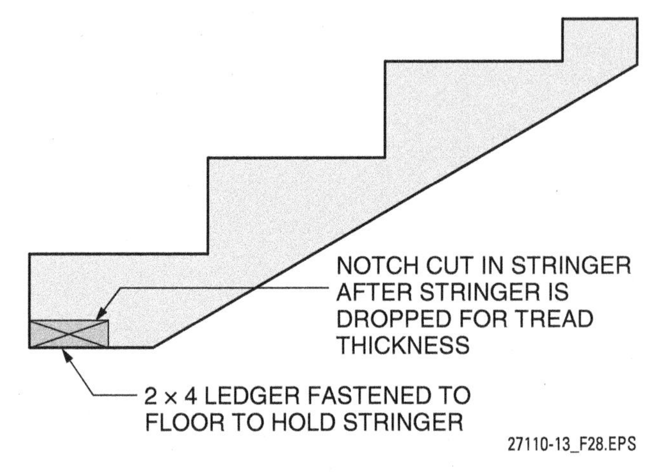

Figure 28 Reinforced stringer mounting.

5.3.0 Concrete Stairway Forms

Figure 29 shows a basic concrete stairway form. Note that the size and position of the riser boards establish the height, depth, and spacing of the treads and risers. When the stairs are wide, a center brace is needed for additional support of the riser board.

Stakes to support the sheathing are laid out and driven into place. The sheathing is then fastened to the stakes. Place riser boards between the sheathing panels. The riser boards should be beveled inward along the bottom edges to allow the cement mason to trowel and/or brush the stair tread. A center brace and sheathing braces are then installed to resist the outward pressure of the concrete. Ensure that the landing and steps slope away from the building ⅛" to ¼" per foot to allow moisture to drain away from the building. An expansion joint is placed between the building wall and stairway to prevent the wall from cracking if the stairway settles or otherwise moves due to expansion and contraction of the concrete.

Compensating for Bottom Riser Height of Stringers

Be sure to allow for the thickness of the finished floor and/or treads when fitting the stringers. One of the most common mistakes is not dropping (adjusting) the bottom riser height of the stringers so the riser height for the top and bottom steps will be the same as the other steps.

Field-Fabricated Housed Stringers

Field-mortising a housed stringer can be done with the use of a specialized European-made jig and router. The layout of a mortised stringer starts with the same procedures described previously, including the procedure with a framing square or pitch board.

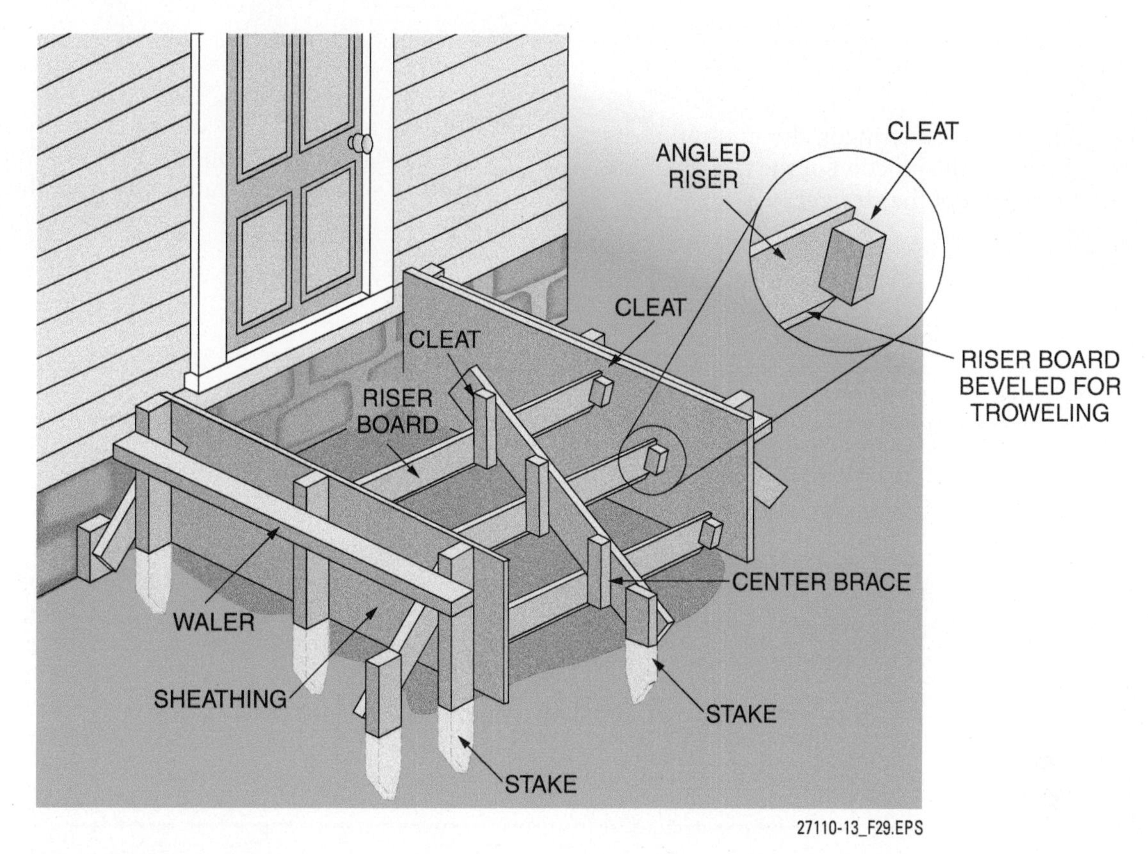

Figure 29 Basic concrete stair form.

Preventing Stairway Squeak

The use of glue or construction adhesive at each tread will help prevent the stairway from squeaking over time.

5.0.0 Section Review

1. When laying out a stringer, the minimum distance between the junction of the rise/run lines and the bottom of the stringer is _____.
   a. 1"
   b. 2"
   c. 4"
   d. none of the above.

2. Long stringers may be reinforced with 2 × 4 or 2 × 6 _____.
   a. trimmers
   b. strongbacks
   c. headers
   d. cripples

3. Treads on a concrete exterior stairway should slope away from the building ½" to ¾" per foot to allow moisture to drain away from the building.
   a. True
   b. False

SUMMARY

A variety of stairs are used in residential and commercial construction. This module covered stair terminology and general building code requirements. It also described the methods used when designing, laying out, cutting, and installing the framing for basic wood stairs. As a professional carpenter, you must be thoroughly familiar with the stair construction information and techniques covered in this module.

Review Questions

1. Because it is difficult to build, which type of stairway is often fabricated off-site?
   a. Straight-run stairway
   b. Platform stairway
   c. Double-L stairway
   d. Winding stairway

2. Which type of stairway uses wedge-shaped treads when changing direction?
   a. Narrow-U
   b. Platform
   c. Long-L
   d. Winding

3. A series of steps that are straight and continuous without breaks is called a _____.
   a. rise
   b. flight
   c. run
   d. covey

4. Geometrical stairways can be potentially dangerous if not properly constructed.
   a. True
   b. False

5. All of the following are types of geometrical stairs *except* _____.
   a. circular
   b. elliptical
   c. wide-U
   d. spiral

6. Which of the following statements about a straight-run stairway is *not* true?
   a. It is continuous from one level to another
   b. It is the easiest stairway to build
   c. It has a landing
   d. It requires a long stairwell

7. Service stairs are the primary means of access in a two-story commercial building.
   a. True
   b. False

8. Per the *International Building Code®*, minimum stairway width for a building occupied by 26 people is _____.
   a. 32"
   b. 36"
   c. 40"
   d. 44"

9. Per the *International Building Code®*, minimum handrail height from top of rail to nosing is _____.
   a. 34"
   b. 36"
   c. 40"
   d. 44"

10. Per the *International Building Code®*, maximum handrail height from top of rail to nosing is _____.
    a. 34"
    b. 38"
    c. 46"
    d. 50"

11. What international code specifies size requirements for residential stairways and handrails?
    a. *International Building Code®*
    b. *International Residential Code®*
    c. ANSI
    d. International Stairway Association

12. A wood stairway that is wider than 30 inches requires one or more _____.
    a. strongbacks
    b. housed stringers
    c. support purlins
    d. center cutout stringers

13. A closed stairway has _____.
    a. a wall on one side
    b. doors at top and bottom
    c. walls on both sides
    d. a landing with a railing

14. The main support for a handrail at the bottom of a stairway is called a(n) _____.
    a. baluster
    b. main newel post
    c. ending newel post
    d. starting newel post

15. The maximum stair riser height specified in the *International Building Code®* is _____.
    a. 6½"
    b. 7"
    c. 7¾"
    d. 8"

Refer to *Review Question Figure 1* to answer Questions 16–19.

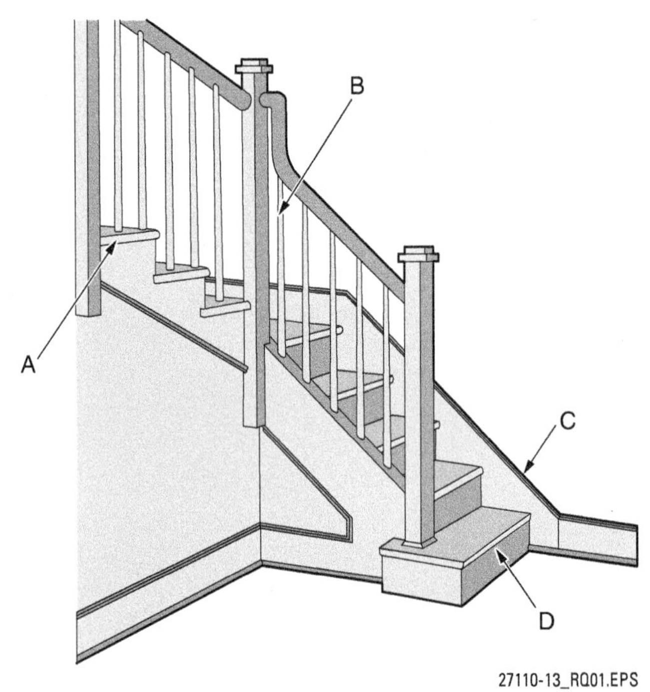

Figure 1

16. The stair part marked A is the _____.
    a. nosing
    b. platform
    c. skirtboard
    d. nosing return

17. The stair part marked B is the _____.
    a. baluster
    b. newel post
    c. handrail
    d. handrail support

18. The stair part marked C is the _____.
    a. nosing
    b. skirtboard
    c. spindle
    d. railing

19. The stair part marked D is the _____.
    a. nosing
    b. skirtboard
    c. spindle
    d. newel post

20. The vertical difference between a stair tread nosing and any structure or ceiling above the stairway is known as the _____.
    a. newel post
    b. skirtboard
    c. crown
    d. headroom

21. The handrail known as a wall rail is used on a(n) _____.
    a. closed stairway
    b. winding stairway
    c. open stairway
    d. balustrade

22. Which of the following is used to establish the rise and run of a staircase?
    a. Tread width + (2 × riser height) = 17" to 18"
    b. Tread width × riser height = 24" to 25"
    c. Tread width + (2 × riser height) = 24" to 25"
    d. Tread width + (2 × riser height) = 70" to 75"

23. For maximum safety, the preferred slope of a stairway is between _____.
    a. 15 and 20 degrees
    b. 30 and 35 degrees
    c. 45 and 50 degrees
    d. 55 and 60 degrees

24. A rule of thumb for establishing the proper angle of a staircase states that the sum of one riser plus one tread should equal between _____.
    a. 14 and 16 inches
    b. 16 and 17 inches
    c. 17 and 18 inches
    d. 18 and 19 inches

25. The item placed between the building wall and stairway to prevent the wall from cracking created by settling is the _____.
    a. crown jabber
    b. expansion joint
    c. skirtboard
    d. stringer

Trade Terms Quiz

Fill in the blank with the correct term that you learned from your study of this module.

1. Circular, elliptical, and spiral stairs are all types of ________.
2. A(n) ________ has solid walls on both sides.
3. The ________ is the main post supporting the handrail of a staircase.
4. The projection of the tread beyond the face of the riser is called the ________.
5. The height of each riser is the ________.
6. The depth of each tread is the ________.
7. When you walk up several flights of stairs, you are likely to pause on the ________ between flights.
8. An open fire-escape stairway is likely to be enclosed by a(n) ________.
9. An enclosed stairway will have a(n) ________ along the wall to serve as a handhold.
10. The spindles of a staircase are also called the ________.
11. The distance between any portion of a step and the ceiling is the ________.
12. A cutout ________ provides the main support for a stairway.
13. The horizontal part of a step is known as the ________.
14. The vertical part of a step is known as the ________.
15. A(n) ________ is a vertical compartment in a building into which stairs are placed.
16. The ________ is the horizontal distance from the face of the first riser to the face of the last riser.
17. The ________ is the vertical dimension of a set of stairs.
18. A(n) ________ is sometimes used against the wall in a closed staircase.
19. The complete handrail, post, and spindle assembly is known as the ________.
20. A stringer with grooves cut into it to hold the risers and treads is a(n) ________.
21. A stairway that can be viewed from within a room is called a(n) ________.
22. A(n) ________ can be used as a pattern for marking cuts for stairs.
23. The degree of incline of a set of stairs is the ________.
24. A ________ is used to ascend from one level of a building to another.

Trade Terms

Baluster	Guard	Newel post	Riser	Total rise
Balustrade	Handrail	Nosing	Skirtboard	Total run
Closed stairway	Headroom	Open stairway	Stairway	Tread
Geometrical stairways	Housed stringer	Pitch board	Stairwell	Unit rise
	Landing	Rise and run	Stringer	Unit run

Cornerstone of Craftsmanship

Owen Carpenter

Construction Technology Instructor
Mountain Home High School Career Academies
Mountain Home, Arkansas

How did you get started in the construction industry?
I got my start in construction by enrolling in high school shop classes. After graduating from high school, I enrolled in college as an Industrial Education major.

Who or what inspired you to enter the Industry?
Teachers were my inspiration for wanting to be a Construction Technology Instructor. Two high school teachers, Gordon Morgan and Doug VanDolah, got me interested in the field. My college professor, Charlie Powers, provided encouragement to continue to strive for my goal.

What do you enjoy most about your career?
I really enjoy helping students get a start in the industry. When a former student thanks me for things they learned while in my classes (that helped them later in life), I take a great deal of pride in that.

Why do you think training and education are important in construction?
Training and education are important in whatever type of work a person chooses. In the construction industry, safety is the most important reason for training and education, followed closely by better performance, which typically equals better pay.

Why do you think credentials are important in construction?
Credentials will never take the place of actual performance. However, a person's credentials are the one thing that will get them into a position to demonstrate their proficiency.

Would you recommend construction as a career to others? Why?
I constantly encourage my students to enter the field of construction if they truly enjoy the satisfaction of being able to step back at the end of the day and take pride in what they have accomplished either alone or as a team.

What does craftsmanship mean to you?
It means putting your best work into every aspect of a project whether it will be seen in the finished product or not. As I tell my students, "If your name is attached to it, make it the very best you are capable of producing."

Trade Terms Introduced in This Module

Baluster: A supporting column or member; a support for a railing, particularly one of the upright columns of a balustrade.

Balustrade: A stair rail assembly consisting of a handrail, balusters, and posts.

Closed stairway: A stairway that has solid walls on each side.

Geometrical stairway: A winding stairway with an elliptical or circular design.

Guard: A rail secured to uprights and erected along the exposed sides and ends of platform stairs, etc.

Handrail: A member supported on brackets from a wall or partition to furnish a handhold.

Headroom: The vertical and clear space in height between a stair tread and the ceiling or stairs above.

Housed stringer: A stair stringer with horizontal and vertical grooves cut (mortised) on the inside to receive the ends of the risers and treads. Wedges covered with glue are often used to hold the risers and treads in place in the grooves.

Landing: A horizontal area at the end of a flight of stairs or between two flights of stairs.

Newel post: An upright post supporting the handrail at the top and bottom of a stairway or at the turn of a landing; also, the main post about which a geometrical staircase winds.

Nosing: The portion of a stair tread that extends beyond the face of the riser.

Open stairway: A stairway that is open on at least one side.

Pitch board: A board that serves as a pattern for marking cuts for stairs. The shortest side is the height of the riser cut; the perpendicular side is the tread width. Mainly used when there is great repetition such as in production housing.

Rise and run: A term used to indicate the degree of stairway incline.

Riser: A vertical board under the tread of a stairway.

Skirtboard: A baseboard or finishing board at the junction of the interior wall and floor. Also called a finished stringer.

Stairway: The complete set of stairs extending from one level of a structure to another level.

Stairwell: A compartment extending vertically through a building into which a stairway is placed.

Stringer: The inclined member that supports the treads and risers of a stairway.

Total rise: The vertical distance of a stairway one floor to the floor above.

Total run: Horizontal length of a stairway.

Tread: The horizontal member of a stairway upon which a person steps.

Unit rise: The vertical distance from the top of one stair tread to the top of the next one above it; also called stair rise.

Unit run: The horizontal distance from the face of one riser to the face of the next riser.

Additional Resources

This module presents thorough resources for task training. The following reference material is suggested for further study.

Basic Stairbuilding, Newtown, CT: Taunton Press, Inc. (Book with companion video or DVD.)
Constructing Staircases, Balustrades & Landings. New York: Sterling Publishing Co., Inc.
For Pros By Pros: Building Stairs. Newtown, CT: Taunton Press, Inc.
Framing Floors and Stairs. Berkeley, CA: Publishers Group West. (Book with companion video or DVD.)
A Simplified Guide to Custom Stairbuilding and Tangent Handrailing. Fresno, CA: Linden Publishing.
Stair Builders Handbook. Carlsbad, CA: Craftsman Book Company.
Staircases. New York: Watson-Guptill Publications.
Stairs: Design and Construction. New York: Birkhauser.
Arcways, Inc. Builders of custom stairways, **http://www.arcways.com**
WM Coffman Resources, LLC. Hardwood stair parts manufacturer, **http://www.wm-coffman.com**
L.J. Smith Stair Systems. Manufacturer of stair products, **http://www.ljsmith.net**

Figure Credits

SkillsUSA, Module opener
The Haskell Company, Figure 11
Weyerhaeuser Wood Products, Figure SA03

Section Review Answer Key

Answer	Section Reference	Objective Reference
Section One		
1. b	1.0.0	1a
Section Two		
1. c	2.0.0	2
Section Three		
1. b	3.1.0	3a
2. b	3.2.0	3b
3. a	3.3.0	3c
4. c	3.4.0	3d
5. d	3.5.0	3e
6. b	3.6.0	3f
Section Four		
1. a	4.1.0	4a
2. d	4.1.2	4b
Section Five		
1. c	5.1.1	5a
2. b	5.2.0	5b
3. b	5.3.0	5c

NCCER CURRICULA — USER UPDATE

NCCER makes every effort to keep its textbooks up-to-date and free of technical errors. We appreciate your help in this process. If you find an error, a typographical mistake, or an inaccuracy in NCCER's curricula, please fill out this form (or a photocopy), or complete the online form at **www.nccer.org/olf**. Be sure to include the exact module ID number, page number, a detailed description, and your recommended correction. Your input will be brought to the attention of the Authoring Team. Thank you for your assistance.

Instructors – If you have an idea for improving this textbook, or have found that additional materials were necessary to teach this module effectively, please let us know so that we may present your suggestions to the Authoring Team.

NCCER Product Development and Revision
13614 Progress Blvd., Alachua, FL 32615

Email: curriculum@nccer.org
Online: www.nccer.org/olf

❑ Trainee Guide ❑ Lesson Plans ❑ Exam ❑ PowerPoints Other ____________________

Craft / Level: Copyright Date:

Module ID Number / Title:

Section Number(s):

Description:

Recommended Correction:

Your Name:

Address:

Email: Phone:

Glossary

Abate: : To reduce or minimize.

Admixtures: Materials that are added to a concrete mix to change certain properties of the concrete such as retarding setting time, reducing water requirements, or making the concrete easier to work with.

Area: The surface or amount of space occupied by a two-dimensional object such as a rectangle, circle, or square.

Baluster: A supporting column or member; a support for a railing, particularly one of the upright columns of a balustrade.

Balustrade: A stair rail assembly consisting of a handrail, balusters, and posts.

Barge rafter: A gable end roof member that extends beyond the gable to support a decorative end piece. Also referred to as a fly rafter.

Benchmark: A point established by the surveyor on or close to the building site. It is used as a reference for determining elevations during the construction of a building.

Bevel cut: A cut made across the sloping edge or side of a workpiece at an angle of less than 90 degrees.

Blocking: A wood block used as a filler piece and a support between framing members.

Board foot: Lumber quantity measure equivalent to a piece of lumber that is 1" thick, 12" wide, and 1' long.

Bottom chord: The lower member of a truss.

Bucks: Well-braced frame placed inside a concrete form to provide an opening for a door or window.

Building envelope: Consists of all building components, such as basement walls, exterior walls, floor, and roof, that separate conditioned space from unconditioned space or outdoor air.

Building wrap: Made of spun, high-density polyethylene fibers randomly bonded into an extremely tough, durable sheet material and used to minimize air infiltration while allowing moisture to escape.

Butt joint: The joint formed when one square-cut edge of a piece of material is placed against another material.

Cantilever: A beam, truss, or slab (floor) that extends past the last point of support.

Career: A profession for which an individual trains and considers to be a lifelong calling.

Casing: Trim around a window or door.

Catalyst: A substance that causes a reaction in another substance.

Closed stairway: A stairway that has solid walls on each side.

Combustible: Capable of easily igniting and rapidly burning; used to describe a fuel with a flash point at or above 100°F.

Compound cut: A simultaneous bevel and miter cut.

Conditioned space: An area of a building in which heating and/or cooling is deliberately supplied to it or is indirectly supplied through uninsulated surfaces of water or heating equipment or through uninsulated ducts.

Contour lines: Imaginary lines on a site plan that connect points of the same elevation. Contour lines never cross each other.

Cripple stud: In wall framing, a short framing stud that fills the space between a header and a top plate or between the sill and the soleplate.

Crosscut: A cut made across the grain in lumber.

Crown: The highest point of the curved edge of a framing member.

Curb: A framework on which a skylight is mounted.

Cured concrete: Concrete that has hardened and gained its structural strength.

Dados: Rectangular grooves that are made part of the way through and across the grain of lumber.

Dead load: The weight of permanent, stationary construction and equipment included in a building.

Deadbolt: A square-head bolt in a door lock that requires a key to move it in either direction.

Dome pan: Metal or fiberglass concrete form used for two-way joist systems.

Door jack: A holder or stand used to hold a door on edge while planing, routing, or installing hinges.

Dormer: Framework extending from the roof surface that provides additional light and ventilation to an attic space.

Double top plate: A plate made of two members to provide better stiffening of a wall. It is also used for connecting splices, corners, and partitions that are at right angles (perpendicular) to the wall.

Dovetail joints: Interlocking wood joints with a triangular shape like that of a dove's tail.

Drying-in: Applying sheathing, windows, and exterior doors to a framed building.

Easement: A legal right-of-way provision on another person's property (for example, the right of a neighbor to build a road or a public utility to install water and gas lines on the property). A property owner cannot build on an area where an easement has been identified.

Elevation view: A drawing providing a view from the front or side of a structure.

End joist: The first and last joists in a floor system; they run parallel to the common joists.

Expansion strip: Resilient material added to a joint to allow expansion and contraction of adjacent materials while minimizing cracking.

Experience modification rate (EMR): A rate computation to determine surcharge or credit to workers' compensation premium based on a company's previous accident experience.

Exterior insulation and finishing system (EIFS): Method of exterior finish consisting of a coat of acrylic copolymers and portland cement applied over a base of exterior sheathing, insulation board, and reinforcing mesh.

False fascia: The board that is attached to the tails of the rafters to straighten and space the rafters and provide a nailing surface for the fascia. Also called subfascia and rough fascia.

Finish carpentry: The portion of the carpentry trade associated with interior and exterior trim, cabinetry, siding, wall finishes, and decorative work.

Fire stop: An approved material used to fill air passages in a frame to retard the spread of fire.

Flashing: Sheet metal strips used to seal a roof or wall against leakage.

Floor truss: An engineered assembly that is used to support floors.

Foundation: The supporting portion of a structure, including the footings.

Front setback: The distance from the property line to the front of the building.

Furring strip: Narrow wood strip nailed to a wall or ceiling as a nailing base for finish material.

Gable: The triangular wall enclosed by the sloping ends of a ridged roof.

Galvanized: Protected from corrosion (rust) by a zinc coating.

Geometrical stairway: A winding stairway with an elliptical or circular design.

Glazing: Material such as glass or plastic used in windows, skylights, and doors.

Green concrete: Concrete that has hardened but has not yet gained its structural strength.

Guard: A rail secured to uprights and erected along the exposed sides and ends of platform stairs, etc.

Gypsum: A chalky material that is a main ingredient in plaster and gypsum board.

Handrail: A member supported on brackets from a wall or partition to furnish a handhold.

Header: A horizontal structural member that supports the load over a window, door, or other wall openings; used to frame floor openings; they run perpendicular to the direction of the joists and are typically doubled.

Header joist: A framing member used in platform framing into which the common joists are fitted. Header joists are also used to support the free ends of joists when framing openings in a floor.

Headroom: The vertical and clear space in height between a stair tread and the ceiling or stairs above.

Hinge jamb: The side of the door jamb into which the hinges are installed.

Hip roof: A roof with four sides or slopes extending toward the center.

Hipped: The external angle formed by the intersection of two adjacent sloping sides.

Housed stringer: A stair stringer with horizontal and vertical grooves cut (mortised) on the inside to receive the ends of the risers and treads. Wedges covered with glue are often used to hold the risers and treads in place in the grooves.

Hydration: The catalytic action water has in transforming the chemicals in portland cement into a hard solid. The water interacts with the chemicals to form calcium silicate hydrate gel.

Jamb: The top (head jamb) and side members of a door or window frame that come into contact with the door or window.

Job hazard analysis: An approach to hazard recognition in which the task to be performed is broken down into its individual parts or steps, and then each step is analyzed for its potential hazard(s). When the hazard is identified, certain actions or procedures are recommended that will correct the hazard before an accident occurs.

Joist: Generally, equally spaced framing members that support floors and ceilings.

Joist hanger: A metal stirrup secured to the face of a structural member, such as a girder, to support and align the ends of joists flush with the member.

Kerf: The width of the cut made by a saw blade. It is also the amount of material removed by the blade In a through (complete) cut or the slot that is produced by the blade in a partial cut.

Kickback: A sharp, uncontrolled grabbing and throwing of the workpiece by a tool as it rejects material being forced into it.

Landing: A horizontal area at the end of a flight of stairs or between two flights of stairs.

Lateral pressure: Sideways pressure against a structure, such as a foundation.

Let-in: Any type of notch in a stud or joist, which holds another piece. The item that is supported by the notch is said to be let in.

Light: Glass insert in a door.

Live load: The total of all moving and variable loads that may be placed upon a structure.

Lockset: The entire lock unit, including locks, strike plate, and trim pieces.

Long pan: Metal or fiberglass concrete form used for one-way joist systems.

Lookout: A structural member used to frame an overhang.

Material safety data sheet (MSDS): Information that details any toxic, chemical, or potentially harmful substances that are contained in a product.

Material takeoff: A list of building materials obtained by analyzing the project drawings (also known as a takeoff).

Millwork: Manufactured wood products such as doors, windows, and moldings.

Miter box: A device used to cut lumber at precise angles.

Miter cut: A cut made at the end of a piece of lumber at any angle other than 90 degrees.

Monolithic: Concrete that is cast as a single unit; i.e., a concrete floor and joists.

Monuments: Physical structures that mark the location of a survey point.

Mortise and tenon joints: Wood joints in which a rectangular cutout or opening receive the tongue that is cut on the end of an adjoining member.

Muntin: A thin framework used to secure panes of glass in a window or door.

Nail set: A punch-like tool used to recess finishing nails.

Newel post: An upright post supporting the handrail at the top and bottom of a stairway or at the turn of a landing; also, the main post about which a geometrical staircase winds.

Nominal size: Approximate or rough size (commercial size) by which lumber, CMUs, etc., is commonly known and sold, normally slightly larger than the actual size (for example, 2 × 4s).

Nosing: The portion of a stair tread that extends beyond the face of the riser.

Occupational Safety and Health Administration (OSHA): An agency of the US Department of Labor whose mission is to set occupational safety and health standards for all places of employment, enforce these standards, ensure that employers provide and maintain a safe workplace for all employees, and provide research and educational programs to support safe working practices.

On center: Distance between the center of one member (typically studs) to the center of the next member. Commonly abbreviated as OC.

Open stairway: A stairway that is open on at least one side.

Open web steel joist: Steel joist, also known as a bar joist, that is used to support roof loads.

Penny: A measure of nail length; commonly abbreviated with the letter d.

Personal protective equipment (PPE): Equipment or clothing designed to prevent or reduce injuries.

Pier: A structural column used to support other structural members, typically girders or beams.

Pitch board: A board that serves as a pattern for marking cuts for stairs. The shortest side is the height of the riser cut; the perpendicular side is the tread width. Mainly used when there is great repetition such as in production housing.

Plan view: A drawing that represents a view looking down on an object.

Plane iron: Flat metal blade that is sharpened on one end to provide cutting edge for a hand plane.

Plastic concrete: Concrete when it is first mixed and is in a semiliquid and moldable state.

Pocket (plunge) cuts: A cut made to remove an interior section of a workpiece or stock (such as a sink cutout for a countertop) or to make square or rectangular openings in floors or walls.

Polyvinyl chloride (PVC): Plastic polymer used for electrical insulation and pipe.

Precast: Concrete structural elements that have been cast at a casting yard and raised into position using a crane.

Property lines: The recorded legal boundaries of a piece of property.

Purlin: A horizontal roof support member parallel to the plate and installed between the plate and the ridgeboard.

R-value: Measure of effectiveness of a material to insulate another material. A higher R value indicates a higher insulating value.

Rabbet cuts: Rectangular cuts made along the edge or end of a board to receive another board.

Rafter: A sloping structural member of a roof frame to which sheathing is secured.

Rail: Horizontal member of a window sash or panel door.

Resins: Protective natural or synthetic coatings.

Ribband: A 1 × 4 nailed to the ceiling joists at the center of the span to prevent twisting and bowing of the joists.

Ridgeboard: Horizontal board (typically a 2× member) placed at the ridge of the roof. Upper ends of the rafters are fastened to it.

Rim joist: Consists of two header joists and the end joists.

Rise and run: A term used to indicate the degree of stairway incline.

Riser diagram: A schematic drawing that depicts the layout, components, and connections of a piping system.

Riser: A vertical board under the tread of a stairway.

Rough carpentry: The portion of the carpentry trade associated with framing and other work that will be covered with finish materials.

Rough opening: Dimensions shown on the drawings that indicate the distance from the inside edge of one trimmer joist to the inside edge of the opposing trimmer joist.

Rough sill: The lower framing member attached to the top of the lower cripple studs to form the base of a rough opening for a window.

Safety culture: The culture created when the whole company sees the value of a safe work environment.

Sash: The part of a window that holds the glass.

Scab: A length of lumber applied over a joint to strengthen it.

Seasoned: Lumber with the appropriate amount of moisture removed to make it usable for construction.

Sheathing: Panel material to which roofing material or siding is secured.

Shim: Narrow, tapered piece of wood or composite material used to plumb or level a door or window.

Shiplap: A method of cutting siding in which each board is tapered and grooved so that the upper piece fits tightly over the lower piece.

Shoring: Temporary bracing used to support above-grade concrete slabs while they set.

Sill: The lowest framing member of a window or exterior door frame.

Sill plate: A horizontal member that supports the framework of a building on the bottom of a wall or box joist. It is also called a sole plate.

Single-layer floor system: Combined subfloor and underlayment; typically installed where direct application of carpet or tile, to the floor is intended

Site-specific safety program: A safety program developed for a job site that identifies and takes into account any specific potential hazards that may be encountered.

Skirtboard: A baseboard or finishing board at the junction of the interior wall and floor. Also called a finished stringer.

Slag: The ash produced during the reduction of iron ore to iron in a blast furnace.

Soleplate: The bottom horizontal member of a wall frame.

Sound transmission class (STC): Rating scale for the transmission of sound through a wall or ceiling of a building.

Span: The distance between structural supports such as walls, columns, piers, beams, or girders.

Specifications: Written document included with a set of prints that clarifies information presented on the prints and provides additional information not easily presented on the prints.

Spreader bar: A rigging device used when lifting large members with a crane.

Stairway: The complete set of stairs extending from one level of a structure to another level.

Stairwell: A compartment extending vertically through a building into which a stairway is placed.

Stick-built framing: Framing method in which framing members are installed one at a time (rather than as a prefabricated assembly).

Stile: Vertical member of a window sash or panel door.

Stringer: The inclined member that supports the treads and risers of a stairway.

Strongback: An L-shaped arrangement of lumber used to support ceiling joists and keep them in alignment.

Swing: The direction of door rotation.

Synthetic: Man-made (not naturally occurring); typically a result of a chemical reaction.

Tag line: Cord or rope attached to a structural member or other large unit that is being positioned by a crane, to assist in guiding the member into proper position.

Tail joist: Short joist that runs from an opening to a bearing.

Threshold: A piece of wood or metal that is set between the door jamb and the bottom of a door opening.

Tiedown: Metal anchoring device used to fasten framing members to one another.

Tilt-up concrete construction: Concrete construction method in which the wall sections of a building are cast horizontally on the job site or precast off-site and are lifted into position using a crane.

Top plate: The upper horizontal framing member of a wall used to carry the roof trusses or rafters.

Topographical survey: An accurate and detailed drawing of a place or region that depicts all the natural and man-made physical features, showing their relative positions and elevations.

Total rise: The vertical distance of a stairway one floor to the floor above.

Total run: Horizontal length of a stairway.

Tread: The horizontal member of a stairway upon which a person steps.

Trimmer joist: A full-length joist that reinforces a rough opening in the floor.

Trimmer stud: A vertical framing member that forms the sides of rough openings for doors and windows. It provides stiffening for the frame and supports the weight of the header.

True: To accurately shape adjoining members so they fit well together.

Two-layer floor system: Floor system in which a layer of underlayment is installed over a layer of subfloor.

Underlayment: A material, such as particleboard or plywood, laid on top of the subfloor to provide a smoother surface for the finished floor.

Unit rise: The amount (in inches) that a common rafter will rise per foot of run; the vertical distance from the top of one stair tread to the top of the next one above it; also called stair rise.

Unit run: The horizontal distance from the face of one riser to the face of the next riser.

Vaulted ceiling: A high, open ceiling that generally follows the roof pitch.

Volume: The amount of space occupied in three dimensions (length, width, and height).

Window: A wall opening that admits light and/or ventilation.

Index

C

F

G

H

I

J

K

L

M

N

O

P

Q

R

S

T

U

V

W

Y

Z